ROCK-FORMING MINERALS

Volume 5B Second Edition

Non-silicates: Sulphates, Carbonates, Phosphates, Halides

L. L. Y. Chang, BS, PhD
Professor of Geology
University of Maryland

R. A. Howie, MA, ScD, FGS, FKC
Emeritus Professor of Mineralogy
University of London

J. Zussman, MA, PhD
Emeritus Professor of Geology
University of Manchester

The Geological Society
London
1998

THE GEOLOGICAL SOCIETY

The Society was founded in 1807 as The Geological Society of London and is the oldest geological society in the world. It received its Royal Charter in 1825 for the purpose of 'investigating the mineral structure of the Earth'. The Society is Britain's national society for geology with a membership of around 8500. It has countrywide coverage and approximately 1500 members reside overseas. The Society is responsible for all aspects of the geological sciences including professional matters. The Society has its own publishing house, which produces the Society's international journals, books and maps, and which acts as the European distributor for publications of the American Association of Petroleum Geologists, SEPM and the Geological Society of America.

Fellowship is open to those holding a recognized honours degree in geology or cognate subject and who have at least two years' relevant postgraduate experience, or who have not less than six years' relevant experience in geology or a cognate subject. A Fellow who has not less than five years' relevant postgraduate experience in the practice of geology may apply for validation and, subject to approval, may be able to use the designatory letters C Geol (Chartered Geologist).

Further information about the Society is available from the Membership Manager, The Geological Society, Burlington House, Piccadilly, London W1V 0JU, UK. The Society is a Registered Charity, No. 210161.

Published by The Geological Society from:
The Geological Society Publishing House
Unit 7, Brassmill Enterprise Centre
Brassmill Lane
Bath BA1 3JN
UK
(*Orders*: Tel. 01225 445046
Fax 01225 442836)

First edition published 1962
Second edition published 1996
Reprinted 1998

British Library Cataloguing in Publication Data
A catalogue record for this book is available from the British Library.

ISBN 1-897799-90-X

Printed by The Alden Press, Osney Mead, Oxford, UK.

Distributors

USA
AAPG Bookstore
PO Box 979
Tulsa
OK 74101-0979
USA
(*Orders*: Tel. (918) 584-2555
Fax (918) 560-2652)

Australia
Australian Mineral Foundation
63 Conyngham Street
Glenside
South Australia 5065
Australia
(*Orders*: Tel. (08) 379-0444
Fax (08) 379-4634)

India
Affiliated East-West Press PVT Ltd
G-1/16 Ansari Road
New Delhi 110 002
India
(*Orders:* Tel. (11) 327-9113
Fax (11) 326-0538)

Japan
Kanda Book Trading Co.
Cityhouse Tama 204
Tsurumaki 1-3-10
Tama-Shi
Tokyo 0206-0034
Japan
(*Orders*: Tel. (0423) 57-7650
Fax (0423) 57-7651)

Contents

Abbreviations and Symbols	iv
Preface to First Edition	v
Preface to Second Edition	vii
Acknowledgements	viii
Sulphates (J.Z)	1
Baryte	3
Celestine	30
Gypsum	40
Anhydrite	74
Carbonates (L.L.Y.C.)	95
Introduction	97
Calcite	101
Magnesite	136
Rhodochrosite	150
Siderite	163
Smithsonite	178
Dolomite	189
Ankerite	219
Aragonite	235
Strontianite	253
Witherite	263
Cerrusite	272
Huntite	277
Nyerereite	282
Malachite and azurite	288
Phosphates (R.A.H.)	295
Apatite	297
Monazite	335
Halides (R.A.H.)	353
Fluorite	355
Halite	369
Index	382

Abbreviations and Symbols

The following abbreviations have been used in the text except where otherwise stated.

Å	Ångstrom units (10^{-10} m)
a	cell edge in the x direction
a_{hex}	hexagonal cell edge
a_{rh}	rhombohedral cell edge
anal.	analysis or analyst
b	cell edge in the y direction
c	cell edge in the z direction
calc.	calculated
C_p	heat capacity
D	specific gravity
d	interplanar spacing
DTA	differential thermal analysis
Eh	redox potential
f	fugacity (of substance denoted by subscript)
Fe*	total iron
G	Gibbs energy
H	hardness (Mohs scale)
H	enthalpy
K_D	distribution coefficient
n	refractive index (for a cubic mineral)
O.A.P.	optic axial plane
P	pressure
pfu	per formula unit
p.p.m.	parts per million
‰	parts per mille (parts per thousand)
R	metal ions
R	reflectance
RE, *REE*	rare earth, rare earth element
$r<v$ (or $r>v$)	optic axial angle in red light is less than (or greater than) that in violet light
S	entropy
subscript$_{ss}$	solid solution
T	temperature
tr.	trace
VHN_{100}	Vickers Hardness Number (for 100 g load)
2V	the optic axial angle
x, y, z	crystallographic axes
Z	number of formula units per unit cell
α, β, γ	least, intermediate and greatest refractive indices; the vibration direction of the fast, intermediate and slow rays; also these rays
α, β, γ	angles between the positive directions of the y and z, x and z, and x and y crystal axes
δ	birefringence
$\varepsilon(\omega)$	extraordinary (ordinary) ray, refractive index
$\sim$	approximately equal to

Preface to First Edition

In writing these volumes the primary aim has been to provide a work of reference useful to advanced students and research workers in the geological sciences. It is hoped, however, that it will also prove useful to workers in other sciences who require information about minerals or their synthetic equivalents. Each mineral has been treated in some detail, and it has thus been necessary to restrict the coverage to a selection of the more important minerals. The principle in this selection is implied in the title *Rock-Forming Minerals*, as, with a few exceptions, only those minerals are dealt with which, by their presence or absence, serve to determine or modify the name of a rock. Some may quarrel with the inclusion or omission of particular minerals; once committed, however, to the discussion of a mineral or mineral series, the less common varieties have also been considered.

Most of the information contained in this text is available in the various scientific journals. An attempt has been made to collect, summarize and group these contributions under mineral headings, and the source of information is given in the references at the end of each section. The bibliography is not historically or otherwise complete, but the omission of reference to work which has been encompassed by a later and broader study does not belittle the importance of earlier investigations; where many papers have been published on a given topic, only a limited number have been selected to illustrate the scope and results of the work they report.

The collection of data and references should bring a saving of time and labour to the research worker embarking on a mineralogical study, but it is hoped also that the presentation of the results of study from many different aspects, and in particular their correlation, will further the understanding of the nature and properties of the minerals. Determinative properties are described and tabulated, but the intended function of this work is the understanding of minerals as well as their identification, and to this end, wherever possible, correlation has been attempted, optics with composition, composition with paragenesis, physical properties with structure, and so on. For each mineral the body of well-established data is summarized, but unsolved and partially solved problems are also mentioned.

The rock-forming minerals are dealt with in five volumes. The silicates are allocated on a structural basis: vol. 1. *Ortho- and Ring Silicates*, vol. 2. *Chain Silicates*, vol. 3. *Sheet Silicates*, vol. 4. *Framework Silicates. Non-Silicates* are grouped chemically in the various sections of volume 5.

With a few exceptions, the treatment of each mineral or mineral group is in five sub-sections. In the *Structure* section, in addition to a brief description of the atomic structure, descriptions of X-ray methods for determining chemical composition and any other applications of X-rays to the study of the mineral are given. The *Chemistry* section describes the principal variations in chemical composition and includes a table of analyses representative, wherever possible, of the range of chemical and paragenetic variation. From most analyses a structural formula has been calculated. The chemistry sections also consider the synthesis and breakdown of the minerals and the phase equilibria in relevant chemical systems, together with differential thermal analysis observations and alteration products. The third section lists *Optical and Physical Properties* and discusses them in relation to structure and chemistry. The fourth section contains *Distinguishing Features* or tests by which each mineral may be recognized and in particular distinguished from those with which it is most likely to be confused. The *Paragenesis* section gives the principal rock types in which the mineral occurs and some typical mineral assemblages: possible derivations of the minerals are discussed and are related wherever possible to the results of phase equilibria studies. The five sub-sections for each mineral are preceded by a condensed table of properties, together with an orientation sketch for biaxial minerals and an introductory paragraph, and are followed by a list of references to the literature. The references are comprehensive to 1959 but later additions extend the coverage for some sections to 1961. In the present text, mineral data are frequently presented in diagrams, and those which can be used determinatively have been drawn to an exact centimetre scale, thus enabling the reader to use them by direct measurement: numbers on such diagrams refer to the number of the analysis of the particular mineral as quoted in the tables.

The dependence of these volumes upon the researches and reports of very many workers will be so obvious to the reader as to need no emphasis but we wish especially to record our indebtedness to those authors whose diagrams have served as a basis for the illustrations and thus facilitated our task. In this connection we would thank also the many publishers who have given permission to use their diagrams, and Mr H. C. Waddams, the artist who has so ably executed the versions used in the present text. *Mineralogical Abstracts* have been an indispensable starting point for bringing many papers to our attention: in by far the majority of cases reference has been made directly to the original papers; where this has not been possible the *Mineralogical Abstracts* reference is also given *e.g.* (M.A. 13–351). Our warmest thanks are due also to our ex-colleagues in the Department of Geology, Manchester University, who have been helpful with discussions and information, and who have tolerated, together with the publishers, repeatedly over-optimistic reports about the work's progress and completion. We wish to thank Miss J. I. Norcott who has executed so efficiently the preparation of the typescript and also Longmans, Green & Co. for their continued co-operation.

Department of Geology, The University, Manchester, 13. *October 1961*

Preface to Second Edition

For this completely new edition of *Rock-Forming Minerals* (volume 5B) we have maintained the general principles and organization adopted for the first edition. The past thirty years, however, has seen an enormous expansion in activity in the Earth Sciences as a whole, and the subjects of mineralogy and petrology have certainly not been exceptions. There have been many improved and new techniques for investigating minerals, producing a new body of data and often a better understanding of their nature, properties and relationships. Such changes are evident in each section under which we treat each major mineral.

Crystal structures are far better known through the development of automatic X-ray diffraction methods supplemented by spectroscopic studies. The chemical variations in minerals are better appreciated through the development of electron-probe microanalysis, which is rapid and is done without separation of minerals from the rock. New techniques have greatly extended the range of pressure and temperature at which phase transformations can be studied in the laboratory. Studies of fluid inclusions and of stable isotope ratios (S, O, H, etc.) have increasingly given information on temperatures of crystallization and genetic sequences.

Whereas we are now not so dependent on optical properties for determining chemical composition, light microscopy remains the basic general tool underpinning all other methods, and the study of submicroscopic features (*e.g.* habits and intergrowths) is enhanced by the use of scanning or transmission electron microscopy.

In the paragenesis sections of this new edition we are able to take account of the results of ocean floor exploration and deep-sea drilling projects. These and other techniques of recent decades have led to better understanding of the processes of mineral ore deposition and hydrocarbon formation.

The above, and other growth areas in mineralogy, have led to the fact that in this new edition the average number of pages devoted to each mineral is more than twice that for the first. The extent of growth is also reflected by the list of references for each mineral, which for this volume we have attempted to bring up to date to the later months of 1994. The minerals covered are as before, but the previous brief notes on malachite and azurite are now extended to full sections, and there are new sections on smithsonite, cerussite and nyerereite.

In order to expedite the completion of new revised volumes, the team of Deer, Howie and Zussman has been joined by L. L. Y. Chang, J. F. W. Bowles and D. J. Vaughan, as authors for carbonates (present volume 5B), oxides and hydroxides (vol. 5A) and sulphides (vol. 5A), respectively. Rcsponsibility for the series *Rock-Forming Minerals* as a whole remains with D. H. and Z., and all three are at present involved with revision of volume 2B, *Double Chain Silicates.*

L. L. Y. Chang, R. A. Howie and J. Zussman *July, 1994*

Acknowledgements

We are grateful to the following for permission to reproduce copyright material: Elsevier Science Ltd, The Boulevard, Langford Lane, Kidlington, OX5 1GB, UK for figs 6 (Reardon and Armstrong 1987), 19 (Sofer 1978), 48 (Huang and Wyllie 1976), 59 (Bruno and Stumm 1992), 63 (Haselton and Goldsmith 1987), 75 and 76 (Irving and Wyllie 1975) and 101 (Bilinski and Schindler 1982) from *Geochim. Cosmochim. Acta*; Economic Geology for figs 60 (Klein and Bricker 1977) and 64 (Takahashi 1960) from *Economic Geology,* Macmillan Magazines Ltd for fig. 106 (Frankis and McKie 1973) from *Nature*; the Mineralogical Association of Canada for fig. 20 (Hawthorne et al. 1975) from *Canadian Mineralogist*; Staubli Verlag AG, Zurich for fig 4 (Burkhard 1973) from *Schweiz. Min. Pet. Mitt.*; Chapman & Hall for fig. 50 (Rao and Murthy 1970) from *J. Mater. Sci.*; Springer-Verlag GmbH, Berlin for figs 35, 36 and 51 (De Capitani and Peters 1981), 113 (Binder and Troll 1989), 122 and 123 (Tromsdorff et al. 1985) from *Contributions to Mineralogy and Petrology*, and figs 21 (Ryabov et al. 1983) and 23 (Evans 1979) from *Phys. Chem. Min.;* Oxford University Press for figs 34 and 83 (Anovitz and Essene 1987) from *Journal of Petrology*; R. Oldenbourg Verlag GmbH, Stuttgart for fig. 108 (Pertlik 1981) from *Zeitschrift fur Krystallographie*; International Union of Crystallography for fig. 8 (Cole and Lancucki 1982) from *Acta Cryst. B*; University of Chicago Press for figs 38, 39 and 95 (Chang 1965), 70 (Goldsmith et al. 1962), 71 (Goldsmith and Graf 1960) and 73 (Goldsmith and Heard 1961) from *Journal of Geology*; American Journal of Science for figs 16 (Yamomoto and Kennedy 1969), 22(Dickson et al 1963), 41 (Wyllie and Boettcher 1969), 58 (French 1971) and 107 (Cooper et al. 1975) from *American Journal of Science*; Mineralogical Society of America for figs 1 (Colville and Staudhammer 1967), 2 and 3 (Blount 1977), 9 (Lager et al. 1984), 12 (Hardie 1967), 13, 14 and 15 (Blount and Dickson 1973), 17 (Oglesby et al. 1976), 31 (Markgraf and Reeder 1985), 33 (Paquette and Reeder 1990), 42 (Kennedy 1947), 43 (Adler and Kerr 1963), 44 (Greenwood 1967), 46 (Figuet et al. 1994), 53 (Huebner 1969), 54 (Abrecht 1988), 56 (Rowland and Jones 1949), 67 (Reeder and Markgraf 1986), 69 (Rosenberg 1967), 72 (Rosenberg and Champress 1989), 82 (Ross and Reeder 1992), 85 (DeVilliers 1971), 89 and 90 (Carlson 1980), 91 (Chang and Brice 1972), 98 (Speer and Hensley-Dunn 1976), 103 (Dollase and Reeder 1986), 104 (Cole and Lancucki 1982), 112 and 114 (Hughes et al. 1989) and 115 (Hughes et al. 1990) from *American Mineralogist*, figs 30, 32, 52, 65, 84 and 86 from R.J. Reeder (ed.), *Reviews in Mineralogy Volume 11: Carbonates: Mineralogy and Chemistry,* 1983, and figs 18, 24, 25, 26 and 27 from Holser, *Marine Minerals,* Mineralogical Society of America Short Course Notes Vol. 6, 1979: American Chemical Society for fig 10 (Weiser et al. 1936) from *Journal of the American Chemical Society*; E. Schweizerbart'sche Verlagsbuchhandlung, Stuttgart for fig. 55 (Nagy et al. 1975) from *Neue Jahrbuch Min. Monatshefte*; Mineralogical Society of Great Britain and Ireland for figs 11 (McConnell et al 1987) and 62 (Birch 1986) from *Mineralogical Magazine*, and figs 79, 94 and 95 from V.C. Farmer (ed.), *The Infrared Spectra of Minerals,* Mineralogical Society Monograph No. 4, 1974; Blackwell Science Ltd, Oxford for fig. 78 (Fairchild 1983) from *Sedimentology*. While every effort has been made to trace to owners of copyright material, in a few cases this has proved impossible, and we would like to take this opportunity to apologize to anyone whose rights we may have unwittingly infringed.

Sulphates

Baryte $BaSO_4$
Celestine $SrSO_4$
Gypsum $CaSO_4 \cdot 2H_2O$
Anhydrite $CaSO_4$

Baryte $BaSO_4$

Orthorhombic (+)

α	1.635–1.637
β	1.636–1.639
γ	1.646–1.649
δ	~0.012
$2V_\gamma$	37°–39°
Orientation:	
$\alpha=z$, $\beta=y$, $\gamma=x$, O.A.P. (010)	
Dispersion:	$r<v$, weak
D	~4.5
H	$2\frac{1}{2}$–$3\frac{1}{2}$

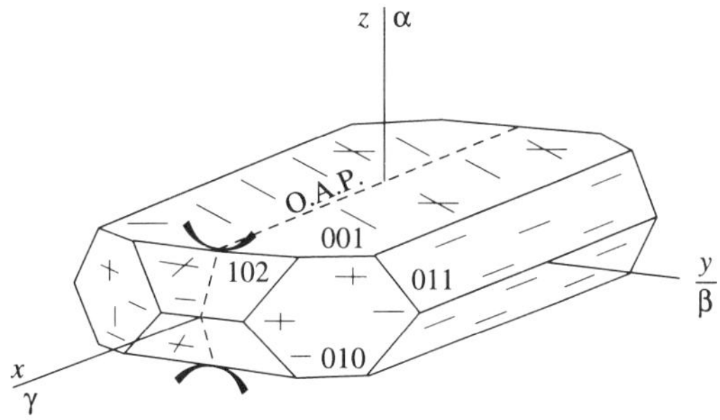

Cleavage:	{001} perfect, {210} very good, {010} good.
Twinning:	Glide twinning on {110} frequent in massive material.
Colour:	White, yellowish brown, grey, pale green, pale blue, red, brown; colourless in thin section.
Pleochroism:	Slight in coloured varieties; absorption $\gamma>\beta>\alpha$.
Unit cell:	*a* 8.88, *b* 5.45, *c* 7.15 Å Z=4. Space group *Pnma*

Insoluble in acids.

Baryte (or barite) is the least soluble sulphate and the most abundant barium mineral in the Earth's crust. Its principal occurrences are as a gangue mineral in hydrothermal metalliferous veins or, mainly dispersed but also as rosettes and nodules, in laminated sedimentary deposits commonly interbedded with cherts and siliceous shales. Baryte is also found as a residual deposit left after the host rock (e.g. carbonate) has dissolved. It is the principal source of barium compounds and is widely used in drilling muds for the petroleum industry, or in other dense fluid media, and also as a filler or extender in the manufacture of papers, cards, plasters, rubber and plastics. Baryte also finds use in glasses and ceramics after conversion to the carbonate, chloride or hydroxide, and in the manufacture of sulphuric acid.

Structure

Although several studies of the unit cell and space group of baryte had been made previously, the first determination of its structure was by James and Wood (1925). Other early structural studies reported were by Rinne *et al.* (1924) and Basche and Mark (1926), and later by Sahl (1963), Colville and Staudhammer (1967), Hill (1977) and Miyake *et al.* (1978).

In the structure of baryte (Fig. 1), the barium atoms lie on mirror planes and link the SO_4 ions in such a way as to be surrounded by 12 oxygens belonging to

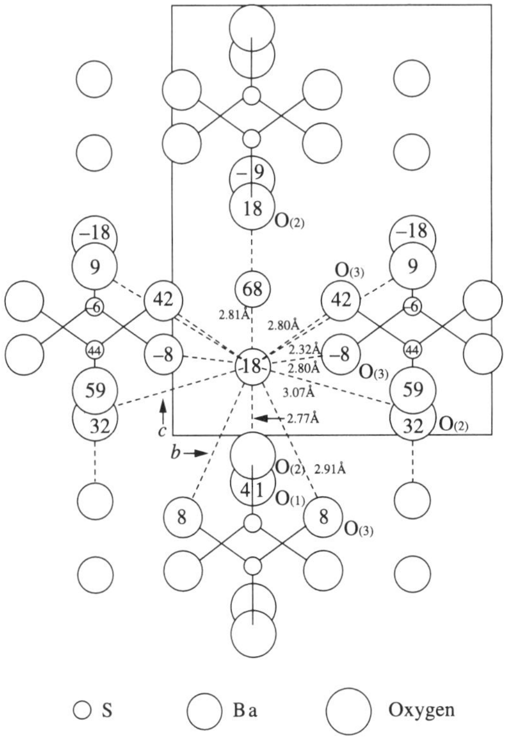

Fig. 1. Structure of baryte projected on (100) showing the coordination of the barium atom, and *x* coordinates as a percentage of *a* (after Colville and Staudhammer, 1967).

seven different SO_4 groups. The nearest six have a mean Ba–O distance of 2.80 Å, but Hill regarded all 12 distances (2.8–3.3 Å) as significant bonds. The SO_4 tetrahedra are not quite regular, having symmetry *Cs*(*m*) and three different S–O distances: S–O(1)m, S–O(2) and 2 × S–O(3). The mean S–O distance is 1.472 Å, and the mean O–S–O bond angle is 109.5° (Colville and Staudhammer, 1967). Hill (1977) noted that the oxygen atoms lying on the mirror planes had higher thermal vibration parameters than others, but found no evidence for a space group with lower symmetry.

Miyake *et al.* (1978) determined the structures of a baryte from Tamagawa, Akita, Japan, a celestine from Clay Center, Ohio, USA, and an anglesite from Osarizawa, Akita, Japan, with compositions respectively $Ba_{0.99}Pb_{0.01}SO_4$, $Sr_{0.986}Ba_{0.013}Pb_{0.001}SO_4$ and $Pb_{0.977}Ba_{0.013}Sr_{0.010}SO_4$. In each case the longest S–O distance was S–O(3), the widest O–S–O angle O(1)–S–O(2) and the average S–O distances 1.478, 1.474 and 1.49 Å respectively.

Miyake *et al.* (1978) correlated their bond length determinations with infrared absorption frequencies and concluded that the electronegativity of the metal and the coulombic force between it and oxygen modify the structure of the SO_4 tetrahedron. They derived stretching and repulsive force constants for S–O for baryte as 6.27 and 0.73, and for celestine as 6.34 and 0.75 md/Å. Comparing them with similar data for sulphates of magnesium, calcium, tin and lead, they show that an increase in field strength of the metal causes an increase in the force constants of the SO_4 tetrahedron.

Table 1. Cell Parameters and Specific Gravities of Baryte, Anglesite and Celestine

Name	*a* (Å)	*b* (Å)	*c* (Å)	*D*	Reference
Baryte	8.878	5.450	7.152	4.480	Swanson *et al.* (1954)
	8.884	5.458	7.153		Colville & Staudhammer (1967)
	8.842	5.4559	7.1569		Hill (1977)
	8.884	5.457	7.157		Miyake *et al.* (1978)
Anglesite	8.480	5.398	6.958	6.323	Swanson *et al.* (1954)
Celestine	8.359	5.352	6.866	3.971	Swanson and Fuyat (1953)

X-ray evidence for the existence of the baryte–celestine isomorphous series was obtained by Sabine and Young (1954); see also Burkhard (1973, 1978). Boström *et al.* (1967) concluded that there is complete ternary solid solution between $BaSO_4$, $SrSO_4$ and $PbSO_4$. Although there is a complete range of solid solution between baryte and celestine, Goldish (1989) found that the cell dimensions deviate from strictly linear relationships; reflection 121 was recommended for estimating composition. Papunen (1967) had used d_{410}, and Soriano *et al.* (1991) used d_{210}.

Renault and Brower (1971) noted that X-ray powder diffraction peaks were broader at intermediate Ba,Sr compositions and associated this with diminished crystallite size, itself related to compositional heterogeneity and the possibility of twin faulting. Compositional changes, and the general paucity of intermediate members of the Ba–Sr series, are probably related to the very low solubility of $BaSO_4$ compared with $SrSO_4$ (see p. 10).

James and Wood (1925) showed that anglesite ($PbSO_4$) is isostructural with baryte and celestine. More recently Chen and Yu (1984) confirmed hokutolite (baryte with appreciable lead content) as having the baryte space group *Pnma*. Sasaki and Minato (1982) determined lattice constants for members of the $BaSO_4$–$PbSO_4$ series, and attributed deviations from linearity to the presence of Sr and Ca (see also Sugitani *et al.*, 1969). Several other minerals and compounds have the baryte structure; among them are $BaSeO_4$, $PbSeO_4$ (kerstenite), $SrSeO_4$, $BaCrO_4$, $KMnO_4$, $KClO_4$ and $(K,Cs)BF_4$ (avogadrite). Anhydrite does not belong to this structural group.

Some determinations of the cell parameters of baryte are listed, along with those for celestine and anglesite, in Table 1. The values for baryte yield axial ratios which agree closely with those derived by Helmhacker (1872) from morphology, using double his value for *a*.

The changes undergone by the structure of baryte on heating up to near the transition to the α-form were studied by Sawada and Takeuchi (1990). They found that for most of the range, *a*, *b* and *c* rise linearly with temperature, but at about 1010 °C (close to the transition temperature ~1090 °C) *a* increases more rapidly whereas *b* and *c* stay almost constant. With regard to the Ba—O bonds, the eight shorter bonds expand more than the four longest, and the shortest bond at room temperature has the highest expansion rate. Sawada and Takeuchi (1990) comment that this shortest bond holds the structure together, especially at temperatures near to the transition, at which point it is severed.

Chemistry

Although there is a complete solid solution series between baryte ($BaSO_4$) and celestine ($SrSO_4$) (Grahmann, 1920; Boström *et al.*, 1967; Brower, 1973), most natural specimens tend to cluster near to the end-member compositions, i.e. <7 mol. per cent Sr in baryte and <4 per cent Ba in celestine (Starke, 1964; Tufar, 1965; Hanor, 1968). There are, however, exceptions; Burkhard (1978) found up to 30 per cent Sr in baryte and similar amounts of Ba in celestine in specimens from sedimentary rocks of the Jura Mountains, and a complete range of Ba/Sr compositions was reported for natural baryte by Touray *et al.* (1988). Members of this series with a preponderance of the Ba 'molecule' are called strontiobaryte, and those near the Sr end, barytocelestine. Some analyses of baryte are given in Table 2. Ideal $BaSO_4$ has 65.70 per cent BaO and 34.30 per cent SO_3.

Anglesite, $PbSO_4$, is isostructural with baryte and has similar cell parameters, so complete solid solution would be expected. Although Takano *et al.* (1970) reported a gap in the series for natural specimens, and indirect suggestions of a gap from infrared and thermal data, Sugitani *et al.* (1969) synthesized the range from $BaSO_4$ to $PbSO_4$ and found that lattice parameters and refractive indices varied linearly across the series. Some examples of appreciable substitution of Pb for Ba are those described by Ohashi (1920; up to 30 per cent $PbSO_4$) and Chen and Yu (1984). The barium-rich members of the Ba–Pb series are called anglesobaryte or hokutolite, and the lead-rich members barytoanglesite. The lead content of barytes coexisting with galena was used by Kajiwara and Honma (1976) to define the oxygen fugacity during ore formation.

Studies of the system $CaSO_4$–$BaSO_4$ (Grahmann, 1920) showed that at room temperature only about 6 per cent $CaSO_4$ can enter into solid solution in the baryte structure, and at the other end 8 per cent. $BaSO_4$ is the limit of solid solution of Ba in anhydrite. The amounts of $CaSO_4$ reported in natural occurrences are generally low, e.g. a range from 0.08 to 0.12 per cent $CaSO_4$ in 42 specimens of baryte examined by Tufar (1978). Slightly higher values were reported by Burkhard (1978). For the system $CaSO_4$–$SrSO_4$, Grahmann (1920) showed that about 12 per cent $CaSO_4$ is the limit of solid solution in celestine, and thus slightly more Ca can be accepted in strontiobaryte than in baryte. Barytocalcite (the name is somewhat misleading) is a barium–calcium carbonate.

Khomyakov (1964) had noted that the ratio *REE* (Ce)/*REE* (Y) was high in baryte compared with that in associated fluorite, calcite and ankerite.

The *REE* contents of baryte from a variety of marine and continental environments were determined by Guichard *et al.* (1979). The ranges reported, in p.p.m., are: La 75–0.2, Sm 4.4–0.02, Eu 3.1–<0.1 and Dy 6.5–<0.002. It was noted that because of diminishing ion size relative to barium, the lighter elements decrease in amount from La towards Eu. Morgan and Wandless (1980) also found that the *REE* content of barytes and anhydrites showed a systematic variation with ionic radius, but that the baryte specimens were anomalously enriched in europium. This was attributed to the latter ion's large radius (similar to that of strontium).

Minor replacements of Ba by other elements may occur, including Fe, Cu, Zn, Ag, Ni, Ra and Hg (Saukov, 1939). Preuss and Ziehr (1966) reported that the mercury contents of baryte were dependent upon the temperatures of formation, those in fluorite veins having a Hg content up to 250 p.p.m., but those in pegmatites containing none. Vanadium can also be a significant trace element in baryte (e.g. Moro *et al.*, 1991). The anion groups $[MnO_4]$, $[SeO_4]$ and $[CrO_4]$ have been substituted for SO_4 in synthetic material. Ryabov *et al.* (1983, 1989) used electron paramagnetic resonance to show that SO_3^- and PO_3^- radicals were present in baryte, celestine and anhydrite.

A study by Srinivasan (1976) of krypton and xenon isotopes in sedimentary baryte of age >3000 m.y. showed that they contain spallogenic Xe produced by

Table 2. Baryte Analyses

	1	2	3	4	5	6	7
SiO_2	2.01	0.48	0.02	—	0.02	—	—
Al_2O_3	0.05	—	0.35	—	0.23	—	—
Fe_2O_3	0.32	0.31	—	—	—	—	tr.
MgO	—	—	tr.	—	tr.	—	—
CaO	0.38	0.35	0.13	0.44	0.16	—	—
SrO	tr.	0.10	1.34	1.68	2.79	14.7	—
PbO	—	—	tr.	—	tr.	—	17.78
BaO	62.49	64.40	63.47	63.10	61.70	48.6	48.95
SO_3	34.09	34.33	34.18	34.40	34.67	36.2	32.24
H_2O^+	} 0.54	0.21	} 0.21	—	} 0.20	—	} 0.60
H_2O^-		0.10		—		—	
Total	99.88	100.28	99.70	[100.21]	99.77	99.5	99.57
α	1.636	1.6357	1.637	1.635	1.637	—	—
β	—	1.6369	1.638	1.638	1.638	—	—
γ	1.647	1.6480	1.649	1.647	1.649	—	—
$2V_\gamma$	—	37° 3′	$36\frac{1}{2}°$	—	$36\frac{1}{2}°$	—	—
D	4.300	4.4644	4.466	—	4.391	3.9	4.62
	Numbers of Ions on the Basis of Four (O)						
Ca	0.0161	0.0144	0.0054	0.0181	0.0067	—	—
Sr	—	0.0225	0.0302	0.0376	0.0622	0.3263	—
Pb	—	—	—	—	—	—	0.1984
Ba	0.9633	0.9753	0.9681	0.9546	0.9296	0.6956	0.7945
Σ	0.979	1.012	1.004	1.010	0.999	1.022	0.993
S	1.007	0.996	0.999	0.997	1.001	0.993	1.002

1 Acicular baryte, Kominato town, Aomori Prefecture, Japan (Hasegawa, 1954).
2 Pale yellow baryte, Trepča mines, near Kosovska, Mitrovica, Serbia (Barić, 1948).
3 Yellow baryte in middle zones of druses, Dobrevo mine, Zletovo, Macedonia (Jurković, 1953).
4 Baryte, siderite veins, South Belozerka district, USSR (Kornilov, 1966). Includes CO_2 0.59. Anal. I. I. Gorda.
5 White baryte in cores of druses, Dobrevo mine, Zletovo, Macedonia (Jurković, 1953).
6 Strontian baryte, Clifton, England (Collie, 1879; see Palache *et al.*, 1951).
7 Anglesobaryte, Shibukuro, Akita Prefecture, Japan (Ohashi, 1920). Anal. Yoshioka mine.

high energy particles in cosmic rays. The $^{124}Xe/^{126}Xe$ ratios were like those for lunar samples and meteorites, and excess ^{131}Xe was derived from the reaction $^{130}Ba(n,\gamma)\rightarrow{}^{131}Xe$ (see also Li *et al.*, 1994).

Experimental Work

Blount (1974) synthesized well-formed crystals of baryte by slow precipitation on combining dilute barium chloride and sodium sulphate solutions in the presence of 0.03 M HCl. The best results were obtained with 0.01 M $BaCl_2$ and 0.01 M Na_2SO_4; higher concentrations and more rapid crystallization led to the formation of smaller crystals. A similar reaction was used by Patel and Bahat (1972) to produce crystals of $BaSO_4$ but using silica gel as a growth medium.

Baryte, celestine and anhydrite have been used (e.g. Denis and Michard, 1983) as good cases on which to test various theories of dissolution encompassing both equilibrium and kinetic aspects. The solubility of baryte in pure water is low (about 2 mg/l), but its variation with temperature and salinity has been the subject of many investigations. For pure water a value of about 10 mg/l was found at 206 °C (Thuggutt, 1936; Palmqvist, 1939), and about 40 mg/l at 500 °C, 1000 bar (Morey and Hesselgesser, 1951).

The solubility of $BaSO_4$ in aqueous solutions of various salts has been measured by many other workers (e.g. Templeton, 1960; Strübel, 1967) and their results were summarized by Monnin and Galinier (1988), who calculated the solubility of baryte in sea-water as about 2×10^{-7}, whereas its measured value is about 3.7×10^{-7} mol. Ba/kg H_2O, and attributed the discrepancy to the presence of a $(Ba,Sr)SO_4$ solid solution which would give a higher Ba content than pure $BaSO_4$.

The thermodynamic equilibrium relationships for $(Ba,Sr)SO_4$ have been difficult to establish. Thermodynamic calculations by Hanor (1969) showed that the presence of strontium has a significant effect on the saturation relationships of baryte in sea-water. When ocean waters are unsaturated with respect to pure $BaSO_4$ they may not be so with respect to $(Ba,Sr)SO_4$.

The solubility of $BaSO_4$ with 8.5 per cent $SrSO_4$ in pure water and in water with 1.0 N NaCl was determined by Gundlach *et al.* (1972) over the temperature range 20–350 °C. In NaCl solution the solubility of Ba from this strontiobaryte is an order of magnitude lower than from pure $BaSO_4$, so Sr is selectively extracted from Sr-containing baryte and thus it dissolves incongruently. Gundlach *et al.* found that the solubility of $SrSO_4$ is greater than that of $BaSO_4$ in NaCl solution at all temperatures studied. In pure water the solubility of $BaSO_4$ becomes greater than that of $SrSO_4$ at temperatures >250 °C, at which point the two solubilities are equal at about 2 p.p.m. They suggested that baryte formed between 80 and 130 °C is relatively richer in Sr because at these temperatures the maximum difference in solubilities for Ba and Sr occurs, and therefore a greater supply of Sr is available in solution.

Uchameyshvili *et al.* (1966) showed that the solubility of baryte in pure water passes through a maximum within the range 100–300 °C, but for solutions of $CaCl_2$, $MgCl_2$ and 2 N NaCl it increases continuously with temperature due to the formation of new phases ($CaSO_4$, $MgSO_4$, Na_2SO_4). They concluded that temperatures >200 °C are unfavourable for the deposition of baryte, and suggested that baryte was preceded by anhydrite in the deposits (Caucasus and Central Asia) studied. The solubility of baryte in saline water was also studied experimentally by Scherp and Strübel (1974).

Thermodynamic calculation by Kritsotakis (1979) gave the solubility of baryte as a function of *T*, pH, NaCl concentration and total sulphate concentration. He compared his results with those of other investigators (Table 3).

From these results Kritsotakis (1979) concluded that hydrothermal mobilization of baryte is only possible when the concentration of sulphate is low and is kept low by means of reducing conditions. From further calculations Kritsotakis and

Table 3. Solubility of Baryte (p.p.m.) in 2 M NaCl Solution

T (°C)	pH	Kritsotakis (1979)	Strübel (1967)	Uchameyshvili *et al.* (1966)
25	7.0	26.0	29.4	—
150	5.8	79.4	81.0	76
300	5.7	162.7	161.9	142

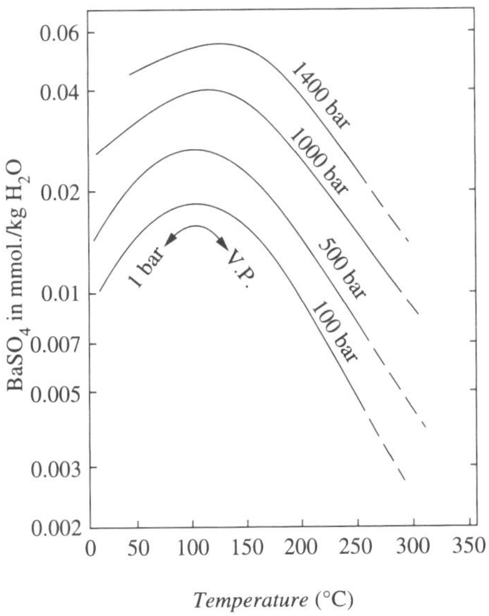

Fig. 2. Solubility of baryte in water as a function of temperature and pressure (after Blount, 1977).

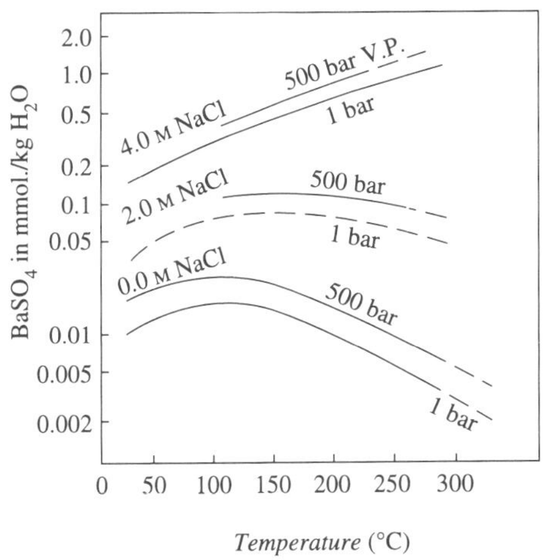

Fig. 3. Solubility of baryte in water and 0.2 and 0.4 mol./kg NaCl–H_2O solutions as a function of temperature and at 1 and 500 bar pressure (after Blount, 1977).

Platen (1980) investigated the mobilization of baryte by reduction with gaseous CH_4, H_2, CO, NH_3 and H_2S and solid C for hydrothermal, metamorphic and anatectic P–T–f–pH conditions. They concluded that the effective reduction of baryte could occur only at temperatures <200 °C and with low pH and high gas fugacities.

Blount (1977) determined the variation of the solubility of baryte with increasing NaCl concentration and increasing pressure (Figs. 2 and 3). At low NaCl concentration the solubility curve shows a maximum at about 100 °C, which moves to a higher temperature with increase in either P or NaCl concentration. Thus with low Cl concentration, Ba-saturated waters will precipitate $BaSO_4$ with either rising or falling temperature. In more saline waters the solubility of baryte just increases with T and P, so precipitation would occur on migration upwards.

Hanor (1968) developed models for the precipitation of baryte and celestine from aqueous solutions which resulted in a similar frequency distribution of compositions in the Ba–Sr series as is found in natural specimens, i.e. clusters near the end-members and few intermediates, even though a complete solid solution series had been verified. The complete series from $BaSO_4$ to $SrSO_4$ was synthesized by Brower (1973) by thermal recycling from $BaSO_4$ + NaCl brines. The paucity of intermediate compositions in nature may be related to the very different solubilities of $BaSO_4$ and $SrSO_4$ ($K_{BaSO_4} \sim 10^{-10}$, $K_{SrSO_4} \sim 10^{-6.4}$) and the fact that the solubility of $SrSO_4$ decreases whereas that of $BaSO_4$ increases with increasing temperature (see also Hanor, 1973).

Malinin and Urusov (1983) concluded that solid solution in the $BaSO_4$–$SrSO_4$ system is far from ideal (though closer to it as temperature increases). They also explain the lack of appreciable Sr in natural baryte as a result of the host $BaSO_4$ being much less soluble than strontium sulphate, the minor component; the build up of strontium in the residual solution explains the observed gradual increase in strontium content of the baryte during its formation.

Oscillatory zoning in $(Ba,Sr)SO_4$ was studied experimentally by Putnis *et al.* (1992) involving the counter diffusion of (Ba^{2+},Sr^{2+}) and SO_4^{2-} ions through a porous silica gel medium. When the end-members have significantly different solubilities, zoning occurs through the coupling of composition gradients in the solid and the liquid, and does not require a non-ideal solution with a miscibility gap or the alteration of imposed intensive parameters.

Felmey *et al.* (1993) investigated $(Ba,Sr)SO_4$ precipitates ranging from 0.05 to 0.90 $SrSO_4$, and took particular care to attain equilibrium and to determine compositions of near-surface material using X-ray photoelectron spectroscopy. They found that their results could not be explained by the presence of only one solid solution phase, whether ideal or non-ideal. They concluded that small quantities of a phase close to $BaSO_4$ in composition were unavoidably present which control the aqueous Ba concentrations, and that Sr substitutes into a (Ba,Sr) solid solution phase in which the Ba component does not attain equilibrium with the aqueous phase.

As well as having uses, baryte can be a nuisance, for example where it develops as a scale deposit in offshore oil wells pressurized by sea-water (sea-water contains ~2400 g/l SO_4^{2-} and the formation waters ~1000 mg/l Ba^{2+}). In this context, Prieto *et al.* (1990) studied the nucleation of baryte in a system involving double diffusion of Ba^{2+} and SO_4^{2-} in an Na-rich aqueous solution through a porous medium (silica gel). They found that the location of baryte precipitation is determined not only by the need to exceed supersaturation, but also by the need to achieve a Ba^{2+}/SO_4^{2-} ratio close to unity.

In further experiments on the growth of baryte crystals in a silica gel, Prieto *et al.* (1992) found that nucleation occurred when the supersaturation exceeded a threshold value which depends on the supersaturation rate. The supersaturation decreases during crystal growth but remains high. They suggested that similarly high supersaturations are present in natural growth conditions.

Christy and Putnis (1993) found that the dissolution of baryte cleavage fragments in water and NaCl brines (over 44–85 °C) followed first-order kinetics, with an activation energy (E_a) of 24.9 ± 10.0 kJ/mol. and a dissolution constant extrapolated to 25 °C of $3.30 \pm 1.31 \times 10^{-3}$ l/m^{2}s. The latter indicated that desorption rather than volume was the rate-determining step. The growth of baryte from supersaturated aqueous solutions followed a second-order rate law with $E_a = 22.0 \pm 14.3$ kJ/mol. and $k_{25\,°C} = 162 \pm 65$ l^2/m^2s.

The solubilities of members of the $BaSO_4$–$PbSO_4$ solid solution series were investigated by Kornicker *et al.* (1991) who found that at 25 °C dissolution was first congruent and then incongruent, whereas at 60 °C it was incongruent from the start. During incongruent dissolution the concentration of Ba decreased and that of Pb increased.

When gently heated, some crystals of baryte decrepitate, liberating H_2S; the variety 'hepatite' does this to a marked extent and consequently it has also been known as '*stinkstein*'. Specimens of baryte may also evolve a strong sulphurous smell on crushing to a powder. Rankin and Shepherd (1978) found that fluid inclusions in a dark grey-brown baryte contained between 600 and 4100 p.p.m. by volume H_2S, and even this amount was detectable by smell. Decrepitation can also involve the emission of CO_2 and H_2O which are trapped in fluid inclusions (Dons, 1956; Touray *et al.*, 1966).

The thermal transition to a high-temperature α-form occurs on heating at 1149 °C (Grahmann, 1920); the transformation temperature was given as 1090 °C by Sawada and Takeuchi (1990). Mohazzabi and Searcy (1976) investigated the thermal decomposition of baryte using the Langmuir method.

A common alteration product from baryte is witherite, $BaCO_3$ (Hancox, 1934), which may be formed through the action of potassium carbonate solutions. Among those minerals which have been recorded as pseudomorphs after baryte are calcite, dolomite, siderite, rhodochrosite, cerussite, witherite, malachite, magnetite, fluorite, limonite, pyrite, marcasite, chalcopyrite and talc.

Optical and Physical Properties

Baryte commonly occurs in well-formed crystals showing tabular {001} and less frequently prismatic [100] or [010] habits. The most common forms are {001}, {210}, {101}, {011}, {211}, {111}, {010} and {100}.

Baryte has been used by several researchers to study the general relationships between morphology, crystal structure and conditions of growth. Earlier studies were by Braun (1932), Niggli (1941) and Seager (1959), who found that for baryte, anglesite and celestine there was only poor agreement between the observed frequency of forms and the prediction (Donnay and Harker, 1937) that the morphological importance of a form will be proportional to its reticular lattice density. The space group was, however, correctly determined for each mineral from its morphology, and it was noted that the frequency of certain forms varies systematically through the sequence baryte–anglesite–celestine, the order of cationic radii. Schneer (1970) suggested that the morphological probability distribution is a function of the crystal structure, and that the structure is a Fourier transform of the morphological probability distribution (see also Schwarz and Follner, 1981).

Dowty (1976) developed a 'template fraction' model which achieved a better fit between observed and predicted morphological importance than that obtained by either the reticular density or the periodic bond chain (PBC) model (Hartman and Perdok, 1955). Hartman and Strom (1989) revised their earlier work for several compounds with the baryte structure, analysing PBCs and also direct chains for various slices of the structure. They also predicted morphology by calculating attachment energies based on an electrostatic point charge model. These considerations correctly gave {210}, {101} and {001} as the main forms, and they suggested that the observed importance of {011} in addition is due to external growth factors.

Some topographical features of baryte crystals were discussed and illustrated by Seager (1953), and were related to mechanisms of crystal growth (see also Sweet, 1930). Changes of habit during growth, observed with the aid of colour zoning, were discussed by Seager and Davidson (1952). Patel and Koshy (1969) examined crystals of baryte by interferometry and noted growth spirals on {011}.

The kinetics of crystal growth for baryte were studied by Fernandez-Diaz *et al.* (1990) who measured the induction period (τ) for nucleation of baryte from supersaturated solutions using a method of light scattering. They found a linear relationship between $\log \tau$ and $(\log S)^{-2}$ (S = supersaturation), a result in accord with homogeneous nucleation and diffusion-controlled growth at concentrations of baryte $>10^{-3}$ M. They also derived the crystal–solution interfacial tension as 73 mJ/m^2.

Baryte also occurs as globular concretions and as fibrous, lamellar, granular and earthy aggregates, in some cases stalactitic. Clusters of platy crystals are found, usually containing sand grains and coloured pink, which assume rosette shapes and are called 'desert roses' (Tarr, 1933). Crystals of baryte have a vitreous to resinous (and sometimes pearly) lustre, and are colourless or white when pure. Yellow, red and brown varieties result from the presence of impurities, principally iron oxides and hydroxides, sulphides and organic matter. Many specimens are blue, probably as a result of exposure to radiation from radium: it is known that exposure of baryte to radium radiation turns it blue, and the presence of Ra in natural specimens has been verified. Crystals of baryte have been described (Sweet, 1930) which change colour from pale yellow to blue on exposure to light: the change in colour is attributed to the diffusion of photoelectrons.

Some specimens of baryte are fluorescent (white, yellow, orange or green) in ultraviolet light and are subsequently phosphorescent; thermoluminescence also occurs. The fluorescent property is at first destroyed by heating but it may be restored by stronger heating (Przibram, 1956). Pale green fluorescence was ascribed to traces of copper by Lovering (1953), who also noted that this fluorescence was inhibited by the presence of iron. Izyumskiĭ (1971) examined the luminescence and trace element contents of several hundred specimens and concluded that the 435 nm absorption band is due to interband transitions, and that only magnesium takes part in the absorption and luminescence transitions.

The refractive indices given on page 3 relate to crystals of baryte which have little or no substitution in the formula $BaSO_4$, and are appropriate to room temperature and the sodium D line. Refractive indices and birefringence are lowered by the substitution of Sr for Ba, and increased by the substitution of Pb, while the optic axial angle is increased by either of these replacements.

The relationships between optical properties and composition for a series of synthetic members of $BaSO_4$–$SrSO_4$ series were determined by Burkhard (1973) and are illustrated in Fig. 4. Similar results were obtained for naturally occurring members of the baryte–celestine series (Burkhard, 1978). The effect of Sr may well be swamped by the effects of small amounts of Pb and other substituents. Kolb (1911) made a detailed study of the temperature and wavelength dependence of the optical properties of baryte, celestine, anglesite and anhydrite; his results for baryte, celestine and anglesite are summarized in Table 4. Dispersion of birefringence with temperature and wavelength, and temperature coefficients of wavelength dispersion, were also determined by Kolb (Table 5) (see also Jaffray, 1943a,b).

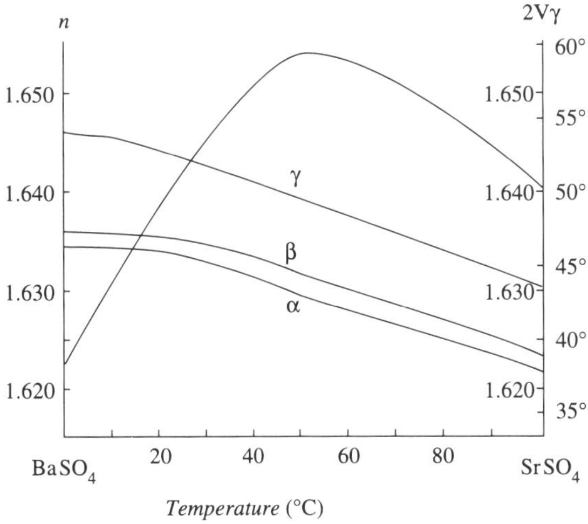

Fig. 4. Variation of α, β, γ and $2V_\gamma$ with composition in the series $BaSO_4$–$SrSO_4$ (after Burkhard, 1973).

Table 4. Optical Constants of Baryte, Celestine and Anglesite (Na_D, 20 °C) (Kolb, 1911)

	$BaSO_4$	$SrSO_4$	$PbSO_4$
α	1.636	1.621	1.878
β	1.637	1.623	1.883
γ	1.648	1.631	1.895
$2V_\gamma$	37° 02′	50° 25′	68° 28′

Table 5. Optical Properties of Baryte (Kolb, 1911)

(a) Temperature variation at constant wavelength (D_2 line)

	−53 °C	20 °C	100 °C	570 °C
α	1.6369	1.6360	1.6348	1.6252
β	1.6383 (−50 °C)	1.6374	1.6364	1.6283
γ	1.6496	1.6480	1.6464	1.6345
		20 °C	115 °C	520 °C
$2V_\gamma$		37° 02′	42° 22′	66° 31′

(b) Wavelength variation at constant temperature

		B	D_2	F	G
	α	1.6326	1.6363	1.6427	1.6481
at 16–19 °C	β	1.6337	1.6373	1.6440	1.6494
	γ	1.6446	1.6484	1.6551	1.6607
at 20 °C	$2V_\gamma$	36° 26′	37° 02′	37° 45′	38° 23′

Infrared spectra. The infrared spectrum of baryte exhibits several significant bands (Vierne and Brunel, 1969; Miyake *et al.*, 1978). The latter's band assignments were ν_1 980, ν_2 439, 470, ν_3 1075, 1110, 1170 and ν_4 600, 625 cm^{-1}; stretching and bending S–O force constants were given as 6.27 and 0.46 mol./Å respectively, and the repulsive force constant for O–O as 0.73 mol./Å. Data were also given by Caillère and Pobeguin (1962) and Griffith (1969), both noting the shift of the ν_1 band to a lower frequency in the order baryte–celestine–anhydrite.

Baryte is brittle, shows perfect {001} cleavage, and less perfect cleavage on {210} and {010}: the {210} cleavages intersect at an angle of ~78°. The average hardness of baryte is about 3 on the Mohs scale but hardness is known to vary with direction in the crystal: on an (001) face the hardness along [100] is approximately twice as great as that in a direction 30° from [100] (Tertsch, 1936). The density of pure chemical $BaSO_4$ is 4.48 g/cm^3 at 26 °C (Swanson *et al.*, 1954), but that of natural baryte is somewhat variable because of chemical substitutions. Replacement of Ba by Sr gives a lower density and by Pb a higher density. Assuming straight-line relationships, the ideal densities of strontiobaryte and anglesobaryte may be obtained by interpolation between the values given in Table 1 for the end-members baryte, celestine and anglesite.

Williams *et al.* (1972) studied dislocations in baryte by means of etch pits and reported the various slip systems as: {201} [101], {401} [010], {110} $\langle\bar{1}10\rangle$, {001} [001], {101} $\langle\bar{1}01\rangle$ or [010], {110} [001] and {201} [001]. Phakey and Aly (1970), using topography, reported pure edge dislocations along [100] with Burgers vector $\mathbf{b}=[010]$, and pure screw dislocations with $\mathbf{b}=[1\bar{1}0]$, as well as mixtures of the two.

Linear compatibilities in different directions were determined for baryte by Bridgman (1949), and the ratios of elasto-optical constants of baryte, determined by an ultrasonic method, were given by Narasimhamurty (1961) as $P_{31}/P_{11}=0.73$ and $P_{12}/P_{22}=1.33$.

The dielectric constant of baryte has been determined by Errera and Brasseur (1933), and its magnetic anisotropy by Krishnan *et al.* (1933).

There have been several studies by electron paramagnetic resonance (e.g. Ryabov *et al.*, 1989; Das *et al.*, 1990) which indicate the presence in baryte of SO_3^- paramagnetic radical ion centres.

Distinguishing Features

Baryte is not easily distinguished from celestine except by its greater density, which also helps to distinguish it from other minerals, e.g. aragonite, albite, calcite and gypsum. In addition, albite is harder than baryte, gypsum is softer, and calcite effervesces while baryte is insoluble in dilute HCl. Baryte gives a green flame colouration. The cleavages of minerals in the baryte group are characteristic, and are distinct from the three orthogonal cleavages of anhydrite. Baryte and celestine have a lower refringence and birefringence than anglesite.

Paragenesis

The principal occurrences of baryte are as (a) a gangue mineral in hydrothermal veins bearing metalliferous ores, (b) veins, (c) cavity-filling concretions and (d) disseminated in sedimentary rocks which may be limestones, sandstones, shales or clays. Gage (1944) reported baryte in Coal Measure grits; Fowler (1944) described baryte, celestine and gypsum in cavities in dolomites of the Lower Magnesian Limestone. Isolated crystalline masses of baryte occur in the Fuller's Earth at Nutfield, Surrey (Wilson *et al.*, 1922), and fibrous nodules are found embedded in kaolin near Bologna, Italy (Tomba, 1954). Baryte-bearing septarian nodules in Lower Jurassic clay were described by Taylor (1950).

Concretions of baryte in sandstone sometimes take the form of rosettes known as 'sand barite', 'rock roses' or 'desert roses'. According to Tarr (1933), the baryte acts as a cement for the sand grains and has itself been deposited through reaction between a soluble sulphate and water carrying $BaCl_2$. The source of barium is believed to be mainly barium silicates, but also detrital baryte and barium adsorbed on colloids. Baryte also occurs as an authigenic marine precipitate, in residual deposits when continental host rocks are weathered away, and as a secondary mineral in both hydrothermal and diagenetic environments. Minerals which commonly accompany baryte include fluorite, quartz, calcite, dolomite, siderite, galena, pyrite and sphalerite. Among the major reserves of baryte are those of Nevada and Arkansas, USA, and Chaillac, Massif Central, France. A useful review of the formation of baryte in marine environments was presented by Church (1979).

Whereas many earlier investigations concluded that baryte–fluorite veins were derived from mantle fluids, the determination of strontium isotope ratios points to a crustal derivation rather than a uniform mantle source. Even so, the possible sources of Ba^{2+} and SO_4^{2-} leading to the formation of baryte are many and varied. The baryte in hydrothermal veins, fissures, etc., can receive its SO_4^{2-} from sulphides, cap-rock or other adjacent sulphate-bearing strata, or from sea-water. The barium can have its origins in micas, feldspar, etc., in granitic or other igneous rocks, in dolomites or limestones, or in clays or shales.

Similar sources for barium can apply to sedimentary baryte which can occur in either marine or lacustrine conditions; there is also the possibility that barium is extracted from sea-water via marine organisms. The waters involved in the transport of ions required to form baryte can be meteoric, connate or groundwater, as well as sea-water.

Much use is now made of isotope ratios for sulphur, oxygen and strontium in baryte. These data, and also fluid inclusion studies, can usefully supplement chemical, mineralogical and geological observations to at least indicate whether or not two or more deposits are genetically related, and in many occurrences point in each case to likely source rocks and/or formation waters for the barium and sulphate constituents. A recent example of such studies is that by Hall *et al.* (1991) on baryte mineralization at Aberfeldy, Scotland (see p. 17). Halas and Wolacewicz (1981) described a method for extracting SO_2 from sulphates (anhydrite, baryte or celestine) for sulphur isotope analysis. This involves reacting the sulphate with $NaPO_3$ *in vacuo* at 850 °C, and reducing the SO_3 so obtained to SO_2 on copper heated to 750 °C. Another method of preparing baryte for chemical and isotope analysis recommended by Breit *et al.* (1985) involves using $NaCO_3$ to replace SO_4 with CO_3. The SO_4 in solution is separated and used for S and O isotope analysis; the carbonate phase can be used to provide the strontium for isotope analysis. An early use of S and O isotopes was the study by Harzer *et al.* (1964) of baryte in the Erzebirge, Saxony, for which they found that there was enrichment of the heavier isotopes ^{34}S and ^{18}O in the secondary compared with primary baryte. The principal factors affecting $\delta^{34}S$ [a] are the source of the sulphur, the temperature and the proportion of oxidized and reduced species in solution.

There are many examples throughout the world of baryte occurring with lead and zinc sulphides. In the UK such hydrothermal deposits are found in the North Pennine and South Pennine ore fields (e.g. Dunham, 1934, 1990). Ford and King (1965) described baryte with galena and fluorite in Carboniferous Limestone (Derbyshire), where each phase of mineralization produced solution cavities in porous dolomite in which these minerals accumulated; they suggested that the

[a] For definition of $\delta^{34}S$, see p. 86.

fluids responsible were brines from deep groundwaters. Schulz (1968) described baryte in the form of crystal platelets associated with fluorite, quartz, calcite, sphalerite, galena, marcasite and clay minerals in the Upper Wetterstein limestone, Bleiberg Gailtaler Alps, and suggested its formation via submarine hydrothermal solutions. For the formation of the baryte from central Kentucky (USA), Plummer (1971) suggested an ore fluid–groundwater interaction involving the reaction:

$$Ba^{2+} + H_2S + CaCO_3 + 2O_2 = Ca^{2+} + BaSO_4 + CO_2 + H_2O.$$

Heinrich and Vian (1967) found that baryte in carbonatites from many localities show a rather wide range of Sr content, and considered that they formed with decreasing temperature from hydrothermal fluids.

Studies of fluid inclusions in baryte from the North Pennine ore field (Sawkins, 1966) indicated that baryte formed from solutions with Na/K ratios between 15.3 and 46.0. The fluorite–baryte zoning there was explained as due to the mixing of juvenile hydrothermal solutions with Ba-rich connate waters. Sawkins drew attention to the similarity of North Pennine and Mississippi Valley deposits. Temperatures for baryte formation were estimated as ranging from 130 °C down to <50 °C. Similar temperatures (165–80 °C) were derived by Doroshenko (1966) from his studies of gas–liquid inclusions in baryte; somewhat higher temperatures (126–297 °C) were proposed by Rife (1971) for baryte from Cartersville, northwest Georgia, USA, compatible with solutions derived from a buried pluton (but see also Rife, 1972; Roedder, 1972). For the lead–zinc–baryte–fluorite–calcite mineralization in the South Pennine (England) ore field, measurements of sulphur and oxygen isotopes gave wide ranges: $\delta^{34}S$ +4 to 23‰ and $\delta^{18}O$ +9 to −26‰ (Robinson and Ineson, 1979). These were also attributed to different degrees of mixing of freshwater sulphate (low $\delta^{34}S$) with connate sea-water sulphate ($\delta^{34}S \sim +19$‰).

Honma and Okamura (1975), using the lead content (120–290 p.p.m. $PbSO_4$) and fluid inclusion data, deduced the conditions of formation of baryte from the Oe mine, Hokkaido, Japan, to be T 180–230 °C, f_{O_2} $10^{-40.3}$–$10^{-35.7}$ atm, f_S $10^{-13.8}$–$10^{-11.6}$ atm, and f_{CO_2} $10^{-1.5}$–$10^{-0.4}$ atm.

Rafter and Mizutani (1967) determined $\delta^{18}O$ for 40 samples of baryte from Tasmania and England, and these showed a narrow spread for any one mining area but significant differences between them. They also noted a correlation between $\delta^{34}S$ and $\delta^{18}O$, a relationship much studied and developed by later workers.

Perry *et al.* (1971) found that sedimentary baryte of the Swaziland system of South Africa, dated at >3000 m.y., had $\delta^{34}S$ ratios enriched by only 2.5‰ with respect to contemporary sulphides. They concluded that the P_{O_2} of the Earth's atmosphere must have been very low, and that local oxidation had occurred in a photosynthetic layer of the ocean.

Determinations of sulphur and oxygen isotopes in baryte (Solomon *et al.*, 1971) showed fairly uniform results over a wide area of the northern Pennines. They concluded that SO_4^{2-} had come from trapped Lower Carboniferous sea-waters which had mixed with warm brines heated by rising through granitic plutons. Hirst and Smith (1974) invoked the mixing of $BaCl_2$-bearing brines derived from Coal Measures with SO_4^{-2}-bearing brines from the Permian to explain mineralization in the Lower Magnesian Limestone of Ferryhill, County Durham, England.

Sakai and Matsubaya (1971) compared sulphur and oxygen isotope ratios for baryte, anhydrite and gypsum in the black ore deposits of Japan. Whereas $\delta^{34}S$

was similar (+21 to +24‰) for all three minerals, the δ^{18}O values increased in the order baryte, anhydrite, gypsum (see p. 88).

On the basis of experiments on the solubility of baryte in water and NaCl solution, Scherp and Strübel (1974) inferred that for baryte formation in areas of West Germany which they studied, chloridic (Ba,Sr)-bearing Na and (Na,Ca) solutions which originated in meteoric waters in the fault system reacted with fossil waters from Ba-rich strata.

Results obtained by Kusakabe and Robinson (1977) on oxygen and sulphur isotope exchange between baryte and various solutions yielded reasonable temperatures when tested against natural occurrences, e.g. for the Derbyshire ore field $T < 150\,^{\circ}$C, and for the Yauricocha deposit, Peru, $T = 350 \pm 20\,^{\circ}$C, a result consistent with a magmatic origin for the ore fluids.

Buchanan *et al.* (1981) estimated from studies of fluid inclusions that the formation of the baryte associated with ore deposits at Catanzaro, southern Italy, was from solutions ranging from 0 to 19.5 wt. per cent NaCl at minimum temperatures around 190–235 °C. In the case of baryte–sulphide deposits in northeast British Columbia, MacIntyre (1982) concluded that these were formed when brines discharged along fault zones within the Kechica trough and accumulated in sea-floor depressions. With regard to studies of fluid inclusions, Ulrich and Bodnar (1988) warn that for baryte even a small amount of heating beyond the true homogenization temperature can lead to stretching of the inclusions and misleading results.

Coats *et al.* (1980) studied interbanded baryte rock and quartz–celsian rock (as well as sulphide-bearing carbonate rock, muscovite schist and graphitic mica schist) in the region of Aberfeldy (Scotland) and concluded that it is of synsedimentary origin involving the introduction of a metal-rich hydrothermal brine into a euxinic basin. Willan and Coleman (1983) found a bimodal distribution of δ^{34}S at +35 and +31‰ for baryte of the Aberfeldy lead–zinc–baryte deposit. They concluded that most of the sulphur was derived during periods of rapid mixing of hot brine with sea-water enriched in ^{34}S by the bacterial reduction of sulphate within the enclosed Dalradian basin for which δ^{34}S ~ +35‰. When mixing was less rapid, more light sulphate was derived by the slow oxidation of exhalative reduced sulphur (δ^{34}S < +28‰). Pouit and Bois (1986) pointed out the similarity of the Zn, Pb and Ba deposits at Arrens, Pyrenées, and Meggen, Germany, to those at Aberfeldy. Zimmerman (1970) had noted similarities in the sedimentary features of baryte deposits of the Arkansas barytes district to those of Meggen. A combined S, O and Sr isotope study was undertaken by Hall *et al.* (1991) of the stratiform baryte mineralization at Aberfeldy. They described the most common assemblage as baryte+pyrite, but baryte+magnetite also occurs. The δ^{34}S values were respectively +40 and +35 to +38‰, δ^{18}O +14 and +16‰, and ^{87}Sr/^{86}Sr 0.715 and 0.714. These results suggested that reduced hydrothermal Ba^{2+} solutions had mixed with Dalradian anoxic sea-water, and that oxygenic sea-water had not been significantly involved. The S and O isotopic compositions were found to be linearly related by $\delta^{18}O = (-0.61\delta^{34}S) + 38.68$, as plotted in Fig. 5, the negative slope of which contrasts with the positive slope for sulphides precipitated by bacterial reduction. A positive slope would also arise if there was equilibrium fractionation of sulphur during oxidation within the hydrothermal system, as is the case for the Kuroko deposits described by Ohmoto and Skinner (1983).

From their studies of S isotopes of Lower Carboniferous vein deposits of the
hydrothermal system, as is the case for the Kuroko deposits described by Ohmoto and Skinner (1983).

From their studies of S isotopes of Lower Carboniferous vein deposits of the

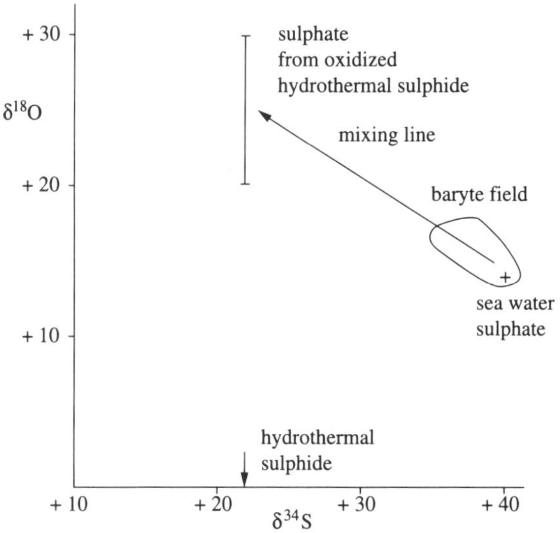

Fig. 5. **$\delta^{18}O$ and $\delta^{34}S$ for sulphates derived from sea-water, oxidized hydrothermal sulphide and mixtures (after Hall *et al.*, 1991).**

British Isles, Pattrick and Russell (1989) deduced that whereas the sulphur of sulphides was derived from underlying strata, the sulphur of barytes ($\delta^{34}S$ +12 to +25‰) was mainly from groundwater or surface water (e.g. continental lakes mixed with rising hydrothermal fluids) (see also Russell, 1986; Andrew, 1993).

From Permian and Triassic Italian marine sulphates, Cortecci *et al.* (1981) found that $\delta^{34}S$ values agreed with those presented by Claypool *et al.* (1980) in their age–isotope curve. There was no consistent relationship, however, between oxygen and sulphur isotope values, explained probably by a post-depositional change in oxygen isotope ratios.

Values of $\delta^{34}S$ (+20.6 to +35.4‰) which were unusually high for Precambrian sulphur were found for baryte in a metavolcanic–sedimentary sequence in Namaqualand, South Africa (von Gehlen *et al.*, 1983). The range observed was taken to indicate varying contributions of sulphur from exhalative sulphide with low $\delta^{34}S$, and that derived from evaporites in the east with high $\delta^{34}S$.

Curves showing the combined variations of $\delta^{34}S$ and $\delta^{18}O$ for oceanic sulphate with geological time were plotted by Claypool *et al.* (1980) and have been used by many subsequent researchers to help determine the provenance of SO_4^{2-} in baryte in various deposits. It is thought that isotopic fractionation occurred during processes of sulphate removal, e.g. by bacterial reduction or evaporite formation. The $\delta^{34}S$ for present ocean sulphate is ~20‰ compared with the average of ~1‰ for sulphides in basic magmas. Lower $\delta^{34}S$ values are found in sulphates derived by the oxidation of hydrothermal sulphur compared with that derived from sea-water sulphate.

Cecile *et al.* (1983) measured $\delta^{34}S$ and $\delta^{18}O$ for 17 samples of stratiform baryte from western Canada. Thick accumulations of baryte in organic-rich strata and also thin beds in organic-poor strata give data which can provide a useful baryte age curve, whereas thin beds and nodules hosted in organic-rich strata are enriched in heavy isotopes through bacterial activity which removes the light isotopes from the aqueous sulphate supply. Extremely high values of $\delta^{34}S$ were obtained for baryte in Lower Cambrian black shales (Wang *et al.*, 1993), and were attributed to a longer period of reduction by bacteria.

Veins of baryte and fluorite constitute an important group of deposits south of the French Massif Central (Jebrak *et al.*, 1985). Study of the *REE* in the fluorite, and oxygen ($\delta^{18}O$ +11 to +15‰) and sulphur isotopes in the baryte, indicate that the veins result from circulated hydrothermal fluids which leached the basement rocks and precipitated the baryte and fluorite when they encountered Jurassic marine waters.

The use of strontium isotope ratios, $^{87}Sr/^{86}Sr$, to identify the source of strontium in a baryte, and thus the origin of an ore deposit, is illustrated by the work of Kesler and Jones (1981), Kesler *et al.* (1983) and Barbieri *et al.* (1987). In the latter, the Sr ratios for baryte and associated fluorite from northwestern Sicily were found to differ, indicating that the fluorite and baryte were derived from different fluids. From their isotopic study of baryte, celestine and gypsum from the Mesozoic basins of northeastern Mexico, Kesler and Jones (1981) concluded that at least two distinct solutions existed and mixed locally during the formation of the celestine and baryte deposits. The baryte contained more radiogenic strontium and indicated that its strontium and probably barium may have originated in basement-derived arkoses of the Cretaceous basin.

In a study by Barbieri *et al.* (1984b) of baryte deposits from Sardinia, Sr isotope data gave evidence of several different origins. In the Cambrian stratiform baryte the $^{87}Sr/^{86}Sr$ ratio of 0.70867 ± 3 was similar to that of Cambrian sea-water. For hydrothermal baryte the ratios (0.70990–0.71837) suggested derivation from a source with high ratio, probably the Lower Palaeozoic marine shales. For karst baryte (which contains less strontium than hydrothermal baryte) the ratio 0.70947 was taken to indicate that it was derived partially from the non-carbonate fraction of the Cambrian wall-rocks.

Dill and Carl (1987) studied vein baryte from the northeast Bavarian basement and used the Sr isotope ratios (0.707117–0.712561) as direct evidence for the environment of fluids percolating through granites and their overlying rocks (see also Dill and Nielsen, 1987). In a further study of fluorite–baryte mineralization in southern Germany, Dill (1988) used *REE* contents, strontium and sulphur isotope determinations and fluid inclusion methods to identify three types of vein deposit: endogenous (Permian), exogenous accumulation of fluorite/baryte by the erosion of pre-Permian source rocks, and 'epigenetic unconformity-related', so called because their concentrations of Ba and F had been influenced by clastic Permo-Triassic sequences lying unconformably above. The ^{87}Sr isotope was found to be more abundant in baryte derived from the crystalline basement or granitic terrains than in baryte hosted by cover sediments.

Cortecci *et al.* (1989) used S, O, C and Sr isotope determinations and fluid inclusion methods to investigate baryte–iron oxide–pyrite deposits from the Apuane Alps, northern Tuscany, Italy. They found that stratiform baryte had $\delta^{34}S$ +28.8 to +32.1‰, $\delta^{18}O$ +12.7 to +15.6‰ and $^{87}Sr/^{86}Sr$ 0.7087, values consistent with the derivation of both the sulphate and the strontium from Cambrian marine waters. For both epigenetic karst and vein baryte they found $\delta^{34}S$ +15.3 to +26.4‰, $\delta^{18}O$ +6.6 to +12.5‰ and $^{87}Sr/^{86}Sr$ 0.7094 to 0.7140, values different from those of stratiform baryte. They suggested that both were formed by the oxidation of reduced sulphur from sulphide ores in warm or hot mixtures of dilute water solutions and saline brines of different $\delta^{18}O$ values.

Archaean sulphates have generally been considered to have formed by the biological oxidation of volcanic sulphides (Lambert *et al.*, 1978). In the case of baryte–celestine intergrowths in plutons of the Kirkland Lake igneous complex

(Abitibi greenstone belt, Ontario, Canada), however, Hattori (1989) used Sr isotope ratios and $\delta^{34}S$ (+6‰) to show that the intergrowths formed as a Ba–Sr solid solution at temperatures >400 °C during oxidizing hydrothermal activity in late Archaean times during emplacement of the alkaline complex.

The strontium isotope ratios for baryte and celestine from the Mesozoic marine basin in central western Argentina range from 0.705 to 0.708 and were taken by de Brodtkorb *et al.* (1989) to indicate a sedimentary syndiagenetic origin. A further report (de Brodtkorb and Barbieri, 1993) concluded that although some of the barium may have volcanogenic sources, the initial Sr ratios are within the range for contemporaneous sea-water, and the deposits in nearshore environments may have formed through the interaction of barium in clay minerals derived from basement rocks and hypersaline sea-water. Stratabound deposits of baryte in several other parts of the world were reviewed in the volume edited by de Brodtkorb (1989).

The Mississippi Valley-type lead–zinc–fluorite–baryte deposits in the Carboniferous of the North Pennines and Peak District of England are reviewed by Ixer and Vaughan (1993), and baryte in the deposits associated with the granites of Devon and Cornwall by Alderton (1993).

Demaiffe and Dejonghe (1990) determined $^{87}Sr/^{86}Sr$ for baryte associated with fluorite and calcite in Belgian lead–zinc deposits. The isotope ratio was >0.710, which is the accepted value for sea-water at the time of deposition of the host rock. It was suggested therefore that the Sr and Ba had a crustal origin.

For the origin of baryte from vein deposits ($^{87}Sr/^{86}Sr$ in the range 0.71042–0.71541), Radvanec *et al.* (1990) ruled out mantle-derived basic or ultrabasic rocks, and also evaporites or Gemeric Permian marine limestones, and concluded that the baryte was derived from the regional amphibolite facies metamorphism of crustal rocks.

In several reported occurrences of baryte, isotopic data have indicated that its formation involved the remobilization of previous sulphate deposits, e.g. dissolved cap-rock sulphate (von Gehlen *et al.*, 1962), sedimentary gypsum (Lambert *et al.*, 1978; Kesler *et al.*, 1988; Parafiniuk, 1989) and anhydrite-rich evaporites (Frimmel and Papesch, 1990).

Baryte (mainly filling veins) in a variety of forms, and with increase in Sr content over the time of deposition, occurs together with quartz, goethite and hematite at Purroy, Zaragoza, Spain (Soriano *et al.*, 1991).

Marine baryte. A thorough review of marine baryte was presented by Church (1979). He reports that very fine-grained baryte makes up 1–3 per cent of many pelagic sediments from the eastern equatorial Pacific, and in southern latitudes baryte seems to be restricted to the East Pacific Rise. It occurs (a) with abundant biological debris, (b) together with abundant Mn and Fe phases (concretionary and dispersed), and (c) with altered volcanic debris (obtaining Ba from the weathering of basalt) consisting mostly of montmorillonite but also some palygorskite and zeolites. Some zeolites may compete with baryte for the incorporation of barium. This baryte of volcanic association is distinct chemically from baryte of other volcanic origins (Goldberg *et al.*, 1969; Guichard *et al.*, 1979).

In a comparative study, Goldberg *et al.* (1969) found that whereas the thorium and uranium contents of marine baryte were 35 and 3 p.p.m. respectively, for continental baryte thorium was ~100 times less and uranium was undetectable. For marine baryte the $^{87}Sr/^{86}Sr$ ratio (0.7093) was close to that of sea-water. For the continental baryte studied, strontium isotope ratios fell into two groups:

~0.7107 and ~0.7060. The latter low value was taken to indicate an amount of strontium of deep-seated origin.

The amount of suspended marine baryte (particle diameter ~1 μm) varies from about 40 μg/kg at the surface to about 10 μg/kg at the bottom of the ocean water column, and it appears to have formed by a process of organic degradation of Ba-enriched biological material, the barium then forming baryte in SO_4-rich environments. Church (1979) estimated that suspended oceanic baryte could contribute one-half of all sedimentary baryte, the remainder growing in surface sediments after burial.

Church (1979) also discussed sulphur and oxygen isotope ratios, which for marine baryte are of course those of sea-water, in contrast to those for continental and borderland baryte. Sulphur isotope fractionation is caused by the bacterial reduction of sulphate, which removes heavy S preferentially as pyrite, leaving baryte to precipitate from the reservoir of lighter sulphur. For oxygen isotopes, the fine-grained (<5 μm) baryte is richer in light oxygen and perhaps also in light sulphur, whereas the more coarse fraction has values similar to sea-water. Borderland baryte tends to be richer in the heavy isotope of both sulphur and oxygen. The light isotopes are removed preferentially by exchange or by the evolution of water, and the baryte precipitates from the residual 'heavy' reservoir.

Bogoch and Shirav (1978) described a small baryte deposit from the Judean desert, Israel, as sedimentary rather than hydrothermal in origin as had been considered previously. Amongst the suggested indicators of a low-temperature sedimentary origin for baryte are the very fine grain size (Mills *et al.*, 1971) and the low trace element content (Shawe *et al.*, 1969).

For baryte in Devonian carbonate rocks of northeastern British Columbia, Morrow *et al.* (1978) found that $\delta^{34}S$ ranged for most samples from +24.2 to +30.1‰, and suggested mineralization by the mixing of reducing sea-water with meteoric groundwater near the seaward edge of a large coastal aquifer causing the contemporaneous solution of dolomite and the precipitation of baryte and fluorite. This was supported by the observation of solution–collapse textures and by the observed concentration of Mn (40–65 p.p.m.) and Sr (~1 per cent); Mn content is much lower than that for baryte precipitated from hydrothermal solutions.

Pak *et al.* (1980) investigated the sulphur isotope compositions for baryte from the stratabound Lower Devonian ore deposits of Graz (Austria) and found values of +23.1 to +27.8‰ which corresponded with Devonian sea-water sulphate.

For baryte occurring with sphalerite, galena and zinc carbonates in epimetamorphic dolomites (Calabria, Italy), Barbieri *et al.* (1984a) found that the strontium content ranged from 5700 p.p.m. for synsedimentary deposits to 32 000 p.p.m. where baryte had been remobilized by metamorphism. The Sr isotope ratios, 0.707029–0.707103, were consistent with direct precipitation from sea-water during the mid- to late Triassic.

The isotopic abundances of sulphur ($\delta^{34}S$ +28.0 to +47.4‰) and oxygen ($\delta^{18}O$ +19.3 to +30.4‰) in baryte in bedded deposits in Devonian rocks of the Valley and Ridge Province, Appalachian mountains, USA, showed enrichment in the heavy isotopes relative to co-eval sea-water (Nuelle and Shelton, 1986), leading to the conclusion that Ba-bearing fluids were derived from submarine springs and deposited baryte in submarine euxinic basins. The sulphur in baryte in veins and vugs was generally heavier than in host nodules, suggesting that ^{34}S was remobilized preferentially during diagenesis.

A syngenetic–diagenetic stratiform baryte deposit in Alcañices (Spain) was described by Moro *et al.* (1991). They suggested that the barium was probably transported to the basin as a chloride complex by exhalative hydrothermal fluids passing through deep fractures or submarine vents and interacting with sulphate of sea-water.

In the Aptian–Albian Formation (southeast France), three main horizons rich in baryte nodules are closely related to sedimentary discontinuities. It was suggested (Bréhéret and Delamette, 1989) that the crystallization of baryte was linked genetically to pauses in sedimentation (see also Dill and Nielsen, 1987) but not to local hydrothermal activity.

For the authigenic baryte which occurs in the Tertiary marine sedimentary rocks of Lefkas Island, Greece, Stamatakis and Hein (1993) concluded that the barium was derived from the decomposition of organic matter and from the diagenesis of silica and carbonate tests; the sulphate ions were derived largely from sea-water.

Submarine baryte. Baryte in a slab from the sea-floor was found by Cortecci and Longinelli (1972) to be enriched in ^{18}O (+16 to +20‰) compared with oceanic sulphate (~9.5 per cent). They concluded that it formed by reaction between ascending magmatic solutions containing barium and the interstitial water of sediments, the SO_4 of which was probably enriched in ^{18}O by bacterial activity.

An unusual submarine baryte concretion, consisting of baryte, opal and volcanic debris, was found in the Lau Basin, southwest Pacific (Bertine and Kenne, 1975). Sulphur and isotope data were used to deduce that the opal and baryte were precipitated at 117–134 °C when hydrothermal solutions came into contact with cooler ocean water.

Baryte is a common gangue mineral in massive submarine volcanogenic sulphide deposits, including those known as 'Kuroko-type'. Examples of such occurrences are those described by Kusakabe and Chiba (1979, 1983), Eldridge *et al.* (1983), Farrell and Holland (1983), Vidal (1987) and Marumo (1989). Watanabe and Sakai (1983) tabulated oxygen and sulphur isotope data for baryte from Kuroko deposits and compared them with other parageneses. Most $\delta^{18}O$ ranged narrowly from +6.4 to +10.9‰, and most $\delta^{34}S$ from +19.0 to +25.5‰, showing that the ore fluids were predominantly Miocene sea-water. In contrast, baryte from hydrothermal vein sulphate shows a mostly lower $\delta^{18}O$ (0 to +14‰) and a higher and more scattered (due to mixing) $\delta^{34}S$ (+16 to +29‰). Sulphur isotope data for baryte and anhydrite associated with the massive sulphide deposits of the East Pacific Rise show that sulphate in ambient sea-water rather than hydrothermal fluid is the source of their sulphur.

More recently, in the discovery and investigations of sea-floor chimneys, spires and mounds, baryte has been found as an important constituent. Baryte and iron oxide-bearing chimneys on the flank of a rhyolite lava dome (Sumusu Rift, Izu-Bonin Arc, northwestern Pacific) are interpreted as an end product of Kuroko-type mineralization (Urabe and Kusakabe, 1990). The ^{18}O content is slightly higher than for Kuroko deposits because of the extensive mixing of sea-water with the hydrothermal solutions of the rift. A similar mode of formation was suggested by Peter and Scott (1988) for sea-floor baryte from the Southern Trough of the Guyayamas Basin, Gulf of California. The baryte-rich chimneys described by Grasty *et al.* (1988) were found to be unusually radioactive (0.1 per cent Th, 0.2 per cent U), and they used the level of radioactivity to estimate that the residence time of the fluids which gave rise to these deposits was about 20 years.

Baryte in residual deposits. An example of residual baryte is provided by that which occurs in a sulphide–baryte–carbonate lode at Roseberry, Tasmania

(Solomon *et al.*, 1969). For this baryte average values of $\delta^{34}S$ and $\delta^{18}O$ were respectively +38.1 and +10.7‰, and it was suggested that it was residual from adjacent sulphide ores. Kaiser *et al.* (1987) reported residual baryte deposits derived from the weathering of Palaeozoic dolostones (Missouri, USA). The early baryte mineralization, however, was caused by the mixing of brines from faults and joints in the dolostones with oxidizing sulphate-bearing dilute fluid.

Magmatic and other baryte. A primary magmatic origin was suggested for crystals of baryte occurring in vesicles in a rhyolite porphyry (Frank and Moynihan, 1947), and also for a baryte vein cutting granite in southeastern Missouri, USA (Tarr, 1932). An unusual primary magmatic occurrence of baryte and celestine in diorite was reported from Ratagain, Scotland (May *et al.*, 1993).

A deposit of baryte in magnetite and hematite in Namaqualand, South Africa, was believed to be due to metasomatic replacement of quartzite (Mathias, 1940). Baryte, in particular the lead-bearing variety, also occurs as a hot spring deposit. It is sometimes pseudomorphous after calcite, barytocalcite and witherite, and often occurs as a petrifying mineral of fossils.

Baryte was identified in chondritic interplanetary dust particles and is believed to have formed by precipitation from mildly acidic aqueous fluid below 417 K and at low oxygen fugacity (Rietmejer, 1990).

References

Alderton, D. H. M., 1993. Mineralisation associated with the Cornubian granite batholith. In *Mineralisation in the British Isles*: Eds R. A. D. Pattrick and D. A. Polya. London (Chapman & Hall), pp. 270–354.

Andrew, J., 1993. Mineralization in the Irish Midlands. In *Mineralization in the British Isles*: Eds R. A. D. Pattrick and D. A. Polya. London (Chapman & Hall), pp. 208–69.

Barbieri, M., De Vivo, B., Perrone, V. and **Turco, E.,** 1984a. Strontium geochemistry of the San Donato Unit barite mineralization (Calabria, Italy). *Chem. Geol.*, **45**, 279–88.

Barbieri, M., Masi, U. and **Tolomeo, L.,** 1984b. Strontium geochemical evidence for the origin of the barite deposits from Sardinia, Italy. *Econ. Geol.*, **79**, 1360–5.

Barbieri, M., Bellanca, A., Neri, R. and **Tolomeo, L.,** 1987. Use of strontium isotopes to determine the sources of hydrothermal fluorite and barite from northwestern Sicily (Italy). *Chem. Geol.*, **66**, 273–8.

Barić, L., 1948. La baryte des mines de Trepča. *Bull. Mus. Hist. Nat. Pays Serbe, Sér. A. Min. Géol., Beograd*, **1**, 71–9 (M.A. **11**-64).

Basche, W. and **Mark, H.,** 1926. Über die Struktur von Verbindungen des Typus $MeXO_4$. *Z. Krist.*, **64**, 1–70.

Bertine, K. K. and **Keene, J. B.,** 1975. Submarine barite–opal rocks of hydrothermal origin. *Science*, **188**, 150–2.

Blount, C. W., 1974. Synthesis of barite, celestite, anglesite, witherite and strontianite from aqueous solutions. *Amer. Min.*, **59**, 1209–19.

Blount, C. W., 1977. Barite solubilities and thermodynamic quantities up to 300 °C and 1400 bars. *Amer. Min.*, **62**, 942–57.

Bogoch, R. and **Shirav, M.,** 1978. Petrogenesis of a Senonian barite deposit, Judean Desert, Israel. *Min. Deposita*, **13**, 383–90.

Boström, K., Frazer, J. and **Blankenburg, J.,** 1967. Subsolidus phase relations and lattice constants in the system $BaSO_4$–$SrSO_4$–$PbSO_4$. *Arkiv. Min. Geol.*, **4**, 477–85.

Braun, F., 1932. Morphologische, genetische und paragenetische Trachtstudien an Baryt. *Neues Jahrb. Min.*, **65A**, 173–222.

Bréhéret, J.-G. and **Delamette, M.,** 1989. Les nodules barytiques d'âge crétacé moyen dans le domain vocontien (SE France), marqueurs de discontinuitiés sédimentaires en série marneuse de bassin. *Comptes Rend. Acad. Sci. Paris, Ser. II*, **308**, 1369–74.

Breit, G. N., Simmons, E. C. and **Goldhaber, M. B.,** 1985. Dissolution of barite for the analysis of strontium isotopes and other chemical and isotopic variations using sodium carbonate. *Chem. Geol.*, **52**, 333–6.

Bridgman, P. W., 1949. Linear compressions to 30,000 kg./cm.^{2}, including relatively incompressible substances. *Proc. Amer. Acad. Arts Sci.*, **77**, 187–234.
Brower, E., 1973. Synthesis of barite, celestite and barium–strontium sulfate solid solution crystals. *Geochim. Cosmochim. Acta*, **37**, 155–8 and 2688–9.
Buchanan, L. J., **de Vivo, B.**, **Kramer, A. K.** and **Lima, L.**, 1981. Fluid inclusion study of the Fiumarella barite deposit (Catanzaro – S. Italy). *Min. Deposita*, **16**, 215–26.
Burkhard, A., 1973. Optische und röntgenographische Untersuchungen am system $BaSO_4$–$SrSO_4$ (Baryt–Coelestin). *Schweiz. Min. Petr. Mitt.*, **53**, 185–97.
Burkhard, A., 1978. Baryt–Celestin und ihre Mischkristalle aus Schweizer Alpen und Jura. *Schweiz. Min. Petr. Mitt.*, **58**, 1–95.
Caillère, S. and **Pobeguin, Th.**, 1962. Comparison des propriétés thermiques et du comportement dans l'infrarouge de quelques carbonates, nitrates, sulfates et sélénites simple. *Bull. Soc. Franc. Min. Crist.*, **85**, 48–57.
Cecile, M. P., **Shakur, M. A.** and **Krouse, H. R.**, 1983. The isotopic composition of western Canadian barites and the possible derivation of oceanic sulphate $\delta^{34}S$ and $\delta^{18}O$ age curves. *Can. J. Earth Sci.*, **20**, 1528–35.
Chen, C. J. and **Yu, S. C.**, 1984. A substructural study of hokutolite. *Mem. Geol. Soc. China (Taiwan)*, 229–37 [M.A. **37**-2935].
Christy, A. G. and **Putnis, A.**, 1993. The kinetics of barite dissolution and precipitation in water and sodium chloride brines at 44–85 °C. *Geochim. Cosmochim. Acta*, **57**, 2161–8.
Church, T. M., 1979. 'Marine barite' in *Marine Minerals*, Vol. 6 of Short Course Notes, (Min. Soc. Amer.), Washington DC, 175–209.
Claypool, G. E., **Holser, W. T.**, **Kaplan, I. R.**, **Sakai, H.** and **Zak, I.**, 1980. The age curves of sulphur and oxygen isotopes in marine sulfate and their mutual interpretation. *Chem. Geol.*, **28**, 99–260.
Coats, J. S., **Smith, C. G.**, **Fortey, N. J.**, **May, F.** and **McCourt, W. J.**, 1980. Stratabound barium–zinc mineralization in Dalradian schist near Aberfeldy, Scotland. *Trans. Inst. Min. Metall.*, **89(B)**, 110–32.
Colville, A. and **Staudhammer, K.**, 1967. A refinement of the structure of barite. *Amer. Min.*, **52**, 1877–80.
Cortecci, G. and **Longinelli, A.**, 1972. Oxygen isotope variations in a barite slab from the sea bottom off southern California. *Chem. Geol.*, **9**, 113–7.
Cortecci, G., **Reyes, E.**, **Berti, G.** and **Casati, P.**, 1981. Sulfur and oxygen isotopes in Italian Marine sulfates of Permian and Triassic ages. *Chem. Geol.*, **34**, 65–79.
Cortecci, G., **Lattanzi, P.** and **Tanelli, G.**, 1989. Sulphur, oxygen and carbon isotope geochemistry of barite–iron and oxide–pyrite deposits from the Apuane Alps (northern Tuscany, Italy). *Chem. Geol.*, **76**, 249–57.
Das, R., **Venkataraman, B.**, **Mineeva, R. M.** and **Bershov, L. V.**, 1990. Electron spin-lattice relaxation of SO_3^- radial ion in a single crystal of barite. *Phys. Chem. Min.*, **17**, 24–6.
de Brodtkorb, M. K., Ed., 1989. *Non-metalliferous Stratabound Ore-Fields.* London (Chapman & Hall), 322 pp.
de Brodtkorb, M. K. and **Barbieri, M.**, 1993. Jurassic barite and celestite deposits of Nequén Province, Argentina. *Proc. 8th IAGOD symp.* Stuttgart (Schweizerbart'sche Verlag), pp. 243–54.
de Brodtkorb, M. K., **Schalamuk, I. B. A.** and **Arnetrano, S.**, 1989. Barite and celestite stratabound ore fields in Argentina. In *Non-Metalliferous Stratabound Ore-Fields.* Ed. M. K. de Brodtkorb, London (Chapman & Hall), pp. 41–68.
Demaiffe, D. and **Dejonghe, L.**, 1990. Géochemie isotopique du strontium des barites, anhydrites, calcites et fluorites de Belgique. *Ann. Soc. Géol. Belg.*, **113**, 231–40.
Denis, J. and **Michard, G.**, 1983. Dissolution d'une solution solides étude theoretique et experimentale. *Bull. Min.*, **106**, 309–19.
Dill, H., 1988. Geologic setting and age relationship of fluorite–barite mineralisation in southern Germany with special reference to the late Paleozoic unconformity. *Min. Deposita*, **23**, 16–23.
Dill, H. and **Carl, C.**, 1987. Sr isotope variation in vein barites from the NE Bavarian basement: relevance for the source of elements and genesis of unconformity-related barite deposits. *Min. & Petr.*, **36**, 27–39.
Dill, H. and **Nielsen, H.**, 1987. Geochemical and geological constraints on the formation of unconformity-related vein baryte deposits of Central Europe. *J. Geol. Soc.*, **144**, 97–105.
Donnay, J. D. H. and **Harker, D.**, 1937. A new law of crystal morphology extending the law of Bravais. *Amer. Min.*, **22**, 446–67.
Dons, J. A., 1956. Barite which decrepitates at room temperatures. *Norsk. Geol. Tidssk.*, **36**, 241–66.
Doroshenko, Y. P., 1966. Crystallomorphological peculiarities and barite forming temperatures from fluorite deposits of Transbaikal. *Min. Sborn. Lvov.*, **20**, 458–9 [M.A. **19**–143].
Dowty, E., 1976. Crystal structure and crystal growth: I. The influence of internal structure on morphology. *Amer. Min.*, **61**, 448–59.
Dunham, K. C., 1934. The genesis of the North Pennine ore deposits. *Quart. Journ. Geol. Soc.*, **90**, 689–720.
Dunham, K. C., 1990. Geology of the Northern Pennine orefield (vol. 1, Tyne to Stainmore). 2nd edn.

Mem. Geol. Surv. Gt. Britain.

Eldridge, C. S., Barton, P. B. and **Ohmoto, H.,** 1983. Mineral textures and their bearing on formation of the Kuroko orebodies. In *Econ. Geol. Monograph 5. The Kuroko and Related Volcanogenic Sulfide Deposits*: Eds H. Ohmoto and B. J. Skinner. (Econ. Geol. Publ. Co.), pp. 241–81.

Errerra, J. and **Brasseur, H.,** 1933. Ionenpolarisation in Kristallen. *Phys. Zeit.*, **34**, 368–73.

Farrell, C. W. and **Holland, H. D.,** 1983. Strontium isotope geochemistry of the Kuroko deposits. In *Econ. Geol. Monograph 5. The Kuroko and Related Volcanogenic Sulfide Deposits*: Eds H. Ohmoto and B. J. Skinner. (Econ. Geol. Publ. Co.), New Haven, CT, pp. 302–19.

Felmey, A. R., Rai, D. and **Moore, D. A.,** 1993. The solubility of $(Ba,Sr)SO_4$ precipitates: thermodynamic equilibrium and reaction path analysis. *Geochim. Cosmochim. Acta*, **57**, 4345–63.

Fernandez-Diaz, L., Putnis, A. and **Cumberbatch, T. J.,** 1990. Barite nucleation kinetics and the effect of additives. *Eur. J. Min.*, **2**, 495–501.

Ford, T. D. and **King, R. J.,** 1965. Layered epigenetic galena–barite deposits in the Golconda Mine, Brassington, Derbyshire, England. *Econ. Geol.*, **60**, 1686–1701.

Fowler, A., 1944. A deep bore in the Cleveland Hills. *Geol. Mag.*, **81**, 193–206.

Frank, A. J. and **Moynihan, C. S.,** 1947. Occurrence of barite at Pilot Knob, Missouri. *Amer. Min.*, **32**, 681–3.

Frimmel, H. E. and **Papesch, W.,** 1990. Sr, O and C isotope study of the Brixlegg barite deposit, Tyrol (Austria). *Econ. Geol.*, **85**, 1162–71.

Gage, M., 1944. Barite deposit near Stockton mine, Ngakanau survey district, Westport district. *New Zealand J. Sci. Tech., Sect. B*, **25**, 181–2 (M.A. **9**–252).

Goldberg, E. D., Somayajulu, B. L. K., Galloway, J., Kaplan, I. R. and **Faure, G.,** 1969. Difference between barites of marine and continental origins. *Geochim. Cosmochim. Acta*, **33**, 287–9.

Goldish, E., 1989. X-ray diffraction analysis of barium-strontium sulphate (barite–celestite) solid solutions. *Powder Diffraction*, **4**, 214–6.

Grahmann, W., 1920. Über Barytocölestin und das Verhältnis von Anhydrit zu Cölestin und Baryt. *Neues Jahrb. Min.*, **1**, 1–23.

Grasty, R. L., Smith, C. W., Franklin, J. M. and **Jonasson, I. R.,** 1988. Radioactive orphans in barite-rich chimneys, axial caldera, Juan de Fuca Ridge. *Can. Min.*, **26**, 627–36.

Griffith, W. P., 1969. Raman spectroscopy of minerals. *Nature*, **224**, 264–7.

Guichard, F., Church, T. M., Treuil, M. and **Jaffrezic, H.,** 1979. Rare earths in barites: distribution and effects on aqueous partitioning. *Geochim. Cosmochim. Acta*, **43**, 983–7.

Gundlach, H., Stoppel, D. and **Strübel, G.,** 1972. Zur hydrothermalen Löslichkeit von Baryt. *Neues Jahrb. Min., Abh.*, **116**, 321–8.

Halas, S. and **Wolacewicz, W. P.,** 1981. Direct extraction of sulfur dioxide from sulfates for isotopic analysis. *Anal. Chem.*, **53**, 686–9.

Hall, A.T., Boyce, A.J., Fallick, A. E. and **Hamilton, P. J.,** 1991. Isotopic evidence of the depositional environment of Late Proterozoic stratiform barite mineralisation, Aberfeldy, Scotland. *Chem. Geol.*, **87**, 99–114.

Hancox, E. G., 1934. Witherite and barytes. *Mining Mag.*, **51**, 76–9.

Hanor, J. S., 1968. Frequency distribution of compositions in the barite–celestite series. *Amer. Min.*, **53**, 1215–22.

Hanor, J. S., 1969. Barite saturation in sea water. *Geochim. Cosmochim. Acta*, **33**, 894–8.

Hanor, J. S., 1973. Synthesis of barite, celestite and barium–strontium sulfate solid solution crystals. *Geochim. Cosmochim. Acta*, **37**, 2685–7.

Hartman, P. and **Perdok, W. G.,** 1955. On the relations between structure and morphology of crystals, III. *Acta Cryst.*, **8**, 525–9.

Hartman, P. and **Strom, C. S.,** 1989. Structural morphology of crystals with the barite $(BaSO_4)$ structure: a revision and extension. *J. Cryst. Growth*, **97**, 502–12.

Harzer, D., Pilot, J. and **Starke, R.,** 1964. Schwefel- und Sauerstoffisotopenverhältnisse von Baryten des sächsichen Erzgebirges. *Bergakademie*, **16**, 709–13 (M.A. **17**–346).

Hasegawa, S., 1954. Barite from Kominato town, Aomori Prefecture. *Journ. Japan. Assoc. Min. Petr. Econ. Geol.*, **38**, 86–91 (M.A. **12**–477).

Hattori, K., 1989. Barite–celestine intergrowths in Archaean plutons: the product of oxidizing hydrothermal activity related to alkaline intrusions. *Amer. Min.*, **74**, 1270–7.

Heinrich, E. W. and **Vian, R. W.,** 1967. Carbonatitic barytes. *Amer. Min.*, **52**, 1179–89.

Helmhacker, K., 1872. *Svárov, Denkschrift, Acad. Wien.*, **32**(2), 1.

Hill, R. J., 1977. A further refinement of the barite structure. *Can. Min.*, **15**, 522–6.

Hirst, D. M. and **Smith, F. W.,** 1974. Controls of barite mineralization in the Lower Magnesian Limestone of the Ferryhill area, County Durham. *Trans. Inst. Min. Met.*, **83(B)**, 49–55.

Honma, H. and **Okamura, S.,** 1975. Lead content of barite from Oe mine, Hokkaido. *Mining Geol. (Japan)*, **25**, 397–400.

Ixer, R. A. and **Vaughan, D. J.,** 1993. Lead–zinc–fluorite–baryte deposits of the Pennines, North Wales and the Mendíps. In *Mineralization in the British Isles*: Eds R. A. D. Pattrick and D. A. Polya. London (Chapman & Hall), pp. 355–418.

Izyumskiĭ, S. I., 1971. Luminescence and absorption spectra and band scheme of baryte. *Zap. Vses. Min. Obshch.*, **100**, 512–6 (M.A. **23**–2310).

Jaffray, J., 1943a. Sur la biréfringence du gypse et de la barytine dans le rouge extrême et le très proche infrarouge. *Compt. Rend. Acad. Sci. Paris*, **216**, 409–10.

Jaffray, J., 1943b. Sur les variations avec la température de la biréfringence du gypse et de la barytine. *Compt. Rend. Acad. Sci. Paris*, **216**, 814–6.

James, R. W. and **Wood, W. A.**, 1925. The crystal structure of barytes, celestine and anglesite. *Proc. Roy. Soc.*, **109A**, 598–620.

Jebrak, M., **Smejkal, V.** and **Albert, D.**, 1985. Rare earth and isotopic geochemistry of the fluorite–barite vein deposits from the western Rouergue district (France). *Econ. Geol.*, **80**, 2030–4.

Jurković, I., 1953. Baryt aus der Grube Dobreva bei Zletovo. *Jugoslav. Akad. Znan. Umjet.*, 221–33 (M.A. **12**–475).

Kaiser, C. J., **Kelly, W. C.**, **Wagner, R. J.** and **Shanks, W. C.**, 1987. Geological and geochemical controls of mineralization in the southeast Missouri barite district. *Econ. Geol.*, **82**, 719–34.

Kajiwara, Y. and **Honma, H.**, 1976. Lead content of barite coexisting with galena – an indicator of oxygen fugacity during formation. *Mining Geol. (Japan)*, **25**, 397–400.

Kesler, S. E. and **Jones, L. M.**, 1981. Sulfur- and strontium-isotope geochemistry of celestite, barite and gypsum from the Mesozoic basins of north-eastern Mexico. *Chem. Geol.*, **31**, 211–24.

Kesler, S. E., **Ruiz, J.** and **Jones, L. M.**, 1983. Strontium isotope geochemistry of fluorite mineralisation (Coahuila, Mexico). *Isotop. Geosci.*, **1**, 65–75.

Kesler, S. E., **Jones, L. M.** and **Ruiz, J.**, 1988. Strontium and sulphur isotope geochemistry of the Galeana Barite District, Nuevo Leon, Mexico. *Econ. Geol.*, **83**, 1907–17.

Khomyakov, A. P., 1964. Distribution of rare earths in carbonate–haematite veins of Western Tannu-ola. *Geochem. Internat.*, **1**, 40–3.

Kolb, R., 1911. Vergleich von Anhydrit, Cölestin, Baryt und Anglesit in bezug auf die Veränderung ihrer geometrischen und optischen Verhältnisse mit der Temperatur. *Z. Krist.*, **69**, 14–61.

Kornicker, W. A., **Presta, P. A.**, **Paige, C. R.**, **Johnson, D. M.**, **Hileman, O. E.** and **Snodgrass, W. J.**, 1991. The aqueous dissolution kinetics of the barium/lead sulfate solid solution series at 25° and 60 °C. *Geochim. Cosmochim. Acta*, **55**, 3531–41.

Kornilov, N. A., 1966. Barite from iron ores of the Krivoy Rog type. *Dokl. Acad. Sci. U.S.S.R., Earth Sci. Sect.*, **168**, 134–6.

Krishnan, K. S., **Guha, B. C.** and **Bannerjee, S.**, 1933. Investigations on magnecrystallic action. Part I. Diamagnetics. *Phil. Trans.*, **231**, **A**, 235–62.

Kritsotakis, K., 1979. Der Einfluss der Löslichkeit auf die hydrothermale Mobilisation des Baryts. *Neues Jahrb. Min., Abh.*, **135**, 132–47.

Kritsotakis, K. and **Platen, H. v.**, 1980. Reduktive Barytmobilisation. *Neues Jahrb. Min., Abh.*, **137**, 282–306.

Kusakabe, M. and **Chiba, H.**, 1979. Oxygen isotope geothermometry applicable to sulphate minerals from the Kuroko deposits. *Mining Geol.*, **29**, 257–64.

Kusakabe, M. and **Chiba, H.**, 1983. Oxygen and sulfur isotope composition of barite and anhydrite from the Fukazawa deposit, Japan. In *Econ. Geol. Monograph 5. The Kuroko and Related Volcanogenic Sulfide Deposits*: Eds H. Ohmoto and B. J. Skinner. New Haven, CT (Econ. Geol. Publ. Co.), pp. 292–301.

Kusakabe, M. and **Robinson, B. W.**, 1977. Oxygen and sulfur isotope equilibria in the $BaSO_4$–HSO_4^-–H_2O system from 110 to 350 °C and applications. *Geochim. Cosmochim. Acta*, **41**, 1033–40.

Lambert, I. B., **Donnelly, T. H.**, **Dunlop, J. S. R.** and **Groves, D. I.**, 1978. Stable isotopic compositions of early Archaean sulphate deposits of probable evaporitic and volcanogenic origins. *Nature*, **276**, 808–11.

Li, B., **Lee, J. T.** and **Manuel, O. K.**, 1994. Anomalous ^{131}Xe in barites. *Earth & Planet. Sci. Letters*, **123**, 71–9.

Lovering, J. F., 1953. Mineralization of the Ashfield Shale, Wianamatta Group. *J. and Proc. Roy. Soc. New South Wales*, **87**, 163–70.

MacIntyre, D. G., 1982. Geologic setting of recently discovered stratiform barite–sulphide deposits in northeast British Columbia. *Can. Min. & Met. Bull.*, **75**, 99–113.

Malinin, S. D. and **Urusov, V. S.**, 1983. The experimental and theoretical data on isomorphism in the $(Ba,Sr)SO_4$ system in relation to barite formation. *Geochem. Internat.*, **20**, 70–80.

Marumo, K., 1989. The barite ore fields of Kuroko-type of Japan. In *Non-metalliferous Stratabound Ore-fields*: Ed. M. K. de Brodtkorb. London (Chapman & Hall), pp. 201–31.

Mathias, M., 1940. The occurrence of barite in an iron ore deposit in Namaqualand. *Trans. Roy. Sco. South Africa*, **28**, 207–17 (M.A. **8**–239).

May, F., **Peacock, J. D.**, **Smith, D. I.** and **Barber, A. J.**, 1993. *Geology of the Kintail district. Memoir* *for 1:50,000 sheet 72W* and part of *71E (Scotland)*. British Geological Survey, 77 pp.

Mills, J. W., Carlson, C. L., Fewkes, R. H., Handlen, L. W., Jayprakash, G. P., Johns, M. A., Margasti, J. M., Neitzel, T. W., Ream, L. R., Sanford, S. S. and **Todd, S. G.**, 1971. Bedded barite deposits of Stevens County, Washington. *Econ. Geol.*, **66**, 1157–63.

Miyake, M., Minato, I., Morikawa, H. and **Iwai, S.**, 1978. Crystal structures and sulphate force constants of barite, celestite and anglesite. *Amer. Min.*, **63**, 506–10.

Mohazzabi, P. and **Searcy, A. W.**, 1976. Kinetics and thermodynamics of decomposition of barium sulphate. *J. Chem. Soc. Farad. Trans.*, **72**, 290–5.

Monnin, C. and **Galinier, C.**, 1988. The solubility of celestite and barite in electrolyte solutions and natural waters at 25°C: a thermodynamic study. *Chem. Geol.*, **71**, 283–96.

Morey, G. W. and **Hesselgesser, J. M.**, 1952. The solubility of some minerals in superheated steam at high pressures. *Econ. Geol.*, **46**, 821–35.

Morgan, J. W. and **Wandless, G. A.**, 1980. Rare earth element distribution in some hydrothermal minerals: evidence for crystallographic control. *Geochim. Cosmochim. Acta*, **44**, 973–80.

Moro, M. C., Cembranos, M. L. and **Pérez del Villar, L.**, 1991. Las mineralizaciones estratiformes de barita Paleozoicas del Sinforme de Alcañices (Zamora, España). *Mem. Not. Publ. Mus. Lab. Min. Geol., Univ. Coimbra*, **112**, 591–618.

Morrow, D. W., Krouse, H. R., Ghent, E. D., Taylor, G. C. and **Dawson, K. R.**, 1978. A hypothesis concerning the origin of barite in Devonian carbonate rocks of northeastern British Columbia. *Can. J. Earth Sci.*, **15**, 1391–1406.

Narasimhamurty, T. S., 1961. Ultrasonic methods of determining elasto-optic constants of uniaxial and triaxial crystals. *Acta Cryst.*, **14**, 1176–9.

Niggli, P., 1941. *Lehrbuch der Mineralogie und Kirstallchemie.* Berlin (I. Borntraeger).

Nuelle, L. M. and **Shelton, K. L.**, 1986. Geologic and geochemical evidence of possible bedded barite deposits in Devonian rocks of the Valley and Ridge Province, Appalachian Mountains. *Econ. Geol.*, **81**, 1408–30.

Ohashi, R., 1920. Note on the plumbiferous barytes from Shibukuro Prefecture of Akita, Japan. *Min. Mag.*, **19**, 73–6.

Ohmoto, H. and **Skinner, B. J.**, 1983. The Kuroko and related volcanogenic massive sulfide deposits: Introduction and summary of new findings. *Econ. Geol. Mon.*, **5**, 1–8.

Pak, E., Schroll, E. and **Weber, L.**, 1980. Zur Schwefelisotopenzusammensetzung der Pb-Zn-Vererzung des Grazer Paläozoikums (Ostalpen). *Min. Deposita*, **15**, 315–25.

Palache, C., Berman, H. and **Fondel, C.**, 1951. *Dana's system of mineralogy. Vol. 2.* New York (Wiley), 1124 pp.

Palmqvist, S., 1939. On the solubility relations of barium sulphate. *Kungl. Fysiograf. Sällsk. Lund. Förh.*, **8**, 21–7 (M.A. 9-213).

Papunen, H., 1967. On the barytes of Finland. *Bull. Comm. géol. Finland*, **229**, 45–69.

Parafiniuk, J., 1989. Strontium and barium minerals in the sulphur deposits from Tarnobrzeg region (S.E. Poland). *Archiv. Min.*, **43**, 41–60.

Patel, A. R. and **Bahat, H. L.**, 1972. Growth of single crystals of $BaSO_4$ and $SrSO_4$ from gels. *J. Cryst. Growth*, **12**, 288–90.

Patel, A. R. and **Koshy, J.**, 1969. Growth spirals on baryte. *Min. Mag.*, **37**, 296–8.

Pattrick, R. A. D. and **Russell, M. J.**, 1989. Sulphur isotopic investigation of Lower Carboniferous vein deposits of the British Isles. *Min. Deposita*, **24**, 148–53.

Perry, E.C., Monster, J. and **Remier, T.**, 1971. Sulfur isotopes in Swaziland System: Barites and evolution of the Earth's atmosphere. *Science*, **171**, 1015–6.

Peter, J. M. and **Scott, S. D.**, 1988. Mineralogy, composition and fluid-inclusion microthermometry of seafloor hydrothermal deposits in the southern trough of Guaymas Basin, Gulf of California. *Can. Min.*, **26**, 567–87.

Phakey, P. P. and **Aly, A. M.**, 1970. Direct observations of dislocations in natural barite ($BaSO_4$). *Phil. Mag.*, **22**, 2117–25.

Plummer, L. N., 1971. Barite deposition in central Kentucky. *Econ. Geol.*, **66**, 252–8.

Pouit, G. and **Bois, J.-P.**, 1986. Arrens Zn (Pb), Ba Devonian deposit, Pyrenées, France: an exhalative sedimentary type deposit similar to Meggen. *Min. Deposita*, **21**, 181–9.

Preuss, E. and **Ziehr, H.**, 1966. Zur Verbreitung des Quicksilber in ostbayrischen Flusspatlagerstätten. *Geol. Rundschau*, **55**, 400–14.

Prieto, M., Putnis, A. and **Fernandez-Diaz, L.**, 1990. Factors controlling the kinetics of crystallization: supersaturation evolution in a porous medium. Application to barite crystallization. *Geol. Mag.*, **127**, 485–95.

Prieto, M., Putnis, A., Arribas, J. and **Fernandez-Diaz, L.**, 1992. Ontogeny of baryte crystals grown in a porous medium. *Min. Mag.*, **56**, 587–98.

Przibram, K., 1956. *Irradiation Colours and Luminescence.* London (Pergamon Press), 332 pp.

Putnis, A., Fernandez-Diaz, L. and **Prieto, M.**, 1992. Experimentally produced oscillatory zoning in

the (Ba,Sr)SO_4 solid solution. *Nature*, **358**, 743–5.
Radvanec, M., **Zák, K.** and **Grecula, P.**, 1990. Isotopic ratio of $^{87}Sr/^{86}Sr$ in baryte of Gemericum. *Min. Slov.*, **22**, 219–24 (in Slovak) (M.A. 91M/3019).
Rafter, T. A. and **Mizutani, Y.**, 1967. Oxygen isotopic composition of sulphates. Part 2. Preliminary results on oxygen isotopic variation in sulphates and the relationship to their environment and to their $\delta^{34}S$ values. *New Zealand J. Sci.*, **10**, 816–40.
Rankin, A. H. and **Shepherd, T. J.**, 1978. H_2S-bearing fluid inclusions in baryte from the North Pole deposit, Western Australia. *Min. Mag.*, **42**, 408–10.
Renault, J. and **Brower, E.**, 1971. X-ray line broadening in the barium sulfate–strontium sulfate series. *Amer. Min.*, **56**, 1481–6.
Rietmeijer, F. J. M., 1990. Salts in two chondritic porous interplanetary dust particles. *Meteoritics*, **25**, 209–13.
Rife, D. L., 1971 and 1972. Barite fluid inclusion geothermometry, Cartersville mining district, northwest Georgia. *Econ. Geol.*, **66**, 1164–7, and **67**, 822–4.
Rinne, F., **Hentschel, H.** and **Schiebold, E.**, 1924. Zum Feinbau von Anhydrit und Schwerspat. *Z. Krist.*, **61**, 164–76.
Robinson, B. W. and **Ineson, P. R.**, 1979. Sulphur, oxygen and carbon isotope investigations of lead–zinc–barite–fluorite–calcite mineralization, Derbyshire, England. *Trans. Inst. Min. Metall.*, **88**, B107–17.
Roedder, E., 1972. Barite fluid inclusion geothermometry, Cartersville mining district, northwest Georgia. *Econ. Geol.*, **67**, 821–2.
Russell, M. J., 1986. Reflections on the genesis of the Irish Carboniferous lead + zinc ± baryte deposits. In *Geology in the Real World*. Eds R. W. Nesbitt and I. Nicol. London (Inst. Min. Metall.), pp. 363–8.
Ryabov, I.D., **Bershov, L. V.**, **Sperianskiy, A. V.** and **Ganeev, I. G.**, 1983. Electron paramagnetic resonance of PO_3^{2-} and SO_3^- radicals in anhydrite, celestite and barite: the hyperfine structure and dynamics. *Phys. Chem. Min.*, **10**, 21–6.
Ryabov, I. D., **Bershov, L. V.** and **Ganeev, I. G.**, 1989. Electron paramagnetic resonance of PO_2^{2-} and SO_2^- radicals in barite. *Phys. Chem. Min.*, **16**, 374–7.
Sabine, P. A. and **Young, B. R.**, 1954. Cell size and composition of the baryte–celestine isomorphous series. *Acta Cryst.*, **7**, 630 (abstract).
Sahl, K., 1963. Die Verfeinerung der Kristallstrukturen von $PbCl_2$ (Cotunnit), $PbSO_4$ (Anglesit) und $BaSO_4$ (Baryt). *Beitr. Min. Petr.*, **9**, 111–32.
Sakai, H. and **Matsubaya, O.**, 1971. Sulfur and oxygen isotopic ratios of gypsum and barite in the black ore deposits of Japan. *Soc. Mining Geol. Japan Spec. Issue*, **2**, (Proc. IMA-IAGOD Meetings '70, Joint Symp. vol.), 80–3.
Sasaki, N. and **Minato, H.**, 1982. Relationship between lattice constants and strontium and calcium contents of hokutolite. *Min. J. (Japan)*, **11**, 62–71.
Saukov, A. A., 1939. Mercury in barytes. *Doklady Acad. Sci. USSR*, **22**, 254–6 (M.A. 7–558).
Sawada, H. and **Takeuchi, Y.**, 1990. The crystal structure of barite, $BaSO_4$, at high temperatures. *Z. Krist.*, **191**, 161–71.
Sawkins, F. J., 1966. Ore genesis in the North Pennine orefield, in the light of fluid inclusion studies. *Econ. Geol.*, **61**, 385–401.
Scherp, A. and **Strübel, G.**, 1974. Zur Barium–Strontium-Mineralisation. *Min. Deposita*, **9**, 155–68.
Schneer, C. J., 1970. Morphological basis for the reticular hypothesis. *Amer. Min.*, **55**, 1466–88.
Schulz, O., 1968. Sedimentäre Barytgefüge im Wettersteinkalk der Gailtaler Alpen. *Tscherm. Min. Petr. Mitt.*, **12**, 1–16.
Schwarz, H. and **Follner, H.**, 1981. A new investigation of correspondence between structure and crystal morphology of barite. *Bull. Min.*, **104**, 215–7.
Seager, A. F., 1953. The surface structure of crystals. *Min. Mag.*, **30**, 1–25.
Seager, A. F., 1959. The morphology of the baryte group. *Min. Mag.*, **32**, 63–86.
Seager, A. F. and **Davidson, W. F.**, 1952. Changes in habit during the growth of baryte crystals from the north of England. *Min. Mag.*, **29**, 885–94.
Shawe, D. R., **Poole, F. G.** and **Brobst, D. A.**, 1969. Newly discovered barite deposits in East Northumberland Canyon, Nye County, Nevada. *Econ. Geol.*, **64**, 245–54.
Solomon, M., **Rafter, T. A.** and **Jensen, M. L.**, 1969. Isotope studies on the Roseberry, Mount Farrell and Mount Lyell ores, Tasmania. *Min. Deposita*, **4**, 172–99.
Solomon, M., **Rafter, T. A.** and **Dunham, K. C.**, 1971. Sulphur and oxygen isotope studies in the northern Pennines in relation to ore genesis. *Trans. Inst. Min. Metall.*, **80(B)**, 259–75.
Soriano, J. B., **Ráfales, J. G.** and **Martinez, J. G.**, 1991. X-ray diffraction analysis of strontium in barites. *Powder Diffraction*, **6**, 70–3.
Srinivasan, B., 1976. Barites: anomalous xenon from spallation and neutron-induced reactions. *Earth & Planet. Sci. Lett.*, **31**, 129–41.
Stamatakis, M. G. and **Hein, J. R.**, 1993. Origin of barite in Tertiary marine sedimentary rocks from Lefkas Island, Greece. *Econ. Geol.*, **88**, 91–103.

Starke, R., 1964. Die Strontiumgehalte der Baryte. *Freiberger Forschungsh., C*, **150**, 86 pp.

Strübel, G., 1967. Zur Kentniss und genetischen Bedeutung des Systems $BaSO_4$–NaCl–H_2O. *Neues Jahrb. Min., Mh.*, 233–338.

Sugitani, Y., **Hasegawa, N.**, **Harada, K.** and **Nagashima, K.**, 1969. Variation of lattice constants and refractive indices in the system, barium sulphate–lead sulphate. *J. Min. Soc. Japan*, **9**, 266–72.

Swanson, H. E. and **Fuyat, R. K.**, 1953. Standard X-ray diffraction powder patterns. *Nat. Bur. Stds. U.S.A., Circ.* **539**, **2**, 61.

Swanson, H. E., **Fuyat, R. A.** and **Ugrinic, G. H.**, 1954. Standard X-ray diffraction powder patterns. *Nat. Bur. Stds. U.S.A., Circ.* **539**, **3**, 65.

Sweet, J. M., 1930. Notes on British barytes. *Min. Mag.*, **22**, 257–70.

Takano, B., **Yanagisawa, M.** and **Watanuki, K.**, 1970. Structure gap in $BaSO_4$–$PbSO_4$ solid solution series. *Min. J. (Japan)*, **6**, 159–71.

Tarr, W. A., 1932. A barite vein cutting granite of southeastern Missouri. *Amer. Min.*, **17**, 443–8.

Tarr, W. A., 1933. The origin of the sand barites of the Lower Permian of Oklahoma. *Amer. Min.*, **18**, 260–72.

Taylor, J. H., 1950. Baryte-bearing nodules from the Middle Lias of the English east midlands. *Min. Mag.*, **29**, 18–26.

Templeton, C. C., 1960. Solubility of barium sulfate in sodium chloride solutions from 25° to 95 °C. *J. Chem. Eng. Data*, **5**, 514–6.

Tretsch, H., 1936. Schleifhärtenanisotropie am Baryt. *Z. Krist.*, **95**, 296–307.

Thugutt, St. J., 1936. Sur la solubilité de la barytine dans l'eau distillée. *Arch. Min. Soc. Sci. Varsovie*, **12**, 75–81 (M.A. **6**–473).

Tomba, A. M., 1954. Le inclusioni argillose di alcuni gessi e baritine del Bolognese. *Rend. Soc. Min. Ital.*, **10**, 428–52 (M.A. **12**–472).

Touray, J.-C., **Lantelme, F.** and **Vogler, M.**, 1966. Étude par spectrométrie de masse de la décrépitation des inclusions fluides. *Comptes Rend. Acad. Sci. Paris*, **262**, 1–4.

Touray, J.-C., **Beng, C.** and **Bouhel, S.**, 1988. Caractérisation, par analyse à la microsonde électronique et microspectrométrie Raman, de barytocelestites de différentes compositions. *Comptes Rend. Acad. Sci. Paris*, II, **306**, 1353–7.

Tufar, W., 1965. Geochemische Untersuchungen an österreichischen Baryten. *Tschermaks Min. Petr. Mitt.*, **9**, 242–51.

Tufar, W., 1978. Calcium–Gehalte in ostalpinen Baryten. *Osterr. Akad. Wiss. Mat. nat. Kl. Anz.*, **115**, 34–41.

Uchameyshvili, N. Y., **Malinin, S. D.** and **Khitarov, N. I.**, 1966. Solubility of barite in concentrated chloride solutions of some metals at elevated temperatures in relation to problems of the genesis of barite deposits. *Geochem. Internat.*, **3**, 951–63.

Ulrich, M. R. and **Bodnar, R. J.**, 1988. Systematics of stretching of fluid inclusions II. Barite at 1 atm. confining pressure. *Econ. Geol.*, **83**, 1037–46.

Urabe, T. and **Kusakabe, M.**, 1990. Barite silica chimneys from the Sumusu Rift, Iza-Bonin Arc: possible analog to hematitic chert associated with Kuroko deposits. *Earth & Planet. Sci. Lett.*, **100**, 283–90.

Vidal, C. E., 1987. Kuroko-type deposits in the Middle Cretaceous marginal basin of Central Peru. *Econ. Geol.*, **82**, 1409–30.

Vierne, R. and **Brunel, R.**, 1969. Spectres de réflexion infrarouge de minéraux monocristallins ou en poudre. I. *Bull. Soc. franc. Min. Crist.*, **92**, 409–19.

von Gehlen, K., **Nielsen, H.** and **Ricke, W.**, 1962. S-Isotopen Verhältnisse in Baryt und Sulfiden aus hydrothermalen Gängen im Schwarzwald und jüngeren Barytgängen in Suddeutschland und ihre genetische Bedeutung. *Geochim. Cosmochim. Acta*, **26**, 1189–207.

von Gehlen, K., **Nielsen, H.**, **Chunnett, I.** and **Rozendaal, A.**, 1983. Sulphur isotopes in metamorphosed Precambrian Fe–Pb–Zn–Cu sulphides and baryte at Aggeneys and Gamsberg, South Africa. *Min. Mag.*, **47**, 481–6.

Wang, Z., **Chu, X.** and **Li, Z.**, 1993. Original explanation of the high $\delta^{34}S$ values of a barite deposit. *Sci. Geol. Sinica.*, **28**, 191–2 (M.A. 94M/606).

Watanabe, M. and **Sakai, H.**, 1983. Stable isotope geochemistry of sulfates from the neogene ore deposits in the Green Tuff Region, Japan. In *Econ. Geol. Monograph 5. The Kuroko and Related Volcanogenic Sulfide Deposits.* New Haven, CT (Econ. Geol. Publ. Co.), pp. 282–91.

Willan, R. C. R. and **Coleman, M. L.**, 1983. Sulphur isotope study of the Aberfeldy barite, zinc, lead deposit and minor sulphide mineralization in the Dalradian metamorphic terrain, Scotland. *Econ. Geol.*, **78**, 1619–56.

Williams, J. O., **Tennakoon, D. T. B.**, **Thomas, J. M.** and **Jacobs, P. W. M.**, 1972. Dislocations in orthorhombic minerals. *J. Chem. Soc. Farad. Trans.*, **1**, 1987–91.

Wilson, G. V., **Eastwood, T.**, **Pocock, R. W.**, **Wray, D. A.** and **Robertson, T.**, 1922. Special reports on the mineral resources of Great Britain, vol. II. Barytes and witherite. *Mem. Geol. Surv. G.B.*

Zimmerman, R. A., 1970. Sedimentary features in the Meggen barite–pyrite–sphalerite deposit and a comparison with the Arkansas barite deposits. *Neues Jahrb. Min., Abh.*, **113**, 179–214.

Celestine $SrSO_4$

Orthorhombic (+)

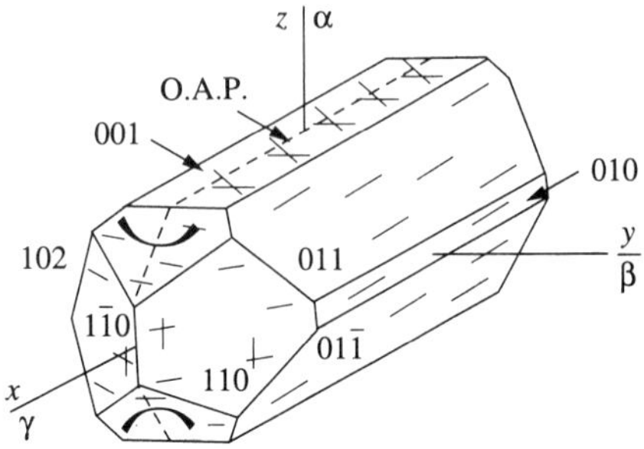

α	1.619–1.622
β	1.622–1.624
γ	1.630–1.632
δ	~0.009
$2V_\gamma$	~50°
Orientation:	
$\alpha = z$, $\beta = y$, $\gamma = x$, O.A.P. (010)	
Dispersion:	$r < v$, moderate
D	~3.98
H	$3–3\frac{1}{2}$

Cleavage:	{001} perfect, {210} good, {010} poor.
Twinning:	Very rare.
Colour:	Colourless, white, pale blue, orange, reddish, greenish, yellow, yellow-brown, brownish; colourless in thin section.
Pleochroism:	Blue crystals weakly pleochroic; indigo, lavender-blue, blue-green, violet; $\gamma > \beta > \alpha$.
Unit cell:	*a* 8.36, *b* 5.35, *c* 6.87 Å Z = 4. Space group *Pnma*

Slightly soluble in hot concentrated acids.

Celestine, $SrSO_4$, is a member of the baryte group of minerals and is the principal source of strontium. The crimson flame colouration produced by strontium makes it an important constituent in the manufacture of fireworks, flares, etc., for which purpose it is converted into the nitrate. Strontium is used also, in its carbonate form, for special glasses (colour television tubes) and for ferrite magnets; celestine itself is used as a filler for white and coloured paints. Celestine occurs mainly in sedimentary beds adjacent to anhydrite or gypsum deposits. The name is derived from the Latin *celestis*, celestial, in allusion to the sky-blue colour often exhibited by its crystals (though this colour is apt to fade in sunlight).

Structure

The structure of celestine was determined, along with those of baryte and anglesite, by James and Wood (1925) and Basche and Mark (1926). The three minerals are isostructural, and in celestine Sr occupies those sites filled by barium in baryte (see structure of baryte, p. 4), but in slightly different positions. A refinement of the structure of a celestine (Waterville, Ohio, USA; BaO between 1.0 and 0.5 per cent; *a* 8.377, *b* 5.350, *c* 6.873 Å) was carried out by Garske and Peacor (1965). They found the Sr atoms in a quite irregular 12-fold coordination with oxygens: Sr–O distances ranged from 2.48 to 3.25 Å, but the S–O tetrahedron was close to ideal. Brunel and Vierne (1970) concluded, however, from the infrared spectra that the SO_4 tetrahedron in celestine is less ideal than that in baryte. Five of the tetrahedral edges are each shared with one Sr–O polyhedron, and the sixth, its largest edge, is shared with two.

The structure of a celestine (Clay Center, Ohio, USA; $Sr_{0.986}Ba_{0.013}Pb_{0.001}SO_4$) was determined by Miyake *et al.* (1978). Its Sr–O distances ranged from 2.536 to 3.263 Å. As for baryte, S, O(1) and O(2) lie on the mirror plane, whereas O(3) is in a general position; the symmetry of the S–O tetrahedron is therefore *Cs*(*m*). The average cation–oxygen distances are 2.831 Å in celestine and 2.952 Å in baryte, and the average S–O distances are 1.474 and 1.478 Å respectively. Miyake *et al.* (1978) also correlated their bond length determinations with infrared absorption frequencies (see p. 34). The cell parameters of the celestine studied were given in the alternative setting *Pbnm* as *a* 6.870, *b* 8.371, *c* 5.355 Å. A refinement of the structure of a synthetic celestine (*a* 8.389, *b* 5.365, *c* 6.885 Å) was reported by Simonov and Troyanov (1987).

A comparison of the cell dimensions of celestine, baryte and anglesite is made in Table 1, p. 5. In accordance with the relative ionic sizes, celestine has the smallest unit cell. The use of X-ray powder patterns to determine composition in the baryte–celestine series is discussed on p. 5.

Measurements of interfacial angles at different temperatures (Kolb, 1911) gave the result $a{:}b{:}c = 0.7785{:}1{:}1.2821$ at 62 °C, and 0.7794:1:1.2780 at 550 °C. (The *a* parameter of the morphological cell is half that of the true unit cell.)

Chemistry

The principal substituent for strontium in celestine is barium, in accordance with the complete solid solution series between celestine and baryte (Grahmann, 1920; Starke, 1964). Although the complete range has been synthesized (e.g. Brower, 1973), in natural specimens intermediate compositions appear to be rare. Most baryte has <7 per cent $SrSO_4$ and most celestine has <4 per cent $BaSO_4$ (Hanor, 1968). Burkhard (1978), however, recorded celestine and baryte with up to 30 per cent of $BaSO_4$ and $SrSO_4$ respectively. Ternary solid solution involving $PbSO_4$, as well as $BaSO_4$ and $SrSO_4$, was reported by Boström *et al.* (1967). Specimens with appreciable Ba content are called barytocelestine.

At ordinary temperatures there is only limited solid solution of $CaSO_4$ in $SrSO_4$ (they have different crystal structures), and in natural specimens the presence of more than 2 or 3 per cent is relatively unusual.

Khomyakov (1964) compared the *REE* content of celestine (and baryte) with fluorite, calcite and ankerite and found that the sulphates have the highest ratio of cerium to yttrium *REE*, a result consistent with the higher coordination number for Ba and Sr compared with Ca.

Some chemical analyses of celestines are presented in Table 6.

There have been many determinations of the solubility of celestine in pure water and in NaCl solutions, and these were summarized by Reardon and Armstrong (1987), in whose own experiments equilibrium was approached from both higher and lower temperatures over the range 0–90 °C. They found that in pure water maximum solubility occurred at 20 °C. Figure 6(a) shows a solubility curve representative of their results and those of Strübel (1966). Reardon and Armstrong (1987) derived the relationship:

$$-\log K = -35.3106 + 0.00422837T + 318312/T^2 + 14.919586 \log T.$$

Table 6. Celestine Analyses

	1	2	3	4	5	6	7	8	9
SiO_2	—	—	—	0.20	0.32	0.64	—	—	0.19
Al_2O_3	—	—	—	—	—	—	tr.	—	0.16
Fe_2O_3	—	—	—	—	0.15	—	1.03	1.26	—
Mn_3O_4	—	—	—	0.77	—	—	—	—	—
MgO	—	—	—	—	—	0.30	—	—	0.10
CaO	0.02	0.14	0.40	—	0.46	1.12	1.30	1.22	0.26
SrO	56.38	56.17	56.20	54.30	54.67	53.76	45.66	31.89	28.27
BaO	—	—	—	1.78	0.78	0.42	8.47	24.48	32.14
SO_3	43.48	43.48	43.40	42.77	43.04	43.95	42.09	39.03	38.97
P_2O_5	—	—	—	—	tr.	—	—	—	—
ign. loss	—	—	—	—	0.12	—	—	1.81	—
Total	99.88	99.79	100.00	99.82	99.54	100.19	98.55	99.69	100.09
α	—	—	—	—	1.6223	1.621	—	—	—
β	—	—	—	—	1.6238	1.624	—	—	—
γ	—	—	—	—	1.6313	1.630	—	—	—
$2V_\gamma$	—	—	—	—	—	$49\frac{1}{2}°$	—	—	—
D	—	—	—	—	3.9841	4.01	—	—	—
	Number of ions on the basis of four (O)								
Ca	0.001	0.005	0.013	0.007	0.015	0.036	0.044	0.044	0.009
Sr	1.001	0.998	0.997	0.976	0.974	0.936	0.833	0.623	0.560
Ba	—	—	—	0.022	0.009	0.005	0.104	0.323	0.431
Σ	1.002	1.003	1.010	1.005	0.998	0.977	0.981	0.990	1.000
S	0.999	0.999	0.997	0.995	0.992	0.990	0.994	0.987	1.000

1 Celestine from geode in ammonite mould, limestone, Cabo Mondego, Portugal (Pinto *et al.*, 1991).
2 Celestine, Asaka gypsum mine, Koriyama City, Japan (Matsubara *et al.*, 1992).
3 Celestine, Djebel Kelbouch, Tunis (Termier, 1902). Anal. M. Pisani.
4 Pink celestine, Djebel Mezzouna, Tunis (Solignac, 1931). Anal. M. H. Faivre.
5 Celestine, cracks in phosphatic nodules, Cretaceous rocks, Trichy, India (Jayaraman, 1940a).
6 Celestine nodules, anhydrite–dolomite rock, Bolshoi Porog Vymsk district, southern Timan (Serdyuchenko, 1947).
7 Celestine, Caramanico, Italy (Onorato, 1924).
8 Barytocelestine, Chipping Sodbury, Gloucestershire, England (see Guppy and Thomas, 1931). Anal. W. Pollard.
9 Barytocelestine, Greiner, Switzerland (Zepharovich, 1886; see Palache *et al.*, 1951).

The solubility of celestine in water is greater than that of baryte but less than that of anhydrite.

The solubility of celestine at 25 °C in various strengths of NaCl solution follows the curve shown in Fig. 6(b), which represents the experimental results obtained by Reardon and Armstrong (1987) and also Strübel (1966), and those obtained theoretically by Reardon and Armstrong using the Pitzer ion interaction model (see also Kushnir, 1985). Monnin and Galinier (1988) used the Pitzer model to evaluate the solubilities of celestine and baryte in aqueous solutions of NaCl, KCl, $CaCl_2$, $MgCl_2$ and Na_2SO_4 at 25 °C. Their results agreed well with measured values.

Celestine crystals were grown by Patel and Bahat (1971) using a flux evaporation method, and from a silica gel using strontium chloride and sodium sulphate (Patel and Bahat, 1972). Well-formed crystals of celestine were obtained by Blount (1974) using 0.02 M $SrCl_2$ and 0.02 M Na_2SO_4 in 0.03 M HCl solution at around 95 °C.

Brower (1973) synthesized representatives of the complete series from $BaSO_4$ to $SrSO_4$ by recrystallization from brines by thermal cycling (see also p. 10, and Starke, 1964; Boström *et al.*, 1967; Hanor, 1973).

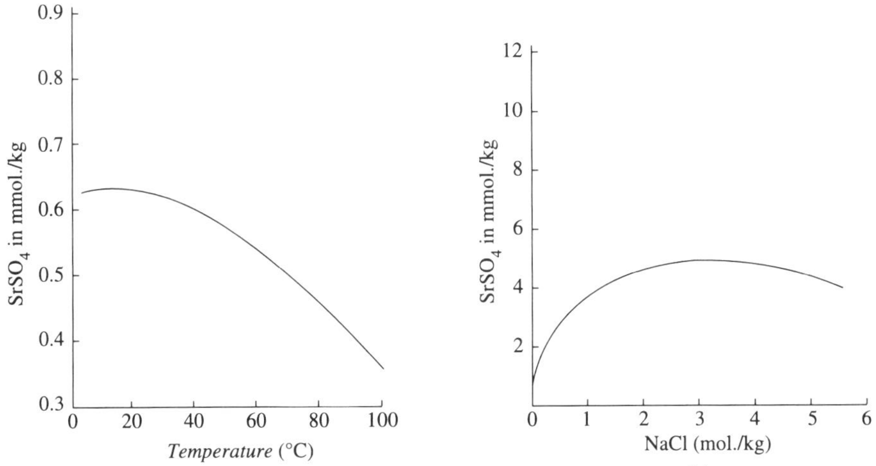

(a) (b)

Fig. 6. (a) Solubility of celestine as a function of temperature. (b) Solubility of celestine at 25 °C in NaCl solutions. Curves derived from data by Reardon and Armstrong (1987) and Strübel (1966) (after Reardon and Armstrong, 1987).

Crystals of celestine containing 1.53 per cent Hg were prepared from $SrCl_2$ and H_2SO_4 in a concentrated solution of $HgSO_4$ (Carrobbi, 1950).

A common alteration product of celestine is strontianite, and among others, some of which may be pseudomorphous, are calcite, witherite, quartz, chalcedony, baryte and sulphur.

Optical and Physical Properties

Celestine occurs in fibrous or rounded aggregates but it is also often found as well-formed crystals with tabular {001} habit, sometimes with large {210} faces and elongated parallel to [100] or [010]; other forms commonly exhibited are {011}, {101}, {210}, {100} and {102}, and fibrous varieties have been reported as being elongated parallel to *z* (Mitchell and Pharr, 1961; Ul'yanova and Petrov, 1971).

The relative development of different forms has been studied by Seager (1959), who noted that the frequency of occurrence of certain forms varies systematically through the sequence baryte–anglesite–celestine, the order of the cationic radii. The space group of celestine was predicted correctly from its morphology (by Seager, 1959) by means of the Donnay–Harker hypothesis (Donnay and Harker, 1937).

In some cases the colour of blue celestine has been associated with the presence of impurities, including organic material and colloidal sulphur. Friend and Allchin (1939, 1940) found that more iron was present in colourless than in blue and red specimens, and that the coloured specimens contained colloidal gold, the colour being a result of Tyndall scattering. Bernstein (1979) found that two orange celestines contained about 50 p.p.m. of copper, and that blue celestines contained more sodium and potassium. Przibram (1956), however, regarded the blue of celestine as a radiation colour, because on heating the colour disappears and weak thermoluminescence is exhibited; on exposure to irradiation the blue colour is restored. Bernstein (1979) reviewed the several other mechanisms suggested, including Fe^{2+}–Fe^{3+} charge transfer and 'colour centres'. The latter are of three types: (a) *F*-centres produced by alkali ions substituting for strontium, forming electron traps, or by ionization of Sr to $Sr^{+} + e^{-}$ with the electron filling an SO^{2-} vacancy; (b) PO^{3-} substituting for SO_4^{2-}; and (c) the presence of monovalent sulphur-containing radicals plus O^{-}.

Bernstein's own studies found that the colour of blue celestine is lost at around 200 °C and reappears with X-radiation, and that the blue celestine gives a thermoluminescence maximum at about 200 °C. Investigation by electron spin resonance showed the presence of SO_4^{-}, SO_3^{-}, SO_2^{-} and O^{-} in blue celestine which were thought to have been formed by the action of ionizing radiation on SO_4^{2-}, releasing energetic electrons that decompose other SO_4^{2-} groups. The centres SO_3^{-}, SO_2^{-} and O^{-} produce the blue colour by absorption in the visible spectrum, while SO_4^{2-} absorbs in the ultraviolet, and the colour centres are stabilized by the presence of trace elements, principally K^{+} substituting for Sr^{2+}. Some celestine specimens show fluorescence in ultraviolet light.

Gaft *et al.* (1985) studied the photoluminescence of many specimens of celestine, baryte and anhydrite over the range 400–750 nm and from 300 to 77 K, as well as X-ray luminescence spectra at 300 K. The inherent lattice luminesence centre of SO_4^{2-} appeared at 360 nm for all sulphates, but was absent if impurities were present in large amounts. Centres of Gd^{3+}, Ce^{3+} and O_2 were observed in celestine.

A study by Raman microspectroscopy (Touray *et al.*, 1988) of a range of barytocelestines showed that the peak ($\nu\ SO_4^{2-}$) corresponding with stretching of S–O in the (SO_4) group shifts linearly with substitution of Sr for Ba from 987.5 cm^{-1} for baryte to 1000 cm^{-1} for celestine (see also Matossi and Kindler, 1934; Griffith, 1969).

The density of pure $SrSO_4$ was given by Swanson and Fuyat (1953) as 3.971 g/cm^3 at 26 °C,[1] but its value is affected by chemical substitution (Ba, Pb, etc.); assuming linear relationships, the densities for intermediate compositions can be calculated from those of the end-members of solid solution series, listed in Table 1, p. 5. Linear compressibilities in different directions have been determined for celestine by Bridgman (1949); its magnetic anisotropy has been determined by Krishnan *et al.* (1933) and its dielectric constant by Errera and Brasseur (1933).

The refractive indices of celestine are compared with those of baryte and anglesite in Table 4, p. 13.

The variation of optical properties across the series $BaSO_4$–$SrSO_4$ is presented in Fig. 4, p. 13, and the temperature and wavelength dependence of the optical properties of celestine are indicated in Table 6A.

Distinguishing Features

Celestine is similar to baryte in many respects but it has a considerably lower density; its solubility in water, though small, is greater than that of baryte. Albite is harder than celestine, gypsum is softer, and calcite effervesces with dilute HCl: celestine gives a crimson flame colouration. The cleavages of minerals in the baryte group are characteristic, and are distinct from the three orthogonal cleavages

[1] Mukae (1957) gave the specific gravity of spectroscopically pure celestine as 3.961.

Table 6A. Optical Properties of Celestine (Kolb, 1911)

(a) Temperature variation at constant wavelength (D_2 line)

	−52 °C	**20 °C**	**103 °C**	**530 °C**
α	1.6225	1.6216	1.6202	1.6133
β	1.6240	1.6232	1.6221	1.6159
γ	1.6316 (−55 °C)	1.6304	1.6295 (100 °C)	1.6203 (550 °C)
		20 °C	100 °C	215 °C
$2V_\gamma$		50° 25′	55° 39′	63° 04′

(b) Wavelength variation at constant temperature

		B	**D_2**	**F**	**G**
	α	1.6180	1.6215	1.6273	1.6322
at 16–19 °C	β	1.6198	1.6232	1.6292	1.6341
	γ	1.6266	1.6305	1.6367	1.6418
at 20 °C	$2V_\gamma$	49° 40′	50° 25′	51° 36′	52° 44′

of anhydrite. Celestine and baryte have lower refringence and birefringence than anglesite.

Paragenesis

Celestine occurs mainly in sedimentary rocks, particularly dolomites, dolomitic limestone and marls, either as a primary precipitate from aqueous solutions or, more usually, by the interaction of gypsum or anhydrite with Sr-rich waters. Beds of celestine are therefore commonly found immediately above or below gypsum or anhydrite deposits. De Brodtkorb (1989) gave brief descriptions of the important deposits of celestine in England, Mexico, Spain, West Germany, Iran, Canada and Algeria. Celestine was described by Taylor (1937) from Louisiana (USA) salt plugs and by Stewart (1949) from the anhydrite zone of the evaporites of Eskdale, Yorkshire. Varve-like banding in celestine from Clay Center, Ohio, USA, due to variable substitution of Ba for Sr, was reported by Roedder (1969), possibly associated with annual variation in compositions of surface waters diluting warm brines.

The celestine associated with gypsum in the Cretaceous clay beds of Trichy (India) encrusts casts of ammonites; it was suggested that the Sr in the clay is from radiolaria, and that the celestine is formed by the action of $Sr(HCO_3)_2$ on gypsum (Jayaraman, 1940b). For celestine partly replacing anhydrite in the Main Dolomite of the German Zechstein, Theilig and Pensold (1964) suggested the aragonite of shells as the source of strontium. Bernstein *et al.* (1987) noted that ancantharians are the only marine organisms to use strontium (as celestine) as a major skeletal component. Below a depth of 900 m most acantharians will have dissolved, contributing to the pool of dissolved strontium. They consider that these organisms play an important role in the strontium budget of the oceans.

One of the principal world occurrences of celestine is in the region of Yate, near Bristol, where it is found near limestone as stratiform beds or horizons of nodular or disseminated celestine, and in veins and sheets (or filling fractures and vugs) in Triassic Keuper Marl or Coal Measures and shales. Nickless *et al.*

(1975, 1976) concluded that the celestine is derived from anhydrite and gypsum, and suggested that the strontium was made available when aragonite expelled this element on conversion to calcite. Little or no celestine was found in regions away from the limestone. They argued that if the celestine had precipitated directly by the evaporation of sea-water it would have occurred together with halite, following gypsum; they pointed out that in modern sabkhas, celestine occurs only in minor amounts. Confirmation that the celestine at Yate was not precipitated directly from sea-water was provided by the strontium isotope ratios ($^{87}Sr/^{86}Sr$ 0.7095–0.7109), which are significantly higher than those for Triassic sea-water (Wood and Shaw, 1976).

The sulphur and strontium isotope geochemistry of celestine, baryte and gypsum of the Mesozoic basins of northeastern Mexico was studied by Kesler and Jones (1981). They found that whereas the isotope ratios for gypsum confirmed its Cretaceous sedimentary origin, celestine contained lighter sulphur and less radiogenic strontium, and was probably formed by mixing two solutions and involving nearby Cretaceous limestones.

Kushnir (1985) investigated experimentally the Sr concentrations in brackish and saline waters which are needed to convert gypsum into celestine, and concluded that veins and bands of fine-grained celestine can be produced in $CaSO_4$-bearing rock only from Sr-rich (>50 mg/l) deep saline waters.

Celestine occurs as small crystals in dolomicrite or in high intertidal facies, and as larger prismatic crystals in pores in the low intertidal to subtidal facies in marls and carbonates in the Wenlock of the Oslo region, Norway (Olaussen, 1981). It was suggested that here the celestine formed from pore-water enriched in SO^{2-} by solution of calcium sulphates in the high supratidal environment, and again that the strontium was released during the transformation of aragonite to calcite.

The association of celestine (Table 6, anal. 2) with gypsum at the Asaka mine (Koriyama City, Japan) was considered by Matsubara *et al.* (1992) to result from the hydration of a strontian anhydrite, the $CaSO_4$ being converted to gypsum but the $SrSO_4$ content remaining and crystallizing as celestine. The replacement of anhydrite by celestine had also been described by Salter and West (1966).

De Brodtkorb *et al.* (1982) reported that the precipitation of celestine from sea-water occurs at the boundary between calcium carbonate and calcium sulphate precipitation. Epigenetic stratabound deposits also occur through the mobilization of celestine.

In a study of the fluorite deposits of north Tunisia, Bouhlel (1985) found both strontium-rich baryte and barium-rich celestine (the latter in pockets or late fractures) in stratoid bodies or karsts. Temperatures of formation were estimated as 150–200 °C.

Baker and Bloomer (1988), investigating cores from deep-sea drilling (Lord Howe Rise, southwest Pacific Ocean), found that with increasing depth below sea-bottom strontium content increases (through its expulsion from carbonate) and SO_4 decreases (through the action of reducing bacteria). The increase of strontium dominated and celestine precipitated below depths of ~ 100 m. The celestine has quite high contents of $BaSO_4$ and $CaSO_4$ (4.3–7.8 and 1.7–6.4 mol. per cent respectively).

The lower zone of the Tatum Dome (Mississippi, USA) cap-rock contains considerable celestine and strontianite. Saunders *et al.* (1988) suggested that these replaced calcite and that strontium was derived from oilfield brines.

For the baryte and celestine deposits in central western Argentina, de Brodtkorb *et al.* (1989) took the $^{87}Sr/^{86}Sr$ ratios of 0.705–0.708 to indicate a sedimentary syndiagenetic origin.

Celestine and baryte occur (though not together) associated with the sulphur deposits of southeast Poland. Parafiniuk (1989) found that both were enriched in the heavy isotope of sulphur (^{34}S) and concluded that they had crystallized from sulphate-rich solutions which remained after the bacterial reduction of gypsum had preferentially removed ^{32}S.

For the Jurassic sulphate deposits of Neuquén Province, Argentina, de Brodtkorb and Barbieri (1993) concluded that celestine precipitated first directly from sea-water, later crystallized during diagenesis, and recrystallized through Eocene intrusive activity.

In addition to the occurrences in sediments, celestine can be a primary mineral in hydrothermal veins, and is also found filling veins and cavities in basic eruptive rocks. The mineral has been found in a diorite as interstitial grains associated with hornblende, biotite and apatite (Nicholls, 1951).

References

Baker, P. A. and **Bloomer, S. H.,** 1988. The origin of celestite in deep-sea carbonate sediments. *Geochim. Cosmochim. Acta*, **52**, 335–9.

Basche, W. and **Mark, H.,** 1926. Über die Struktur von Verbindungen des Typus $MeXO_4$. *Z. Krist.*, **64**, 1–70.

Bernstein, L. R., 1979. Colouring mechanisms in celestite. *Amer. Min.*, **64**, 160–8.

Bernstein, R. E., Betzer, P. R., Feely, R. A., Byrne, R. H., Lamb, M. F. and **Michaels, A. F.,** 1987. Acantharian fluxes and strontium to chlorinity ratios in the North Pacific Ocean. *Science*, **237**, 1490–4.

Blount, C. W., 1974. Synthesis of barite, celestite, anglesite, witherite and strontianite from aqueous solutions. *Amer. Min.*, **59**, 1209–19.

Boström, K., Frazer, J. and **Blankenburg, J.,** 1967. Subsolidus phase relations and lattice constants in the system $BaSO_4$–$SrSO_4$–$PbSO_4$. *Arkiv. Min. Geol.*, **4**, 477–85.

Bouhlel, S., 1985. Composition chimique, fréquence et distribution des minéraux de la série barytine–célestite dans les gisements de fluorine de Hammam Jédidi et Hammam Zriba-Jébel Guébli (Tunisie nord-orientale). *Bull. Min.*, **108**, 403–20.

Bridgman, P. W., 1949. Linear compressions to 30,000 kg./cm.^{2}, including relatively incompressible substances. *Proc. Amer. Acad. Arts Sci.*, **77**, 187–234.

Brower, E., 1973. Synthesis of barite, celestite and barium–strontium sulphate solid solution crystals. *Geochim. Cosmochim. Acta*, **37**, 155–8 and 2688–9.

Brunel, R. and **Vierne, R.,** 1970. Spectres de réflexion infrarouge de minéraux monocristallins ou en poudre, II. *Bull. Soc. Franc. Min. Crist.*, **93**, 328–40.

Burkhard, A., 1978. Baryte–Coelestin und ihre Mischkristalle aus Schweizer Alpen und Jura. *Schweiz. Min. Petr. Mitt.*, **58**, 1–95.

Carobbi, G., 1950. Celestina e apatite di stronzio sintetiche contenenti piccole quantità di mercurio. *Atti (Rend.) Accad. Naz. Lincei, Cl. Sci. fis. mat. nat., ser. 8*, **8**, 87–93. (M.A. **11**–193).

de Brodtkorb, M. K., 1989. Celestite: worldwide classical ore fields. In *Non-metalliferous Stratabound Ore-Fields*: Ed. M. K. de Brodtkorb. London (Chapman & Hall), pp. 17–39.

de Brodtkorb, M. K. and **Barbieri, M.,** 1993. Jurassic barite and celestite deposits of Nequén Province, Argentina. *Proc. 8th IAGOD Symp. Scheweizerbart'sche Verlag. Stuttgart*, 243–54.

de Brodtkorb, M. K., Ramos, V., Barbieri, M. and **Ametrano, S.,** 1982. The evaporitic celestite–barite deposits of Neuquen, Argentina. *Min. Dep.*, **17**, 423–6.

de Brodtkorb, M. K., Schalamuk, I. B. A. and **Ametrano, S.,** 1989. Barite and celestite stratabound ore fields in Argentina. In *Non-Metalliferous Stratabound Ore-Fields*: Ed. M. K. de Brodtkorb, London (Chapman & Hall), pp. 41–68.

Donnay, J. D. H. and **Harker, D.,** 1937. A new law of crystal morphology extending the law of Bravais. *Amer. Min.*, **22**, 446–67.

Errera, J. and **Brasseur, H.,** 1933. Ionenpolarisation in Kristallen. *Phys. Zeit.*, **34**, 368–73.

Friend, J. N. and **Allchin, J. P.,** 1939. Colour of celestine. *Nature*, **144**, 633.

Friend, J. N., 1940. Colloidal gold as a colouring principle in minerals. *Min. Mag.*, **25**, 584–96.

Gaft, M. L., Bershov, L. V., Krasnaya, A. R. and **Yaskolko, V. Ya.,** 1985. Luminescence centres in anhydrite, barite, celestite and their synthesized analogs. *Phys. Chem. Min.*, **11**, 255–60.

Garske, D. and **Peacor, D. R.,** 1965. Refinement of the structure of celestite $SrSO_4$. *Z. Krist.*, **121**, 204–10.

Grahmann, W., 1920. Über Barytocölestin und das Verhältnis von Anhydrit zu Cölestin und Baryt. *Neues Jahrb. Min.*, **1**, 1–23.
Griffith, W. P., 1969. Raman spectroscopy of minerals. *Nature*, **224**, 264–7.
Guppy, E. M. and **Thomas, H. H.**, 1931. Chemical analyses of igneous rocks, metamorphic rocks and minerals. *Mem. Geol. Surv. Gr. Britain.*
Hanor, J. S., 1968. Frequency distribution of compositions in the barite–celestite series. *Amer. Min.*, **53**, 1215–22.
Hanor, J. S., 1973. Synthesis of barite, celestite and barium–strontium sulfate solid solution crystals. *Geochim. Cosmochim. Acta*, **37**, 2685–7.
James, R. W. and **Wood, W. A.**, 1925. The crystal structure of barytes, celestine and anglesite. *Proc. Roy. Soc.*, **109**, **A**, 598–620.
Jayaraman, N., 1940a. A chemical and mineralogical study of the celestine from the phosphatic nodules of the Cretaceous rocks of Trichy. *Quart. Journ. Indian Inst. Sci.*, **3**, 11–20 (M.A. **9**–34).
Jayaraman, N., 1940b. Origin of celestine in the Cretaceous beds of Trichy. *Quart. Journ. Indian Inst. Sci.*, **3**, 25–27 (M.A. **9**–34).
Kesler, S. E. and **Jones, L. M.**, 1981. Sulphur- and strontium-isotopic geochemistry of celestite, barite and gypsum from the Mesozoic basins of north-eastern Mexico. *Chem. Geol.*, **31**, 211–24.
Khomyakov, A. P., 1964. Distribution of rare earths in carbonate–haematite veins of Western Tannu-ola. *Geochem. Internat.*, **1**, 40–3.
Kolb, R., 1911. Vergleich von Anhydrite, Cölestin, Baryt und Anglesit in bezug auf die Veränderung ihrer geómetrischen und optischen Verhältnisse mit der Temperatur. *Z. Krist.*, **69**, 14–61.
Krishnan, K. S., **Guha, B. C.** and **Bannerjee, S.**, 1933. Investigations on magnecrystallic action. Part I. Diamagnetics. *Phil. Trans.*, **231**, **A**, 235–62.
Kushnir, S. V., 1985. The epigenetic celestite formation mechanism for rocks containing $CaSO_4$. *Geochem. Internat.*, **23**, 1–9.
Matossi, F. and **Kindler, H.**, 1934. Die Temperaturabhängigkeit der Ultrarotabsorption von Kristallen. II. Beobachtungen an Sulfaten und Glimmern. *Z. Physik.*, **92**, 312–36. (M.A. **6**–275).
Matsubara, S., **Kato, A.** and **Hashimoto, E.**, 1992. Celestine from the Asaka gypsum mine, Koriyama City, Fukushima Pref. Japan. *Min. J. (Japan)*, **16**, 16–20.
Mitchell, R. S. and **Pharr, R. F.**, 1961. Celestite and calciostrontianite from Wise County, Virginia. *Amer. Min.*, **46**, 189–95.
Miyake, M., **Minato, I.**, **Morikawa, H.** and **Iwai, S.**, 1978. Crystal structures and sulphate force constants of barite, celestite and anglesite. *Amer. Min.*, **63**, 506–10.
Monnin, C. and **Galinier, C.**, 1988. The solubility of celestite and barite in electrolyte solutions and natural waters at 25 °C: a thermodynamic study. *Chem. Geol.*, **71**, 283–96.
Mukae, M., 1957. Celestine from the Udô mine, Shimané Prefecture. *J. Min. Soc. Japan*, **3**, 165–6 (M.A. **14**–113).
Nicholls, G. D., 1951. The Glenelg–Ratagain igneous complex. *Quart. Journ. Geol. Soc.*, **106**, 309–44.
Nickless, E. F. P., **Booth, S. J.** and **Mosley, P. N.**, 1975. Celestite deposits of the Bristol area. *Trans. Inst. Min. Met.*, **B84**, 62–4.
Nickless, E. F. P., **Booth, S. J.** and **Mosley, P. N.**, 1976. The celestite resources of the area north-east of Bristol. *Inst. Geol. Sci., Min. Assess. Rept.*, **25**, 84 pp.
Olaussen, S., 1981. Formation of celestite in the Wenlock, Oslo region, Norway – evidence for evaporitic depositional environments. *J. Sed. Pet.*, **51**, 37–46.
Onorato, E., 1924. La celestina di Caramanico. *Atti. (Rend.) R. Accad. Lincei, Roma, Cl. Sci. fis. mat. nat. Ser. 5*, **33**, 259–61 (M.A. **2**–563).
Palache, C., **Berman, H.** and **Frondel, C.**, 1952. *Dana's System of Mineralogy. Vol. 2*, New York (Wiley) 1124 pp.
Parafiniuk, J., 1989. Strontium and barium minerals in the sulphur deposits from Tarnobrzeg region (S.E. Poland). *Arkiv. Min.*, **43**, 41–50.
Patel, A. R. and **Bahat, H. L.**, 1971. Growth of barite group crystals by the flux evaporation method. *J. Cryst. Growth*, **11**, 166–70.
Patel, A. R. and **Bahat, H. L.**, 1972. Growth of single crystals of $BaSO_4$ and $SrSO_4$ from gels. *J. Cryst. Growth*, **12**, 288–90.
Pinto, J. M. S., **Pereira, L. C. G.** and **da Silva, F. G.**, 1991. Sobre a ocorrência de celestite (em ḿoldes de ammonites) nos calcárus Margosos do cabo Mondego (Portugal central). *Memoriás e Noficias*, Publ. Mus. Lab. Min. Geol. Univ. Coimbra, **112**, 383–94.
Przibram, K., 1956. *Irradiation Colours and Luminescence.* London (Pergamon Press), 332 pp.
Reardon, E. J. and **Armstrong, D. K.**, 1987. Celestite solubility in water, seawater and NaCl solution. *Geochim. Cosmochim. Acta*, **51**, 63–72.
Roedder, E., 1969. Varvelike banding of possible annual origin in celestite crystals from Clay Center, Ohio, and in other minerals. *Amer. Min.*, **54**, 796–810.
Salter, D. L. and **West, I. M.**, 1966. Calciostrontianite in the basal Purbeck beds of Durlston Head,

Dorset. *Min. Mag.*, **35**, 146–50.

Saunders, J. A., Prikryl, J. D. and **Posey, H. H.**, 1988. Mineralogic and isotopic constraints on the origin of the strontium-rich cap rock, Tatum Dome, Mississippi, U.S.A. *Chem. Geol.*, **74**, 137–52.

Seager, A. F., 1959. The morphology of the baryte group. *Min. Mag.*, **32**, 63–86.

Serdyuchenko, D. P., 1947. Celestine from Permian beds in southern Timan. *Doklady Acad. Sci. USSR*, **58**, 1141–2.

Simonov, M. A. and **Troyanov, S. I.**, 1987. The refined crystal structure of synthetic celestite $SrSO_4$. *Vestnik. Mosk. Univ. Ser. 4, Geol.*, **1**, 77–9 (in Russian) (M.A. **87**M/5578).

Solignac, M. M., 1931. La célestite du Sud tunisien. *Bull. Soc. franc. Min.*, **54**, 64–74.

Starke, R., 1964. Die Strontiumgehalte der Baryte. *Freiberger Forschungsh. C.*, **150**, 86 pp.

Stewart, F. H., 1949. The petrology of the evaporites of the Eskdale No. 2 boring, east Yorkshire. *Min. Mag.*, **28**, 621–75.

Strübel, G., 1966. Die hydrothermale Löslichkeit von Cölestin im System $SrSO_4$–NaCl–H_2O. *Neues Jahrb. Min., Monat.*, 99–108.

Swanson, H. E. and **Fuyat, R. K.**, 1953. Standard X-ray diffraction powder patterns. *Nat. Bur. Stds. U.S.A. Circ.* **539**, **2**, 61.

Taylor, R. E., 1937. Water-insoluble residues in rock salt of Louisiana salt plugs. *Bull. Amer. Assoc. Petroleum Geol.*, **21**, 1268–1310.

Termier, M. P., 1902. Sur la célestine du Djebel Kelbouch et du Djebel Bezina (Tunisie). *Bull. Soc. franc. Min.*, **25**, 178–80.

Theilig, F. and **Pensold, G.**, 1964. Uber das Vorkommen von Cölestin im Hauptodolomit des norddeutschen Zechsteins. *Chem. Erde.*, **23**, 215–8.

Touray, J.-C., Beny, C. and **Bouhel, S.**, 1988. Charactérisation, par analyse à la microsonde électronique et microspectrométrie Raman, de barytocelestites de différentes compositions. *C. R. Acad. Sci. Paris*, **306**, 1353–7.

Ul'yanova, T. P. and **Petrov, T. G.**, 1971. Formation conditions of celestite spherulites in the strontium chloride–potassium sulphate system. *Zap. Vses. Min. Obshch.*, **100**, 324–7 (M.A. **23**–2958).

Wood, M. W. and **Shaw, H. F.**, 1976. The geochemistry of celestites from the Yate area near Bristol (U.K.). *Chem. Geol.*, **17**, 179–93.

Gypsum $CaSO_4{\cdot}2H_2O$

Monoclinic (+)

α	1.519–1.521
β	1.523–1.526
γ	1.529–1.531
δ	~0.01
$2V_\gamma$ [1]	~58°
O.A.P.	(010), $\gamma{:}z \sim 52°$
Dispersion:	$r > v$, strong
D	2.30–2.37
H	2

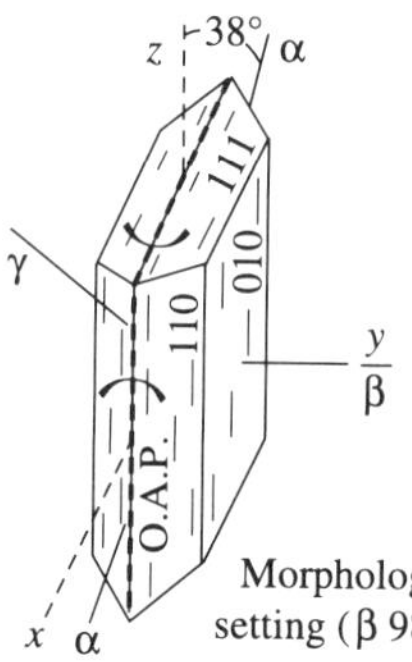

Morphological setting (β 98°58')

Cleavage:	{010} perfect, {100} and {011} distinct.
Twinning:	Very common on {100}; less common on {101}.
Colour:	Usually white or colourless, sometimes grey, red, yellow, brown, blue; colourless in thin section.
Unit cell[2]:	*a* 5.68, *b* 15.18, *c* 6.29 Å, β 113° 50′ Z = 4. Space group *A2/a*

Soluble in HCl.

Gypsum is the most abundant sulphate mineral; it often occurs in extensive masses of great thickness, in association with limestones and shales and in evaporite deposits. Among the various forms in which gypsum occurs are: selenite, a clear crystalline variety well known for its use in a testing instrument with the polarizing microscope (gypsum plate or sensitive tint); alabaster, a fine-grained massive variety used as an ornamental stone; and satin-spar, a fibrous variety. Gypsum is taken as one of the standard minerals (H = 2) for the Mohs scale of hardness. An important commercial use of gypsum is in the manufacture (by its partial dehydration) of Plaster of Paris. It is also used as a fertilizer, a filler in paper and paint, in muds for oil well drilling, and as a retarding agent in cement. The name gypsum is from the Greek *gupsos*, which was probably applied to the mineral itself or to the product of calcination. Selenite is from the Greek *selenites*, moon-stone; alabaster, from *alabastros*, is said to have been named after a town in Egypt.

Hemihydrate ($CaSO_4{\cdot}\frac{1}{2}H_2O$) and γ-$CaSO_4$ (sometimes called 'soluble anhydrite') are low-temperature (below about 110 °C) dehydration products of gypsum; γ-$CaSO_4$ is a metastable phase which is readily re-hydrated.

[1] $2V_\gamma = 0°$ at 91 °C; above 91 °C, O.A.P. ⊥ (010) and dispersion is horizontal with $r < v$.
[2] Except in the sketch above, all indices for gypsum refer to the X-ray cell given here.

Structure

Euhedral crystals of gypsum often occur in the habit depicted above, and in most morphological studies the clinodome was taken as the form {111} and as prism {110}; this description yields axial ratios 0.6910:1:0.4145, and *β* 98° 58′ (Des Cloizeaux, 1886). This setting of the crystal was adopted by Gossner (1937), as well as by Onorato (1929) for his X-ray study of the crystal structure, which showed that the cell is *F*-centred with *a* 10.49, *b* 15.18, *c* 6.29 Å, *β* 98° 58′ (axial ratios 0.6895:1:0.4132), and contains 8[$CaSO_4 \cdot 2H_2O$]. A complete structure determination by Wooster (1936) was carried out with the choice of a simpler cell which is *A*-centred, space group *A*2/*a* with *a* 6.59, *b* 15.18, *c* 10.49 Å, *β* 151° 33′, and contains only 4 formula units. In the Bravais lattice for gypsum the *b* cell edge is uniquely determined by the diad axis, but the four shortest repeat distances in the *xz* plane (~5.68, 6.29, 6.52, 10.49 Å) may be taken in pairs in six different ways to describe alternative unit cells. These include the two mentioned above and are illustrated in Fig. 7. All except the *F*-cell have equal volumes and contain 4 formula units, but four of them are *A*-centred and one (chosen by Bragg, 1937) is *I*-centred. With the *A*-centred cell chosen by Strunz (1942) the edges of the face (010) of the crystal illustrated on p. 40 define the *x* and *z* axial directions, and the clinodome has indices {011}. Onorato (1929) gave, in addition to the morphological cell, the cell with *β*~147° shown in Fig. 7.

The cell chosen in the present text is that selected by Jong and Bouman (1939), because this uses the smallest and second smallest of the four possible vectors. It has *a* 5.68, *b* 15.18, *c* 6.29 Å, *β* 113° 50′, Z=4, space group *A*2/*a*. Corresponding

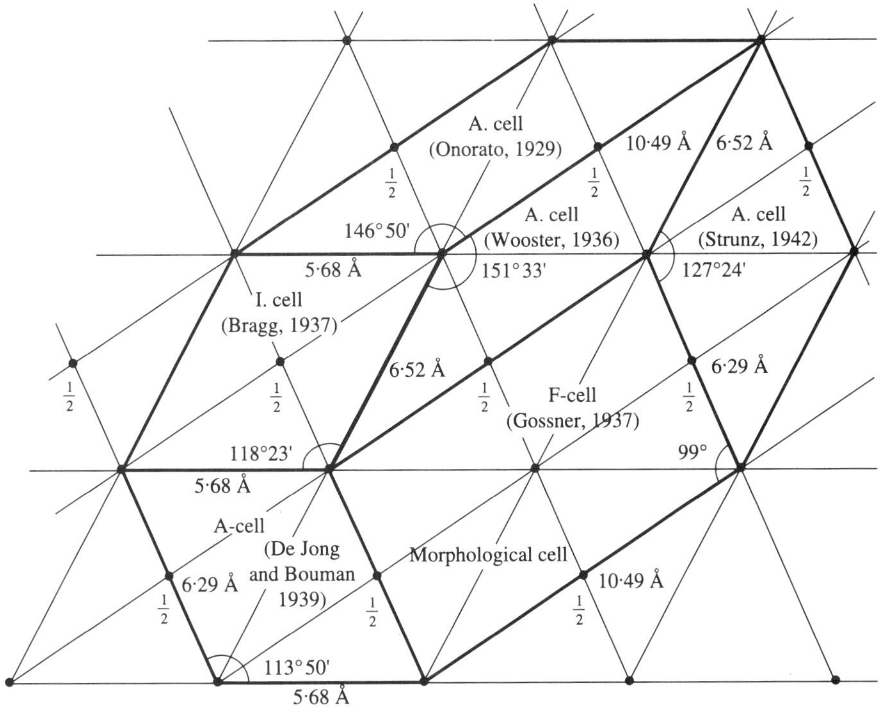

Fig. 7. The Bravais lattice of gypsum projected on (010), showing six ways of choosing unit cells defined by two out of the four shortest vectors.

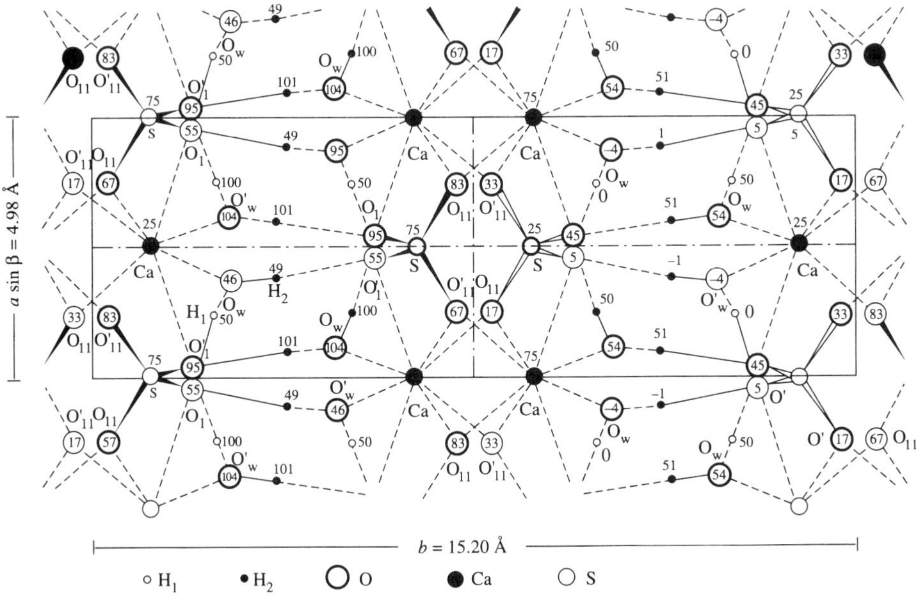

Fig. 8. Half of the unit cell of gypsum projected on (001) (*I*-cell) with *z* coordinates given as percentages of *c* (after Cole and Lancucki, 1974).

values given by the Powder Diffraction File (No. 33-311) for synthetic gypsum are *a* 5.6776, *b* 15.2079, *c* 6.2845 Å, β 114.09°.

A face indexed as (*hkl*) for the morphological *F*-centred cell becomes $-\frac{1}{2}(h+l)$, *k*, *l* for the cell with shortest translations.

In the structure of gypsum, determined by Wooster (1936), two sheets of SO_4 ions parallel to (010) are bonded together by Ca ions, and these double sheets are separated by layers of water molecules (see Fig. 8). Each Ca ion is coordinated by eight oxygens, six belonging to SO_4 tetrahedra and two to water molecules. The Ca–O distances range from 2.552 to 2.366 Å. The water molecules lie near the centres of triangles formed by a calcium ion and two oxygens (of sulphate groups) to which they are hydrogen bonded. Each water molecule links a Ca ion to an oxygen in the same double sheet at 2.81 Å and to an oxygen in a neighbouring sheet at 2.88 Å. The relatively weak hydrogen bonding between layers explains the perfect (010) cleavage of gypsum. Also, the direction [001] is that in which Ca and O ions are closest together; this corresponds approximately with the direction of fibrous cleavage, small expansion coefficient and high refractive index.

The expansion coefficient is large for the direction perpendicular to the perfect cleavage, and on heating this results in a rapid decrease in refractive index for this vibration direction and the consequent change in optic axial plane (Wooster, 1936).

Piezoelectric tests have confirmed that gypsum is holosymmetric and not hemisymmetric as had been suggested from the evidence of etch figures (Bellanca, 1938). X-ray studies on different forms of gypsum (Raman and Jayaraman, 1954) showed that in satin-spar the fibres are elongated parallel to the *z* axis, while a variety called 'fascicular gypsum' consists of rods elongated parallel to *y*; alabaster is an irregular granular aggregate.

The cell parameters of gypsum from different localities have been shown to vary, the largest difference between specimens of a selection by Nagy (1928) being about 0.25 per cent. The cells of gypsum and of pharmacolite, $CaHAsO_4{\cdot}2H_2O$, were compared by Gossner (1937).

In the structure of gypsum there are channels of water molecules giving a somewhat zeolitic character and perhaps also facilitating the formation of anhydrite as a secondary mineral.

The earlier determinations of the structure of gypsum showed the SO_4 tetrahedron as being regular but the spectroscopic data indicated lower symmetry. This was confirmed by Cole and Lancucki (1974) who re-worked the neutron diffraction data of Atoji and Rundle (1958), and also carried out a three-dimensional X-ray diffraction study. They found that while the S–O distances are equal (1.459 Å), for one O–O pair the distance is 2.323 Å and O–S–O is 105.5°, and for the other pair corresponding values are 2.411 Å and 111.5°. They also found that there are two distinct hydrogen bond lengths at 2.816 and 2.89 Å rather than the one determined previously as 2.820 Å. An earlier study by infrared spectroscopy (Seidl *et al.*, 1969) and a re-working of the neutron data by Denne and Jones (1969) had also indicated two kinds of hydrogen bond.

The results of a three-dimensional neutron diffraction study (Pedersen and Semmingsen, 1982) also showed irregularity in the SO_4 tetrahedron. Again, the S–O distances were found to be equal (1.472 Å) but one of the O–S–O angles was appreciably smaller (106.3°) than the other three (111.1, 111.0, 111.1°); the O–O edges were 2.429, 2.432, 2.426 and 2.357 Å. The shortest edge is shared with the CaO_8 square bipyramid.

Using nuclear magnetic resonance spectroscopy Pedersen and Holcomb (1963) determined the proton–proton distance in gypsum to be 1.62 Å compared with 1.58 Å obtained by Atoji and Rundle (1958), McGrath and Silvidi (1961) and Pake (1948). The later study by Pedersen and Semmingsen (1982) showed that in the water molecule the two O–H distances are different (0.942 and 0.959 Å), and the H–O–H bond angle is 107.5°. The water molecule donates two hydrogen bonds (1.856 and 1.941 Å) to sulphate oxygens.

Single crystals of gypsum were 50 per cent deuterated (Hutton and Pedersen, 1969) and studied by proton and deuteron magnetic resonance. The orientations of the water molecules so determined agreed with those obtained by neutron diffraction. The quadrupole coupling constant and asymmetry factor were determined as 118.3 ± 0.7 kHz and 0.877 ± 0.009, respectively, and the hyperfine structure indicated non-random distribution of D among the water molecules; the mol. fractions indicated were DDO 0.34, HDO 0.14 and HHO 0.52.

Pedersen and Semmingsen (1982) used the $I2/a$ unit cell with a 5.679, b 15.202, c 6.522 Å, β 118.43°, Z = 4, as did Innorta *et al.* (1980) who gave a 5.6774, b 15.1872, c 6.5274 Å, β 118.39°. Gypsum/anhydrite ratios were measured using the intensity ratios of the peaks at $d = 3.49$ Å for anhydrite and $d = 4.27$ Å for gypsum.

Caution is needed if X-ray diffractometer intensities are used quantitatively because the well-developed (010) faces of gypsum crystals can lead to marked enhancement of the 020 peak (Grattan-Bellew, 1975).

Bassanite (hemihydrate, $CaSO_4{\cdot}\frac{1}{2}H_2O$). Gay (1965) showed that heating gypsum above 75–80 °C yielded hemihydrate with its z axis parallel to that of the gypsum. At these temperatures some γ-$CaSO_4$ was also produced, and the amount increased with further heating (100–110 °C). At 150 °C some anhydrite was detectable, also with the z axis parallel to that of gypsum. Gay observed that ‘chains’ of Ca and SO_4 ions occur in each of the three structures.

Table 7. Cell Parameters and Symmetries Determined for $CaSO_4 \cdot nH_2O$

Reference	*a* (Å)	*b* (Å)	*c* (Å)	β°		Z	*n*
Gaitelli (1933)	11.94	6.83	12.70	90.66	$C2$	12	0.5
Caspari (1936)		6.82	6.24		$P3m1$	3	0.5
Flörke (1952)	11.5	6.8	12.7		$C222$	6	0.5
Flörke (1952)		6.8	12.7		$P3_21$	6	0.5
Gay (1965)	11.88	6.85	12.60	90	*I* mon.	12	0.5
Frik and Kuzel (1982)	12.06	6.933	12.670		orth.	12	0.48
Frik and Kuzel (1982)		13.865	12.718		hex.	24	0.52
Busheev and Borisov (1982)	12.028	6.927	12.674	90.21	$I2$	12	0.67
Busheev and Borisov (1982)		6.977	12.617		$P3_121$	12	0.50
Abriel (1983)		6.968	6.410		$P3_121$	3	0.80
Lager *et al.* (1984)	12.062	6.930	12.660	~90	$I2$	12	0.5
Abriel and Nesper (1993)		6.937	6.345		$P3_121$	3	0.5
Abriel and Nesper (1993)	12.015	6.937	12.690	~90	$I2$	12	0.5
Lager *et al.* (1984)		6.9694	6.3033		$P6_222$	3	0.0[a]

[a] $n = 0$, i.e. γ-$CaSO_4$.

There have been many attempts to determine the structures of hydrates of calcium sulphate other than gypsum. Recent work points to the existence of a range of water content perhaps up to 0.8 H_2O per mol. $CaSO_4$, and there have been discussions about whether the well-known hemihydrate is just a member of a series $CaSO_4 \cdot nH_2O$ or has a unique structure at $n = \frac{1}{2}$. The various cell parameters and symmetries given for hemihydrate are listed in Table 7, which has been arranged to show that they all have much in common. Although none have a corresponding conclusive, directly determined, crystal structure, the nearest to this is the work of Lager *et al.* (1984), which determined the structure of γ-$CaSO_4$ (also known as soluble anhydrite) and related the structure of hemihydrate to this. Their description of hemihydrate is as monoclinic $I2$ with *a* 12.062, *b* 12.660 (diad axis), *c* 6.930 Å, $\gamma \sim 90°$. Because γ is $\sim 90°$ this is dimensionally pseudo-orthorhombic; moreover, since $a \sim \sqrt{3}c$, it is orthohexagonal and is closely related to a trigonal cell with rhombus base 6.93 and height 12.66 Å. All of the cells listed in Table 7 have one axis of ~12.7 Å, except in some cases where it is halved. Approximately perpendicular to this, all have either a rhombus network with dimension ~6.9 Å or a rectangular network with 6.9 Å × 12 Å ($6.9 \times \sqrt{3}$).

Frik and Kuzel (1982) had suggested that the cell and symmetry depended upon water vapour pressure ($n <$ or > 0.50). For $n = 0.52$ the 6.9 Å parameter is doubled; Lager *et al.* (1984) suggested that this may represent a finely twinned monoclinic hemihydrate.

Lager *et al.* (1984) used neutron powder diffraction data from $CaSO_4 \cdot 0.5D_2O$ to show that the water molecules occupy two different levels in the channels of a distorted γ-$CaSO_4$ framework. They also found that hemihydrates grown in saturated NaCl solution contain 0.50 wt. per cent Na_2O, which may help to stabilize the structure and affect the refractive indices and cell parameters; in addition, it may inhibit the movement of water molecules through the channels. They suggest that when 2Na replaces Ca, half of the sodium may go into the framework and half reside in the channels.

Busheev and Borisov (1982) determined the structure of $CaSO_4 \cdot 0.67H_2O$. They described the water molecules as in 9-fold coordination around Ca and having long-range ordering, giving a 38 Å repeat along the channel axis. They argued that because of steric hindrance all water sites could not be occupied, and that

0.67 H_2O was the limit and produced monoclinic distortion, whereas 0.50 H_2O gives the more ideal trigonal structure.

The study by Lager *et al.* (1984) supported the results obtained by Flörke (1952) and Gay (1965), but provided no support for the trigonal structure. They also had difficulty reconciling the structure found with the possibility of 0.67 H_2O (Busheev and Borisov, 1982), and still more so with 0.8 H_2O (Abriel, 1983). In the latter structure Ca ions are surrounded by six SO_4 groups, as in anhydrite, and one water molecule. It was suggested that within the channels parallel to *z*, four water molecules are distributed randomly over five sites.

Recent work (Abriel and Nesper, 1993) shows that hemihydrate has a superstructure with high pseudosymmetry, and that the substructure is trigonal $P3_221$ with *a* 6.937, *c* 6.345 Å, Z = 3. This 'average' structure has a statistical distribution of 1.5 H_2O molecules over three positions in the channels which run between the basic Ca and SO_4 polyhedral framework. An ordered arrangement of H_2O molecules due to hydrogen bonding leads to the stable hemihydrate structure with $a' = a\sqrt{3}$, $b' = b$, $c' = 2c$, space group *I*2, $\beta = 90°$ and Z = 12.

Morris (1963) concluded that the so-called α- and β-hemihydrates possessed the same structure but exhibited differences in stacking order, which could affect properties such as density, water uptake, particle size and shape, and thermal behaviour. A similar conclusion was reached by others, e.g. Fowler *et al.* (1969), Bartram (1969) from infrared spectroscopy, and Miyazaki and Takagi (1970).

γ-$CaSO_4$. The structure of γ-$CaSO_4$ ('soluble anhydrite') was determined by Lager *et al.* (1984) using time-of-flight neutron powder diffraction data. The structure (Fig. 9a,b) contains chains of SO_4 tetrahedra showing edges with CaO_8 polyhedra. These chains are related by six-fold screw axes to form channels (diameter ~4 Å) parallel to *z* (Fig. 9c). On the channel walls at intervals of *c*/6, oxygens of SO_4 groups are arranged in a nearly planar configuration equidistant (~2.78 Å) from the channel centre. The unit cell was described as having *a* 6.9694, *c* 6.3033 Å, $P6_222$, Z = 3.

The *c* parameters of γ-$CaSO_4$ and hemihydrate (~6.3 or 2 × 6.3 Å) are echoed in the unit cells of both gypsum *c* ~ 6.3Å and anhydrite *c* ~ 6.24 Å.

Chemistry

Gypsum shows very little variation in chemical composition. Some chemical analyses of gypsum are presented in Table 8. To calculate their formulae, CO_2 was assumed to be present in calcium carbonate and an appropriate amount of CaO was subtracted in each case; SiO_2, Al_2O_3, Fe_2O_3, MgO, NaCl and Cl were assumed to be present in impurity minerals. Spectrographic analyses of primary and secondary gypsum from the Permian deposits of Tartary, central Asia, showed that Mg, Si, Ca, Ti, Fe and Sr were present in all samples, B, Al, Mn and Cu were frequently found, and Li, Be, Y, Zr and Bi occurred sporadically as traces (Miropolsky and Borovick, 1943); trace amounts of Ba have also been found in gypsum.

Average amounts of strontium in specimens of gypsum from central and southern Italy ranged from 1013 to 2437 p.p.m. (Barbieri *et al.*, 1976). Some of the strontium found in gypsum is likely to occur in inclusions of celestine. The ratio of boron isotopes $^{11}B/^{10}B$ for gypsum was found to range between 4.03 and 4.19 (Grégoire, 1987).

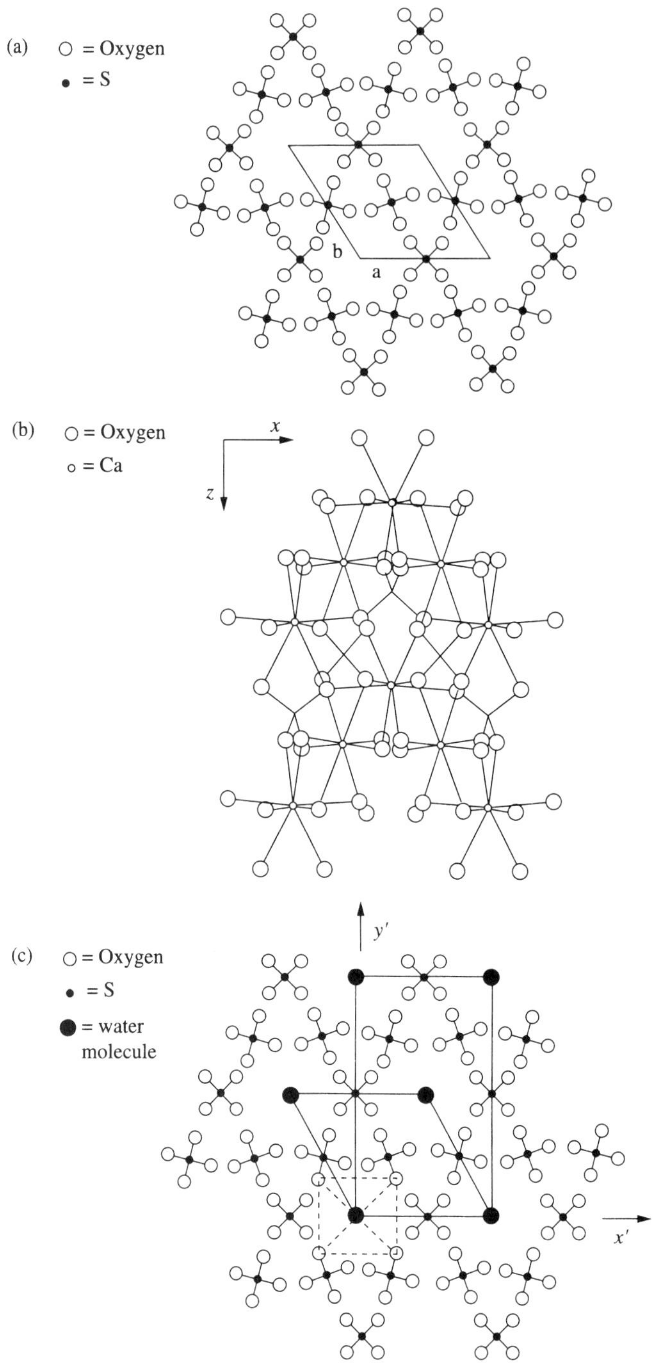

Fig. 9. (a) Structure of γ-$CaSO_4$ projected along z axis showing only the sulphate tetrahedra, and the hexagonal channels parallel to z. (b) Structure of γ-$CaSO_4$ illustrating chains parallel to z formed by sulphate tetrahedra sharing edges with CaO_8 polyhedra. (c) Projection along z axis of structure of γ-$CaSO_4$ showing relationship between the hexagonal and monoclinic ($\gamma = 90°$, orthohexagonal) cells of γ-$CaSO_4$ and hemihydrate. Possible positions of water molecules in hemihydrate are indicated '●'. All after Lager *et al.* (1984).

Table 8. Gypsum Analyses

	1	2	3	4	5
SiO_2	1.46	0.68	—	—	—
Al_2O_3 } Fe_2O_3 }	0.16	0.18	0.11	—	—
MgO	1.14	0.28	0.09	—	—
CaO	33.64	32.76	32.08	33.30	32.36
SO_3	45.88	45.88	45.37	45.30	46.00
CO_2	2.04	0.28	0.95	1.26	0.28
H_2O^+	16.04	20.23	20.10	19.74	20.82
H_2O^+	0.01	0.01	—	—	—
Total	100.71	100.99	98.88	99.69	99.62
α	—	—	—	1.519	1.521
β	—	—	—	—	1.526
γ	—	—	—	1.530	1.531
$2V_\gamma$	—	—	—	57°–58°	55.5°–56°
D	—	—	—	2.37	2.30
	Numbers of ions on the basis of four (O) in the anhydrous 'molecule'				
Ca	0.974	1.006	0.981[a]	0.999	0.995
S	1.009	0.998	1.007	1.000	1.002
O	4.000	4.000	4.000	4.000	4.000
H_2O	1.567	1.956	1.982	1.937	2.015

[a]Includes Sr 0.001, (Na, K) 0.003.

1 Gypsum, St Louis limestone, Sangamon Co., Illinois, USA (Saxby and Lamar, 1957). Anal. L. D. McVicker (includes NaCl 0.34).
2 Gypsum, St Louis limestone, Sangamon Co., Illinois, USA (Saxby and Lamar, 1957). Anal. L. D. McVicker (includes NaCl 0.69).
3 Gypsum, Nentershausen (Flörke, 1952) (includes SrO 0.08, (Na_2O+K_2O) 0.06, Cl 0.04).
4 Coarse-grained gypsum, Zaleschiki, Ukraine (Zolotukhin, 1954) (includes 0.09 residue).
5 Fine-grained gypsum, Zaleschiki, Ukraine (Zolotukhin, 1954) (includes 0.16 residue).

In a study of the influence of gold content on the colour of minerals, reddish brown crystals of gypsum were examined by Friend and Allchin (1940). Gold, varying between 0 and 95 p.p.m., was found in different specimens, but no correlation with colour was observed; the iron content varied between 5 and 230 p.p.m., and the colour appeared to be governed by the state of oxidation of the iron.

The various names commonly used to describe components and phases of the system $CaSO_4$–H_2O are summarized in Table 9.

Dehydration of gypsum. The main point of interest in the chemistry of gypsum concerns the products of its dehydration. Four principal phases have been reported in the system calcium sulphate–water: $CaSO_4{\cdot}2H_2O$ (in mineral form, gypsum), $CaSO_4{\cdot}\frac{1}{2}H_2O$ (hemihydrate or, as a mineral, bassanite), γ-$CaSO_4$ ('soluble anhydrite') and β-$CaSO_4$ (anhydrite). There have been some reports of two polymorphs of hemihydrate, but it seems that they have identical structures but different degrees of structural disorder. The existence of a high-temperature form of anhydrite was suggested by Grahmann (1920) and is discussed on p. 76.

The relationship between hemihydrate and γ-$CaSO_4$ has been much studied. Recent work implies that hydrates of $CaSO_4{\cdot}nH_2O$ with n between 0 and 0.5 have

Table 9. Nomenclature in the System $CaSO_4$–H_2O

Formula	Description	Symmetry
$CaSO_4 \cdot 2H_2O$	Gypsum (selenite, satin-spar, calcium sulphate dihydrate)	Monoclinic $A2/a$
$CaSO_4 \cdot \frac{1}{2}H_2O$	Hemihydrate (Plaster of Paris, bassanite)	Monoclinic $I2$ (pseudo-trigonal $P3_121$)
$CaSO_4$ (metastable)	γ-$CaSO_4$ (soluble anhydrite, anhydrite, dehydrated hemihydrate)	Hexagonal $P6_222$
$CaSO_4$ (stable to 1200 °C)	Anhydrite (β-$CaSO_4$[a], insoluble anhydrite, dead-burned gypsum)	Orthorhombic *Amma*
$CaSO_4$ (stable above 1200 °C)	α-$CaSO_4$[a]	Trigonal $P31c$

[a]The use of α and β for high- and low-temperature forms, respectively, is contrary to general mineralogical usage but is retained here because of its prevalence in the literature.

distorted forms of the structure of γ-$CaSO_4$, the latter being the phase produced by the complete dehydration of gypsum but below the temperatures required to convert it to anhydrite, which involves a distinct structural change. The term γ-$CaSO_4$ is to be preferred therefore to γ-anhydrite.

Chemical compositions and thermal experiments appear to show that although n can vary, a water content of $n = \frac{1}{2}$ yields an energetically more favourable structure. Laboratory preparation of components with $n > \frac{1}{2}$ (up to 0.8) has been reported (see p. 49).

Of the four phases in the system $CaSO_4$–H_2O, hemihydrate and γ-$CaSO_4$ exist only metastably. Thus, under equilibrium conditions the reaction gypsum = anhydrite + water occurs without intermediate compounds.

When heated in air, however, gypsum is converted slowly to hemihydrate at or below about 70 °C, and rapidly at or above 90 °C; commercial Plaster of Paris consists mainly of hemihydrate and is made by heating gypsum at 190–200 °C. Heating gypsum at higher temperatures produces anhydrite and this change is irreversible. At the stage of dehydration where γ-$CaSO_4$ or hemihydrate is produced, re-hydration to gypsum readily occurs, and the re-hydration of hemihydrate to produce interlocking crystals of gypsum is the process of setting of Plaster of Paris.

Among the many thermogravimetric and differential thermal analysis (DTA) studies of the dehydration of gypsum are those by Weiser *et al.* (1936), Flörke (1952), Barshad (1952), West and Sutton (1954) and Caillère and Pobeguin (1962). These and more recent work provide evidence of the distinctiveness of the hemihydrate, as for example the results illustrated in Fig. 10.

A combined thermogravimetric analysis (TGA), DTA and X-ray study by Khalil and Gad (1972) showed that hemihydrate formed between 100 and 200 °C when the heating rate was 10 °C/min, but complete conversion below 100 °C occurred when heating was very slow. A few minutes heating above 250 °C, or longer between 100 and 130 °C, resulted in the formation of γ-$CaSO_4$ (soluble anhydrite); further heating to around 360 °C gave anhydrite. Transition to the high-temperature form of anhydrite took place at 1230 °C and decomposition at 1300–1350 °C.

Thermal changes in gypsum have also been followed by measuring changes in dielectric constant ε with rising temperature (Egerer, 1981).

Ball and Norwood (1969) concluded that at temperatures >110 °C the dehydration of synthetic $CaSO_4 \cdot 2H_2O$ at $P_{H_2O} \sim 10^{-5}$ and 45 mm pressure was

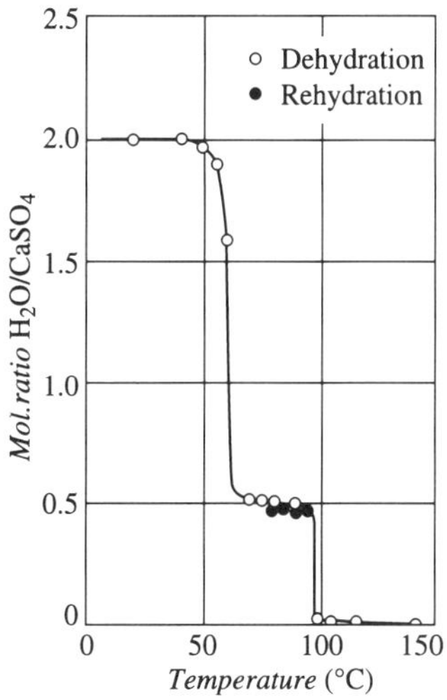

Fig. 10. Isobaric dehydration and rehydration of a sample of gypsum, showing transition to the metastable hemihydrate at about 60 °C and the formation of γ-$CaSO_4$ ('soluble anhydrite') at about 100 °C (after Weiser *et al.*, 1936).

controlled by diffusion, but at lower temperatures both nucleation and the formation and migration of boundaries are important. For nucleation, diffusion and boundary control, at 17 mm pressure, activation energies were given as 34.6, 10.0 and about 20 kcal/mol. respectively. Taylor and Baines (1970) found that the reaction rate for the thermal transformation of gypsum to anhydrite increased at pressures <1 bar, and Combe *et al.* (1970) concluded that the rate of reaction for the hydration of hemihydrate is diffusion controlled.

For the dehydration of hemihydrate (described as the β-form), experiments by Ball and Norwood (1969, 1970) indicated a diffusion-controlled mechanism with an activation energy of 6.4 kcal/mol. at 10^{-5} mm pressure and 49 kcal/mol. at 42 mm pressure. Somewhat different results (15.6 and 36 kcal/mol.) were obtained for hemihydrate described as the α-form (Ball and Urie, 1970).

It has been suggested elsewhere that α- and β-hemihydrates have the same crystal structure but differ in 'packing' and surface properties (p. 45). The α-form is well crystallized and is normally prepared from saturated solutions, whereas the β-form can be prepared by the dehydration of gypsum, i.e. in the absence of solution.

The equilibrium temperature for the transformation gypsum$\rightarrow CaSO_4 \cdot \frac{1}{2}H_2O + 1\frac{1}{2}H_2O$, determined by topotactic dehydration of single crystals of gypsum (Šatava, 1974), agreed with temperatures reported from solubility experiments and free energy calculations, i.e. close to 97 °C.

Experiments on the pressure-dependence of the metastable dehydration of gypsum to hemihydrate (α-form) by McConnell *et al.* (1987) gave the P–T relationship expressed in Fig. 11. In these experiments care was taken to avoid the appearance of anhydrite or re-hydration to gypsum. The slope of the curve in Fig. 11 is 81.1 bar/K, a result similar to that (85.5) calculated from ΔS 12.22 cal/mol./K and ΔV 5.99 cm^3. At 1 bar the curve extrapolates to just below 100 °C. McConnell *et al.* noted that this metastable reaction is likely to be dominant in mechanical experiments on gypsum (e.g. Murrell and Ismail, 1976) at relevant temperatures.

The transformation of gypsum to anhydrite was observed (Kosztolanyi *et al.*,

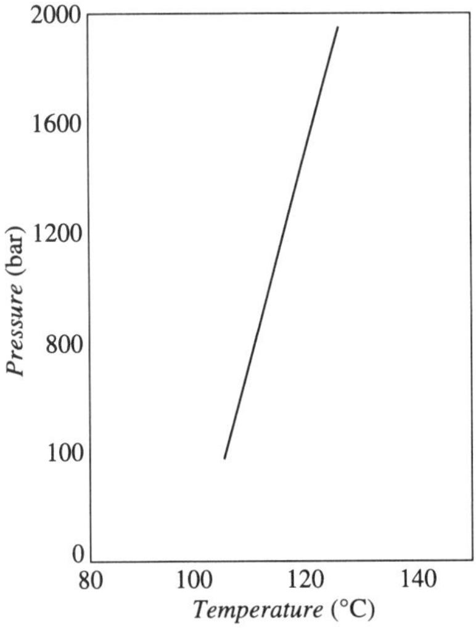

Fig. 11. Experimentally determined univariant *P–T* relationship for the coexistence of gypsum and hemihydrate (after McConnell *et al.*, 1987).

1987) using microthermometry and Raman microspectroscopy to study inclusions of gypsum in quartz (i.e. in a closed system). Because of the increase of pressure, dehydration did not begin until about 100 °C and was complete at 200 °C. The transformation occurred via the intermediate, hemihydrate.

Thermogravimetric experiments (in dry nitrogen) by Putnis *et al.* (1990) showed the continuous loss of water from gypsum, with an activation energy of 90.3 kJ/mol. (~21 kcal/mol.) and no significant change in this value with degree of hydration. Their *in situ, in vacuo* studies by infrared spectroscopy, however, showed three distinct phases, gypsum, hemihydrate and γ-$CaSO_4$, and no other intermediate water content (see also Dubrovinsky *et al.*, 1991).

Rull *et al.* (1989) followed the transformation of gypsum into hemihydrate by infrared spectroscopy using the symmetric stretching bands in the region 4000–3000 cm^{-1}. Isothermal data (at 115–135 °C) were consistent with random growth by a first-order mechanism, and an Arrhenius plot gave a rate constant and thus an activation energy of 24.6 kcal/mol., a result consistent with those obtained by TGA methods.

McAdie and Kuntze (1961) described a method for determining small quantities of gypsum and soluble anhydrite in the presence of hemihydrate (Plaster of Paris). Their approach requires the measurement of weight gain on hydration in aqueous alcohol, followed by dehydration to obtain the total amount of combined water.

Solubility of gypsum and anhydrite. Because of their importance to the cement and plaster industries, and also to the understanding of evaporite deposits, the chemical properties, solubilities, transformations and phase equilibrium relationships of calcium sulphate and its hydrates have been widely studied. Among early laboratory studies of the $CaSO_4$–H_2O system were those by Hall *et al.* (1926), Hill (1937), Posnjak (1938), Kelly *et al.* (1941), D'Ans *et al.* (1955) and Bock (1961).

The temperature of the equilibrium transition of gypsum to anhydrite in pure water determined by solubility measurements has been given mostly as in the region of 40–42 °C, and for the transition of metastable gypsum to metastable

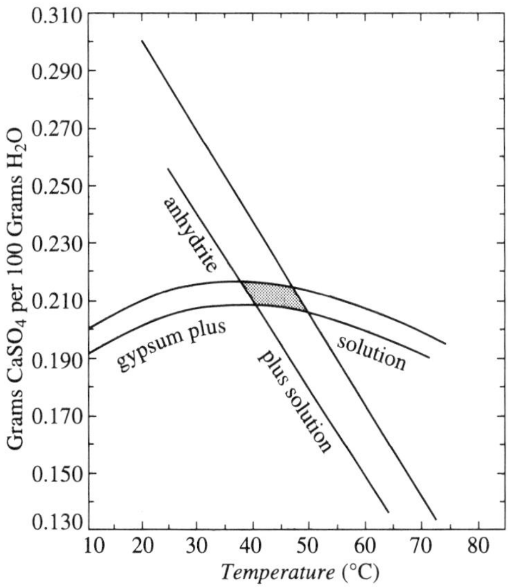

Fig. 12. The solubilities of gypsum and anhydrite in the system $CaSO_4$–H_2O at 1 bar. Two solubility curves are given for each mineral which bracket most of the results published, including Hill (1937), Posnjak (1938), Bock (1961) and Zen (1965, calculated) (after Hardie, 1967).

hemihydrate as about 97 °C. For the gypsum–anhydrite temperature, however, other studies have suggested higher values, e.g. 58 °C (Hardie, 1967), and the range of values is illustrated in Fig. 12. Some of the variations in results are thought to arise through experiments on solutions in which equilibrium was not approached from both undersaturated and oversaturated conditions. Also, the reactions of anhydrite with solutions tend to be very sluggish.

Cherry (1968) investigated various factors that could affect the calculated value of the solubility product for gypsum and gave a range from 1.9 to 2.4×10^{-5}, rather than a single value, for gypsum saturation at 25 °C and at low pressure.

Because gypsum is commonly associated with halite, more geologically applicable experiments were based on the ternary system $CaSO_4$–NaCl–H_2O, e.g. Macdonald (1953), Bock (1961), Zen (1965) and Blount and Dickson (1973). Bock (1961) determined the solubility of gypsum in pure water as 0.207 g $CaSO_4 \cdot 2H_2O$/100 g water at 25 °C, 0.222 g/100 g at 30°C and 0.212 g/100 g at 50 °C. Corresponding solubilities in water with approx. 1 mol. NaCl/100 g were 0.614, 0.600 and 0.599. The intersection of the solubility curve with that for anhydrite gave the temperature for transition of gypsum to anhydrite as 42 °C, but it decreased to 30, 25 and 20 °C with NaCl at 3.28, 4.3 and 5.3 mol./100 g water.

Zen (1965) obtained similar results for solubilities in the system $CaSO_4$–NaCl–H_2O and, approaching equilibrium from undersaturated and oversaturated sides, derived the transition temperature as 46 ± 25 °C. He found that anhydrite was not precipitated from any composition in the ternary system, whereas gypsum is formed from all solutions of all compositions at 35 °C, and from most compositions at 70 °C. He concluded therefore that when anhydrite occurs in evaporite beds it originated as gypsum and has been subsequently subjected to metamorphism (see also Conley and Bundy, 1958; Ostroff, 1964).

Hardie (1967) concluded that in sea-water saturated with halite and gypsum, the gypsum should dehydrate at about 18 °C. Hardie worked with solutions of

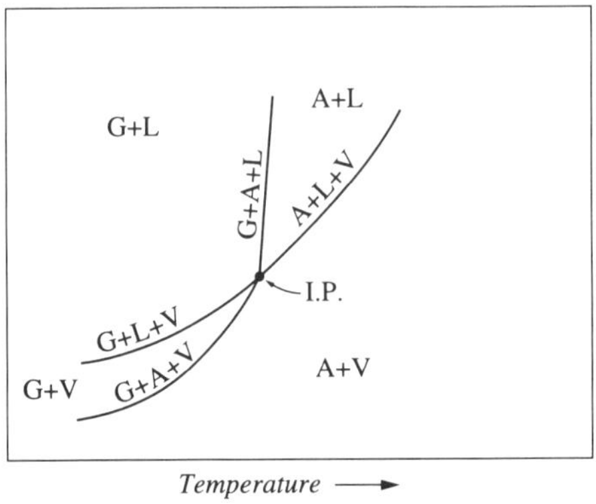

Fig. 13. Schematic *P–T* phase relationships in the system $CaSO_4$–H_2O. G, gypsum; A, anhydrite; L, water solution; V, vapour (after Blount and Dickson, 1973).

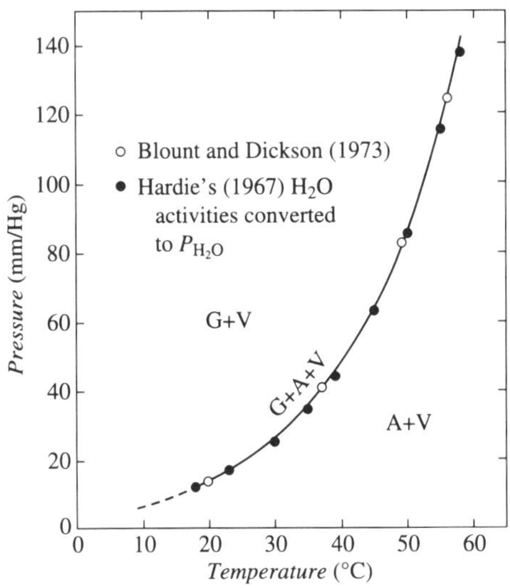

Fig. 14. Vapour pressure–temperature plot of gypsum–anhydrite–vapour equilibrium at 1 bar total pressure. The upper end of the curve coincides with the invariant point (I.P.) on Fig. 13 (after Blount and Dickson, 1973).

Na_2SO_4 and H_2SO_4, and extrapolating to zero concentration found the transition temperature to be 58 °C.

D'Ans (1968) defended his value of 42 °C against Hardie's 58 °C, suggesting that results could be affected by the presence of small amounts of metastable γ-$CaSO_4$. The effect of pressure on the equilibrium temperature in the $CaSO_4$–H_2O system was calculated by MacDonald (1953) to be 85.4 bar/°C, and by Zen (1965) to be 71 bar/°C.

For the system $CaSO_4$–H_2O, Fig. 13 shows schematically four three-phase univariant equilibria meeting at the four-phase invariant point. Blount and Dickson (1973) used their own experimental results and those of Hardie (1967) to plot the *P–T* equilibrium curve (Fig. 14) for the reaction gypsum + vapour = anhydrite + vapour.

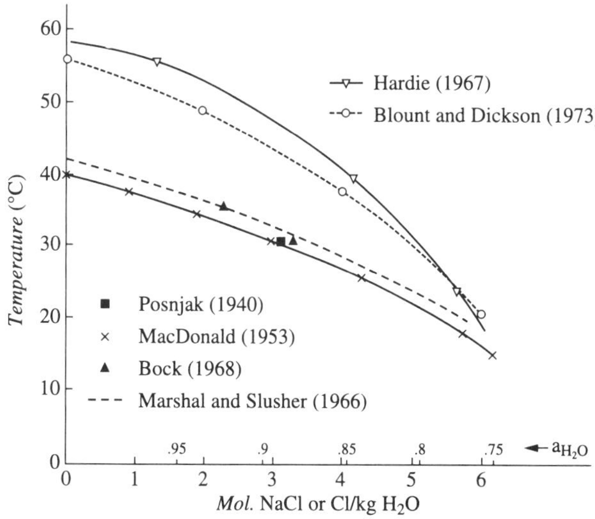

Fig. 15. Variation of the equilibrium temperature of the gypsum, anhydrite, solution and vapour assemblage (at 1 bar total pressure) with NaCl concentration (and a_{H_2O}) (after Blount and Dickson, 1973).

Blount and Dickson (1973) also explored the effects of both temperature and NaCl concentration (Fig. 15) on the gypsum + anhydrite + H_2O equilibrium. They overcame the sluggishness of the reaction by using higher temperatures and extrapolating downwards. The transition temperatures obtained were 56, 48, 36 and 20 °C for solutions with 0, 2, 4 and 6 mol./kg NaCl respectively, and the pressure coefficients of 70–80 bar/°C agreed with those obtained from thermochemical calculations. The curves from their data and those of Hardie (1967), passing through approximately 58 °C for pure water, are clearly significantly above those obtained from data by Posjnak (1940), MacDonald (1953) and others.

Christoffersen and Christoffersen (1976) determined the solubility of gypsum at 25 °C by several methods and adopted the value of 1.52×10^{-2} M, agreeing with that of Bock (1961) and close to many others in the literature, e.g. Marshall and Slusher (1966) and Madgin and Swales (1956), which range between 1.47 and 1.52×10^{-2} M. They also concluded that the rate-controlling process is diffusion and gave the rate constant as 4×10^{-6} cm^2/s (see also Ridge and Surkevicius, 1962; Gobran and Miyamoto, 1985).

Innorta *et al.* (1980) determined solubility curves for gypsum defined by:

$$S(\text{wt. per cent } CaSO_4) = 8.69 \times 10^{-6} T^2 - 1.73 \times 10^{-3} T + 0.2624,$$

and for anhydrite:

$$S = 1.14 \times 10^{-5} T^2 - 3.667 \times 10^{-3} T + 0.3517.$$

Lebedev and Lekhov (1990) studied the kinetics of the dissolution of gypsum in water at 5–25 °C. They found that the activation energy was 42–46 kJ/mol. and that the saturation level for calcium in solution with gypsum dissolving is always close to 0.0146 mmol./cm^3 at 25 °C.

Conley *et al.* (1938) and Conley and Bundy (1958) used data from Hill (1934) and Hill and Wills (1938), as well as their own experimental results, to explore systems involving Na_2SO_4, K_2SO_4 and Mg_2SO_4, in addition to $CaSO_4$–H_2O. In these systems gypsum occurs at appropriate temperatures stably or metastably along with mirabilite ($Na_2SO_4 \cdot 10H_2O$), syngenite ($K_2SO_4 \cdot CaSO_4 \cdot H_2O$) or other double-salt hydrates.

Gypsum is soluble in HCl. In Clerici solution it becomes cloudy, increases in density and eventually sinks in the liquid; thallium sulphate is formed (Pehrman, 1935).

Miscellaneous reactions. The treatment of powdered gypsum with a solution of sodium carbonate produced calcite, but when the solution was applied to cleavage plates, pseudomorphous vaterite, as well as calcite, was formed (Flörke and Flörke, 1961).

When gypsum particles are reacted with solutions of $SrCl_2$, $BaCl_2$ or $Pb(NO_3)_2$, celestine, baryte and anglesite respectively are precipitated (Kaushansky and Yariv, 1990).

Alteration. Among the alteration products of gypsum, many of them pseudomorphous, are celestine, anhydrite, halite, polyhalite, calcite, aragonite, malachite, chalcedony, opal and quartz.

Holser (1966) reported that at Baja California, Mexico, due to the action of a bittern which has reached 60 times the salt concentration of sea-water, gypsum was being replaced by polyhalite without any intermediate formation of anhydrite or bassanite.

Hemihydrate (bassanite). Razouk *et al.* (1960) noted that because of structural similarities, γ-$CaSO_4$ takes up water easily to form hemihydrate, but because anhydrite has a different structure it cannot be easily hydrated. To form hemihydrate or gypsum from anhydrite it may need to go through a process of solution and re-crystallization (see also Gevork'yan and Povarennykh, 1980).

McConnell (1965) determined thermogravimetrically that hemihydrate loses reversibly 0.416 mol. H_2O per mol. $CaSO_4$ in the temperature range 40–100 °C. The mean heat of absorption was reported as 14.955 kcal/mol. H_2O and ΔH as 7.477 kcal for the dehydration reaction, a value similar to that obtained by Kelly *et al.* (1941) using a completely different technique.

Although hemihydrate (bassanite) is theoretically unstable at all temperatures at 1 bar, but nevertheless occurs naturally (e.g. Bundy, 1956), Yamamoto and Kennedy (1969) decided to investigate the effect of pressure on the reactions:

$$CaSO_4 \cdot 2H_2O = CaSO_4 \cdot \tfrac{1}{2}H_2O + 1\tfrac{1}{2}H_2O \quad (1)$$

and

$$CaSO_4 \cdot \tfrac{1}{2}H_2O = CaSO_4 + \tfrac{1}{2}H_2O \quad (2)$$

and obtained the equilibrium diagram shown in Fig. 16. For reaction 1 at 1 bar the equilibrium temperature was obtained as 99 °C. The two curves intersect at 85 °C and 2 kbar, indicating that below this pressure hemihydrate does not exist stably and suggesting that natural bassanite is metastable under the conditions reported for its occurrence. Nancollas *et al.* (1973) gave 103 °C as the temperature of transition from dihydrate to hemihydrate in salt-free solutions at 4 bar.

Because of its wide industrial use as plaster, there have been many studies of the mechanisms, thermodynamics and kinetics of the hydration of $CaSO_4 \cdot \tfrac{1}{2}H_2O$ to the dihydrate, e.g. those by Ridge and Surkevicus (1962) and Kuzel (1987). [Tables containing thermochemical data relating to the hydrates of calcium sulphate have been presented by Ladd and Lee (1965), Gardner and Glueckauf (1970) and others.] Reported values of ΔH range from about 16 000 to 21 000

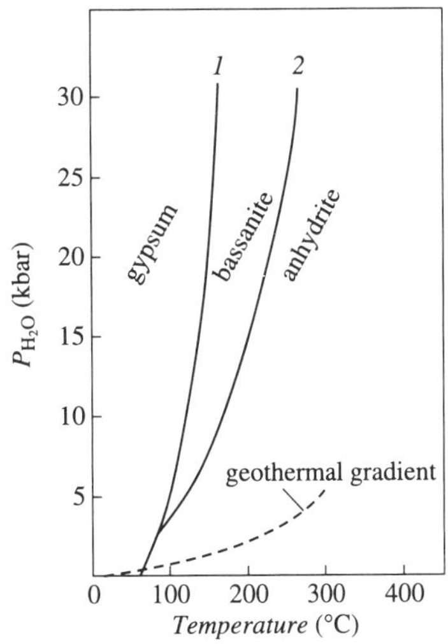

Fig. 16. Equilibrium diagram for the system $CaSO_4$–H_2O (after Yamamoto and Kennedy, 1969).

J/mol. For the hydration of γ-$CaSO_4$ to $CaSO_4{\cdot}2H_2O$, values range between 27 500 and 30 187 J/mol.

Kuzel (1987) prepared hemihydrate with composition $CaSO_4{\cdot}0.53H_2O$ by the dehydration of the dihydrate in concentrated nitric acid at 36 °C, followed by rehydration at 35 per cent relative humidity. The extent of hydration achieved depended on the water vapour pressure, up to a limit of 0.62 H_2O at 92 per cent relative humidity. Kuzel concluded that the hydration of γ-$CaSO_4$ to hemihydrate did not occur by a continuous increase in channel water, but rather by a single step hydration to $CaSO_4{\cdot}0.53H_2O$.

Cation partitioning. Kushnir (1980) investigated the distribution of several ions between gypsum and brine and found the order of preference for gypsum to be $Sr \gg Na \sim K > Mg > Cl$. Kushnir found that the concentrations of ions in gypsum increased with increasing growth rate up to a limiting value, and also that the preference of Sr for gypsum increased with increasing brine concentration and decreasing temperature. A mechanism involving adsorption and desorption of cations on surface sites was suggested, and also that some cations locate at interstitial rather than structural sites.

The transformation of gypsum into anhydrite occurs via dissolution and precipitation so that the ions are re-partitioned between the new phase and the solution. Kushnir (1982) found that the partitioning of potassium and sodium between anhydrite and solution was not temperature dependent but was affected by brine concentration. For magnesium and strontium, however, the partitioning into anhydrite decreased with increasing temperature. $D_{Sr}[(Sr/Ca)_{solid}/(Sr/Ca)_{brine}]$ for anhydrite fell from 0.40 at 20 °C to 0.02 at 170 °C. This may provide a method for estimating temperatures and thus the depth of burial of gypsum at a particular stage of dehydration. Kushnir (1982) noted that bassanite may form as an intermediate phase and that strontium appears to concentrate in bassanite to a greater extent than in either anhydrite or gypsum. It was also noted that during the dehydration of gypsum rock the water released contains high strontium concentrations and can subsequently precipitate celestine (see p. 36).

Crystal growth. Barta *et al.* (1971) grew crystals of gypsum in silicate acid gels and found that at 22 °C the best results were obtained with pH 7–8 and when the crystals were on the boundary of the gel and a 1 M solution of $(NH_4)_2SO_4$. The calcium was provided by a $CaCl_2$ solution. To simulate the growth of gypsum in lake or sea-floor muds, Cody (1976) grew crystals in gels and pastes of Wyoming bentonite in the presence of sodium or other soluble chloride. Only gypsum (tabular $\perp z$ and hemihydrate (no anhydrite) were formed. Cody concluded that primary anhydrite could occur only in extremely saline conditions.

Fowler *et al.* (1969) carried out a microscopic study of the growth of crystals of hemihydrate produced on sheets of selenite by heating from 50 to 150 °C under high vacuum. They recognized three steps:

$$CaSO_4{\cdot}2H_2O \rightarrow CaSO_4{\cdot}\tfrac{1}{2}H_2O + 1\tfrac{1}{2}H_2O$$
$$CaSO_4{\cdot}\tfrac{1}{2}H_2O \rightarrow \gamma\text{-}CaSO_4 + \tfrac{1}{2}H_2O$$
$$\gamma\text{-}CaSO_4 \rightarrow \underset{\text{anhydrite}}{CaSO_4}$$

It was noted that the transition from gypsum to hemihydrate involves a large change in specific volume (from 74.5 to 52.8 cm^3/mol.).

Optical and Physical Properties

Most gypsum occurs as massive rock gypsum; when this is fine-grained and white or lightly coloured it finds use as an ornamental stone known as alabaster, but it is often darkened by impurities of clays, iron oxides and other minerals. Granular or compact masses of gypsum are sometimes called sand or seed gypsum; kopi is an earthy variety. Aggregates of crystals usually permeated by sand grains are sometimes found in the form of rosettes which are similar to those of baryte; both are commonly referred to as 'desert roses'.

Gypsum often occurs as euhedral transparent crystals, in which form it is known as selenite. Crystals of selenite commonly adopt a tabular habit {010}, showing the additional forms {120}, {$\bar{1}$11} and {011}; sometimes a lenticular appearance results from the presence of curved faces. The {010} faces may exhibit a pearly lustre. The order of importance of forms predicted by the periodic bond chain (PBC) theory for gypsum is {010}, {120}, {0$\bar{1}$1}, {$\bar{1}$11}, but in practice {0$\bar{1}$1} features less commonly, probably because of the adsorption of water molecules on faces and edges of crystals (Simon and Bienfait, 1965). Another distinctive habit is that of translucent fibrous aggregates (fibres parallel to *z* axis), the surfaces of which show a pearly sheen; this variety is known as satin-spar. Mellis (1964) showed that in some cases the fibres are parallel to *y*.

Crystal growth. There have been many experiments based on the growth of crystals of gypsum to establish which conditions favour particular habits, curvature of faces and twinning. Deicha (1946) concluded that curved crystals result when impurities such as NaCl are present in solution (but see below). Ridge (1958) found that the crystal habits of gypsum formed by the hydration of Plaster of Paris at different temperatures varied from very fine-grained at 0 °C to swallow-tail crystals at 20 °C and sheaf-like aggregates at 46 °C.

The effects of cation impurities as well as temperature were reported by Rinaudo *et al.* (1989). Crystals of gypsum were grown at temperatures of 25, 35 and 45 °C from solutions to which trace amounts of Mg, Cd and Na had been added. Those containing Mg or Cd had curved faces, whereas those with Na were twinned but undeformed (see also Rinaudo *et al.*, 1985). Curved crystals were also produced from solutions with added calcium chloride (Rinaudo *et al.*, 1988). It was suggested that the curvature is due to crack formation and mechanical twinning which have been induced by stresses associated with the cation impurities. The inducement of twinning without deformation in the case of sodium impurity was attributed by Franchini-Angela and Rinaudo (1989) to the epitaxial growth of glauberite.

Crystals grown in different media also appear to develop different morphologies (Van Rosmalen *et al.*, 1976). Equant crystals formed when grown in pure agar, but acicular crystals formed in HCl. Crystals were also acicular when grown in silica gel in the presence of inorganic compounds, but changed to a more stubby morphology in the presence of organic compounds. Many other observations and experiments (e.g. Barcelona and Atwood, 1978; and others mentioned below) point to the modifying influence of organic compounds on the morphology of gypsum crystals grown from solutions. A generalized approach to the presentation of growth curves of slightly soluble salts was devised by Van Leeuwen and Blomen (1979) and gave good results when applied to gypsum and baryte.

The crystal habit of gypsum also appears to be influenced by the pH conditions in which it grew. Cody (1979) found that lens-shaped crystals occurred in the presence of dissolved organic material under alkaline conditions, whereas prismatic crystals developed in acid conditions with or without organic material. Kirov (1980) described acidity as the main modifier of habit, with decreasing pH leading to increasing elongation and fibrous development.

The influence of organic compounds on unusual growth orientations of gypsum crystals from the Miocene evaporites of Poland was described by Babel (1990).

Guillou *et al.* (1991) studied gypsum from the Casmance area, Senegal, where it is precipitated on concentration and re-dissolved with new influxes of sea-water. The crystals displayed the tabular {001} habit associated with an organic-rich environment. Their slightly rounded faces show striations, and on examination by scanning electron microscopy were found to be composed of fibres (~ 1 μm in diameter) stacked parallel to [100], the largest dimension of the (001) face.

Sabouraud (1976) found triangular cavities in synthetic gypsum grown at low temperature from $CaSO_4 + NaCl + H_2O$ similar to those which occur in natural gypsum.

Gypsum crystals grown from a mixture of fine-grained synthetic calcite and sea-water (Kastner, 1970) contained hour-glass patterns of calcite inclusions. In directions of slow crystal growth particles had been pushed aside instead of being incorporated; in directions of rapid growth clear areas developed.

The activation energy for the crystallization of gypsum from saturated solution was determined by Liu and Nancollas (1970) as 15 ± 0.5 kcal/mol. and by Nancollas *et al.* (1970), who noted that the rate of reaction was proportional to the square of the relative supersaturation, as 14 ± 2 kcal/mol.

Because gypsum may be the main constituent of unwanted scale deposits, there have been various studies of crystal-growth inhibition. Edinger (1973) found that HSO_4, NO_3 and Na^+ are preferentially adsorbed on (111) and reduce its growth rate, as do OH^- on (010) and H^+, Sr^+ and Ag^+ on (110).

In studying the effects of inhibitors on the growth of gypsum, Weijnen *et al.* (1987) calculated the interfacial tension using single-layer and double-layer Ising models and obtained good agreement with experimental values. They also found that whereas in gypsum grown from pure solutions with pure water (011) developed less than ($\bar{1}11$), the reverse was true in the presence of organic compounds. They noted that the best inhibitors are phosphonates and polycarboxylates.

Table 10. Variation of Refractive Indices of Gypsum with Temperature and Wavelength (Hutchinson and Tutton, 1913)

Wavelength (mμ)	α		β		γ	
	12 °C	105 °C	12 °C	105 °C	12 °C	105 °C
Li 671	1.5178	1.5154	1.5201	1.5158	1.5270	1.5243
Na 589	1.5207	1.5184	1.5230	1.5188	1.5299	1.5274
Tl 535	1.5231	1.5209	1.5255	1.5213	1.5325	1.5300
H_β 486	1.5262	1.5239	1.5285	1.5243	1.5355	1.5330

Optical properties. Gypsum is usually white or colourless, but it is sometimes grey, brown or pale shades of red, yellow or blue. The refractive indices of gypsum at different temperatures and for different wavelengths of light are listed in Table 10. For sodium light at room temperature $2V_\gamma$ is about 58°, the optic axial plane is (010) and γ:z is +51° 52′. Change of 2V with temperature is large, however, and with increasing temperature 2V decreases and is zero at about 91 °C for sodium light. At higher temperatures 2V opens out in the plane perpendicular to (010) (Hutchinson and Tutton, 1913). At constant temperature gypsum shows strong inclined dispersion of the bisectrix and dispersion of 2V with $r > v$: because of the change of optic axial plane, dispersion of the bisectrix is horizontal above 91 °C and dispersion of 2V has $r < v$. The extinction angle γ:z varies with wavelength over a range of about 1° and is at its maximum for a wavelength near that of sodium light. The angle γ:z also varies with changing temperature, increasing by about 2′ or 3′ for a rise of 1 °C (Berek, 1912). The variation of the birefringence of gypsum with wavelength and temperature has been studied by Jaffray (1943); at 17 °C (γ–α) is 0.00873 at λ 8500 Å and 0.00970 at λ 3500 Å. Cleavage plates of selenite may show asterism in the form of bands of light perpendicular to the traces of (100), ($\bar{1}$11) and (122) on (010) (Rutherford, 1940). A bluish fluorescence exhibited by some gypsum, anhydrite and many other minerals was attributed by Przibram (1962) to traces of organic substances.

The polarized infrared spectrum of gypsum was reported by Vierne and Brunel (1969) as having principal bands at 8.68–8.75 (ν_3, valence antisymmetric, x direction), 14.83–16.55 (ν_4, deformation antisymmetric, x direction), and 21.4 μm (ν_4, deformation symmetric). The results indicated a non-regular SO_4 tetrahedron, as was confirmed subsequently by X-ray structure determination. Bands associated with the water molecules in gypsum occurred at 2.83 (valence symmetric), 2.95 (valence antisymmetric) and 6.15 μm (deformation) (see also Razouk *et al.*, 1960; Caillère and Pobeguin, 1962; Ross, 1974). Benstead (1976) used laser Raman spectroscopy and reported spectral bands for gypsum at wave numbers (cm^{-1}) 414 and 494 (ν_2), 622, 624 and 676 (ν_4), 1010 (ν_1, very strong) and 1114, 1138 and 1144 (ν_3) (see also Griffith, 1969; Berenblut *et al.*, 1973; Benstead and Prakash, 1968). For hemihydrate the corresponding values (cm^{-1}) were: ν_2 438, 494; ν_4 602, 636, 676; ν_1 1018 and ν_3 1132, 1150, 1174; almost identical results were obtained for γ-$CaSO_4$ ('soluble anhydrite').

The Raman spectra of gypsum were recorded at pressures ranging from 0.001 to 7 kbar and $d\nu/dP$ values obtained for bands from 3600 to 100 cm^{-1} (Couty *et al.*, 1983). The internal modes of SO_4 tetrahedra were found to be more sensitive to pressure than to temperature, and coupled H_2O–SO_4 modes to be more affected by pressure and temperature than $CaSO_4$ modes. For the stretching vibration of

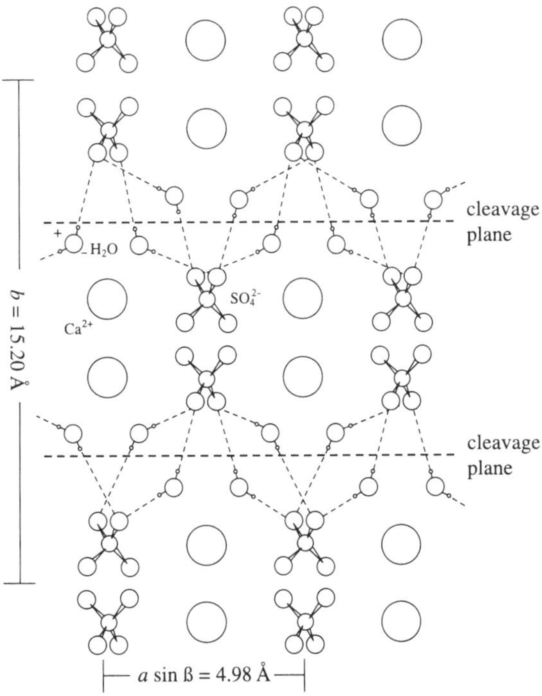

Fig. 17. The structure of gypsum viewed along the *z* axis (*I*-cell) showing the way in which the (010) cleavage planes intersect hydrogen bonds between water molecules and SO_4 tetrahedra (after Oglesby *et al.*, 1976).

water molecules at 3500 cm^{-1} $d\nu/dP \sim 2.11$ cm^{-1}/kbar, but for that at 3400 cm^{-1} $d\nu/dP \sim 0.10$ cm^{-1}/kbar. This large difference was explained by the stronger hydrogen bonding of one of the hydrogen atoms outside the double H_2O planes in the structure (Fig. 17).

Iishi (1979) determined the static dielectric constants of gypsum to be: $\varepsilon_0^{min} = y = 5.98$, $\varepsilon_0^{max} = 11.65$, ε_0^{max}: $z \sim 12°$ (*A*-cell).

Bassanite, the natural form of $CaSO_4 \cdot \frac{1}{2}H_2O$, occurs as microscopic needles. Synthetic hemihydrate is uniaxial with positive sign, ω 1.558, ε 1.586 and density 2.74 g/cm^3, while γ-$CaSO_4$ has ω 1.505, ε 1.548 and density 2.61 g/cm^3 (Gaubert, 1933).

Twinning. Contact twins on (100) are often found, the appearance of which gives rise to the names 'swallow tail' or 'arrow-head'. Translation gliding on {010} in the direction [001] is easily produced.

The investigation of the conditions in which twins are or are not formed in the growth of gypsum has importance for both industrial processes and the understanding of palaeo- and recent environments. In gypsum prepared by the action of sulphuric acid on calcium chloride, Kern and Rehn (1960) observed that the most abundant kind of twin, particularly at high supersaturation, was interpenetrant on ($\bar{1}$01); less common were (100) interpenetrant and (100) contact, while contact twins on ($\bar{1}$01) were not observed at all. In the interpenetrant ($\bar{1}$01) and (100) twins the original composition plane is (010).

Gypsum produced by the slow evaporation of saturated solutions of $CaSO_4$

above 36 °C (Bright and Ridge, 1964) contained acicular crystals twinned on (100) with re-entrant angle 132°; some were twinned on (101) (re-entrant 104°), which is generally uncommon in natural gypsum. The (101) twins are, however, common in the gypsum of the Paris Basin, which is believed to have formed through laguno-lacustrine evaporation, and in the absence of the sodium ion (Cody and Cody, 1989a). They found that these features are favoured by the presence in natural evaporite conditions of α-amylase, which results from certain microbial blooms. They also found that low pH tended to produce (101) twins flattened on (010), similar to those formed in the presence of acid mine waters (see also Cody, 1979). Some earlier studies had associated the (101) twins with supersaturation and the absence of Na^+ ions, but a re-examination by Cody and Cody (1989b) showed that the twins were not on (101) but were habit-modified (100) twins with a very similar re-entrant angle.

Rinaudo *et al.* (1985) concluded from an X-ray topographic study that twinning was affected by the nature of the gel in which crystals of gypsum were grown, and also that twinning occurred more in the larger crystals.

Mechanical properties. The perfect {010} cleavage of crystalline gypsum results in platy cleavage fragments which, because of further cleavages {100} and {011}, often show a lozenge-shaped outline (angle 114°), with edges parallel to those of the two smallest *a* and *c* lattice parameters (Fig. 7, p. 41). The topography of cleavage faces of selenite was studied using interferometry by Tolansky (1945).

The (010) cleavage plane of gypsum is indicated in Fig. 17, which shows how it passes through hydrogen bonds in the structure. The specific surface energy of the cleavage was determined by Oglesby *et al.* (1976) as 358 ergs/cm^2 in air, and was little changed at 10^{-5} bar and over the temperature range −35 to 65 °C.

The hardness of gypsum is generally given as H $1\frac{1}{2}$–2; using a Knoop microhardness tester, Winchell (1945) found variation according to direction from 32–45 on the Knoop scale. Taylor (1949) gave the Vickers hardness of gypsum as approximately 60.

Heard and Rubey (1966) carried out triaxial compression tests at 5 kbar on sealed cylinders of polycrystalline gypsum. The observed decrease in strength (from 2.70 to 0.25 kbar) when the temperature was increased from 100 to 150 °C was attributed to the dehydration process. A useful summary of textures and deformation behaviour for gypsum and anhydrite is that by Kern and Richter (1985).

Single crystals of gypsum were sheared in different directions on (010) and showed both elastic and ductile behaviour prior to failure (Williams, 1988). The ultimate shear strength τ_u ranged from 0.58 to 1.36 MPa, and the minimum τ_u was found to be in the direction [103], which is the preferred direction for translation glide. On the secondary cleavages much higher values of τ were found, ranging from 14.3 MPa on (011) to 18.5 MPa on (100). For (010), the static and dynamic modulus of rigidity ranged respectively from 0.07 to 0.16 and 2.02 to 2.85 GPa. The compressibility (V/V_0) of gypsum was given by Vaidya *et al.* (1973) as 0.9880 and 0.9289 at 5 and 40 kbar respectively.

Distinguishing Features

Gypsum is easily distinguished from anhydrite because the latter mineral has higher refringence and birefringence. In addition, anhydrite has characteristic pinacoidal cleavages and a higher density. Hydromagnesite also has higher birefringence than gypsum.

Shearman (1979) described a field test useful for identifying gypsum in soils and sediments. When heated on a metal plate, grains of gypsum dehydrate and turn white in a few minutes whereas most others are unaltered. A simple and rapid staining test for sulphates was given by Hounslow (1979). A 10 per cent solution of $Hg(NO_3)_2$ is mixed with 1:1 nitric acid until the hydrolysis precipitate formed just dissolves; with a sulphate, this solution gives a yellow stain or precipitate.

Paragenesis

Gypsum is the most common of the sulphate minerals. Its main occurrences are as sedimentary deposits associated with limestones, shales, marls and clays, and in evaporite deposits. The principal modes of formation of gypsum are direct deposition by the evaporation of brine, by hydration of anhydrite, by oxidation of sulphides and by the action of acid sulphate solutions on calcium-bearing rocks. In clays and marls the acid solution may be produced by the weathering of sulphides, and in metalliferous veins by the oxidation of sulphides. Gypsum/anhydrite deposits exhibit a range of different textures including lamellar (sometimes in clastic or algal varves replacing carbonates), nodular, pegmatitic (lensoid selenite crystals) and massive. Important world localities for the occurrence of gypsum/anhydrite include the USA, Canada, China, France (Paris and Rhone basins), Germany (Zechstein) and the UK (Permo-Trias).

Evaporites. Sea-water contains about 3.5 per cent by weight of dissolved materials, about 80 per cent of which is sodium chloride and about 4 per cent calcium sulphate. The usual sequence of deposition of salts from sea-water has been shown experimentally to be: calcium carbonate–calcium sulphate–sodium chloride–sulphates and chlorides of magnesium–sodium bromide and potassium chloride. If all the salt of a 1000 m column of water were precipitated it would make only 15 m of salt deposits, of which about 0.4 m would be calcium sulphate, 11.6 m halite and the remainder potassium and magnesium-bearing salts. It is not essential, however, to have deep basin conditions for the formation of calcium sulphate; it can occur in a lagoonal environment, and in salt flat or sabkha conditions like those of the present day in areas such as the Trucial Coast and the Gulf of California. In many areas gypsum has been dissolved in percolating waters (solubility is increased by the presence of NaCl or $CaCO_3$), which in the dry season are drawn to the surface by capillary action, are evaporated, and leave gypsum deposited as crystals, sometimes in aggregates described as 'desert roses'. In experiments on soils using artificial columns, Jafarzadeh and Burnham (1992) found that gypsum crystals with a great variety of crystal habits were deposited by capillary rise and by the movement of water downwards. A general review of the chemistry, mineralogy and origins of gypsum 'crusts' of the main arid areas of the world was presented by Watson (1983).

Gypsum precipitates continuously from the Dead Sea, and this mineral accounts for the majority of the recent deposits in the area. Most of the gypsum is decomposed at the bottom and transformed into calcite (Neev, 1963).

Holser (1979) discussed the precipitation of gypsum from actual sea-water in which sodium chloride is accompanied by salts of magnesium, potassium, etc. Figure 18 illustrates how the solubility of gypsum rises sharply with increasing

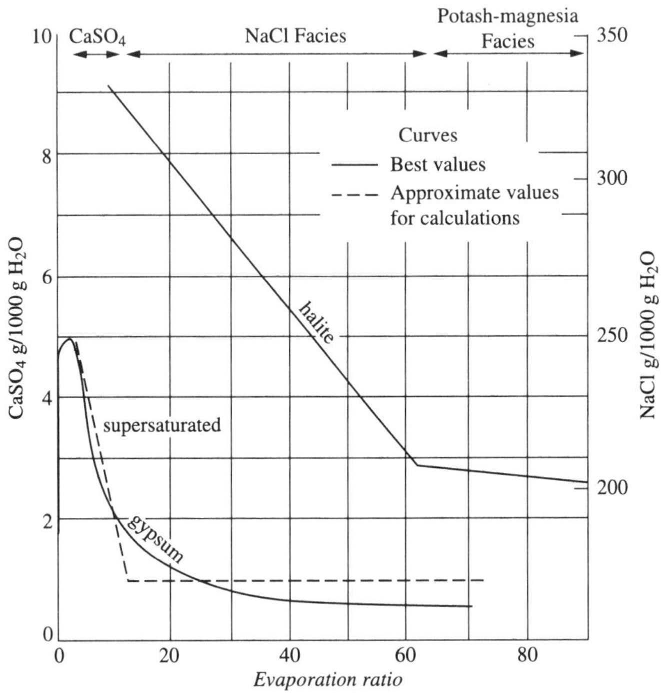

Fig. 18. Solubilities of gypsum and halite in concentrated sea-water (plotted by Holser, 1979, from published data). Evaporation ratio = mass of water in original sea-water/mass of water in brine.

salt concentration but then falls to a low and almost constant level. Holser points out that because of its very low solubility in concentrated brines, all calcium has been precipitated as gypsum before the halite facies, but new sea-water mixed with the brine will precipitate its own gypsum. Also, sea-water concentrated to saturation with gypsum will precipitate more gypsum if it dissolves halite.

In evaporites, the calcium sulphate sometimes occurs as gypsum, sometimes as anhydrite, and very often both minerals together. Different opinions have been expressed with reference to particular localities as to whether the gypsum or the anhydrite was the first to precipitate from solution. Some geological evidence indicates that the original material of most gypsum beds was anhydrite. This view is apparently supported where gypsum beds are observed to grade into anhydrite at depth (see for example Goodman (1952) on lagoonal deposits of gypsum and anhydrite in Nova Scotia), and by textural relationships which indicate that gypsum is formed later than anhydrite; in both cases, however, the anhydrite itself may have resulted from the dehydration of gypsum. Later opinions have allowed for the possibility that either gypsum or anhydrite may occur as the primary deposit: Stewart (1953), for example, found in the Permian evaporites of northeastern England, pseudomorphs of anhydrite after gypsum at a depth of about 1150 m (see also Kulstad *et al.*, 1956). Gypsum occurs with halite and anhydrite in the Triassic rocks of Eskdale, Yorkshire, England, and the vertical distribution of the minerals was discussed in terms of cycles of evaporation by Raymond (1953).

Experimental work on the solubility of gypsum and anhydrite (see p. 52) shows that gypsum should be deposited first from saturated solution in pure water when evaporation takes place below about 58 °C; above this temperature anhydrite is the stable form. The presence of sodium chloride is shown to lower the transition temperature, so that in concentrated NaCl solution anhydrite can form at lower temperatures. Bundy (1956), in a study of the gypsum–anhydrite deposits in southwestern Indiana, USA, suggested that in most evaporite deposits gypsum was precipitated first and was later converted to anhydrite; subsequent removal of the overburden and exposure regenerated gypsum from the secondary anhydrite. Conley and Bundy (1958) went further in stating that primary precipitation of anhydrite from sea-water is improbable. From the results of experimental studies, they concluded that the hydration of anhydrite in nature takes place mainly by the action of activating cations which occur in groundwater (see p. 53). These ions are also present in sea-water and cause the formation of gypsum rather than anhydrite. Moreover gypsum can exist metastably at temperatures up to 98 °C, the hemihydrate transition temperature, and at temperatures well above the transition point predicted for solutions with various salt concentrations. Conley and Bundy found that finely ground gypsum in contact with saturated NaCl solution remained unchanged at room temperature over a period of six months, and at 60 °C no alteration to anhydrite had taken place in three weeks.

It thus appears that anhydrite is nearly always a secondary mineral produced by the dehydration of gypsum. Douglas and Goodman (1957) showed experimentally that gypsum remains stable under pressures equivalent to a load of about 5500 m of rock; it would appear therefore that the dehydration of gypsum is largely a thermal effect. The conversion of gypsum to anhydrite involves a decrease in volume of the solid phase; sometimes halite fills the resultant voids and a halite–anhydrite assemblage is formed (Stewart, 1949). The water released by the dehydration of gypsum may result in the local solution, redistribution and deposition of the soluble salts of the surrounding evaporites.

An example of modern evaporite deposition is that of the Bocana de Virrila, Peru. In this marine estuary a horizontal gradient of salinity has been established, with gypsum being deposited near the head and halite at the mouth (Morris and Dickey, 1957). Large gypsum deposits are also found in saline lakes and salt pans, where the usual order of precipitation prevails, but the deposition may be interrupted at times by influxes of water, which may cause re-solution of some of the deposited salts (see for example De La Hunty and Low (1958) on the gypsum deposits of Western Australia). The unique several hundred square miles of nearly pure gypsum of the 'White Sands' of New Mexico were observed by remote sensing (infrared) by Lindberg and Smith (1973).

Fontes and Gonfiatini (1967) determined oxygen and hydrogen isotope ratios for the water of crystallization of gypsum. They found that while for oxygen the heavy isotope is higher by 4‰ in the gypsum than in solution, the heavy isotope of hydrogen (deuterium) is lower by 15‰.

Much use has been made of the determination of sulphur ($^{34}S/^{32}S$) and oxygen ($^{18}O/^{16}O$) isotope ratios in tracing the mode of formation of gypsum. If an occurrence of gypsum is attributed to its primary crystallization from a brine, its sulphur and oxygen isotope ratios would be expected to match those for sea-water of the appropriate age. Isotope–age curves for sea-water were given for example by Claypool *et al.* (1980). Present day sea-water has $\delta^{34}S \sim 20.5$‰ (for definition of $\delta^{34}S$, see p. 86). Account may need to be taken of the fractionation occurring on precipitation, if slow enough to equilibrate, which may itself enrich the heavy isotope in the crystals by 2 or 3‰. Gypsum deposited in a shallow basin by rapid evaporation will, however, maintain the isotope ratio of its water. Interpretation may also need to take account of mixing sea-water and freshwater.

For sulphur derived from magmatic or mantle sources $\delta^{34}S$ is very low (0–1‰),

[a] For definition of $\delta^{34}S$, see p. 86.

so that gypsum incorporating sulphur from oxidized sulphide with such origins would have low $\delta^{34}S$.

The bacterial reduction of sulphate is known to remove ^{34}S preferentially as sulphide, leaving the sulphate solution enriched in ^{32}S (Fritz *et al.*, 1989). Stemmerik *et al.* (1988) used sulphur isotope ratios to distinguish between gypsum deposited from residual brines (more ^{34}S) and that derived from re-oxidized sulphur which had originally been organic or in sulphide.

Gypsum, anhydrite and baryte in Kuroko-type deposits were studied by Sakai and Matsubaya (1971), who found that the sulphur isotope ratios for all three were similar (21–24‰), but that $\delta^{18}O$ increased in the order barytes–anhydrite–gypsum. They suggested that the water of crystallization of the gypsum was of meteoric origin for some Kuroko-type deposits but not for others.

Sabouraud *et al.* (1972) found in gypsum from a Mediterranean deposit a Cl/Br ratio of 311, similar to the value for oceanic water and normal for gypsum precipitating before halite. The ratio 151 obtained from a New Caledonian marine evaporite deposit was explained as 'plausibly' due to enrichment in Br of biological origin.

The $^{34}S/^{32}S$ and $^{18}O/^{16}O$ isotope ratios decrease in passing from the base to the top of the Parisian evaporite beds (Fontes and Létolle, 1976). The mean $\delta^{34}S$ value of +16.4‰ could not be attributed to Tertiary marine sulphate ($\delta^{34}S$ +21 to +22‰), so it was concluded that the SO_4^{2-} resulted from pre-existing and non-marine evaporite layers altered by solution erosion.

Sofer (1978) showed how a combination of oxygen and hydrogen isotope ratios of the water molecules of gypsum can help to distinguish formation via the evaporation of brine from the processes of hydration of anhydrite (Fig. 19) or the *in situ* oxidation of sulphides. For some selenite crystals from Sicily, Censi (1984) reported δ^2H values of +5 to +6‰ and $\delta^{18}O$ values of +17 to +27‰, confirming that they were of primary origin.

The $\delta^{34}S$ values for some Italian marine gypsum of various ages were given by Cortecci *et al.* (1981) as +9.3 to +13.8, +27, +25 and +17‰ for Upper Permian, Lower Trias, Mid Trias and Upper Trias deposits respectively, fitting broadly with the isotope–age curve determined by Claypool *et al.* (1980). The lack of a consistent relationship between oxygen and sulphur isotope ratios was attributed to the post-depositional change of oxygen isotope abundances.

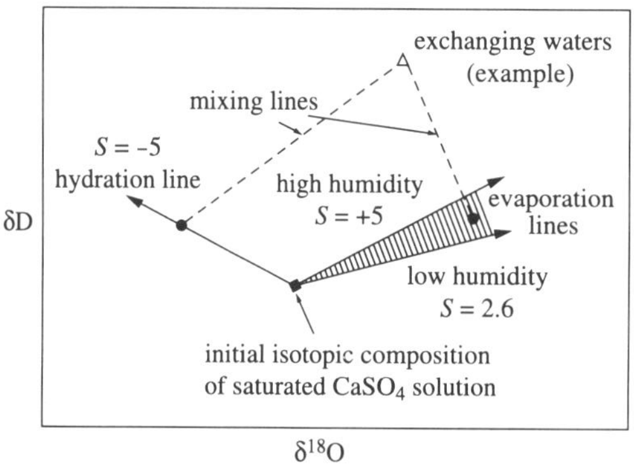

Fig. 19. Illustration of the trends of the isotope ratios for hydrogen and oxygen during processes of evaporation, hydration and mixing of water (after Sofer, 1978). S = slope.

A study by Geisler (1981) of gypsum in modern solar salt-works showed alternating clear and inclusion-rich layers related to changes in salinity. When salinity was high, crusts of gypsum crystals were preserved; when salinity was low, an algal and bacterial layer at the bottom of the salt-pans caused breakdown into a fine sand. It was suggested that these observations might help interpret the formation of some 'gypsarenites' in certain natural evaporite basins (e.g. Messinian in the Mediterranean).

The values of $\delta^{34}S$ of marine gypsum of different geological ages were determined by Chen *et al.* (1981) to be +18.8, +25.3 to +29.8 and +21.2 to +27.7‰ for Sinian, Lower Cambrian and Middle Ordovician deposits respectively.

A number of explanations have been offered for the formation of veins of satin-spar gypsum. Some involve crystal growth from the centre to the margins and others suggest growth from the vein walls to the centre. Growth by the contraction of a gel has also been suggested, and not uncommonly the formation of gypsum from anhydrite. Machel (1985) suggested that the hydration of anhydrite is not a pre-requisite for the formation of satin-spar veins, and that the calcium and sulphate ions may be derived from adjacent evaporite beds or from external sources. In those deposits studied by Machel, the fibres were generally flattened parallel to (010) and elongated parallel to z. It appeared that while the veins opened under tensile stress, the crystals grew from the centre keeping pace with dilation. Machel describes fibrous anhydrite with similar features.

Strontium content in gypsum from three Upper Miocene vaporitic basins ranged from 140 to 320 p.p.m. (Berrettoni *et al.*, 1986). Their Sr isotope ratios $^{87}Sr/^{86}Sr$ range from 0.70877 to 0.70919, indicating a diagenetic interaction with percolating waters which affected the surrounding shales.

Kelepertzis (1989) described the occurrence of gypsum in the Thiapes area of Milos island. He postulated its formation from plagioclase by the action of hydrothermal solutions rich in SO_4^{2-} and H_2S-bearing gases and of low pH, forming gypsum as well as the kaolinite deposits in the area:

$$\underset{\text{anorthite component of plagioclase}}{CaAl_2Si_2O_8} + H_2SO_4 + 3H_2O \rightarrow \underset{\text{gypsum}}{CaSO_4{\cdot}2H_2O} + \underset{\text{kaolinite}}{Al_2Si_2O_5(OH)_4}$$

Values of $\delta^{34}S$ and $\delta^{18}O$ (+15.4 to +17.4 and +11.2 to +12.79‰ respectively) for gypsum of the Swiss Gipskeuper (Rick, 1990) were consistent with a Triassic marine origin.

Gypsification of anhydrite. Whereas anhydrite is usually secondary, gypsum can be either primary or secondary. Gypsification of anhydrite occurs frequently along contacts of evaporites with carbonate rocks. It proceeds along anhydrite cleavages, with the textures showing that the gypsum is secondary. For gypsum to form by the hydration of anhydrite there must be space available for the required amount of water. Anhydrite requires about 50 per cent of its own volume of water to form gypsum, but the resulting gypsum occupies about 9 per cent less volume than the equivalent amount of anhydrite plus water (Farnsworth, 1924). Space for gypsification is provided by the removal of solutions through penecontemporaneous vein formation, and it has been suggested that transverse veins of gypsum are enlarged by the separation of their walls through the forces exerted by directional crystal growth (Bundy, 1956). In the gypsum–anhydrite cap-rock of the Sulphur Salt Dome, Louisiana, USA, transverse gypsum veins may have altered directly from transverse anhydrite veins (Goldman, 1952). Shearman *et al.* (1972) suggested that in many secondary gypsum deposits the additional volume is in the form of gypsum in veins cutting associated rock, water having moved up under pressure producing hydraulic fracture aided by the relief of overburden.

Hydrothermal gypsum deposits in Japan (altered dacite and shale), in which the gypsum is associated with anhydrite and was probably derived from it by hydration, were described by Iwao (1956). Gypsum occurs as an alteration product of anhydrite, which itself has replaced calcite in limestone at Calumet Island, Quebec; it was suggested that the sulphate is of magmatic origin (Osborne, 1941).

Holliday (1970) reviewed the processes by which secondary gypsum can form from anhydrite, i.e. (a) by solution followed by precipitation, (b) by direct hydration or (c) by stepwise hydration via bassanite. The hydration textures in some alabastrine rocks suggest formation by process (b), but for most secondary gypsum, particularly where it is clear that there has been no increase in volume, process (a) is the most likely.

Where large volumes of anhydrite have been altered to gypsum by hydration, masses of gypsum are found with relict nodules of anhydrite. This type of formation is found in the gypsum–anhydrite deposit in Somaliland: in this case, however, it was suggested that the gypsification of anhydrite is merely a superficial weathering phenomenon (Warden and Pallister, 1958).

Icart *et al.* (1965) found that the strontium content of the evaporite deposit of Tarascan (Ariège, France) was unusually high and, because strontium replaces calcium more readily in anhydrite than in gypsum, they argued that the gypsum in the deposit resulted from the slow hydration of anhydrite. The Lower Eocene anhydrite–secondary gypsum rocks (St Andrews, Jamaica) were described by Holliday (1971) as having formed in an arid supratidal (sabkha) environment.

In a deep-sea drilling project west of the Mid-Atlantic Ridge, gypsum and halite were found in a vein in a basalt sill 625 m below the sea-floor (Drever *et al.*, 1979). The gypsum had a $\delta^{34}S$ value of $+19.4‰$, indicating a sea-water source for the SO_4^{2-}. Nevertheless, the gypsum might be secondary. Anhydrite may have precipitated first at $T > 60\,°C$ and then converted to gypsum with falling temperature, or the anhydrite could have formed by the reaction of SO_4^{2-} from sea-water with calcium from the basalt.

Oxidation of sulphides. Vinogradov and Stephanov (1964) studied the oxidation products coating a large pyrite crystal in the sequence jarosite + goethite (outermost)–jarosite + gypsum–gypsum + pyrite (innermost). They found that the reactions involved had caused fractionation of the sulphur isotopes, with the enrichment of ^{34}S in the order pyrite–jarosite–gypsum. Faure and Felder (1981) investigated secondary prismatic crystals of gypsum in soil (Brown Hills, Transantarctic Mountains). The $\delta^{34}S$ value of $+11.5‰$ matched that of Permian marine evaporites, but the absence of such a source in the region, and the $^{87}Sr/^{86}Sr$ ratio of 0.7517, led the authors to the conclusion that the gypsum was formed by the oxidation of sulphides in the igneous basement rocks of the Brown Hills.

Gypsum filling fractures in an Archaean granite (Atikokan, Ontario) was described by Kamineni (1983). Its mean $\delta^{34}S$ value of $7.1‰$ is much lower than that for marine sulphates of appropriate age. It was suggested that the gypsum was derived locally by the interaction of pore water with epidote, plagioclase feldspar and pyrite, with the pyrite providing the sulphur and the alteration of epidote and feldspar supplying the calcium.

Von Gehlen *et al.* (1986) concluded from the $\delta^{34}S$ values (6.2–6.3‰) of gypsum in veins in southern Schwarzwald that it was not derived from Mesozoic evaporites; rather it was formed from solutions at or below 40 °C and its sulphur involved contributions from oxidized sulphide. Whereas the formation of massive deposits of gypsum generally requires arid conditions and high rates of evaporation, these constraints may not apply to the formation of scattered crystals and crusts. Bain (1990) reported large crystals of gypsum ocurring in the glaciolacustrine deltaic sequences in Ohio, USA, which formed diagenetically by the oxidation of sulphides by Ca-rich oxidizing groundwater.

Volcanic origin. Gypsum is found in deposits of native sulphur and can be produced in volcanic regions by the action of sulphurous vapours on calcium-bearing minerals. For gypsum at the lower and upper part of the crater neck of Mount Zimin, Kamchatka, Vinogradov *et al.* (1965) deduced from sulphur isotope ratios that whereas the lower gypsum was formed from sulphur of deep origin, the upper deposits involved the near-surface oxidation of H_2S. A sublimate of gypsum occurs inside the crater of Nyiragongo (Herman *et al.*, 1960).

Miscellaneous. Gypsum and anhydrite within pyrite deposits (Urals; Stolyarov, 1965) were accounted for by the reaction of post-magmatic, relatively hypothermal SO_4^{2-}-bearing solutions, with calcium carbonate already present or with Ca-bearing deeply circulating groundwater.

Pisarchik *et al.* (1977) found that gypsum and anhydrite from red beds overlying halide–carbonate rocks (Upper Lena Formation, Siberian platform) have $\delta^{34}S$ ranging between +5.4 and 9.75‰, significantly less than the underlying beds (~26.7‰). They attributed this to the influx of terrigenous sulphates containing more ^{32}S than those of marine origin.

Among the minerals which may be found in association with gypsum are halite (sometimes pseudomorphous), celestine, calcite, aragonite, dolomite, pyrite, sulphur and quartz. Since gypsum and anhydrite are interconvertible, pseudomorphs of one after the other commonly occur.

Occurrences of gypsum are dealt with further in the section on anhydrite (pp. 83–90).

Bassanite. The mineral bassanite, calcium sulphate hemihydrate, is rare and nearly always a product of alteration. It was found in cavities of leucite tephrite blocks ejected from Vesuvius (Spencer, 1913), and with gibbsite in fumaroles of this volcano. It has also been reported associated with gypsum or anhydrite in Indiana, USA (Bundy, 1956), and in Nova Scotia (Goodman, 1957). The latter author gave the mineral the name vibertite, and previously similar artificial material had been named miltonite (Milton, 1942), but the name bassanite has priority (Fleischer, 1958). Bassanite was found to be stable in restricted soil profiles of the Australian arid zone (Apokodje, 1984). The white chalky coating sometimes found on crystals of gypsum may be bassanite.

References

Abriel, W., 1983. Calcium Sulfat Subhydrat, $CaSO_4{\cdot}0.8H_2O$. *Acta Cryst.*, *C*, **39**, 956–8.

Abriel, W. and **Nesper, R.**, 1993. Bestimmung der Kristallstruktur von $CaSO_4(H_2O)_{0.5}$ mit Röntgenbeugungsmethoden und mit Potentialprofil–Rechnungen. *Z. Krist.*, **205**, 99–113.

Apokodje, E. G., 1984. The occurrence of bassanite in some Australian arid-zone soils. *Chem. Geol.*, **47**, 361–4.

Atoji, M. and **Rundle, R. E.**, 1958. Neutron diffraction study of gypsum, $CaSO_4{\cdot}2H_2O$. *J. Chem. Phys.*, **29**, 1306–11.

Babel, M., 1990. Crystallography and genesis of the giant intergrowths of gypsum from the Miocene evaporites of Poland. *Arch. Min.*, **44**, 103–35.

Bain, R. J., 1990. Diagenetic, non-evaporitic origin for gypsum. *Geology*, **18**, 447–50.

Ball, M. C. and **Norwood, L. S.**, 1969. Studies in the system calcium sulphate–water. Part I. Kinetics of dehydration of calcium sulphate dihydrate. *J. Chem. Soc. A*, 1633–7.

Ball, M. C. and **Norwood, L. S.**, 1970. Studies in the system calcium sulphate–water. Part III. Kinetics of dehydration of α-calcium sulphate hemihydrate. *J. Chem. Soc. A*, 1476–9.

Ball, M. C. and **Urie, R. G.**, 1970. Studies in the system calcium sulphate–water. Part II. The kinetics of dehydration of β $CaSO_4 \cdot \frac{1}{2}H_2O$. *J. Chem. Soc. A*, 528–30.

Barbieri, M., Masi, U. and **Tolomeo, L.**, 1976. Distribuzione dello stronzio nei gessi e nelle anidriti delle formazioni evaporitiche dell' Italia Centrale. *Rend. Soc. Italiana Min. Petr.*, **32**, 551–60.

Barcelona, M. J. and **Atwood, D. K.**, 1978. Gypsum–organic interactions in natural seawater: effect of organics on precipitation kinetics and crystal morphology. *Marine Chem.*, **6**, 99–115.

Barshad, I., 1952. Temperature and heat of reaction calibration of the differential thermal analysis apparatus. *Amer. Min.*, **37**, 667–94.

Barta, Č., Ženlička, J. and **René, V.**, 1971. Growth of $CaCO_3$ and $CaSO_4 \cdot 2H_2O$ crystals in gels. *J. Cryst. Growth*, **10**, 158–62.

Bartram, D. A., 1969. Infrared absorption spectra of α- and β-calcium sulphate hemihydrates. *Nature*, **223**, 494.

Bellanca, A., 1938. Piezoelettricita polare e simmetria di alcune specie cristalline. (Nota I.) *Periodico Min. Roma*, **9**, 323–31 (M.A. **7**–246).

Benstead, J., 1976. Uses of Raman spectroscopy in cement chemistry. *J. Amer. Ceram. Soc.*, **59**, 140–3.

Benstead, J. and **Prakash, S.**, 1968. Investigation of the calcium sulphate–water system by infrared spectroscopy. *Nature*, **219**, 60–1.

Berek, M., 1912. Die Bestimmung von Auslöschungsrichtungen doppelbrechender inaktiver Kristallplatten mit Hilfe von Halbschattenvorrichtungen in einfarbigen Licht. *Neues Jahrb. Min.*, Bl. Bd. **33**, 583.

Berenblut, B. J., Dawson, P. and **Wilkinson, G. R.**, 1973. A comparison of the Raman spectra of anhydrite ($CaSO_4$) and gypsum ($CaSO_4 \cdot 2H_2O$). *Spectrochim. Acta*, **29A**, 29–36.

Berrettoni, C., Tolomeo, L. and **Veneri, F.**, 1986. Strontium isotopic compositions and contents in gypsum from the Upper Miocene evaporitic basins of the Umbria-Marche-Ranagna series (Italy). *Period. Min.*, **55**, 255–9.

Blount, C. W. and **Dickson, F. W.**, 1973. Gypsum–anhydrite equilibria in systems $CaSO_4$–H_2O and $CaSO_4$–NaCl–H_2O. *Amer. Min.*, **58**, 323–31.

Bock, E., 1961. On the solubility of anhydrous calcium sulphate and of gypsum in concentrated solutions of sodium chloride at 25 °C, 30 °C, 40 °C and 50 °C. *Can. J. Chem.*, **39**, 1746–51.

Bragg, W. L., 1937. *The Atomic Structure of Minerals.* Cornell University Press, 292 pp.

Bright, J. E. and **Ridge, M. J.**, 1964. Twinned crystals of precipitated gypsum. *Min. Mag.*, **33**, 347–8.

Bundy, W. M., 1956. Petrology of gypsum–anhydrite deposits in southwestern Indiana. *Journ. Sed. Petr.*, **26**, 240–52.

Busheev, N. N. and **Borisov, V. M.**, 1982. X-ray diffraction investigation of $CaSO_4 \cdot 0.67H_2O$. *Russ. J. Inorg. Chem.*, **27**, 604–9.

Caillère, S. and **Pobeguin, Th.**, 1962. Comparison des propriétés thermiques et du comportement dans l'infrarouge de quelques carbonates, nitrates, sulfates et sélénates simples. *Bull. Soc. franc. Min. Crist.*, **85**, 48–57.

Caspari, J. E., 1936. Calcium sulphate hemi-hydrate and the anhydrites. I. Crystallography. *Proc. Roy. Soc. Lond.*, **155**, A, 41–48.

Censi, P., 1984. Composizione isotopica dell'acqua di crystallizzazione di cristalli di gesso selenitico siciliano: interpretazione delle variazioni osservate. *Min. Petr. Acta Bologna*, **28**, 139–53.

Chen, J., Zhao, R., Huo, W., Yao, Y., Pan, S., Shao, M. and **Hai, C.**, 1981. Sulphur isotopes of some marine gypsum. *Scientia Geol. Sinica*, 273–8 (M.A. 82M/1437).

Cherry, J. A., 1968. Chemical equilibrium between gypsum and brackish and slightly saline waters at low temperature and pressures. *Chem. Geol.*, **3**, 239–47.

Christoffersen, J. and **Christoffersen, R.**, 1976. The kinetics of dissolution of calcium sulphate dihydrate in water. *J. Cryst. Growth*, **35**, 79–88.

Claypool, G. E., Holser, W. T., Kaplan, J. R., Sakai, H. and **Zak, I.**, 1980. The age curves of sulphur and oxygen isotopes in marine sulfate and their mutual interpretation. *Chem. Geol.*, **28**, 99–260.

Cody, A. M. and **Cody, R. D.**, 1989a. Evidence for micro-biological induction of {101} Montmartre twinning of gypsum ($CaSO_4 \cdot 2H_2O$). *J. Cryst. Growth*, **98**, 721–30.

Cody, A. M. and **Cody, R. D.**, 1989b. SEM and polarization analyses updating early light microscope studies related to {010} twin formation in gypsum. *J. Cryst. Growth*, **98**, 731–8.

Cody, R. D., 1976. Growth and early diagenetic changes in artificial gypsum crystals grown within bentonite muds and gels. *Bull. Geol. Soc. Amer.*, **87**, 1163–8.

Cody, R. D., 1979. Lenticular gypsum: occurrence in nature, and experimental determinations of effects of soluble green plant material on its formation. *J. Sed. Petr.*, **49**, 1015–28.

Cole, W. F. and **Lancucki, C. J.**, 1974. A refinement of the crystal structure of gypsum $CaSO_4 \cdot 2H_2O$. *Acta Cryst.*, **B30**, 921–9.

Combe, E. C., Smith, D. C. and **Braden, M.**, 1970. Kinetics of hydration of autoclaved calcium sulphate hemihydrate. *J. Appl. Chem.*, **20**, 287–92.

Conley, R. F. and **Bundy, W. M.,** 1958. Mechanism of gypsification. *Geochim. Cosmochim. Acta*, **15**, 57–72.

Conley, J. E., Gabriel, A. and **Partridge, E. P.,** 1938. Equilibria in the system potassium sulphate–magnesium sulphate–calcium sulphate–water at 100 °C. *J. Phys. Chem.*, **42**, 587–616.

Cortecci, G., Reyes, E., Berti, G. and **Casati, P.,** 1981. Sulphur and oxygen isotopes in Italian marine sulfates of Permian Triassic ages. *Chem. Geol.*, **34**, 65–79.

Couty, R., Velde, B. and **Besson, J. M.,** 1983. Raman spectra of gypsum under pressure. *Phys. Chem. Min.*, **10**, 89–93.

D'Ans, J., 1968. Der Ubergangspunkt Gips–Anhydrit. *Kali Steinsaltz*, **5**, 109–11 (M.A. **20**–2083).

D'Ans, J., Bretschneider, D., Eick, H. and **Freund, H. E.,** 1955. Untersuchungen uber die Calciumsulfate. *Kali Steinsaltz*, **1**, 17–38.

De La Hunty, L. E. and **Low, G. H.,** 1958. The gypsum deposits of Western Australia. *Dept. of Mines: Mineral Resources of Western Australia, Bull.*, **6**.

Deicha, G., 1946. Remarque sur la genèse des faces courbes de cristaux. *Bull. Soc. franc. Min.*, **69**, 47–50.

Denne, W. A. and **Jones, D. W.,** 1969. Hydrogen location by occupation analysis: a test on the neutron diffraction data for gypsum, $CaSO_4{\cdot}2H_2O$. *Z. Krist.*, **130**, 314–7.

Des Cloizeaux, M., 1886. Note sur la détermination des paramètres du gypse et sur les incidences des formes observées dans ce minéral. *Bull. Soc. Franc. Min.*, **9**, 175.

Douglas, G. V. and **Goodman, N. R.,** 1957. The deposition of gypsum and anhydrite. *Econ. Geol.*, **52**, 831–7.

Drever, J. I., Lawrence, J. R. and **Antweiler, R. C.,** 1979. Gypsum and halite from the Mid-Atlantic Ridge. D.S.D.P. site 395. *Earth Planet. Sci. Lett.*, **42**, 98–102.

Dubrovinsky, L. S., Rubin, S. G., Iskhakova, C. D. and **Urosov, R. S.,** 1991. Prediction of the thermodynamic properties and relative stabilities of anhydrite. *Dokl. Earth Sci. Sect.*, **310**, 216–8.

Edinger, S. E., 1973. The growth of gypsum. *J. Cryst. Growth*, **18**, 217–24.

Egerer, F., 1981. The dielectric thermal analysis. *Bull. Min.*, **104**, 763–7.

Farnsworth, M., 1924. Effects of temperature and pressure on gypsum and anhydrite. *U.S. Bureau of Mines, Rept. Invest.* **2654**.

Faure, G. and **Felder, R. P.,** 1981. Isotopic composition of strontium and sulphur in secondary gypsum crystals, Brown Hills, Transantarctic Mountains. *J. Geochem. Explor.*, **14**, 265–70.

Fleischer, M., 1958. New mineral names. Vibertite. *Amer. Min.*, **43**, 791.

Flörke, O. W., 1952. Kristallographische und röntgenometrische Untersuchungen im System $CaSO_4$–$CaSO_4{\cdot}2H_2O$. *Neues Jahrb. Min., Abh.*, **84**, 189–240.

Flörke, W. and **Flörke, O. W.,** 1961. Vateritbildung aus Gips in Sodalösung. *Neues Jahrb. Min., Monat.*, 179–81.

Fontes, J.-C. and **Gonfiantini, R.,** 1967. Fractionnement isotopique de l'hydrogène dans l'eau de cristallisation du gypse. *C. R. Acad. Sci. Paris*, **265D**, 4–6.

Fontes, J.-C. and **Létolle, R.,** 1976. ^{18}O and ^{34}S in the upper Bartonian gypsum deposits of the Paris basin. *Chem. Geol.*, **18**, 285–95.

Fowler, A., Howell, H. G. and **Schiller, K. K.,** 1969. The dihydrate–hemihydrate transformation in gypsum. *J. Appl. Chem.*, **18**, 366–72.

Franchini-Angela, M. and **Rinaudo, C.,** 1989. Influence of sodium and magnesium on the growth morphology of gypsum, $CaSO_4{\cdot}2H_2O$. *Neues Jahrb. Min., Abh.*, **160**, 105–15.

Friend, J. N. and **Allchin, J. P.,** 1940. Colloidal gold as a colouring principle in minerals. *Min. Mag.*, **25**, 584–96.

Frik, M. and **Kuzel, H. J.,** 1982. Röntgenographische und thermoanalytiche Untersuchungen an Calciumsulfate-Halbhydrat. *Fortsch. Min.*, **60**, 80–1.

Fritz, P., Basharmal, G. M., Drimmie, R. J., Ibsen, J. and **Qureshi, R. M.,** 1989. Oxygen isotope exchange between sulphate and water during bacterial reduction of sulphate. *Chem. Geol.*, **79**, 99–105.

Gallitelli, P., 1933. Richerche sul solfato di calcio semidrato e sull' anidrite solubile. *Periodico Min. Roma*, **4**, 132–71 (M.A. **5**–314).

Gardner, A. W. and **Glueckauf, E.,** 1970. Thermodynamic data of the calcium sulphate solution process between 0 and 200 °C. *Trans. Farad. Soc.*, **66**, 1081–6.

Gaubert, P., 1933. Sur l'hemihydrate de sulphate de calcium et ses produits de déhydration. *C. R. Acad. Sci. Paris*, **197**, 72–5.

Gay, P., 1965. Some crystallographic studies in the system $CaSO_4$–$CaSO_4{\cdot}2H_2O$. II. The hydrous forms. *Min. Mag.*, **35**, 354–62.

Geisler, D., 1981. Genèse et évolution des gypses des marais salants de Salin-de-Giraud (Camargue). *Bull. Min.*, **104**, 625–9.

Gevork'yan, S. V. and **Povarennykh, A. S.,** 1980. A study of the dehydration processes in certain sulphates by infrared spectroscopy. *Min. Zhurn.*, **2**, 33–9 (M.A. **81**–444).

Gobran, G. R. and **Miyamoto, S.,** 1985. Dissolution rate of gypsum in aqueous salt solutions. *Soil Sci.*, **140**, 89–93.

Goldman, M., 1952. Deformation, metamorphism and mineralisation in gypsum–anhydrite cap rock of Sulphur Salt Dome, Louisiana. *Geol. Soc. Amer., Mem. 50*, 169 pp.
Goodman, N. R., 1952. Gypsum and anhydrite in Nova Scotia. *Nova Scotia Dept. Mines, Mem. 1*, 75 pp.
Goodman, N. R., 1957. Gypsum in Nova Scotia and its associated minerals. In *The Geology of Canadian Industrial Mineral Deposits*: Can. Inst. Mining Metall., p. 110.
Gossner, B., 1937. Über Sulfate und Phosphate mit ähnlichem Kristallgitter. *Z. Krist.*, **96**, 488–92.
Grahmann, W., 1920. Über Barytocölestin und das Verhältnis von Anhydrit zu Cölestin und Baryt. *Neues Jahrb. Min.*, **1**, 1.
Grattan-Bellew, P. E., 1975. Effects of preferred orientation on X-ray diffraction patterns of gypsum. *Amer. Min.*, **60**, 1127–9.
Grégoire, D. C., 1987. Determination of boron isotope ratios in geological materials by ICPMS. *Anal. Chem.*, **59**, 2479–84.
Guillou, J. J., **Debenay, J. P.**, **Pages, J.** and **Villemin, G.**, 1991. Cristallogenèse du gypse fibreux et évolution en monocristaux tabulaires (Ria de la Casamance, Sénégal). *C. R. Acad. Soc. Paris, II*, **313**, 399–405.
Hall, R. E., **Robb, J. A.** and **Coleman, C. E.**, 1926. The solubility of calcium sulfate at boiler water temperatures. *J. Amer. Chem. Soc.*, **48**, 927–38.
Hardie, L. A., 1967. The gypsum–anhydrite equilibrium at one atmosphere pressure. *Amer. Min.*, **52**, 171–200.
Heard, H. C. and **Rubey, W. W.**, 1966. Tectonic implications of gypsum dehydration. *Geol. Soc. Amer. Bull.*, **77**, 741–60.
Herman, P., **Vanderstappen, R.** and **Hubaux, A.**, 1960. Sublimés du Nyiragongo (Kivu). *Bull. Acad. Roy. Sci. d'Outre Mer, Brussels*, **7**, 964–73.
Hill, A. E., 1934. Ternary systems. XIX. Calcium sulfate, potassium sulfate and water. *J. Amer. Chem. Soc.*, **56**, 1071–8.
Hill, A. E., 1937. The transition temperature of gypsum to anhydrite. *J. Amer. Chem. Soc.*, **59**, 2242–4.
Hill, A. E. and **Wills, J. H.**, 1938. Ternary systems. XIV. Calcium sulfate, sodium sulfate and water. *J. Amer. Chem. Soc.*, **60**, 1647–55.
Holliday, D. W., 1970. The petrology of secondary gypsum rocks: a review. *J. Sed. Petr.*, **40**, 734–44.
Holliday, D. W., 1971. Origin of lower Eocene gypsum–anhydrite rocks, southeast St Andrew, Jamaica. *Trans. Inst. Min. Met.*, **80(B)**, 305–15.
Holser, W. T., 1966. Diagenetic polyhalite in recent salt from Baja California. *Amer. Min.*, **51**, 99–109.
Holser, W. T., 1979. Mineralogy of Evaporites. In *Marine Minerals*: Min. Soc. Amer. Short Course Notes, **6**, 211–94.
Hounslow, A. W., 1979. Modified gypsum/anhydrite stain. *J. Sed. Petr.*, **49**, 636–7.
Hutchinson, A. and **Tutton, A. E. H.**, 1913. Über die Temperatur der optischen Einaxigkeit von Gyps. *Z. Krist.*, **52**, 218–24.
Hutton, G. and **Pedersen, B.**, 1969. Proton and deuteron magnetic resonance in partly deuterated crystals. III. Gypsum. *J. Phys. Chem. Sol.*, **30**, 235–42.
Icart, J.-C., **Monchoux, P.** and **Tollon, F.**, 1965. Le strontium dans les évaporites de Tarascon (Ariège). *Bull. Soc. franc. Min. Crist.*, **88**, 351–2.
Iishi, K., 1979. Phonospectroscopy and lattice dynamical calculations of anhydrite and gypsum. *Phys. Chem. Min.*, **4**, 341–59.
Innorta, G., **Rabbi, E.** and **Tomadin, L.**, 1980. The gypsum–anhydrite equilibrium by solubility measurements. *Geochim. Cosmochim. Acta*, **44**, 1931–6.
Iwao, S., 1956. Hydrothermal gypsum deposits of the Wanibuchi Mine, Japan, with special reference to alteration of wall rocks. *Jap. Journ. Geol. Geogr.*, **27**, 105–32.
Jafarzadeh, A. A. and **Burnham, C. P.**, 1992. Gypsum crystals in soils. *J. Soil. Sci.*, **43**, 409–20.
Jaffray, J., 1943. Sur les variations avec la température de la biréfringence du gypse et de la barytine. *C. R. Acad. Sci. Paris*, **216**, 814–6.
Jong, W. F. de and **Bourman, J.**, 1939. Das reziproke und das Brasvaissche Gitter von Gips. *Z. Krist.*, **100**, 275–6.
Kamineni, D. C., 1983. Sulfur-isotope geochemistry of fracture-filling gypsum in an Archaean granite near Atikokan, Ontario, Canada. *Chem. Geol.*, **39**, 263–72.
Kastner, M., 1970. An inclusion hourglass pattern in synthetic gypsum. *Amer. Min.*, **55**, 2128–30.
Kaushansky, P. and **Yariv, S.**, 1990. The effect of percolating aqueous salt solutions on gypsum. *Chemie der Erde*, **50**, 203–20.
Kelepertzis, A. E., 1989. Formation of sulfates at the Thiapes area of Milos Island: possible precursors of kaolinite mineralization. *Can. Min.*, **27**, 241–5.
Kelly, K. K., **Southard, J. C.** and **Anderson, C. F.**, 1941. Thermodynamic properties of gypsum and its dehydration products. *U.S. Bur. Mines, Tech. Paper*, **625**.
Kern, H. and **Richter, A.**, 1985. Microstructures and textures in evaporites. In *Preferred Orientation in Deformed Metals and Rocks: an Introduction to Modern Texture Analysis*: Ed. H. R. Wenk, New York and London (Acad. Press), pp. 317–33.

Kern, R. and **Rehn, B.**, 1960. Étude expérimentale de la formation des macles de croissance du gypse. *C. R. Acad. Sci. Paris*, **251**, 1300–2.

Khalil, A. A. A. and **Gad, G. M.**, 1972. Thermochemical behaviour of gypsum. *Build. Intern.*, **5**, 145 (M.A. **24**–1572).

Kirov, G. K., 1980. Growth forms of gypsum crystals grown by isothermal evaporation of solutions. *Geochim. Min. Petr.*, **12**, 18–28 (M.A. **34**–4534).

Kosztolanyi, C., **Mullis, J.** and **Weidmann, M.**, 1987. Measurements of the phase transformation temperature of gypsum–anhydrite, included in quartz, by microthermometry and Raman microprobe techniques. *Chem. Geol.*, **61**, 19–28.

Kulstad, R. D., **Fairchild, P.** and **McGregor, D.**, 1956. Gypsum in Kansas. *Bull. Kansas Geol. Surv.*, **113**, 1–110 (M.A. **14**–113).

Kushnir, J., 1980. The coprecipitation of strontium, magnesium, sodium, potassium and chloride ions with gypsum. An experimental study. *Geochim. Cosmochim. Acta*, **44**, 1471–82.

Kushnir, J., 1982. The partitioning of seawater cations during the transformation of gypsum to anhydrite. *Geochim. Cosmochim. Acta*, **46**, 433–46.

Kuzel, H.-J., 1987. Hydratationswärmen von α-Calciumslfaten. *Neues Jahrb. Min., Abh.*, **156**, 155–74.

Ladd, M. F. C. and **Lee, W. H.**, 1965. The thermodynamics of crystalline hydrates. *J. Phys. Chem.*, **69**, 1840–3.

Lager, G. A., **Armbruster, Th.**, **Rotella, F. J.**, **Jorgensen, J. D.** and **Hinks, D. G.**, 1984. A crystallographic study of the low temperature dehydration products of gypsum, $CaSO_4{\cdot}2H_2O$: hemihydrate $CaSO_4{\cdot}0.50H_2O$ and α-$CaSO_4$. *Amer. Min.*, **69**, 910–8.

Lebedev, A. L. and **Lekhov, A. V.**, 1990. Dissolution in water at 5–25 °C. *Geochem. Internat.*, **27**, 85–94.

Lindberg, J. D. and **Smith, M. S.**, 1973. Reflectance spectra of gypsum sand from the White Sands National Monument and basalt from a nearby lava flow. *Amer. Min.*, **58**, 1062–4.

Liu, S.-T. and **Nancollas, G. H.**, 1970. The kinetics of crystal growth of calcium sulfate dihydrate. *J. Cryst. Growth*, **6**, 281–9.

McAdie, H. G. and **Kuntze, R. A.**, 1961. Determination of gypsum and its dehydration products in their mixtures. *Mat. Res. Stand.*, **1**, 27–9.

McConnell, J. D. C., 1965. Study of the reaction $CaSO_4{\cdot}\frac{1}{2}H_2O$ (β-hemihydrate) = $CaSO_4$ (β-soluble anhydrite) + H_2O in the temperature range 20–100 °C. *Min. Mag.*, **34**, 327–45.

McConnell, J. D. C., **Astill, D. M.** and **Hall, P. L.**, 1987. The pressure dependence of the dehydration of gypsum to bassanite. *Min. Mag.*, **51**, 453–7.

MacDonald, G. J. F., 1953. Anhydrite–gypsum equilibrium relations. *Amer. J. Sci.*, **251**, 884–98.

McGrath, J. W. and **Silvidi, A. A.**, 1961. Structure of water in solid hydrated compounds. *J. Chem. Phys.*, **34**, 322–5.

Machel, H.-G., 1985. Fibrous gypsum and fibrous anhydrite in veins. *Sedimentology*, **32**, 443–54.

Madgin, W. M. and **Swales, D. A.**, 1956. Solubilities in the system $CaSO_4$–NaCl–H_2O at 25° and 35°. *J. Appl. Chem.*, **6**, 482–7.

Marshall, W. L. and **Slusher, R.**, 1966. Thermodynamics of calcium sulfate dihydrate in aqueous sodium chloride solutions, 0–110°. *J. Phys. Chem.*, **70**, 4015–27.

Mellis, O., 1964. Die Umkristallisationerscheinungen im Fasergips. *Neues Jahrb. Min., Mh.*, 327–36.

Milton, C., 1942. Note on the occurrence of calcium sulphate hemi-hydrate ($CaSO_4{\cdot}\frac{1}{2}H_2O$) in thin sections of rocks. *Amer. Min.*, **27**, 517–8.

Miropolsky, L. M. and **Borovick, S. A.**, 1943. Spectrum analysis data on gypsum from Permian deposits of Tartary. *Doklady Acad. Sci. USSR*, **38**, 33–36.

Miyazaki, H. and **Takagi, N.**, 1970. Difference between α and β forms of calcium sulphate hemihydrate. *J. Chem. Soc. Japan*, **73**, 1766–9 (M.A. **22**–886).

Morris, R. C. and **Dickey, P. A.**, 1957. Modern evaporite deposition in Peru. *Bull. Amer. Assoc. Petroleum. Geol.*, **41**, 2467–74.

Morris, R. J., 1963. X-ray diffraction identification of the alpha- and beta-forms of calcium sulphate hemihydrate. *Nature*, **198**, 1298–9.

Murrell, S. A. E. and **Ismail, I. A. H.**, 1976. The effect of decomposition of hydrous minerals on the mechanical properties of rocks at high pressures and temperatures. *Tectonophysics*, **31**, 207–58.

Nagy, J., 1928. Untersuchungen über die Atomebenabstände der Gipskristalle. *Z. Phys.*, **51**, 410–2.

Nancollas, G. H., **Reddy, M. M.** and **Tsai, F.**, 1973. Calcium sulfate dihydrate crystal growth in aqueous solution at elevated temperatures. *J. Cryst. Growth*, **20**, 125–34.

Neev, D., 1963. Recent precipitation of calcium salts in the Dead Sea. *Bull. Res. Council, Israel*, **11G**, 153–4 (M.A. **16**–582).

Oglesby, M. L., **Gutshall, P. L.** and **Phillips, J. M.**, 1976. Cleavage surface energy of selenite. *Amer. Min.*, **61**, 295–8.

Onorato, E., 1929. Über den Feinbau des Gipses. *Z. Krist.*, **71**, 277–325.

Osborne, F., 1941. Anhydrite and gypsum at Calumet mines, Calumet Island, Quebec. *Univ. Toronto Studs., Geol. Ser.*, **46**, 75–82.

Ostroff, A. G., 1964. Conversion of gypsum to anhydrite in aqueous salt solutions. *Geochim. Cosmochim.*

Acta, **28**, 1363–72.

Pake, G. E., 1948. Nuclear resonance absorption in hydrated crystals: fine structure of the proton line. *J. Chem. Phys.*, **16**, 327–36.

Pedersen, B. F. and **Holcomb, D. B. F.**, 1963. NMR in hydrate crystals: structural information from broadened fine-structure lines. *J. Chem. Phys.*, **38**, 61–9.

Pedersen, B. F. and **Semmingsen, D.**, 1982. Neutron diffraction refinement of the structure of gypsum, $CaSO_4{\cdot}2H_2O$. *Acta Cryst.*, **B38**, 1074–7.

Pehrman, G., 1935. Die Einwirkung der Clerici'schen Lösung auf Gips. *Zentr. Min., Abt. A*, 54–5.

Pisarchik, Yu. K., Golubchina, M. N. and **Toksubayev, A. J.**, 1977. Isotopic composition of sulfur in calcium sulfates from the Cambrian Upper Lena formation (Siberian Platform). *Geochem. Intern.*, **14**, 182–5.

Posnjak, E., 1938. The system $CaSO_4$–H_2O. *Amer. J. Sci., Ser. 5*, **35A**, 247.

Posnjak, E., 1940. Deposition of calcium sulphate from sea water. *Amer. J. Sci.*, **238**, 559–68.

Przibram, K., 1962. Uber die Fluoreszenz organischer Spuren in anorganischen Stoffen und ihre Verbreiterung in der Natur. *Geochim. Cosmochim. Acta*, **26**, 1045–54.

Putnis, A., Winkler, B. and **Fernandez-Diaz, L.**, 1990. In situ IR spectroscopic and thermogravimetric study of the dehydration of gypsum. *Min. Mag.*, **54**, 123–8.

Raman, Sir C. V. and **Jayaraman, A.**, 1954. X-ray studies on polycrystalline gypsum. *Proc. Ind. Acad. Sci.*, **40**, 57–60.

Raymond, L. R., 1953. Some geological results from the exploration for potash in Northeast Yorkshire. *Quart. J. Geol. Soc.*, **108**, 283–306.

Razouk, R. I., Salem, A. Sh. and **Mikhail, R. Sh.**, 1960. The sorption of water vapour on dehydrated gypsum. *J. Phys. Chem.*, **64**, 1350–5.

Rick, B., 1990. Sulphur and oxygen isotopic composition of Swiss Gipskeuper (Upper Triassic). *Chem. Geol.*, **80**, 243–50.

Ridge, M. J., 1958. Effect of temperature on the structure of set gypsum plaster. *Nature*, **182**, 1224–5.

Ridge, M. J. and **Surkevicius, H.**, 1962. Hydration of calcium sulphate hemihydrate. I. Kinetics of the reaction. *J. Appl. Chem.*, **12**, 246–52.

Rinaudo, C., Robert, M. C. and **Lefaucheux, F.**, 1985. Growth and characterization of gypsum crystals. *J. Cryst. Growth*, **71**, 803–6.

Rinaudo, C., Franchini-Angela, M. and **Boistelle, R.**, 1988. Gypsum crystallization from cadmium-poisoned solutions. *J. Cryst. Growth*, **89**, 257–66.

Rinaudo, C., Franchini-Angela, M. and **Boistelle, R.**, 1989. Curvature of gypsum crystals induced by growth in the presence of impurities. *Min. Mag.*, **53**, 479–82.

Ross, S. D., 1974. Sulphates and other oxyanions of Group VI. In *The Infrared Spectra of Minerals*: Ed. V. C. Farmer, London (Mineralogical Society), pp. 423–44.

Rull, F., Acosta, A. and **Alia, J. M.**, 1989. Estudio de la deshidratación del yeso mediante espectroscopia infrarroja. *Bol. Soc. Espan. Min.*, **12**, 161–8.

Rutherford, R. L., 1940. Asterism in selenite. *Univ. Toronto Studs., Geol. Ser.*, **44**, 71–4.

Sabouraud, C., 1976. Les conditions de genèse de certaines formes de cavités intracrystalline éclairées par la methode expérimentale. *Bull. Soc. franc. Min. Crist.*, **99**, 74–7.

Sabouraud, C., Bourelly, I. N., Tauray, J.-C. and **Deschamps, N.**, 1972. Détermination par activation neutronique du rapport Cl/Br dans les inclusions fluides de gypsum. *C. R. Acad. Sci. Paris*, **274D**, 3161–3.

Sakai, H. and **Matsubaya, O.**, 1971. Sulfur and oxygen isotopic ratios of gypsum and barite in the black ore deposits of Japan. *Soc. Mining. Geol. Japan. Spec. Issue*, **2**, (Proc. IMA–IAGOD Meetings '70, Joint Symp. Vol.) 80–3.

Šatava, V., 1974. A study of gypsum dehydration in aqueous medium. I. The growth kinetics of calcium sulphate hemihydrate crystals from aqueous solution. II. Kinetics of gypsum dissolution in water. *Silikaty*, **18**, 281–93.

Saxby, D. B. and **Lamar, J. E.**, 1957. Gypsum and anhydrite in Illinois. *Illinois State Geol. Surv., Circ.*, **226**, 22 pp.

Seidl, V., Knop, O. and **Falk, M.**, 1969. Infrared study of water in crystalline hydrates: gypsum, $CaSO_4{\cdot}2H_2O$. *Can. J. Chem.*, **47**, 1361–8.

Shearman, D. J., 1979. A field test for identification of gypsum in soils and sediments. *Q. J. Eng. Geol.*, **12**, 51.

Shearman, D. J., Mossop, G., Dunsmore, H. and **Martin, M.**, 1972. Origin of gypsum veins by hydraulic fracture. *Trans. Inst. Min. Met.*, **81(B)**, 149–55.

Simon, B. and **Bienfait, M.**, 1965. Structure et mécanisme de croissance du gypse. *Acta Cryst.*, **19**, 750–6.

Sofer, Z., 1978. Isotopic composition of hydration water in gypsum. *Geochim. Cosmochim. Acta*, **42**, 1141–9.

Spencer, L. J., 1913. Sixth list of new mineral names. Bassanite. *Min. Mag.*, **16**, 354.

Stemmerik, L., Rouse, J. E. and **Spiro, B.**, 1988. S isotope studies of shallow water, laminated gypsum and associated evaporites, Upper Permian, East Greenland. *Sed. Geol.*, **58**, 37–46.

Stewart, F. H., 1949. The petrology of the evaporites of the Eskdale no. 2 boring, east Yorkshire, Part I. The Lower Evaporite bed. *Min. Mag.*, **28**, 621–75.

Stewart, F. H., 1953. Early gypsum in the Permian evaporites of northeastern England. *Proc. Geol. Assoc.*, **64**, 33–8.

Stolyarov, Y. M., 1965. Anhydrite and gypsum in Urals pyrite deposits. *Dokl. Acad. Sci. USSR Earth Sci.*, **161**, 162–4.

Strunz, H., 1942. Isotypie zwischen $YPO_4 \cdot 2H_2O$ und $CaSO_4 \cdot 2H_2O$. *Naturwiss.*, **30**, 64.

Taylor, E. W., 1949. Correlation of the Mohs's scale of hardness with the Vickers's hardness numbers. *Min. Mag.*, **28**, 718–21.

Taylor, J. B. and **Baines, J. E.**, 1970. Kinetics of the calcination of calcium sulphate dihydrate. *J. Appl. Chem.*, **20**, 121–2.

Tolansky, S., 1945. The topography of crystal faces. II. The topography of cleavage faces of mica and selenite. *Proc. Roy. Soc.*, **184**, **A**, 51–63.

Vaidya, S. N., **Bailey, S.**, **Pasternack, T.** and **Kennedy, G. C.**, 1973. Compressibility of fifteen minerals to 45 Kilobars. *J. Geophys. Res.*, **78**, 6893–8.

Van Leeuwen, C. and **Blomen, L. J. M. J.**, 1979. On the presentation of growth curves for growth from solution. *J. Cryst. Growth*, **46**, 96–104.

Van Rosmalen, G. M., **Marchée, W. G. J.** and **Benna, P.**, 1976. A comparison of gypsum crystals grown in silica gel and agar in the presence of additives. *J. Cryst. Growth*, **35**, 169–76.

Vierne, R. and **Brunel, R.**, 1969. Spectres de réflexion infrarouge de minéraux monocristallins ou en poudre. I. *Bull. Soc. franc. Min. Crist.*, **92**, 409–19.

Vinogradov, V. L. and **Stephanov, V. L.**, 1964. Isotopic fractionation of sulfur in the oxidized zone. *Geochem. Internat.*, **1**, 28.

Vinogradov, V. L., **Borisova, V. N.** and **Hsü, Y.**, 1965. The origin of volcanic sulphates. *Dokl. Acad. Sci. USSR Earth Sci. Sect.*, **158**, 149–51.

Von Gehlen, K., **Grauert, M. B.** and **Nielsen, H.**, 1986. REE mineralogy in southern Schwarzwald veins and isotope studies on gypsum from the central Schwarzwald, F. R. Germany. *Neues Jahrb. Min., Mh.*, 393–9.

Warden, A. J. and **Pallister, J. W.**, 1958. A gypsum–anhydrite deposit in Somaliland. *Mining Mag.*, **98**, 333–7.

Watson, A., 1983. In *Chemical sediments and geomorphology*: Eds. A. S. Goudie and K. Pye, London (Academic Press), pp. 133–61.

Weijnen, M. P. C., **Van Rosmalen, G. M.**, **Bennema, P.** and **Rijpkena, J. J. M.**, 1987. The adsorption of additives of the gypsum crystal surface: a theoretical approach. *J. Cryst. Growth*, **82**, Part I. 509–27; Part II. 528–42.

Weiser, H. B., **Milligan, W. O.** and **Eckholm, W. C.**, 1936. The mechanism of the dehydration of calcium sulphate hemihydrate. *J. Amer. Chem. Soc.*, **58**, 1261–5.

West, R. R. and **Sutton, W. J.**, 1954. D.t.a. of gypsum. *J. Amer. Ceram. Soc.*, **37**, 221–4.

Williams, S. C., 1988. The shear strength of gypsum single crystals on three cleavage planes. *Tectonophysics*, **148**, 163–73.

Winchell, H., 1945. The Knoop hardness tester as a mineralogical tool. *Amer. Min.*, **30**, 583–95.

Wooster, W. A., 1936. On the crystal structure of gypsum, $CaSO_4 \cdot 2H_2O$. *Z. Krist.*, **94**, 375–96.

Yamamoto, H. and **Kennedy, G. C.**, 1969. Stability relations in the system $CaSO_4$–H_2O at high temperatures and pressures. *Amer. J. Sci.*, **267A**, 550–7.

Zen, E.-an, 1965. Solubility measurements in the system $CaSO_4$–NaCl–H_2O at 35°, 50° and 70 °C and at 1 atmosphere pressure. *J. Petr.*, **6**, 124–64.

Zolotukhin, V. V., 1954. On two varieties of gypsum from Zaleschiki. *Min. Sbornik. Lvov Geol. Soc.*, **no. 8**, 253–60 (M.A. **13**–62).

Anhydrite $CaSO_4$

Orthorhombic (+)

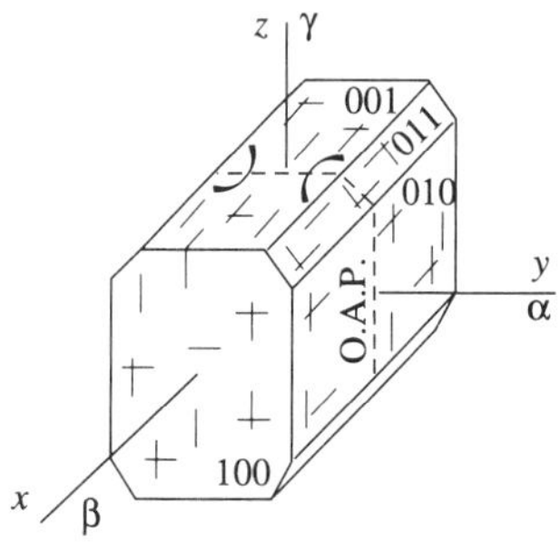

α	1.567–1.574
β	1.574–1.579
γ	1.609–1.618
δ	~0.04
$2V_\gamma$	42°–44°

Orientation:
$\alpha = y$, $\beta = x$, $\gamma = z$, O.A.P. (100)

Dispersion:	$r < v$
D	2.9
H	$3–3\frac{1}{2}$

Cleavage:	{010} perfect, {100} very good, {001} good.
Twinning:	May be simply or repeatedly twinned on {011}.
Colour:	White or colourless when pure, often grey, and more rarely bluish mauve, red or brown; colourless in thin section.
Unit cell:	*a* 6.991, *b* 6.996, *c* 6.238 Å Z=4. Space group *Amma*

Soluble with difficulty in dilute HCl.

Anhydrite is one of the principal minerals of evaporite deposits, and it is also found as an accessory mineral in certain dolomites and limestones. It is used in the manufacture of ammonia, and of ammonium sulphate for sulphuric acid production, and is the main constituent of certain special plasters and cements. Anhydrite is so named because in contrast to gypsum it is a sulphate of calcium which contains no water.

Structure

The structure of anhydrite was determined first by Wasastjerna (1925) and then by Dickson and Binks (1926); the latter workers found basically the same structure but with small though significant differences. Preference for the first structure was indicated by an X-ray study of one projection (Cheng and Zussman, 1963) and was confirmed by work on three projections (Höhne, 1963). Three-dimensional X-ray investigations by Morikawa *et al.* (1975) and Hawthorne and Ferguson (1975) yielded more accurate results.

The unit cell of anhydrite is orthorhombic, with parameters as given by Hawthorne and Ferguson (1975) of *a* 6.993, *b* 6.995, *c* 6.245 Å, and space group *Amma.* Cell parameters and space group have been listed many times in the literature, with some difficulties caused by the near equality of the *a* and *b* cell edges. The above values agree closely with those determined by Swanson *et al.* (1955) for a particularly pure specimen (*a* 6.991, *b* 6.996, *c* 6.238 Å), and the labelling of axes shows best the relationships between the structures of anhydrite and the hydrates of calcium sulphate. Kolb (1911), by the measurement of interfacial angles at various temperatures, deduced that $a{:}b{:}c = 0.9996{:}1{:}0.8911$ at $-64\,^{\circ}$C and 1.0025:1:0.8996 at 500 °C. The optical orientation sketch (p. 74) relates to the *Amma* setting given above. To compare or relate different reports of crystallographic, optical and other directional physical properties correctly it is unsafe to refer them to 'the longest' and 'the intermediate' cell edge because these are so similar in length. If possible they should be distinguished as 'perpendicular to the centred cell face' or parallel to it, or else 'in the direction of the structure's glide translation' or perpendicular to it. Failing this structural information, which may not be easily obtainable, consistency could at least be obtained by referring directions to the optical properties, i.e. α, β and γ, or the optic axial plane and acute bisectrix.

Although the unit cell of anhydrite has $a \sim b$, the structure is not pseudo-tetragonal. It is very different from that of the baryte group of sulphates and shows more similarity with the zircon type of structure (see below). The sulphur atoms of SO_4 tetrahedra and the calcium atoms lie on the lines of intersection of two mirror planes. Planes containing evenly spaced Ca and SO_4 ions lie parallel to (100) and (010) (see Figs 20, 21), whereas layering is not so well defined parallel to (001). This is in accord with anhydrite's better cleavage on (100) and (010) than on (001). The calcium atoms are coordinated by eight oxygens in a distorted triangular dodecahedron. Each Ca polyhedron has common vertices with four SO_4 tetrahedra and common edges with two more tetrahedra. It also has a common vertex with four other Ca polyhedra and a common edge with two more of these. The Ca and S polyhedra form alternating edge-sharing chains parallel to z (Fig. 20b) which may be compared with similar chains parallel to [001] in gypsum (*A2/a* orientation, see pp. 41, 42). This relationship is exhibited in the oriented dehydration reaction converting gypsum to anhydrite.

The SO_4 tetrahedron departs only very slightly from regularity. Hawthorne and Ferguson (1975) found slight distortion from an ideal tetragonal disphenoid, with all S–O bonds equal; Morikawa *et al.* (1975) proposed two S–O distances, 1.474 and 1.475 Å. There are four distinct Ca–O distances ranging between 2.34 and 2.56 Å, and those edges of the CaO_8 polyhedra which are shared with SO_4

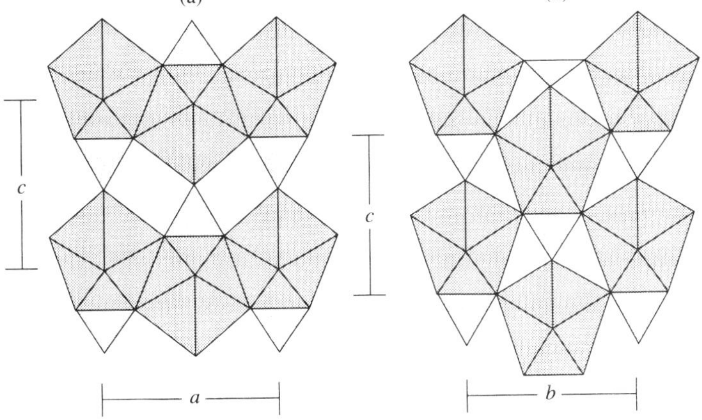

Fig. 20. (a) The structure of a single layer (half-cell) of anhydrite viewed along y showing Ca dodecahedra and SO_4 tetrahedra sharing edges to form chains parallel to z, which are linked along x by edge-sharing dodecahedra. (b) The structure of anhydrite viewed along x showing the edge-sharing chains parallel to z linked along y by corner-sharing between polyhedra (after Hawthorne and Ferguson, 1975).

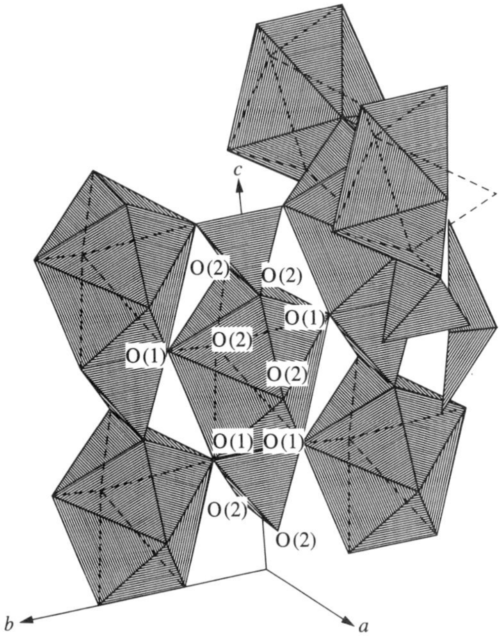

Fig. 21. Perspective view of the crystal structure of anhydrite (Ryabov *et al.*, 1983).

tetrahedra are the shortest. The two Ca–S distances are almost equal (3.102 and 3.143 Å), and their sum constitutes the *c* parameter (6.245 Å).

Kirfel and Will (1980, 1981) carried out X-ray and neutron diffraction refinements of the structure giving special attention to the bonding electron distribution, particularly around the SO_4^{2-} anion. Their results indicated considerable population of the 3d orbitals of the sulphur atoms.

The structures of anhydrite and zircon were compared by Nyman *et al.* (1984). The chains of CaO_8 and SO_4 polyhedra along *z* in anhydrite are similar to those of ZrO_8 and SiO_4 in zircon. In the *x* direction of anhydrite the chains are connected by edge sharing between bisdisphenoids, as in zircon, but along *y* they are joined by corner sharing not edge sharing. The symmetry is thus reduced from tetragonal to orthorhombic.

Corba (1975) found that the 022 X-ray reflection of anhydrite ($2\theta \sim 36°$, Cu$K\alpha$) was useful for the quantitative determination of anhydrite in gypsum plaster. An X-ray method for determining the proportions of anhydrite and gypsum in a mixture uses the intensities of reflections at *d* 3.44 Å for anhydrite and *d* 4.27 Å for gypsum (Innorta *et al.*, 1980).

The metastable γ-$CaSO_4$ obtained by slow dehydration of gypsum or hemihydrate at about 100 °C is sometimes misleadingly called 'soluble anhydrite'. Its structure and relationship to the hydrates of $CaSO_4$ were described in the previous chapter.

At about 1200 °C anhydrite inverts reversibly to a high-temperature form (α-anhydrite) which is trigonal *P*31*c* with *a* 5.064, *c* 7.978, Z = 2, D 2.55. There is a volume increase of 7.3 per cent associated with the $\beta \rightarrow \alpha$ transition (Hanic *et al.*, 1985).

Chemistry

Chemical analyses of anhydrite (e.g. Table 11) show only minor variations in its composition. Small amounts of strontium and barium sometimes occur replacing calcium; other oxides recorded are probably present as impurities; H_2O^+, when it occurs, is mainly due to the presence of gypsum. Pure $CaSO_4$ would contain 41.19 per cent CaO and 58.81 per cent SO_3.

Grahmann (1920) reported complete solid solution between the high-temperature forms of $CaSO_4$, $BaSO_4$ and $SrSO_4$, but at room temperature there was limited solid solution of up to 8 per cent $BaSO_4$ and 42 per cent $SrSO_4$ in $CaSO_4$.

REE contents were determined for anhydrite at the Kachar iron-ore deposit, USSR (Zhuk-Pochekutov *et al.*, 1986). The maximum values noted (p.p.m.) were: La 54, Ce 31, Pr 2.4, Nd 7.7, Eu 2.6, Gd 1.3, Gy 0.7, Ho 0.23, Er 0.46, Tu 0.17, Yb 2.55, Lu 0.44 and Y 5.47.

Table 11. Anhydrite Analyses

	1	2	3	4	5	6	7
SiO_2	0.50	—	2.73	—	0.01	—	—
Al_2O_3 } Fe_2O_3	0.10	0.06	1.06	0.04	0.03	—	—
MgO	0.60	—	0.17	—	0.03	—	—
CaO	40.86	41.13	39.12	41.22	41.08	40.47	40.27
SrO	—	—	0.00	0.01	—	0.71	0.94
BaO	—	—	0.00	—	—	0.05	—
Na_2O } K_2O	—	—	—	0.02	0.07	—	—
SO_3	57.10	58.37	56.63	58.95	58.50	58.94	58.45
CO_2	0.37	0.17	—	—	—	—	—
H_2O^+	0.69	—	0.10	0.01	0.04	—	—
H_2O^-	0.01	—	0.68	—	—	0.40	—
Total	100.61	99.75	100.49	100.25	99.76	100.57	99.66
α	—	—	1.569	—	—	—	—
γ	—	—	1.616	—	—	—	—
D	—	—	2.91–2.935	—	—	2.97	—
	Numbers of ions on the basis of four oxygens (O)						
Ca	1.007[a]	0.990[a]	1.004	0.999	1.001	0.983	0.985
Sr	—	—	—	—	—	0.009	0.013
Na, K	—	—	—	0.001	0.003	—	—
S	0.998	1.003	1.003	1.000	0.999	1.002	1.001

[a]Calculated after subtraction of CaO equivalent of CO_2 present.

1 Anhydrite, Sangamon County, Illinois, USA (Saxby and Lamar, 1957) (includes 0.38 NaCl; SiO_2 figure includes insoluble residue).
2 Anhydrite from gypsum deposit, Yonaibata mine, Kano village, Yama Co., Hukusima Prefecture, Japan (Harada, 1954). Anal. K. Togari (includes FeS_2 0.02).
3 Fibrous anhydrite, dolomite rock, Chuvash Republic, central Russia (Miropolsky, 1942).
4 Anhydrite, Stassfurt, Germany (Flörke, 1952a).
5 Anhydrite, Celle, Germany (Flörke, 1952a).
6 Anhydrite, Kalgoorlie, Western Australia (Simpson, 1930).
7 Anhydrite, Asaka gypsum mine, Koriyama City, Japan (Matsubara *et al.*, 1992).

Chemical analyses of anhydrites by X-ray fluorescence (Belkin *et al.*, 1988) gave values for Rb, Sr, Ba, La and Ce as, respectively, 40–670, 2000–8400, 110–3900, <2–400, and 16–500 p.p.m.

Kagi *et al.* (1993) studied the distribution of rare earth elements between an anhydrite crystal and its saturated solution at different temperatures and concluded that the uptake of these elements occurred in two stages: a non-selective incorporation into the crystal, followed by thermodynamically controlled differential incorporation.

Boron isotope ratios ($^{11}B/^{10}B$) for anhydrite were determined as approximately 4.17 (Gregoire, 1987).

Studies by electron paramagnetic resonance (Ryabov *et al.*, 1983) showed the presence in anhydrite of SO_3^{2-} and two types of PO_3^{2-} radical.

Experimental Work

Crystals of anhydrite may be prepared by slowly cooling fusions of $CaSO_4$ with $CaCl_2$, $BaCl_2$ or NaCl, or of $CaCl_2$ with K_2SO_4, and also by heating gypsum with NaCl or $CaCl_2$ solution in an enclosed tube. In the experimental work by Zen (1965), anhydrite did not precipitate from any composition in the ternary system $CaSO_4$–NaCl–H_2O.

Solubility. Investigations of the systems $CaSO_4$–H_2O, Na_2SO_4–$CaSO_4$–H_2O, K_2SO_4–$CaSO_4$–H_2O and K_2SO_4–$MgSO_4$–$CaSO_4$–H_2O, in which anhydrite has a stability field, and also many of the problems concerning the solubility and stability of anhydrite in relation to the hydrates of $CaSO_4$, have been discussed in the section on gypsum (pp. 50–54). It is clear that the solubility of anhydrite in water decreases with increasing temperature but increases with increasing salinity. Bock (1961) gave some values as 0.275 g $CaSO_4$/100 g water at 25 °C and 0.190 g $CaSO_4$ at 50 °C, these values increasing to 0.751 and 0.520 respectively in water with 1 ml NaCl/1000 g. The range of results obtained by several investigations was illustrated by Blount and Dickson (1969) (see Fig. 12, p. 51).

An increase in pressure acts in the opposite direction to an increase in temperature, i.e. it favours increased solubility (Dickson *et al.*, 1963) as illustrated in Fig. 22. The implications of the *P–T* effects for the occurrence of anhydrite in evaporites are discussed on p. 83. Hardie (1967) studied the effect of changing the water activity, a_{H_2O}, by adding Na_2SO_4 or H_2SO_4, and found that the equilibrium temperature for the transition between gypsum and anhydrite fell from about 58 to 23 °C on reducing a_{H_2O} from 1.00 to 0.77.

Through solubility experiments, calculations and other approaches, estimates have been made of the equilibrium temperature for the transition between gypsum and anhydrite. Early estimates were around 42 °C for pure water, but these were later revised upwards by some workers to around 58 °C. These temperatures are significantly lowered by increase in salinity.

Cody (1976) carried out experiments using bentonite to simulate the environments of lake and sea-floor muds, and found that at temperatures up to 80 °C and with NaCl up to 20 per cent only gypsum or hemihydrate was produced and anhydrite did not form either as a primary precipitate or by replacing gypsum. He concluded that primary anhydrite could occur only in extremely saline conditions. Harvie and Weare (1980) used the Pitzer equations to model mineral solubilities in the system H_2O–Na–K–Mg–Ca–Cl–SO_4, and their results agreed

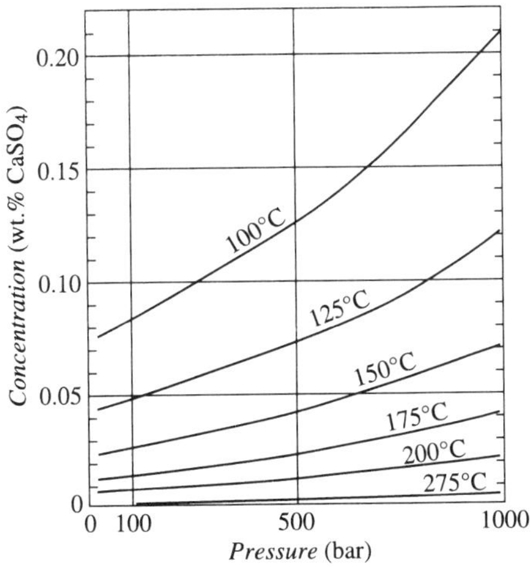

Fig. 22. Variation of solubility of anhydrite in pure water with temperature and pressure (after Dickson *et al.*, 1963).

well with those from experiments. Also, their predicted mineral sequences for the Stassfurt Zechstein II formation involving anhydrite with or without other salts agreed with those observed.

Monnin (1990), using calculated activity coefficients, derived solubility curves for anhydrite in pure water and in sodium chloride solutions up to 6 M. Good agreement was obtained with experimental results up to 200 °C at 1 kbar.

Ryzhenko (1964) found that in solutions with equal contents of carbonates plus CO_2 and sulphates, at 100–200 °C, the association of anhydrite and witherite is forbidden. Precipitation of anhydrite occurs at pH < 6.5 and of baryte and calcite at pH 6.5–10.

Gardner and Glueckauf (1970) derived thermodynamic values from solubility data for anhydrite over the temperature range 100–200 °C.

Hydration. A distinction has to be drawn between the hydration processes for γ-$CaSO_4$ ('soluble anhydrite' or 'dehydrated gypsum') and for anhydrite (β-$CaSO_4$) itself. The hydration of γ-$CaSO_4$ to hemihydrate proceeds readily, but because the structure of anhydrite is very different from that of hemihydrate it cannot be hydrated easily, and it may have to go through a solution and re-crystallization process to do so. The transformation of anhydrite into gypsum involves an increase in volume, but the extent to which this is evident in practice depends upon factors such as temperature and crystal size and shape, and also on the overburden (Reimann, 1991).

Bath *et al.* (1987) studied the gypsum in the Zechstein anhydrite formation (Billingham, northeast England) and found that when anhydrite is hydrated to form gypsum in a closed system, the isotope ratios $^{18}O/^{16}O$ and $^{2}H/^{1}H$ for the water of crystallization vary systematically with depth. The slope of δH versus δO is about -5, agreeing with calculations and with Sofer (1978).

Decomposition. Anhydrite (β-anhydrite) has been reported to invert reversibly on heating to a high-temperature form, the transition temperature being variously given in the range 1193–1232 °C (e.g. Grahmann, 1920; Flörke, 1952b; Gutt and

Smith, 1967). Some doubt was cast upon the validity of two forms (e.g. Gay, 1965), but more recently further confirmation was provided by Hanic *et al.* (1985) who gave the transition temperature as 1228 °C in air and 1212 °C in an atmosphere of SO_2 and oxygen. The high-temperature form (α-anhydrite) is trigonal (see p. 76).

On heating beyond the $\beta \rightarrow \alpha$ transition, decomposition takes place:

$$CaSO_4 \rightarrow CaO + SO_2 + \tfrac{1}{2}O_2.$$

Hanic *et al.* (1985) found that at the sub-eutectic temperature (1365 °C) in the system $CaSO_4$–CaO the decomposition obeys second-order kinetics represented by $G(\alpha) = 1/(1-\alpha) - 1$ with an activation energy, E_a, of 622 ± 7 kJ/mol.

Above the eutectic temperature the phase boundary reaction controls the rate of decomposition and E_a is 373 ± 4 kJ/mol. Anhydrite undergoes a sluggish phase transition at 19.6 kbar with a volume change of 4.1 per cent (Stephens, 1964). The compression $-\Delta V/V_0$ goes from 0.0093 to 0.0310 with an increase of pressure from 5.0 to 19.6 kbar, and then from 0.0720 (at the transition) to 0.0997 at 44 kbar. This phase transition was also reported by Kahn (1975).

Among the minerals which may be found as pseudomorphs after anhydrite are quartz, siderite, calcite, dolomite, gypsum and marcasite.

Optical and Physical Properties

Anhydrite usually occurs in massive aggregates of varying grain size, sometimes as groups of parallel or radiating fibres. The more acicular morphologies are generally attributed to pseudomorphism after gypsum. Larger crystals, when they occur, usually have a thick tabular habit on pinacoidal faces, and may be elongated parallel to *x* or *z*. Twinning (sometimes repeated) occurs on {011} and more rarely on {120}. The crystals show three pinacoidal cleavages, of which {010} is the best. Pure perfect crystals of anhydrite are transparent, but specimens are often white, or coloured blue or violet. This colour disappears on heating and reappears on exposure to radium radiation; thus the colour is thought to be a natural radiation colour, the source of radiation being either external or present as an impurity within the anhydrite (Przibram, 1956; Tocmakchieva, 1980).

In accordance with its relatively high chemical purity, the optical properties of anhydrite do not vary greatly. The dispersion of the optical properties of anhydrite with wavelength and temperature was studied by Kolb (1911), and his results are summarized in Table 12. The dispersion of birefringence with temperature and wavelength, and temperature coefficients of wavelength dispersion, were also determined by Kolb. γ-$CaSO_4$ ('soluble anhydrite' or 'dehydrated hemihydrate') has lower refractive indices: ω 1.501, ε 1.546 (Flörke, 1952a).

Murray (1964) was able to distinguish by reflected light microscopy between bedded, void-filling and replacement anhydrite because abundant included material was present in the latter.

Benstead (1976) reported laser Raman spectroscopy bands for anhydrite occurring at 420 and 500 cm^{-1} (ν_2 SO_4), 611, 630 and 679 cm^{-1} (ν_4), 1022 (ν_1 very strong) and 1116, 1134 and 1116 (ν_3). Wave numbers for corresponding bands are lower for gypsum (e.g. ν_1 1004) and hemihydrate (see also Caillère and Pobeguin, 1962; Benstead and Prakash, 1968; Griffith, 1969). The strong bands at 121 and 150 cm^{-1} for gypsum arise from its water molecules and are of course not present in the infrared spectrum of anhydrite.

Bershov *et al.* (1975) studied natural and irradiated anhydrite crystals by electron

Table 12. Optical Properties of Anhydrite (Kolb, 1911)

(a) Temperature variation at constant wavelength (D_2 line)

	−55°	**20°**	**100°**	**508 °C**
α	1.5698 (−50 °C)	1.5696	1.5694 (105 °C)	1.5674
β	1.5760	1.5757	1.5754	1.5728
γ	1.6144	1.6139	1.6133	1.6079
$2V_\gamma$	—	43° 31′	43° 18′	41° 57′ (570 °C)

(b) Wavelength variation at constant temperature

		B	**D_2**	**F**	**G**
at 16–19 °C	α	1.5667	1.5697	1.5752	1.5798
	β	1.5725	1.5755	1.5811	1.5858
	γ	1.6103	1.6137	1.6198	1.6250
at 20 °C	$2V_\gamma$	43° 13′	43° 31′	43° 57′	44° 13′

paramagnetic resonance (epr) at 77 and 290 K. They provided evidence of stabilized free radicals (e.g. O_2^{3-}, SO_2^-, SO_3^-) in empty pseudo-octahedral sites associated with substituent cations such as B, Y, La, Na and K. Further studies of this nature were carried out by Marfunin *et al.* (1983) who used epr and luminescence spectra to distinguish anhydrites of sedimentary from those of hydrothermal origin. Mariano and Ring (1975) studied the europium-activated cathodoluminescence in a variety of minerals, but found that for anhydrite activation was by Sm^{3+} and Dy^{3+} and not by europium.

Gaft *et al.* (1985) investigated the photoluminescence spectra (400–750 nm) of many specimens of anhydrite, baryte and celestine at temperatures from 300 to 77 K, and also X-ray luminescence spectra at 300 K. The inherent lattice luminescence centre for SO_4^{2-} appeared at 360 nm for all types of sulphate, but was absent if much impurity was present. They concluded that three non-equivalent centres of Gd^{3+} were present in the anhydrites studied.

Iishi (1979) determined the static dielectric constants of anhydrite to be ε_0^x 9.21, ε_0^y 4.94 and ε_0^z 5.40 (orientation *Bbmm*, in which $\alpha = z$, $\beta = y$, $\gamma = x$).

Stress–strain curves for anhydrite at 1 and 2 kbar were determined by Handin and Hager (1957), and Brace (1960), using the method of indentation hardness, obtained similar results.

Texture analysis of anhydrite from an evaporite diapir in the Canadian Arctic Islands was carried out by neutron diffraction texture goniometry (Mainprice *et al.*, 1993). Preferred orientations of (020) and (002) were consistent with [010] glide (cell orientation *Bbmm*, $a < b < c$) in a non-coaxial flow regime. Considerable twinning was observed and this was thought to make an important contribution to the development of the textures observed.

From deformation experiments (1–5 kbar; 25–300 °C), Ramez (1976a,b) found for anhydrite translation gliding [010](001), [$1\bar{2}1$] and [$\bar{1}\bar{2}\bar{1}$]{012}, and twin gliding [$10\bar{1}$]{101}; the frequency and thickness of the latter twinning increased with pressure and temperature. At 300 °C [100] and [001] became respectively normal and parallel to the stress σ_1 because of the twinning on {101}. The indices given by Ramez refer to a unit cell with *a* as the shortest and *c* as the longest dimension; space group *Bbmm*. The rheology of mixtures of anhydrite and halite was studied by Ross *et al.* (1987) to simulate the textural evolution of

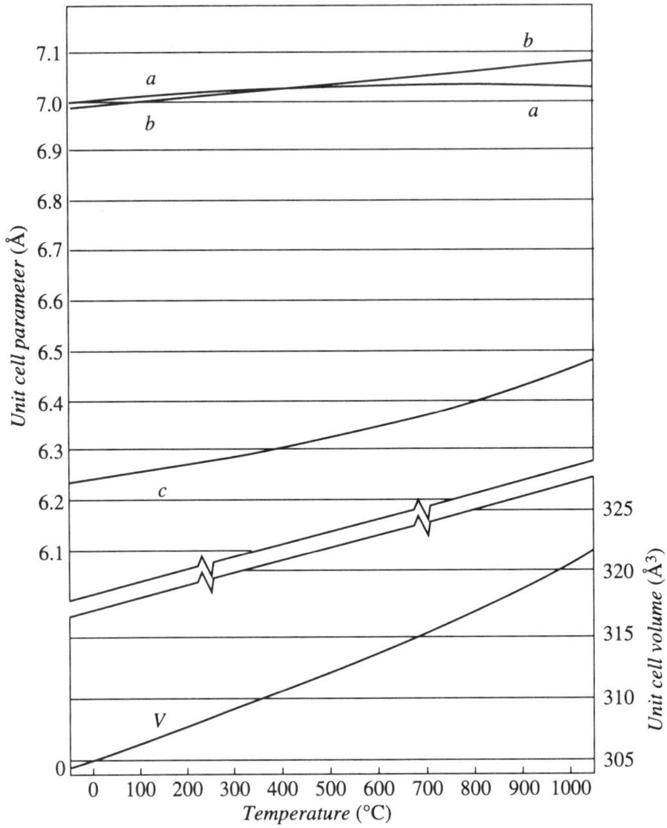

Fig. 23. Variation of cell parameters and cell volume of anhydrite with temperature (after Evans, 1979).

mylonites in silicate rocks. A useful summary of textures and deformation behaviour for anhydrite and gypsum is that by Kern and Richter (1985).

Evans (1979) determined the thermal expansion of an anhydrite from Bancroft, Ontario, by X-ray diffraction between 22 and 1000 °C. The unit cell at 22 °C had *a* 7.003, *b* 6.996, *c* 6.242 Å, *V* 305.9 $Å^3$ (space group *Amma*), and the variations of *a*, *b*, *c* and *V* with temperature are shown in Fig. 23. According to Evans (1979), *a* and *b* became equal at about 410 °C and *a* attains a maximum of 7.027 Å at 730 °C. Khan (1976) had obtained similar results by X-ray diffractometry and by calculation over the lower part of the temperature range, but determined the cross-over point for the *a* parameter as 300 °C. The thermal expansion of anhydrite is seen to be greatest parallel to the chains of SO_4 and Ca polyhedra (pp. 75, 76), which form a more rigid feature of its crystal structure (Evans, 1979; Hanic *et al.*, 1985); in other directions the polyhedra re-orientate themselves.

The anisotropic thermal conductivities of anhydrite were determined by Wooster (1936).

Distinguishing Features

Anhydrite may be distinguished from baryte by its lower density, and in thin section by its pinacoidal cleavages. It is harder and more dense than gypsum and has higher relief and birefringence; it is more dense than calcite. A simple rapid test for a sulphate recommended by Hounslow (1979) mixes a 10 per cent $Hg(NO_3)_2$ solution with 1:1 nitric acid until the hydrolysis precipitate formed just dissolves. The solution then gives a yellow stain or precipitate with a sulphate.

A modified form of Feigl's test for Ca, Ba, Sr and Pb using mercuric nitrate enables a distinction between anhydrite (or gypsum), baryte, celestine and anglesite (Hosking and Polkinghorne, 1958).

Warne (1962) described a staining test using alizarin red S and 30 per cent NaOH which distinguishes between anhydrite and gypsum. Gypsum is stained purple but anhydrite gives no colour.

Paragenesis

Anhydrite occurs in thick beds in sedimentary rock sequences, usually with other evaporite minerals. Anhydrite evaporite deposits are widespread in the USA, Canada and China, and particularly well-known European deposits are those of the Paris and Rhone Basins, the Zechstein in Germany and the Permo-Trias in the UK. Where it occurs associated with dolomites and limestones (e.g. Fowler, 1944; Dunham, 1948; Nagell, 1957), it was suggested that sulphuric acid solution, produced by the alteration of sulphides, had reacted with the calcium in these rocks to produce anhydrite. Anhydrite also occurs disseminated and interstitially as cements in sandstones, and in veins and replacement bodies associated with ore deposits. It is also found in amygdales in extrusive rocks, sometimes with quartz, glauberite and zeolites, as well as in fumarole deposits, where it has formed by the action of sulphurous vapours on calcium-bearing minerals.

Evaporites. Both anhydrite and gypsum occur commonly in evaporite deposits, either separately or together. There has been much discussion (e.g. Stewart, 1953; Conley and Bundy, 1958; Zen, 1965; Hardie, 1967; Yamamoto and Kennedy, 1969; Cody, 1976) as to whether or not anhydrite can occur in evaporites as a primary phase. Experiments and calculations on the solubilities of gypsum and anhydrite in water and saline solutions now point to a temperature of about 58 °C for the gypsum–anhydrite boundary for pure water rather than around 42 °C as estimated earlier, but this temperature is lowered to about 18 °C for saturated sodium chloride brines, making primary precipitation of anhydrite less unlikely as long as salinity is high. Gypsum does, however, appear to be far more common as the primary phase, and most anhydrites have been derived from it. Blount and Dickson (1969) also noted that because an increase in pressure increases the solubility of anhydrite, a highly saline solution would be more effective in precipitating anhydrite at shallow depths. The effect of temperature, however, usually outweighs that of pressure.

The relationships between anhydrite and the hydrates of calcium sulphate were reviewed and summarized by Holser (1979). It is clear that because of their respective solubilities, gypsum and anhydrite are the stable phases at lower and higher temperatures respectively. When heated in the laboratory in the presence of water, gypsum remains metastable up to about 93 °C before transforming to the hemihydrate; on further dehydration it can yield anhydrite, but the latter is very difficult to obtain from solution. Holser illustrated (Fig. 24) the relative ease of transformation between the various forms of calcium sulphate. Figure 25 shows the effects of temperature and salinity on the gypsum–anhydrite equilibrium. Note that increasing the concentration of NaCl lowers the temperature at which

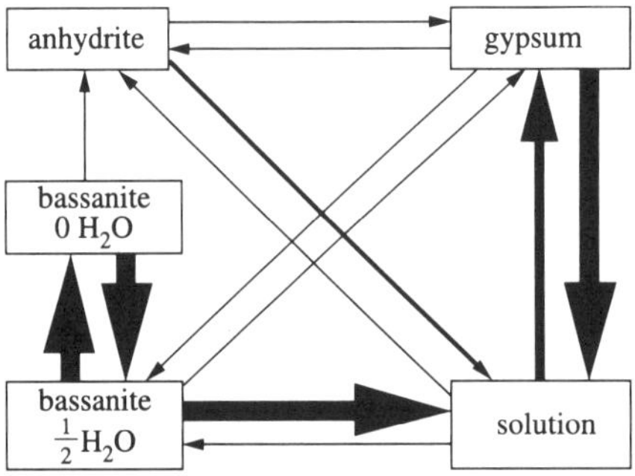

Fig. 24. Sketch showing the relative ease of the transformations between the various forms of calcium sulphate and its hydrates. Thicker lines indicate greater ease of transformation (after Holser, 1979).

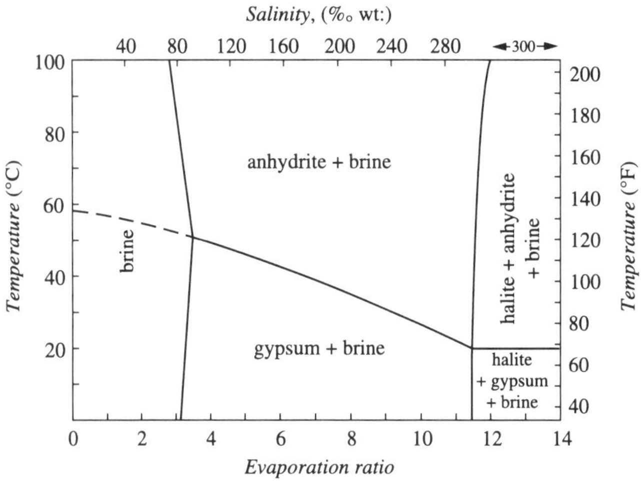

Fig. 25. Equilibrium relationships for gypsum and anhydrite with varying temperature and brine concentration, calculated by Holser using data from Hardie (1967) and others (Holser, 1979).

anhydrite will precipitate, until at the beginning of halite precipitation it would do so at normal sea-surface temperature. Such conditions are particularly pertinent to the primary growth of anhydrite in sabkha muds but, as in the laboratory, gypsum commonly persists metastably above its transition temperature. Holser also summarizes the effect of pressure in that burial raises the temperature and favours anhydrite (Figs. 26, 27), but on uplift and erosion the anhydrite reverts to gypsum (see also Murray, 1964). Although this gypsification process involves an increase in volume, it appears that generally the textures of the anhydrite rock are preserved. Further discussion of anhydrite–gypsum relationships occurs in the section on gypsum (pp. 61–66).

Wardlaw and Schwedtner (1966) described the alternation of thin (<1 cm) layers of anhydrite (plus carbonate) with halite in the lower half of the Devonian Prairie Evaporite Formation, Saskatchewan, similar to that found in the Zechstein, Germany. The alternation was attributed to seasonal changes in basinal waters at the time of deposition. Although Valayashke and Stolle (1965) concluded that

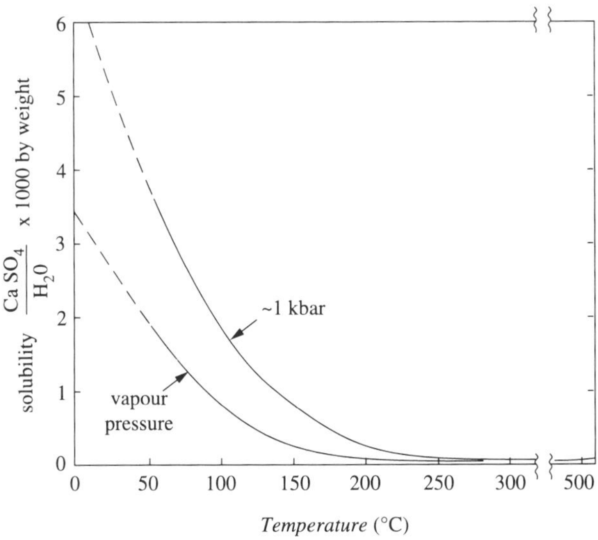

Fig. 26. Effect of pressure on the solubility of anhydrite in pure water. (Holser, 1979, using data by Blount and Dickson, 1969, and others.)

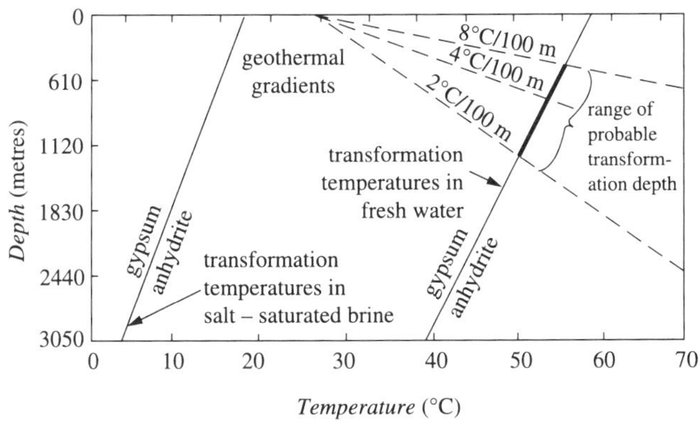

Fig. 27. *P–T* plots of gypsum–anhydrite transformation in freshwater and in saturated NaCl solution, also showing the normal range of geothermal gradients, calculated by Holser from MacDonald (1953) (after Holser, 1979).

the Zechstein II (Stassfurt) deposits of K salts are primary, the anhydrite in the hartsalz (South Harz) is ascribed to base exchange between kieserite and $CaCl_2$-containing brines:

$$CaCl_{2\,liq} + MgSO_{4\,s} \rightarrow CaSO_{4\,s} + MgCl_{2\,liq}$$

According to Schlager and Bolz (1977), although the sulphates in the Messinian and Zechstein evaporites were originally precipitated in supratidal sabkhas and hypersaline shallow waters, much of the sulphates were carried downslope and deposited in the deep water of the basin interior. The concentration of $CaSO_4$ in the water was close to or above saturation level.

Much use has been made of the ratio of the sulphur isotopes ^{34}S and ^{32}S in sulphates to infer their mode of formation. The parameter used is:

$$\delta^{34}S(^0/_{00}) = [^{34}S/^{32}S_{(sample)} - {}^{34}S/^{32}S_{(standard)}] \times 1000/(^{34}S/^{32}S_{(standard)}).$$

Müller *et al.* (1966) gave a range of $\delta^{34}S$ values for evaporite formations in North and South Germany from 11.8 to 21.0$^0/_{00}$, each individual formation covering a narrow range, e.g. 15–21, 12–17 and 18–21$^0/_{00}$. When sulphates are reduced by bacteria the residual sulphates are enriched in ^{34}S, whereas the oxidation of sulphides to sulphate leads to the enrichment of ^{32}S (see also Ault and Kulp, 1959).

Murzeyev (1966) studied the sulphur isotope ratios for Kimmeridgian and Permian anhydrite and gypsum and found that the ratios were unaffected by hydration. They found that the sulphates of the Kimmeridgian were a little lighter and those of the Permian a little heavier than those of the Jurassic ocean.

Pisarchik *et al.* (1977) determined $\delta^{34}S$ as 5.4–9.75$^0/_{00}$ for gypsum from red beds overlying halide and carbonate beds, which themselves had a $\delta^{34}S$ value of 26.75$^0/_{00}$. They concluded that there had been an influx of terrigenous sulphate containing more ^{32}S than that of marine origin. For Lower and Mid-Triassic marine anhydrite, Chen *et al.* (1981) gave $\delta^{34}S$ values as 28.3–32.4 and 14.9–20.9$^0/_{00}$ respectively.

In the sulphide–anhydrite-bearing Archaean carbonate rocks from Bahia (Brazil), Sighinolfi *et al.* (1980) concluded from oxygen isotope studies that the anhydrite was probably of primary origin and the sulphide was produced by sulphate reduction involving Archaean life systems.

For the exchange of oxygen isotopes between anhydrite and water the dependence of α, the fractionation factor, on temperature was given by Chiba *et al.* (1981) as:

$$10^3 \ln \alpha = 3.21 \times (10^3/T)^2 - 4.72.$$

Some α values determined were 1.0178, 0.0066 and 0.0039 at 100, 250 and 350 °C respectively. These values of α are slightly lighter (2$^0/_{00}$) than those determined for baryte, and it was suggested that the stronger SO_4 bonds in anhydrite explain this greater preference for ^{18}O.

Pierre and Rouchy (1986) determined oxygen and sulphur isotope ratios for 40 anhydrite samples so as to fill gaps in the curves given by Claypool *et al.* (1980). They found for Givetian-age anhydrite that $\delta^{18}O = +18.7$ and $\delta^{34}S = +20.4^0/_{00}$, and for Viséan samples the respective values were +15.4 and +15.8$^0/_{00}$ (see also Pierre *et al.*, 1984).

Cameron (1983) argued that sulphate-reducing bacteria lead to sulphides with lower ^{34}S content balanced by an enrichment of ^{34}S in the sulphates of oceans and evaporites, and reported such 'heavy' sulphur ($\delta^{34}S$ 12.3–15.6$^0/_{00}$) in nodular anhydrite from the Gordon Lake Formation, Ontario. Since the formation of this anhydrite is believed to involve continental brines (with lower ^{34}S), the above values should be taken as a minimum for sea-water sulphate at the time of deposition (2300 m.y.).

Anhydrite also occurs in salt domes and plugs (see, for example, Taylor, 1937; Goldman, 1952). The cap-rock at Richton Dome (southeast Mississippi, USA) consists of gypsum-veined anhydrite overlain by calcite. Werner *et al.* (1988) concluded that the anhydrite is a direct residue that accumulated during dissolution of the salt stock, indicating a temperature >60 °C and a salinity >4 M NaCl. The Sr content in the anhydrite, at 751 p.p.m., was similar to that in the salt stock.

In their study of the strontium isotopes of sulphates, Demaiffe and Dejonghe

(1990) concluded that whereas the strontium of baryte from Belgian lead–zinc ore deposits had a crustal origin, that of the massive anhydrite and celestine from evaporites at St Ghislain and Tournai (Belgium) came from sea-water at the time of deposition.

A combined use of sulphur and strontium isotope determinations to establish the ages, sources and diagenetic history of anhydrites (Texas, USA) was described by Posey and Fisher (1989).

In connection with their experimental work on the complex system Na–K–Mg–Ca–SO_4–Cl–H_2O, Eugster *et al.* (1980) noted that although calcium was thought to be unimportant because of its generally low concentration, it can have a significant effect because anhydrite can back-react with brines to produce glauberite, $Na_2SO_4{\cdot}CaSO_4$, and polyhalite, $K_2SO_4{\cdot}MgSO_4{\cdot}2CaSO_4{\cdot}2H_2O$. At a later stage polyhalite can in turn be replaced by anhydrite.

As an alternative to the deposition of gypsum/anhydrite in deep basin conditions, these minerals may have been formed in arid zones in supratidal or sabkha facies sediments similar to those presently being deposited in the Trucial Coast area, Arabian Gulf, and in the Gulf of California (Nickless *et al.*, 1975).

Anhydrite replacing gypsum. In some occurrences of anhydrite, evidence that it has replaced gypsum is provided by crystallographic orientation relationships (see Atoji, 1959; Moiola and Glover, 1965). The replacement of gypsum by anhydrite can also be indicated by its pseudomorphous character, but in some cases the earlier pseudomorph of gypsum is unrecognizable because the texture has been disrupted by subsequent anhydrite growth (see, for example, Holliday (1968) who described both secondary and primary anhydrite in nodular rocks from Spitsbergen).

Both gypsum and anhydrite are widely distributed in Britain, mainly in Permian and Triassic rocks in central and northern England, but also in Upper Jurassic strata in the south. Anhydrite is one of the major constituents of the Lower Evaporites of Fordon, Yorkshire, along with dolomite, polyhalite, kieserite and halite (Stewart, 1963). It was concluded that whereas the anhydrite of the middle cycle had been primary and was subsequently replaced by polyhalite, in the upper and lower cycles of deposition the anhydrite was secondary after gypsum. Kerr and Thomson (1963) suggested that the nodular anhydrite in Permian shelf sediments (Texas and New Mexico, USA) was originally deposited as gypsum.

In the Mississippi cap-rocks of southeastern Saskatchewan, sheets of anhydrite are interpreted as replacing veins of gypsum that have lost water on burial (Kendall, 1975).

Hydrothermal occurrences. It was noted by Blount and Dickson (1973) that although calcium and the sulphate ion are common in waters and hot springs, anhydrite is not very common as a hydrothermal mineral because its formation requires special conditions of temperature, pressure and concentration. Some examples of such occurrences are given below.

Steiner and Rafter (1966) compared the sulphur isotope ratios for hypogene anhydrite (steam wells, Taupo volcanic zone, New Zealand) with those in the H_2S and SO_4^{2-} in the geothermal fluid from which it formed. The values of $\delta^{34}S$ of 16.1, 3 and 22‰ for anhydrite, H_2S and SO_4^{2-}, respectively, indicated that the anhydrite derived its sulphur mainly from the SO_4^{2-} in the fluid.

A comparison of the sulphur and oxygen isotope ratios was made by Sakai and Matsubaya (1971) for anhydrite, gypsum and baryte in the black ore deposits of Japan. Whereas $\delta^{34}S$ values were similar (21–24‰) for all three minerals, the

$\delta^{18}O$ values increased in the order baryte, anhydrite, gypsum. This was explained by the isotopic character of submarine hydrothermal systems, where sulphate from sea-water predominates over that from other sources of sulphur.

Kusakabe and Chiba (1979) studied oxygen isotope ratios for the $CaSO_4$–H_2O and $BaSO_4$–H_2O systems with application to sulphates from Kuroko type deposits. For the anhydrite–gypsum ore body at Shakanai mine, they estimated a temperature of about 200 °C assuming direct precipitation of anhydrite from sea-water. At this temperature 50 per cent of the calcium sulphate originally in the sea-water was precipitated to form the orebody.

Steefel (1987) described a pervasive pre-mineralization nodular anhydrite–Mg-chlorite–clay–sericite assemblage as having formed by the heating and circulation of sea-water within unconsolidated tuffs (Johnson River Prospect, Alaska, USA). The temperature of precipitation of anhydrite was estimated to have been about 150 °C.

Fluid inclusion studies (Ramboz *et al.*, 1988) on epigenetic anhydrites (Atlantis II Deep, Red Sea) showed that the anhydrites had precipitated from successive boiling brines with salinities increasing from 19 to 32 wt. per cent equivalent NaCl. Maximum temperatures of the trapped pure liquids of 390–403 °C were given as compatible with boiling brines at the pressures of the overlying water (223 bar). It was also observed that the cores of the anhydrite crystals had grown from less saline sulphur-bearing fluids and that later growth was from residual solutions that had boiled for longer.

Fluid inclusion studies on hydrothermal anhydrite from the Sabatini volcanic district, Latium, Italy, yielded temperatures of homogenization of 144–304 °C (Belkin *et al.*, 1988). The fluids varied in salinity from 0.5 to 14.0 wt. per cent NaCl equivalent ($\sim$2 M NaCl).

In the Mineral Park Cu–Mo porphyry system, Arizona, USA, studies using oxygen isotopes and fluid inclusions indicated two generations of anhydrite (Lang *et al.*, 1989): anhydrite–molybdenite veins formed from magmatic meteoric waters ($\delta^{18}O$ 6.5–9.2‰), and anhydrite–chalcopyrite veins formed from waters of either magmatic or evaporitic origin, with $\delta^{18}O$ values of 11.8–17.1‰.

Submarine occurrences. Anhydrites have been reported from deep-sea basalts (e.g. Alt *et al.*, 1985). However, studies of the sulphur isotopes show that they are produced by late hydrothermal alteration by fluids isotopically similar to sea-water sulphate.

Anhydrite was reported by Styrt *et al.* (1981) as present in all hydrothermal chimneys, whether active or inactive, in the East Pacific Rise. They found that $\delta^{34}S$ for anhydrite from active vents was close to that for sea-water, and concluded that the anhydrite was derived predominantly from sea-water sulphate. Haymon and Kastner (1981) found anhydrite as a major constituent of active chimneys only, and suggested that it had been removed from inactive chimneys by dissolution (see also Graham *et al.*, 1988).

The study of fluid inclusions in anhydrite from the East Pacific Rise (Kusakabe *et al.*, 1982) gave a wide range of temperatures (180–360 °C), indicating that precipitation occurred during a wide degree of sea-water mixing. Whereas the $\delta^{34}S$ for anhydrite sulphur was similar to the present sea-water value, indicating an origin from sea-water sulphate, the $\delta^{18}O$ value was found to be 2‰ higher; it was suggested that post-depositional partial re-equilibration with solutions at a lower temperature had occurred.

The sulphur isotope ratios of anhydrites associated with the massive sulphide deposits of the East Pacific Rise indicated disequilibrium with coexisting sulphides and showed that their source of sulphur was the sulphate of ambient sea-water and not hydrothermal fluid (Zierenberg *et al.*, 1984). Oxygen isotope ratios for two anhydrites corresponded with $T \sim 222$ and 194 °C.

Thompson *et al.* (1988) investigated minerals of active vents on the Mid-Atlantic Ridge and found that anhydrite was the main sulphate phase and there was little or no baryte. This contrasted with the situation reported for the East Pacific Rise and for the Southern Juan de Fuca Ridge (Paradis *et al.*, 1988).

Metamorphic rocks. The occurrence of anhydrite in a hornfels formed by the intrusion of a gabbro into a pillow lava sequence (Queensland, Australia) was attributed by Phillips (1962) to the action of sulphuric acid arising from the oxidation of pyrite by the gabbro.

Nash (1972) reported mauve twinned anhydrite associated with quartz, microcline feldspar and plagioclase feldspar in a banded gneiss (Swakopmund District, Southwest Africa), and concluded that it was formed in equilibrium with the high-temperature minerals at around 700 °C. Another pale-mauve anhydrite occurring together with quartz, halite, pyrite, sphalerite and kyanite in a mica schist (Labrador) was described by Butchins and Mason (1973) who suggested that it probably formed as primary gypsum in a sulphide-rich pelite; metasomatic alteration could not, however, be ruled out.

Anhydrite (*a* 6.24, *b* 6.99, *c* 7.00 Å, *Ccmm*, α 1.570, γ 1.613) in granitic gneiss (Leventina complex, Ticino, Switzerland) was considered to be of evaporitic origin (Hiss, 1975).

Igneous rocks. Anhydrite is the most common sulphate in igneous rocks, occurring either as phenocrysts or as products of alteration. Primary anhydrite was reported, together with plagioclase, hornblende, biotite, pyroxene and magnetite, in a colourless glass in the lavas of Mount Lamington, Papua New Guinea (Taylor, 1958).

Arculus *et al.* (1983) reported anhydrite in andesite and suggested that the source of sulphur was contamination by crustal sulphides (see also Carrol and Rutherford, 1985). Carrol and Rutherford (1987) described hydrothermal experiments on natural trachyandesite and dacite compositions, which gave anhydrite as a stable phenocryst at log f_{O_2} values of 1.0–1.5 above the fugacity of the Ni–NiO buffer. They noted that this was consistent with anhydrite (an important source of the sulphurous gases emitted on eruption) being observed in the El Chichón (Mexico) pumice (Luhr *et al.*, 1984; Rye *et al.*, 1984). They also suggested that the apparent scarcity of anhydrite in calc-alkaline rocks with appropriate f_{O_2} may be due to their low sulphur content (100–500 p.p.m.) and the easy removal of anhydrite by subaerial weathering.

From a study of anhydrite in orthopyroxene phenocrysts in the Nevada del Ruiz (Colombia) pumice, Fournelle (1990) suggested that the andesitic dacitic melt was not saturated with anhydrite at the prevailing temperature and f_{O_2}, and that the anhydrite is formed by the reaction of pyrrhotite trapped in the phenocrysts with melt, e.g. by:

$$3CaSiO_3 + 3FeS + 6\tfrac{1}{2}O_2 = 3CaSO_4 + Fe_3O_4 + 3SiO_2.$$

Alternatively, SO_2 could originate from a sulphur-saturated magma. From the Noril'sk intrusive, Grinenko and Grinenko (1967) reported a violet anhydrite associated with prehnite, hornblende, serpentine and chlorite in reaction rims on sulphide grains, as well as independently in a picritic dolerite. They concluded from sulphur isotope data that the anhydrite was derived from gypsum and anhydrite in evaporite country rocks, not from oxidation of the sulphides.

References

Alt, J. C., **Saltzman, E. S.** and **Price, D. A.**, 1985. *Industrial reports of the D.S.D.P.*, **83**, 283–8.

Arculus, R. J., **Johnson, R. W.**, **Chappell, B. W.**, **McKee, C. O.** and **Sakai, H.**, 1983. Ophiolite-contaminated andesites, trachybasalts, and cognate inclusions of Mount Lamington, Papua New Guinea: anhydrite–amphibole–bearing lavas and the 1951 cumulodome. *J. Volc. Geotherm. Res.*, **18**, 214–47.

Atoji, M., 1959. Anhydrite obtained by the dehydration of gypsum. *J. Chem. Phys.*, **30**, 341–2.

Ault, W. U. and **Kulp, J. L.**, 1959. Isotopic geochemistry of sulphur. *Geochim. Cosmochim. Acta*, **16**, 201–35.

Bath, A. H., **Darling, W. G.**, **George, I. A.** and **Milodowski, A. E.**, 1987. $^{18}O/^{16}O$ and $^{2}H/^{1}H$ changes during progressive hydration of a Zechstein anhydrite formation. *Geochim. Cosmochim. Acta*, **51**, 3113–8.

Belkin, H. E., **Cavaretta, G.**, **De Viro, B.** and **Tecce, F.**, 1988. Hydrothermal phlogopite and anhydrite from the SH2 well, Sabatini volcanic district, Latium, Italy: fluid inclusions and mineral chemistry. *Amer. Min.*, **73**, 775–97.

Benstead, J., 1976. Uses of Raman spectroscopy in cement chemistry. *J. Amer. Ceram. Soc.*, **59**, 140–3.

Benstead, J. and **Prakash, S.**, 1968. Investigation of the calcium sulphate–water system by infrared spectroscopy. *Nature*, **219**, 60–1.

Bershov, L. V., **Martirosyan, V. O.**, **Marfunin, A. S.** and **Speranskii, A. V.**, 1975. EPR and structure models for radical ions in anhydrite crystals. *Fortschr. Min.*, **52**, 591–604.

Blount, C. W. and **Dickson, F. W.**, 1969. The salinity of anhydrite ($CaSO_4$) in NaCl–H_2O from 100 to 450° and 1 to 1000 bars. *Geochim. Cosmochim. Acta*, **33**, 227–45.

Blount, C. W. and **Dickson, F. W.**, 1973. Gypsum–anhydrite equilibria in systems $CaSO_4$–H_2O and $CaSO_4$–NaCl–H_2O. *Amer. Min.*, **58**, 323–31.

Bock, E., 1961. On the solubility of anhydrous calcium sulphate and of gypsum in concentrated solutions of sodium chloride at 25 °C, 30 °C, 40 °C and 50 °C. *Can. J. Chem.*, **39**, 1746–51.

Brace, W. F., 1960. Behaviour of rock salt, limestone and anhydrite during indentation. *J. Geophys. Res.*, **65**, 1773–88.

Butchins, C. S. and **Mason, R.**, 1973. Metamorphic anhydrite in a kyanite-bearing schist from Churchill Falls, Labrador. *Min. Mag.*, **39**, 488–90.

Caillère, S. and **Pobeguin, Th.**, 1962. Comparison des propriétés thermiques et du comportement dans l'infrarouge de quelques carbonates, nitrates, sulfates et sélénites simples. *Bull. Soc. Franc. Min. Crist.*, **85**, 48–57.

Cameron, E. M., 1983. Evidence from early Proterozoic anhydrite for sulphur isotopic partitioning in Precambrian oceans. *Nature*, **304**, 54–6.

Carroll, M. R. and **Rutherford, M. J.**, 1985. Sulfide and sulfate saturation in hydrous silicate melts. *J. Geophys. Res.*, **90** (*suppl.*), C 601–12.

Carroll, M. R. and **Rutherford, M. J.**, 1987. The stability of igneous anhydrite: experimental results and implications for sulfur behaviour in the 1982 El Chichón trachyandesite and other evolved magmas. *J. Petr.*, **28**, 781–801.

Chen, J., **Zhao, R.**, **Huo, W.**, **Yao, Y.**, **Pan, S.**, **Shao, M.** and **Hai, C.**, 1981. Sulphur isotopes of some marine gypsum. *Scientia Geol. Sinica.*, 273–8 (M.A. **33**–1437).

Cheng, G. C. H. and **Zussman, J.**, 1963. The crystal structure of anhydrite ($CaSO_4$). *Acta Cryst.*, **16**, 767–9.

Chiba, H., **Kusakabe, M.**, **Hirano, S.**, **Matsuo, S.** and **Somiya, S.**, 1981. Oxygen isotope fractionation between anhydrite and water from 100 to 550 °C. *Earth & Planet. Sci. Lett.*, **53**, 55–62.

Claypool, G. E., **Holser, W. T.**, **Kaplan, I. R.**, **Sakai, H.** and **Zak, I.**, 1980. The age curves of sulphur and oxygen isotopes in marine sulfate and their mutual interpretation. *Chem. Geol.*, **28**, 99–260.

Cody, R. D., 1976. Growth and early diagenetic changes in artificial gypsum crystals grown within bentonite muds and gels. *Bull. Geol. Soc. Amer.*, **87**, 1163–8.

Conley, R. F. and **Bundy, W. M.**, 1958. Mechanism of gypsification. *Geochim. Cosmochim. Acta*, **15**, 57–72.

Corba, J., 1975. X-ray quantitative determination of anhydrite in technical grade gypsum plaster. *Silikaty*, **19**, 43–8.

Demaiffe, D. and **Dejonghe, L.**, 1990. Géochemie isotopique du strontium des barites, anhydrites, calcites et fluorites de Belgique. *Ann. Soc. Géol. Belg.*, **113**, 231–40.

Dickson, E. C. S. and **Binks, W.**, 1926. The crystalline structure of anhydrite. *Phil. Mag., ser. 7*, **2**, 114–28.

Dickson, F. W., **Blount, C. W.** and **Tunell, G.**, 1963. Use of hydrothermal solution equipment to determine the solubility of anhydrite in water from 1000 °C to 275 °C and from 1 bar to 1000 bars pressure. *Amer. J. Sci.*, **261**, 61–78.

Dunham, K. C., 1948. A contribution to the petrology of the Permian evaporite deposits of north-eastern England. *Proc. Yorks. Geol. Soc.*, **7**, 217–27.

Eugster, H. P., **Harvie, C. E.** and **Weave, J. H.**, 1980. Mineral equilibria in a six-component seawater system, Na–K–Mg–Ca–SO_4–Cl–H_2O, at 25 °C. *Geochim. Cosmochim. Acta*, **44**, 1335–47.

Evans, H. T., *Jr*, 1979. The thermal expansion of anhydrite to 1000 °C. *Phys. Chem. Min.*, **4**, 77–82.

Flörke, O. W., 1952a. Kristallographische und röntgenographische Untersuchungen im System, $CaSO_4$–$CaSO_4 \cdot 2H_2O$. *Neues Jahrb. Min., Abh.*, **84**, 189–240.

Flörke, O. W., 1952b. Die Hochtemperaturmodificationen von Kalzium-, Strontium-, und Bariumsulphat. *Naturwiss.*, **39**, 478–9.

Fournelle, J., 1990. Anhydrite in Nevado el Ruiz November 1985 pumice: relevance to the sulfur problem. *J. Volc. Geotherm. Res.*, **42**, 189–201.

Fowler, A., 1944. A deep bore in the Cleveland Hills. *Geol. Mag.*, **81**, 193–206.

Gaft, M. L., **Bershov, L. V.**, **Krasnaya, A. R.** and **Yaskolko, V. Ya.**, 1985. Luminescence centres in anhydrite, barite, celestite and their synthesized analogs. *Phys. Chem. Min.*, **11**, 255–60.

Gardner, A. W. and **Glueckauf, E.**, 1970. Thermodynamic data of the calcium sulphate solution process between 0 and 200 °C. *Trans. Farad. Soc.*, **66**, 1081–6.

Gay, P., 1965. Some crystallographic studies in the system $CaSO_4$–$CaSO_4 \cdot 2H_2O$. I. The polymorphism of anhydrous $CaSO_4$. *Min. Mag.*, **35**, 347–53.

Goldman, M. I., 1952. Deformation, metamorphism and mineralization in gypsum–anhydrite cap rock, sulphur salt dome, Louisiana. *Mem. Geol. Soc. Amer.*, **50**.

Graham, U. M., **Bluth, G. J.** and **Ohmoto, H.**, 1988. Sulfide–sulfate chimneys on the East Pacific Rise 11° and 13°N latitudes. Part I. Mineralogy and paragenesis. *Can. Min.*, **26**, 487–504.

Grahmann, W., 1920. Über Barytocölestin und das Verhaltnis von Anhydrit zu Cölestin und Baryt. *Neues Jahrb. Min.*, **1**, 1–23.

Grégoire, D. C., 1987. Determination of boron isotope ratios in geological materials by I.C.P.M.S. *Anal. Chem.*, **59**, 2479–84.

Griffith, W. P., 1969. Raman spectroscopy of minerals. *Nature*, **224**, 264–7.

Grinenko, L. N. and **Grinenko, V. A.**, 1967. Origin of violet anhydrite in the Noril'sk-1 intrusive (as indicated by isotopic analysis of sulfur). *Geochem. Internat.*, **4**, 75–9.

Gutt, W. and **Smith, M. A.**, 1967. The α form of calcium sulphate. *Trans. Brit. Ceram. Soc.*, **66**, 337–345.

Handin, J. and **Hager, R. V.**, 1957. Experimental deformation of sedimentary rocks under confining pressure: tests at room temperature on dry samples. *Bull. Amer. Assoc. Petrol. Geol.*, **41**, 1–50.

Hanic, F., **Galikova, L.**, **Havlica, J.**, **Kaprálik, I.** and **Ambruz, V.**, 1985. Kinetics of the thermal decomposition of $CaSO_4$ in air. *Trans. Brit. Ceram. J.*, **84**, 22–5.

Harada, Z., 1954. Chemical analyses of Japanese minerals (III). *Journ. Fac. Sci. Hokkaido Univ., Ser. IV*, **8**, 289–348.

Hardie, L. A., 1967. The gypsum–anhydrite equilibrium at one atmosphere pressure. *Amer. Min.*, **52**, 171–200.

Harvie, C. E. and **Weare, J. H.**, 1980. The prediction of mineral solubilities in natural waters; the Na–K–Mg–Ca–Cl–SO_4–H_2O system from zero to high concentration at 25 °C. *Geochim. Cosmochim. Acta*, **44**, 981–97.

Hawthorne, F. C. and **Ferguson, R. B.**, 1975. Anhydrous sulphates II. Refinement of the crystal structure of anhydrite. *Can. Min.*, **13**, 289–92.

Haymon, R. M. and **Kastner, M.**, 1981. Hot spring deposits on the East Pacific Rise at 21°N: preliminary description of mineralogy and genesis. *Earth & Planet. Sci. Lett.*, **53**, 363–81.

Hiss, A. M., 1975. Metamorpher Anhydrit im Leventina-Gneiss. *Schweiz. Min. Petr. Mitt.*, **55**, 217–25.

Höhne, E., 1963. A more accurate determination of the crystal structure of anhydrite, $CaSO_4$. *Soviet Physics. Cryst.*, **7**, 559–69.

Holliday, D. W., 1968. Early diagenesis in Middle Carboniferous nodular anhydrite of Spitsbergen. *Proc. Yorks. Geol. Soc.*, **36**, 277–92.

Holser, W. J., 1979. Mineralogy of Evaporites. In *Marine Materials: Min. Soc. Amer. Short Course Notes*, **6**, 211–94.

Hosking, K. F. G. and **Polkinghorne, J. P. R.**, 1958. A note on the identification of 'insoluble' sulphates. *Camborne School Mines Mag.*, **58**, 38–9.

Hounslow, A. W., 1979. Modified gypsum/anhydrite stain. *J. Sed. Petr.*, **49**, 636–7.

Iishi, K., 1979. Phonospectroscopy and lattice dynamical calculations of anhydrite and gypsum. *Phys. Chem. Min.*, **4**, 341–59.

Innorta, G., **Rabbi, E.** and **Tomadin, L.**, 1980. The gypsum–anhydrite equilibrium by solubility measurements. *Geochim. Cosmochim. Acta*, **44**, 1931–6.

Kagi, H., **Dohmoto, Y.**, **Takano, S.** and **Masuda, A.**, 1993. Tetrad effect in lanthanide partitioning between calcium sulfate crystal and its saturated solution. *Chemical Geology (including Isotope*

Geoscience), **107**, 71–82.

Kahn, J. S., 1975. High pressure phase transformation of $CaSO_4$ (anhydrite) during a nuclear explosion. *Science*, **189**, 454–5.

Kendall, A. C., 1975. Anhydrite replacements of gypsum (satin-spar) veins in the Mississippian caprocks of southeastern Saskatchewan. *Can. J. Earth. Sci.*, **12**, 1190–5.

Kern, H. and **Richter, A.**, 1985. Microstructures and textures in evaporites. In *Preferred Orientation in Deformed Metals and Rocks: an Introduction to Modern Texture Analysis*: Ed. H. R. Wenk. New York and London (Academic Press), 610 pp.

Kerr, S. D. and **Thomson, A.**, 1963. Origin of nodular and bedded anhydrite in Permian shelf sediments, Texas and New Mexico. *Bull. Amer. Petrol. Geol.*, **47**, 1726–32.

Khan, A. A., 1976. Computer simulation of thermal expansion of non-cubic crystals: forsterite, anhydrite and scheelite. *Acta Cryst.*, **A32**, 11–6.

Kirfel, A. and **Will, G.**, 1980. Charge density in anhydrite, $CaSO_4$, from X-ray and neutron diffraction measurements. *Acta. Cryst.*, **B36**, 2881–90.

Kirfel, A. and **Will, G.**, 1981. Charge density in anhydrite, $CaSO_4$. II. Static deformation density distributions from a multiple refinement. *Acta Cryst.*, **B37**, 525–32.

Kolb, R., 1911. Vergleich von Anhydrit, Cölestin, Baryt und Anglesit in bezug auf die Veränderung ihrer geometrischen und optischen Verhaltnisse mit der Temperature. *Z. Krist.*, **69**, 14–61.

Kusakabe, M. and **Chiba, H.**, 1979. Oxygen isotope geothermometry applicable to sulphate minerals from the Kuroko deposits. *Mining Geol.*, **29**, 257–64.

Kusakabe, N. M., **Chiba, H.** and **Ohmoto, H.**, 1982. Stable isotopes and fluid inclusion study of anhydrite from the East Pacific Rise at 21°N. *Geochem. Journ.*, **16**, 89–95 (M.A. **36**–1722).

Lang, J. R., **Guan, Y.** and **Eastoe, C. J.**, 1989. Stable isotope studies of sulfates and sulfides in the Mineral Park porphyry Cu–Mo system, Arizona. *Econ. Geol.*, **84**, 650–62.

Luhr, J. R., **Carmichael, I. S. E.** and **Varekamp, J. C.**, 1984. The 1982 eruption of El Chichón volcano, Chiapas, Mexico. Mineralogy and petrology of the anhydrite-bearing pumices. *J. Volc. Geotherm. Res.*, **23**, 69–108.

MacDonald, G. J. F., 1953. Anhydrite–gypsum equilibrium relations. *Amer. J. Sci.*, **251**, 884–98.

Mainprice, D., **Bouchez, J.-L.**, **Casey, M.** and **Dervin, P.**, 1993. Quantitative analysis of naturally deformed anhydrite by neutron diffraction texture goniometry. *J. Struct. Geol.*, **15**, 793–804.

Marfunin, A. S., **Badalov, S. J.**, **Bershov, L. V.** and **Gaft, A. V.**, 1983. The pecularities of composition and the true structure of hypogene and sedimentary anhydrites. *Min. Zhurn.*, **5**, 3–9 (M.A. **34**–4269).

Mariano, A. N. and **Ring, P. J.**, 1975. Europium-activated cathodoluminescence in minerals. *Geochim. Cosmochim. Acta*, **39**, 649–60.

Matsubara, S., **Kato, A.** and **Hashimoto, E.**, 1992. Celestine from the Asaka gypsum mine, Koriyama City, Fukishima Prefecture, Japan. *Min. J.*, **16**, 16–20.

Miropolsky, L. M., 1942. Macrofibrous anhydrite in the Chuvash Republic. *Doklady Acad. Sci. USSR*, **35**, 157–8.

Moiola, R. J. and **Glover, E. D.**, 1965. Recent anhydrite from Clayton Playa, Nevada. *Amer. Min.*, **50**, 2063–9.

Monnin, C., 1990. The influence of pressure on the activity coefficients of the solutes and on the solubility of minerals in the system Na–Ca–Cl–SO_4–H_2O to 200 °C and 1 kbar, and to high NaCl concentration. *Geochim. Cosmochim. Acta*, **54**, 3265–82.

Morikawa, H., **Minato, I.**, **Tomita, T.** and **Iwai, S.**, 1975. Anhydrite: a refinement. *Acta Cryst.*, **B31**, 2164–5.

Müller, G., **Nielsen, H.** and **Ricke, W.**, 1966. Schwefel-isotopen-verhältnisse in Formationswässern und Evaporiten Nordund Süd Deutschlands. *Chem. Geol.*, **1**, 211–20.

Murray, R. C., 1964. Origin and diagenesis of gypsum and anhydrite. *J. Sed. Petr.*, **34**, 512–23.

Murzeyev, P. M., 1966. S^{32}/S^{34} ratio in pyrite and anhydrite from the Baymakliya uplift, southern USSR. *Dokl. Acad. Sci. USSR. Earth Sci.*, **168**, 176–9.

Nagell, R. H., 1957. Anhydrite complex of the Morococha District, Peru. *Econ. Geol.*, **52**, 632–44.

Nash, C. R., 1972. Primary anhydrite in Precambrian gneisses from the Swakopmund District, South West Africa. *Contr. Min. Petr.*, **36**, 27–32.

Nickless, E. F. P., **Booth, S. J.** and **Mosley, P. N.**, 1975. Celestite deposits of the Bristol area. *Trans. Inst. Min. Met.*, **84(B)**, 62–4.

Nyman, H., **Hyde, B. G.** and **Andersson, S.**, 1984. Zircon, anhydrite, scheelite and some related structures containing bisdisphenoids. *Acta Cryst.*, **B40**, 441–7.

Paradis, S., **Jonasson, I. R.**, **Le Cheminant, G. M.** and **Watkinson, D. H.**, 1988. Two zinc-rich chimneys from the Plume Site, Southern Juan de Fuca Ridge. *Can. Min.*, **26**, 637–54.

Phillips, E. A., 1962. A note on the occurrence of anhydrite in a hornfels. *Proc. Roy. Soc. Queensland*, **72**, 121–2.

Pierre, C. and **Rouchy, J.-M.**, 1986. Oxygen and sulfur isotopes in anhydrites from Givetian and Visean evaporites of northern France and Belgium. *Chem. Geol.*, **58**, 245–52.

Pierre, C., Rouchy, J.-M., Laumondais, A. and **Goessens, É.**, 1984. Sédimentologie et géochemie isotopique (^{18}O, ^{34}S) des sulfates évaporitique givétiens et dinantiens du Nord de la France et de la Belgique; importance pour la stratigraphie et la reconstitution des paléomilieux de depot. *C. R. Acad. Sci. Paris*, **299**, Ser. II, 21–6.

Pisarchik, Yu. K., Golubchina, M. N. and **Toksubayev, A. J.**, 1977. Isotopic composition of sulphur in calcium sulfates from the Cambrian Upper Lena formation (Siberian platform). *Geochem. Intern.*, **14**, 182–5.

Posey, H. H. and **Fisher, R. S.**, 1989. A sulfur and strontium isotopic investigation of Lower Permian anhydrite, Palo Duro Basin, Texas, USA. *Applied Geochem.*, **4**, 395–407.

Przibram, K., 1956. *Irradiation Colours and Luminescence.* London (Pergamon Press), 332 pp.

Ramboz, C., Oudin, E. and **Thisse, Y.**, 1988. Geyser-type discharge in Atlantis II Deep Red Sea: evidence of boiling from fluid inclusions in epigenetic anhydrite. *Can. Min.*, **26**, 765–86.

Ramez, M. R. H., 1976a. Mechanisms of intergranular gliding in experimentally deformed anhydrite. *Neues Jahrb. Min., Abh.*, **127**, 311–29.

Ramez, M. R. H., 1976b. Fabric changes in experimentally deformed anhydrite rocks. *Neues Jahrb. Min., Abh.*, **128**, 89–113.

Reimann, M., 1991. Geologisch-lagerstättenkunde und mineralogische Untersuchungen zur Vergipsung und Volumenzunahme der anhydrite verschiedener geologischer Formationen unter natürlichen und labormässigen Bedingungen. *Geol. Jahrb.*, **D97**, 21–125.

Ross, J. V., Bauer, S. J. and **Hansen, F. D.**, 1987. Textural evolution of synthetic anhydrite–halite mylonites. *Tectonophys.*, **140**, 307–26.

Ryabov, I. D., Bershov, L. V., Sperianskiy, A. V. and **Ganeev, I. G.**, 1983. Electron paramagnetic resonance of PO_3^{2-} and SO_3^- radicals in anhydrite, celestite and barite: the hyperfine structure and dynamics. *Phys. Chem. Min.*, **10**, 21–6.

Rye, R. O., Luhr, J. E. and **Wasserman, M. D.**, 1984. Sulfur and oxygen isotopic systematics of the 1982 eruption of El Chichón volcano, Chiapas, Mexico. *J. Volc. Geotherm. Res.*, **23**, 109–23.

Ryzhenko, B. N., 1964. Determination of the second dissociation constant of sulfuric acid, and precipitation of salts in the reciprocal system Ca^{2+}, Ba^{2+}/SO_4^{2-}, CO_3^{2-} under hydrothermal conditions. *Geochem. Internat.*, **1**, 8–12.

Sakai, H. and **Matsubaya, O.**, 1971. Sulfur and oxygen isotopic ratios of gypsum and barite in the black ore deposits of Japan. *Soc. Mining Geol. Japan, Spec. Issue*, **2**, (Proc. IMA-IAGOD Meetings '70, Joint Symp. vol.), 80–3.

Saxby, D. B. and **Lamar, J. E.**, 1957. Gypsum and anhydrite in Illinois. *Illinois State Geol. Surv.*, circ. **226**.

Schlager, W. and **Bolz, H.**, 1977. Clastic accumulation of sulphate evaporites in deep water. *J. Sed. Petr.*, **47**, 600–9.

Sighinolfi, G. P., Kronberg, B. I., Gorgoni, C. and **Fyfe, W. S.**, 1980. Geochemistry and genesis of sulphide–anhydrite bearing Archaean carbonate rocks from Bahia (Brazil). *Chem. Geol.*, **29**, 323–31.

Simpson, E. S., 1930. Contribution to the mineralogy of Western Australia. Series V. *Journ. Roy. Soc. W. Australia*, **16**, 25–40.

Sofer, Z., 1978. Isotopic composition of hydration water in gypsum. *Geochim. Cosmochim. Acta*, **42**, 1141–9.

Steefel, C. I., 1987. The Johnson River Prospect, Alaska: gold-rich sea-floor mineralization from the Jurassic. *Econ. Geol.*, **82**, 894–914.

Steiner, A. and **Rafter, T. A.**, 1966. Sulfur isotopes in pyrite, pyrrhotite, alunite and anhydrite from steam wells in the Taupo volcanic zone, New Zealand. *Econ. Geol.*, **61**, 1115–29.

Stephens, D. R., 1964. The hydrostatic compression of eight rocks. *J. Geophys. Res.*, **69**, 2967–78.

Stewart, F. H., 1953. Early gypsum in the Permian evaporites of northeastern England. *Proc. Geol. Assoc.*, **64**, 33–8.

Stewart, F. H., 1963. The Permian Lower Evaporites of Forden in Yorkshire. *Proc. Yorks. Soc.*, **34**, 1–44.

Styrt, M. M., Brackmann, A. J., Holland, H. D., Clarke, B. C., Pisutha-Arnond, V., Eldridge, C. S. and **Ohmoto, H.**, 1981. The mineralogy and the isotopic composition of sulfur in hydrothermal sulfide/sulfate deposits on the East Pacific Rise, 21°N latitude. *Earth Planet. Sci. Sci. Lett.*, **53**, 382–90.

Swanson, H. E., Fuyat, R. K. and **Ugrinic, G. M.**, 1955. Standard X-ray diffraction patterns. *U.S. Nat. Bur. Stands. Circular 539*, **4**, 67.

Taylor, G. A., 1958. The 1952 eruption of Mount Lamington, Papua. *Bur. Min. Res. Geol. Geophys. Australia, Bull.*, **38**, 117 pp.

Taylor, R. E., 1937. Water-insoluble residues in rock salt of Louisiana salt plugs. *Bull. Amer. Assoc. Petroleum Geol.*, **21**, 1268–1310.

Thompson, G., Humphries, S. E., Schroeder, B. and **Sulanowiska, M.**, 1988. Active vents and massive sulphides at 26°N (TAG) and 23°N (Snakepit) on the mid-Atlantic Ridge. *Can. Min.*, **26**, 697–711.

Tocmakchieva, M., 1980. On the nature of colouring in the anhydrite of Panagyurishte cupro-pyrite ore deposits. *Ann. l'Ecole Super. Mines. Geol., Sofia, Fasc.2. Geol. Appl. & Hydrogeol.*, **25**, 193–207 (M.A. **81**–4410).

Valyashko, M. G. and **Stolle, E.,** 1965. Geochemical analysis of the genesis of the Stassfurt potassium salts. *Geochem. Internat.,* **2**, 709–26.

Wardlaw, N. C. and **Schwedtner, W. M.,** 1966. Halite–anhydrite seasonal layers in the Middle Devonian Prairie Evaporite Formation, Saskatchewan, Canada. *Geol. Soc. Amer. Bull.,* **77**, 331–42.

Warne, S. St J., 1962. A quick field or laboratory staining scheme for the differentiation of the major carbonate minerals. *J. Sed. Petr.,* **32**, 29–38.

Wasastjerna, J. A., 1925. Der Bau der Sulfatgruppe. *Soc. Sci. Fennica, Phys. Math.,* **2**, no. 26.

Werner, M. L., Feldman, M. D. and **Knauth, L. P.,** 1988. Petrography and geochemistry of water–rock interaction in Richton Dome cap rock (southeastern Mississippi, USA). *Chem. Geol.,* **74**, 113–35.

Wooster, W. A., 1936. Thermal conductivity in relation to crystal structure. *Z. Krist.,* **95**, 138–49.

Yamamoto, H. and **Kennedy, G. C.,** 1969. Stability relations in the system $CaSO_4$–H_2O at high temperatures and pressures. *Amer. J. Sci.,* **267A**, 550–7.

Zen, E.-an, 1965. Solubility measurements in the system $CaSO_4$–NaCl–H_2O at 35°, 50°, 70 °C and at 1 atmosphere. *J. Petr.,* **6**, 124–64.

Zhuk-Pochekutov, K. A., Serdobova, L. I., Katayeva, Z. Y., Andreyeva, E. P., Ryabova, T. V. and **Logunova, Ye. V.,** 1986. The rare-earth elements in anhydrite from the Kachar iron-ore deposit. *Geochim. Internat.,* **23**(9), 49.

Zierenberg, R. A., Shanks, W. C. and **Bischoff, J. L.,** 1984. Massive sulphide deposits at 21°N; East Pacific Rise: Chemical composition, stable isotopes, and phase equilibria. *Bull. Geol. Soc. Amer.,* **95**, 922–9.

Carbonates

Calcite $CaCO_3$
Magnesite $MgCO_3$
Rhodochrosite $MnCO_3$
Siderite $FeCO_3$
Smithsonite $ZnCO_3$
Dolomite $MgCa(CO_3)_2$
Ankerite $Ca(Mg,Fe^{2+},Mn)(CO_3)_2$
Aragonite $CaCO_3$
Strontianite $SrCO_3$
Witherite $BaCO_3$
Cerussite $PbCO_3$
Huntite $Mg_3Ca(CO_3)_4$
Nyerereite $Na_2Ca(CO_3)_2$
Malachite $Cu_2(OH)_2CO_3$
Azurite $Cu_3(OH)(CO_3)_2$

Carbonates

Carbonates are a group of minerals in which the essential unit is the $(CO_3)^{2-}$ ion. The unit has a planar arrangement, with the carbon atom at the centre and oxygen atoms at the apices of an equilateral triangle. The CO_3 group is considered to be a rigid structural unit, with the C–O bond length in the range 1.25–1.35 Å and the O–C–O bond angle close to the ideal value of 120° (Zemann, 1981). The linkage of the CO_3 groups is made by other atoms which form carbonates, and the effect of them on the configuration of the CO_3 unit is minimal (Effenberger *et al.*, 1981).

There are approximately 150 known carbonate minerals but many of them are relatively rare; some are hydrated, contain hydroxyl or halogen ions, or are compounds with sulphate, phosphate or silicate radicals. The common rock-forming carbonate minerals may be grouped structurally as calcite-, aragonite- and dolomite-type.

Calcite is the most important and also the most abundant carbonate mineral. It forms extensive sedimentary sequences and massive metamorphosed rock bodies. The calcite structure is characterized by alternate layers of Ca atoms and CO_3 groups along the z axis, and between successive layers the CO_3 groups have opposite orientations. Each CO_3 group is surrounded by six Ca atoms, and each calcium atom has six coordinated oxygen atoms. There are six other carbonate minerals that crystallize with the calcite-type structure. They are listed in Table 13 with their crystallographic and optical data. In this isomorphous group, good correlation exists between metal–oxygen bond length and ionic radius, unit cell volume, polyhedral volume of MO_6, and the axial ratio of c/a (Reeder, 1983). Magnesite, siderite, rhodochrosite and smithsonite are all rock-forming carbonate minerals and each has a distinct mode of occurrence. Otavite and gaspéite are relatively rare. Their properties, given in Table 13, are from Kohls and Rodda (1966), Nickel (1973) and Borodin *et al.* (1979).

Aragonite is the high-pressure form of $CaCO_3$ and is much less common. Listed in Table 14 are carbonate minerals which crystallize with the aragonite-type structure. In contrast to the calcite-type structure, the calcium atoms in the aragonite-type structure have a two-layer sequence parallel to {001} instead of the three-layer sequence. The CO_3 groups, as in calcite, lie perpendicular to the z axis but in two distinct planes, and are surrounded by six calcium atoms. Each calcium atom has nine coordinated oxygen atoms.

In the dolomite-type carbonates, dolomite and ankerite are the most important minerals (Table 15). They have a wide and extensive occurrence in sedimentary strata and metamorphosed carbonate rocks. The dolomite structure is related to

Table 13. Calcite-type Carbonate Materials

Name	Formula	Specific gravity	Cell dimensions (Å) *a*	Cell dimensions (Å) *c*	Refractive indices ω	Refractive indices ε
Calcite	$CaCO_3$	2.71	4.989	17.061	1.658	1.486
Magnesite	$MgCO_3$	3.00	4.632	15.012	1.700	1.509
Siderite	$FeCO_3$	3.97	4.691	15.379	1.875	1.635
Rhodochrosite	$MnCO_3$	3.70	4.768	15.635	1.816	1.597
Smithsonite	$ZnCO_3$	4.43	4.652	15.025	1.850	1.625
Otavite	$CdCO_3$	4.96	4.923	16.787	1.828	1.607
Gaspéite	$NiCO_3$	4.39	4.608	14.805	1.930	1.721

Table 14. Aragonite-type carbonate minerals

Name	Formula	Specific gravity	Cell dimension (Å)			Refractive indices		
			a	*b*	*c*	α	β	γ
Aragonite	$CaCO_3$	2.95	4.960	7.964	5.738	1.531	1.681	1.685
Strontianite	$SrCO_3$	3.7	5.090	8.358	5.997	1.517	1.663	1.667
Witherite	$BaCO_3$	4.3	5.312	8.896	6.428	1.529	1.676	1.677
Cerussite	$PbCO_3$	6.55	5.180	8.492	6.134	1.803	2.074	2.076

Table 15. Dolomite-type carbonate minerals

Name	Formula	Specific gravity	Cell dimension (Å)		Refractive indices	
			a	*c*	ω	ε
Dolomite	$CaMg(CO_3)_2$	2.85	4.807	16.003	1.679	1.500
Ankerite	$Ca(Mg,Fe)(CO_3)_2$	3.01	4.811 –4.831	16.042 –16.166	1.690 –1.750	1.510 –1.548
Kutnahorite	$CaMn(CO_3)_2$	3.12	4.873	16.349	1.727	1.535
Minrecordite	$CaZn(CO_3)_2$	3.45	4.818	16.029	1.750	1.550

the calcite structure in that Mg atoms substitute for Ca atoms in every other layer with no distinct changes in the CO_3 layers of the calcite structure. The other two dolomite-type carbonate minerals, kutnahorite and minrecordite, are not common; they occur closely associated with dolomite. Therefore, further consideration of these two minerals is given in the section on dolomite.

Four other carbonate minerals are considered. Huntite is an alteration product of dolomite- or magnesite-bearing rock, malachite and azurite are secondary minerals in the oxidized zone of copper deposits, especially in regions where limestone is present, and nyerereite is restricted to the carbonate lavas erupted by Volcano Oldoinyo in Tanzania (Dawson, 1962).

References

Borodin, V. L., Lyntin, V. I., Ilyukhin, V. V. and **Belov, N. V.,** 1979. Isomorphic series of calcite–otavite. *Dokl. Akad. Nauk. USSR*, **245**, 1099–1101.

Dawson, J. B., 1962. Sodium carbonate lavas from Oldoinyo Lengai, Tanganyika. *Nature*, **195**, 1075–6.

Effenberger, H., Mereiter, K. and **Zemann, J.,** 1981. Crystal structure refinements of magnesite, calcite, rhodochrosite, siderite, smithsonite, and dolomite with discussion of some aspects of the stereochemistry of calcite-type carbonates. *Z. Krist.*, **156**, 233–43.

Kohls, D. W. and **Rodda, J. L.,** 1966. Gaspéite, $(Ni,Mg,Fe)(CO_3)_2$, a new carbonate from the Gaspé Peninsula, Quebec. *Amer. Min.*, **51**, 677–84.

Nickel, E. H., 1973. An occurrence of gaspéite and pecoraite in the Nullagine region of Western Australia. *Min. Mag.*, **39**, 113–5.

Reeder, R. J., 1983. Crystal chemistry of the rhombohedral carbonates. In *Carbonates: Mineralogy and Chemistry*: Ed. R. J. Reeder, *Reviews in Mineralogy*, **11**, 1–48, Mineralogical Society of America, Washington, DC.

Zemann, J., 1981. Zur Stereochimie der Karbonate. *Fortsch. Min.*, **59**, 95–116.

Calcite and Calcite-type Carbonate Minerals

Calcite $CaCO_3$
Magnesite $MgCO_3$
Rhodochrosite $MnCO_3$
Siderite $FeCO_3$
Smithsonite $ZnCO_3$

Calcite $CaCO_3$

Trigonal (−)

ε	1.486
ω	1.658
δ	0.172
Dispersion:	Very strong.
D	2.711
H	3
Cleavage[1]:	$\{10\bar{1}1\}$ perfect.
Twinning[1]:	$\{01\bar{1}2\}$ and $\{0001\}$ common.
Colour:	Usually white or colourless, sometimes grey, yellow, green or blue. Colourless in thin section.
Unit cell:	a_{hex} 4.989 Å, c_{hex} 17.061 Å, $Z_{hex}=6$. a_{rh} 6.375 Å, α 46° 05′, $Z_{rh}=2$. Space group $R\bar{3}c$

Easily soluble with effervescence in cold dilute HCl.

Calcite is very common and widely distributed in the Earth's crust. It occurs in sedimentary rocks as the major constituent of limestone, in many calcareous metamorphic rocks from marbles to gneisses, and in some alkaline igneous rocks. As a secondary mineral calcite is formed in veins and cavities in rocks of all types.

Calcite is one of the three naturally occurring polymorphic forms of $CaCO_3$; the other two are aragonite and vaterite. Most calcite is relatively pure. Common impurities include magnesium, ferrous iron and manganese. The varieties of crystalline forms of calcite exceed those of all other minerals.

The name is from the Latin *calx* (*calcis*), lime.

Structure

The structure of calcite was determined by Bragg (1914) and has been described as a distorted NaCl structure with Ca atoms and CO_3 groups in place of Na and Cl atoms, respectively. The face-centred rhombohedral unit cell is produced by compressing along one of the three-fold axes of the NaCl unit cell to the degree that the edges which meet it make an angle of 101° 55′ with each other and a_{rh} equals 6.42 Å. The CO_3 group has the form of an equilateral triangle, with oxygen atoms at the corners and a carbon atom in the centre, and is oriented perpendicular to the three-fold axis. The three oxygen atoms are about 2.3 Å apart and the C–O distance is about 1.24 Å. This rhombohedral unit cell contains four $CaCO_3$, in

[1]Cleavage, twinning and morphology are indexed using the hexagonal morphological cell with quadrupled *a* parameter (see Table 16).

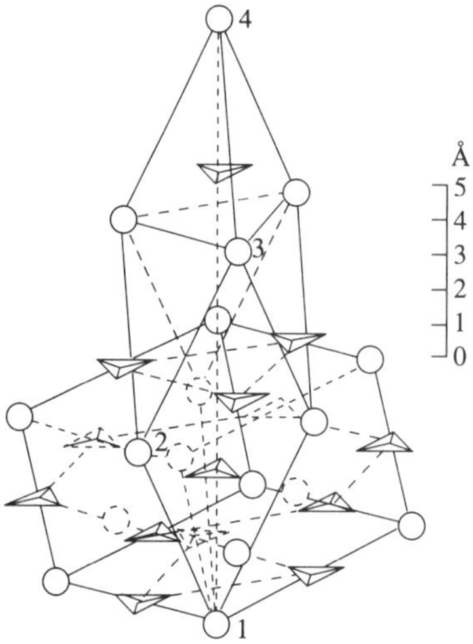

Fig. 28. Face-centred rhombohedral unit cell of calcite containing four $CaCO_3$. Open circles represent Ca atoms and triangles represent CO_3 groups. The elongated rhombohedral unit cell contains two $CaCO_3$ (after Lippmann, 1973).

agreement with the NaCl structure. The shape of the unit cell corresponds well with the cleavage rhomb of calcite.

As shown in Fig. 28, the triangular CO_3 groups have opposite orientations in successive carbonate layers along the three-fold axis. Therefore, a true, face-centred rhombohedral unit cell must have each edge doubled in three dimensions, have eight small pseudo-cells, and contain 32 $CaCO_3$.

A small rhombohedral unit cell which contains two $CaCO_3$ can be chosen to describe the structure of calcite. Ca atoms take the corner and centre positions, and two CO_3 groups lie on the central three-fold axis with opposite orientation. This unit cell has $a_{rh}=6.375$ Å and $\alpha=46°\,05'$; it is illustrated in the elongated rhombohedral cell in Fig. 28. A hexagonal description of the calcite structure was given by Lippmann (1973). As shown in Fig. 29, the Ca atoms and CO_3 groups occupy alternate layers along the z axis, and between successive layers CO_3 groups have opposite orientations. Cell dimensions from the most recent structure refinement are a 4.9896 Å and c 17.0610 Å, with $Z=6CaCO_3$ (Effenberger *et al.*, 1981). A comparison of rhombohedral and hexagonal unit cells of calcite is given in Table 16. In the literature, reports relating to morphology, twinning and cleavage generally use the hexagonal *morphological* cell, while accounts of crystal structure and diffraction patterns relate to the hexagonal *structural* unit cell. The latter cell is also favoured for most microstructural studies involving transmission electron microscopy. If there is a risk of misunderstanding, the cell employed should be stated explicitly.

In the structure, the six Ca atoms are on 6(b) at (0,0,0), six carbon atoms are on 6(a) at $(0,0,\frac{1}{4})$, and 18 oxygens on 18(e) at $(x,0,\frac{1}{4})$ with $x=0.2568$ (Effenberger *et al.*, 1981). The interatomic distances from this refinement are 1.2815 Å for the C–O bond and 2.3598 Å for the Ca–O bond. The edges of the CaO_6 polyhedra inclined to (0001), the edges of the CaO_6 parallel to (0001), and the shortest oxygen–oxygen distance outside the polyhedra are 3.4107, 3.2623 and 3.1883 Å, respectively. Stereoscopic projections of coordinated positions of Ca atoms and

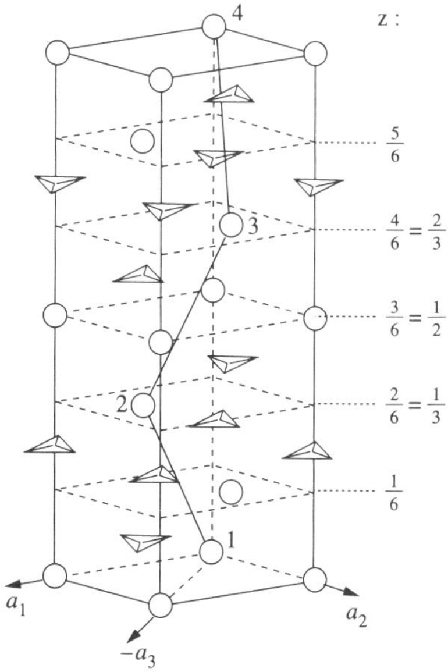

Fig. 29. Hexagonal unit cell of calcite containing six $CaCO_3$. The Ca atoms marked 1, 2, 3 and 4 correspond with those in the face-centred rhombohedral unit cell in Fig. 28 (after Lippmann, 1973).

Table 16. Calcite Unit Cells

		X-ray smallest cell[a]	Cleavage rhomb pseudo cell	Cleavage rhomb true cell[b]
Rhombohedral axes	a_{rh} (Å)	6.37	6.42	12.85
	α_{rh}	46° 05′	101° 55′	101° 55′
	Z_{rh}	2	4	32
	Cleavage rhomb indices	{211}	{100}	{100}
Hexagonal axes	a_{hex} (Å)	~5	~10	~20
	c_{hex} (Å)	~17	~8.5	~17
	Z_{hex}	6	12	96
	Cleavage rhomb indices	$\{10\bar{1}4\}$	$\{10\bar{1}1\}$	$\{10\bar{1}1\}$

[a]Structural cell.
[b]Morphological cell.
For calcite- and dolomite-type carbonates, morphological cell indices are used for twinning and cleavage, and structural cell indices are used for reflections. Unit cell dimensions are calculated from structural cell indices.

CO_3 groups were prepared by Reeder (1983) and are shown in Fig. 30. The vibrational ellipsoids represent the 90 per cent probability electron distribution surfaces. Calculations to determine the proportion of ionic and covalent bonding in the structure of calcite have been reported by Skinner *et al.* (1994) and Lindsay and Jackson (1994).

Using a rigid body model and an isotropic thermal model, high-temperature structure refinement of calcite was carried out by Markgraf and Reeder (1985) in

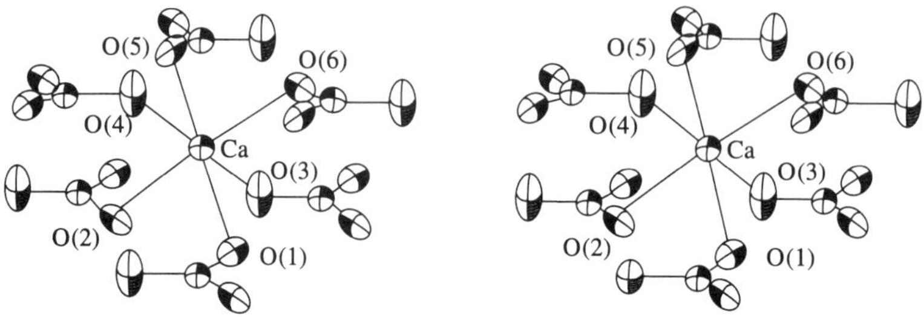

Fig. 30. Stereoscopic view of the coordination of Ca atoms and CO_3 groups. Oxygen atoms are labelled according to the convention described in Reeder (1983).

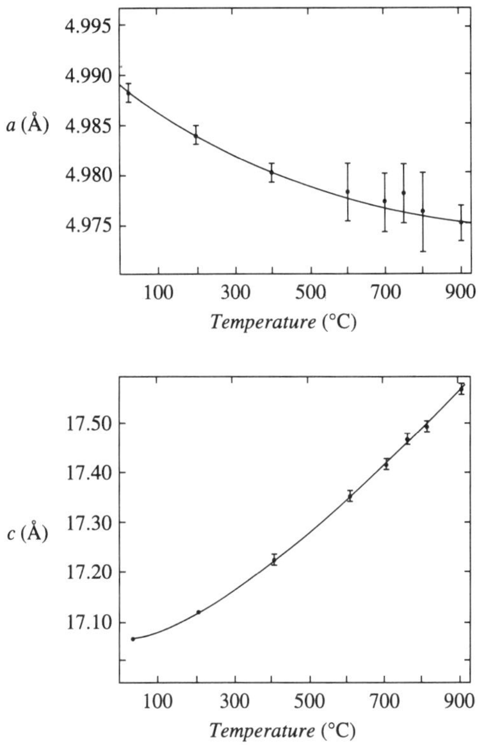

Fig. 31. Variation of cell dimensions with increasing temperature for calcite. Error bars represent two standard deviations (after Markgraf and Reeder, 1985).

the temperature range between 24 and 900 °C. The variation of cell dimensions with increasing temperature is shown in Fig. 31. The negative thermal expansion along the *x* axis is a distinct property of calcite. Both C–O and $(O\text{–}O)_{CO_3}$ show a sharp decrease in the uncorrected bond lengths, which indicates an increased thermal motion, specifically libration. The bond angle of O(1)–Ca–O(2) decreases, whereas those of O(1)–Ca–O(6) and Ca–O–Ca increase from 24 to 800 °C. The CaO_6 octahedra become more distorted with increasing temperature. High-pressure X-ray powder diffraction of $CaCO_3$ by Figuet *et al.* (1994) shows that from 0 to 18 kbar, cell dimensions *a*, *c* and *V* decrease from 4.990 to 4.95 Å, from 17.06 to 16.89 Å and from 368.0 to 359.6 $Å^3$, respectively.

In addition to calcite, there are four other forms of $CaCO_3$ which have been reported to exhibit basic structural features of calcite, designated calcite II and calcite III (Bridgman, 1939) and calcite IV and calcite V (Mirwald, 1979a; Carlson, 1983). Both calcite II and calcite III are high-pressure forms. At temperatures $<200\,^{\circ}C$ the transition pressures are 15 kbar for calcite I$\rightleftharpoons$calcite II and 22 kbar for calcite II$\rightleftharpoons$calcite III (Davis, 1964; Wang, 1968). Both transitions are reversible, and calcite I (the normal calcite) is the end product when the pressure returns to atmospheric conditions. Furthermore, a single crystal of calcite which was subjected to two complete cycles of transition recovered its original orientation. This led Bridgman (1939) to believe that calcite II and calcite III must have structures closely similar to that of calcite I.

Both calcite II and calcite III have lower densities than aragonite, and since both are formed in the stability field of aragonite they are considered to be metastable with respect to aragonite (Jamieson, 1957). A gradual disappearance of the calcite (113) reflection with increasing pressure was noted by Jamieson (1957) and Davis (1964). This reflection is the strongest of a group of reflections proving opposite carbonate ion orientation in successive layers (Mauguin, 1923). Single crystal X-ray data for calcite II were obtained at 21 kbar by Merrill and Bassett (1975). This high-pressure form of $CaCO_3$ is monoclinic, space group $P2_1/c$, with *a* 6.334 ± 0.002, *b* 4.948 ± 0.005, *c* 8.033 ± 0.025 Å, β 107° 54′ and Z=4. The calcite I$\rightleftharpoons$calcite II transition involves a rotation of CO_3 groups by 11°, and a small displacement of adjacent planes of Ca atoms. Liu and Mernagh (1990) found that calcite II was formed more easily under non-hydrostatic than hydrostatic conditions. At room temperature, calcite III was always found to exist at pressures above 20 kbar. High-pressure X-ray powder diffraction studies of both calcite II and calcite III were made by Figuet *et al.* (1994). Cell dimensions of calcite II show no significant changes with pressure between 16 and 18 kbar. For calcite III, *b* changes from 8.37 Å at 23 kbar to 8.27 Å at 68 kbar, whereas *a* decreases from 8.75 Å at 23 kbar to 8.50 Å at 68 kbar. No significant change in *c* could be measured.

Evidence of structural changes of $CaCO_3$ at high temperatures was first given by Boeke (1912). The transition was found to be reversible at $970 \pm 5\,^{\circ}C$, and the crystals grown from $CaCO_3$–CaO melt above the transition temperature have trigonal morphology. Cohen and Klement (1973) have extended the transition to high pressures and showed a boundary line of steep dP/dT slope from 0.25 kbar and 985 °C to 5 kbar and 1000 °C. Further extension of the boundary line to higher pressures was made by Mirwald (1976, 1979a), which confirmed the steep dP/dT slope of Cohen and Klement. At 10 kbar, the slope is about 2 °C/kbar and decreases with increasing pressure. At 25 kbar, it is almost vertical in a *P*–*T* diagram. A triple point with the calcite–aragonite transition boundary has been established at $1030 \pm 20\,^{\circ}C$ and 31 ± 1 kbar. Intensity measurement of the $(11\bar{2}3)$ reflection for the hexagonal structural cell showed a marked decrease in magnitude at temperatures $>700\,^{\circ}C$ (Chang, 1965), even if intensity loss due to the reduced scattering power of Ca is considered. If an extrapolation is made, the temperature of this order–disorder transition in pure $CaCO_3$ would be 955 °C, about 20 °C lower than that originally observed by Boeke (1912). In keeping with Bridgman's nomenclature, Mirwald (1976) named this phase calcite V.

In the course of his differential thermal analysis, Mirwald (1976) also observed breaks in temperature signal (ΔT) in the temperature range 720–820 °C and between 8 and 20 kbar, which suggested the existence of another high-temperature form

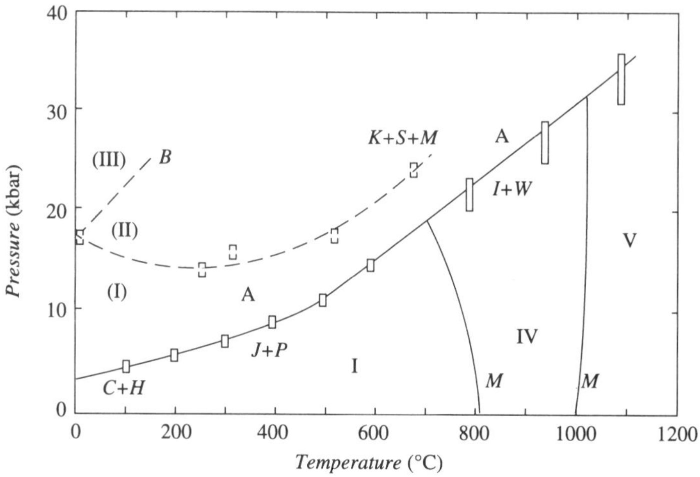

Fig. 32. Generalized phase relationships in the system $CaCO_3$ (after Carlson, 1983). A, aragonite; I–V, polymorphic forms of calcite; C+H, Crawford and Hoersch (1972); J+P, Johannes and Puhan (1971); I+W, Irving and Wyllie (1975); M, Mirwald (1976); K+S+M, Kondo *et al.* (1972); B, Bridgman (1939); dashed lines, metastable equilibrium.

of $CaCO_3$, designated calcite IV. The boundary of the calcite I$\rightleftharpoons$calcite IV transition extends from the triple point with aragonite at 730 ± 20 °C and 18 kbar to 780 °C at 8 kbar. With a linear extrapolation of this boundary line to 1 bar, the transition temperature would be 810 ± 20 °C. Supporting evidence of this transition was given from high-temperature X-ray powder diffraction studies to 900 °C at one atmospheric CO_2 pressure (Mirwald, 1979a). However, high-temperature structure refinement of calcite to 900 °C showed no anomalies in cell dimensions (Markgraf and Reeder, 1985). A generalized phase diagram of $CaCO_3$ is shown in Fig. 32 (Carlson, 1983). The calcite–aragonite equilibrium boundaries used are those of Johannes and Puhan (1971) in the temperature range 400–700 °C and of Crawford and Hoersch (1972) at lower temperatures.

Calcite which commonly occurs in the skeleton of marine organisms and carbonate cements often has appreciable Mg content. Structure refinements of magnesian calcites were made by Althoff (1977) and Paquette and Reeder (1990) on a calcite sample with 10 mol. per cent $MgCO_3$, and two samples with 6.4 and 12.9 mol. per cent $MgCO_3$, respectively. In the structure, the octahedron is slightly distorted, and it appears to be insensitive to the increase of Mg content. The mean Mg^{2+}–O bond length decreases with increasing substitution of Mg for Ca, and the variation with composition is nearly linear. Significant positional disorder exists in the structure. A comparison of some selected crystallographic and geometrical data of magnesian calcite, calcite, magnesite and dolomite is shown in Table 17. The magnesian calcite samples evaluated by Paquette and Reeder (1990) are spine (6.4 mol. per cent $MgCO_3$) and tooth (12.9 mol. per cent $MgCO_3$) materials of a Recent echinoid from Bermuda. Other samples used for comparison are calcite (Markgraf and Reeder, 1985), calcite with 10 mol. per cent $MgCO_3$ (Althoff, 1977), dolomite (Reeder and Markgraf, 1986) and magnesite (Markgraf and Reeder, 1985). The O–*M*–O bond angles within the octahedra are comparable with those in calcite. The variation of cell dimensions with composition for the magnesian calcite has been studied in detail (Goldsmith *et al.*, 1961; Bischoff

Table 17. A comparison of selected crystallographic data of calcite, magnesian calcite, dolomite and magnesite (from Paquette and Reeder, 1990)

	Calcite	Magnesian calcite 6.4 mol. per cent	Magnesian calcite 10.0 mol. per cent	Magnesian calcite 12.9 mol. per cent	Dolomite	Magnesite
Unit cell volume (Å^3)	367.6	362.48	356.60	355.48	320.24	279.3
a (Å)	4.988	4.9673	4.941	4.9382	4.8069	4.635
c (Å)	17.061	16.963	16.864	16.832	16.0034	15.019
c/a ratio	3.420	3.4150	3.413	3.4085	3.329	3.240
Bond length						
M–O bond (Å)	2.3595	2.3466	2.331	2.3282	2.3816	2.1018
C–O bond (Å)	1.281	1.2790	1.276	1.277	1.2858	1.288
Bond angle (°)						
O(1)–M–O(2)	87.44	87.46		87.45	87.69	86.17
O(1)–M–O(6)	92.56	92.54		92.55	92.31	91.83

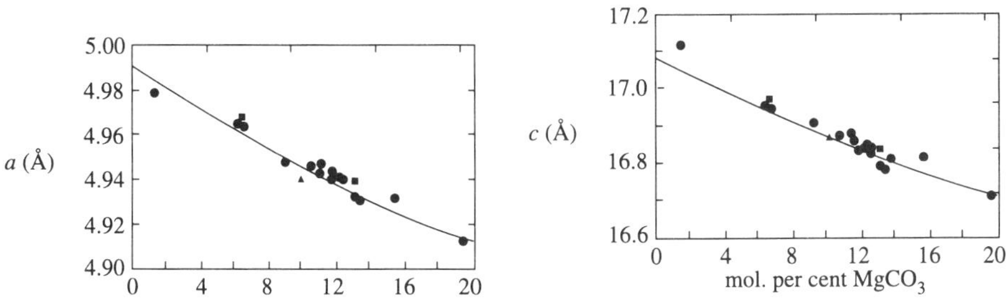

Fig. 33. Variation of unit cell dimensions with $MgCO_3$ content for biogenic magnesian calcites (after Paquette and Reeder, 1990). Symbols used in the figure represent data from three studies: squares, Paquette and Reeder (1990); circles, Bischoff *et al.* (1983); and triangles, Althoff (1977).

et al., 1983; Paquette and Reeder, 1990). In addition to the relationships shown in Fig. 33, quadratic least-squares regressions from synthetic data produced the equations:

$$V = 368.1 - 122x + 131x^2,$$

$$a = 4.996 - 0.50x + 0.56x^2,$$

$$c = 17.069 - 2.27x + 2.1x^2,$$

and

$$c/a = 3.420 - 0.118x + 0.05x^2,$$

where 'x' is the mole. fraction of $MgCO_3$ in calcite (Bischoff *et al.*, 1983).

Vaterite is a metastable form of $CaCO_3$. It is easily precipitated from aqueous solution of calcium salts under favourable conditions of nucleation (Meyer, 1965), but its occurrence in nature is rare (McConnell, 1959; Bentor *et al.*, 1963; Rowland and Webster, 1971). There are reports of biologically produced vaterite in mollusc shells, otoliths of fish and human gallstones (Hall and Taylor, 1971). Vaterite is hexagonal with a 7.16, c 16.98 Å, Z = 12, and space group $P6_3/mmc$ (Kamhi, 1963). In the structure, the planes of the carbonate group are parallel to the c axis and oriented in such a fashion that each oxygen atom is equidistant from the two nearest calcium atoms. Each calcium atom is coordinated by six oxygen atoms at distances of about 2.4 Å, but has two more oxygen atoms at 2.9 Å, close enough to be considered as bonded neighbours. Therefore, the coordination of calcium in vaterite can be regarded as eight-fold. Vaterite changes irreversibly to calcite or aragonite on heating to temperatures $>400\,^{\circ}C$ (DeKeyser and Degueldre, 1950) or on contact with water (Bischoff, 1968). Turnbull (1973), from measurements of the enthalpies of solution of vaterite in 0.1 N HCl under 0.98 bar CO_2 and the solubilities of vaterite and calcite in water under 0.98 bar CO_2, concluded that vaterite is unstable with respect to calcite at 25 °C. Newly precipitated vaterite has been reported to have formed at the expense of carbonate rock in drilling fluids in oil wells in the USA and New Zealand; it occurs as a colourless rim on a light brown core of Ca_3SiO_5, and its origin has been attributed to the solution of original carbonate in the subsurface by natural gas (Friedman and Schultz, 1994).

Chemistry

Most calcite is relatively pure and fairly close to its formula composition. In all, 23 analyses of calcite selected from various geological environments are tabulated in Table 18. Common substitutions for calcium in calcite are magnesium, manganese and ferrous iron, and to a lesser extent strontium and barium.

Magnesian calcite containing up to 30 mol. per cent $MgCO_3$ is an important constituent of skeletons of marine organisms and carbonate cements. According to Chave (1981), the five most common Mg-calcite-secreting organisms are calcareous red algae, foraminifera, bryozoans, echinoderms and barnacles. There is a general trend of decrease in the magnesium content of skeletal Mg-calcite

Table 18. Calcite Analyses (Analyses are Arranged in Order of Sedimentary, Igneous and Metamorphic Calcite)

	1	2	3	4	5	6	7	8
CaO	52.55	55.39	53.72	53.51	50.99	54.64	54.37	53.60
MgO	0.45	0.38	0.33	0.29	0.41	0.40	0.22	1.74
FeO	3.69	0.25	2.75	2.60	1.65	0.60	0.87	0.43
MnO	0.06	—	—	0.06	—	0.51	0.35	—
		0.12 (SrO)						
CO_2[c]	44.03	44.09	44.24	43.91	41.48	44.99	43.66	44.22
Total	100.78	100.23	101.04	100.37	94.53	101.14	99.47	99.99
	Numbers of ions on the basis of six (O)							
Ca	1.873	1.972	1.905	1.912	1.930	1.920	1.954	1.902
Mg	0.022	0.019	0.016	0.014	0.022	0.020	0.010	0.086
Fe	0.103	0.007	0.076	0.073	0.049	0.017	0.026	0.012
Mn	0.002	—	—	0.002	—	0.014	0.010	—
		0.002 (Sr)						
C	2.000	2.000	2.000	2.000	2.000	2.000	2.000	2.000

(c) Calculated from stoichiometry

1–4 Pore-filling, fossil, recrystallized fossil and vein calcite, respectively, in the Avalon Sand, Grand Banks of Newfoundland (Hutcheon *et al.*, 1985).

5–7 Calcite from boreholes and wells of the active geothermal system, Salton Sea, California, USA, at 1176 (195 °C), 1812 (282 °C) and 439 m (190 °C), respectively (McDowell and Paces, 1985).

8 Calcite from the Monterey Formation, California, USA (Burns and Baker, 1986).

Table 18. *Continued*

	9	10	11	12	13	14	15	16
CaO	55.46	50.99	46.53	55.30	55.71	38.79	54.37	55.78
MgO	0.49	4.30	8.10	0.07	0.23	1.87	0.24	0.60
FeO	—	—	—	0.13	0.15	12.18	0.31	0.01
MnO	—	—	—	0.57	0.60	2.49	0.09	0.04
				0.03 (BaO)				
				0.32 (SrO)	0.09 (SrO)			
CO_2[c]	44.05	44.71	45.36	44.07	44.48	41.48	43.18	44.46
Total	100.00	100.00	99.99	100.49	101.26	96.81	98.19	100.89
	Numbers of ions on the basis of six (O)							
Ca	1.976	1.790	1.610	1.970	1.966	1.468	1.958	1.969
Mg	0.024	0.210	0.039	0.003	0.011	0.098	0.030	0.029
Fe	—	—	—	0.004	0.004	0.360	0.009	—
Mn	—	—	—	0.016	0.017	0.075	0.003	0.001
				0.006 (Sr)	0.002 (Sr)			
C	2.000	2.000	2.000	2.000	2.000	2.000	2.000	2.000

9–11 Biogenically produced magnesian calcite (Bischoff *et al.*, 1983).
12, 13 Calcite from carbonatites near Gatineau, Quebec (Hogarth *et al.*, 1985). Samples contain minor amounts of Ce_2O_3.
14 Calcite from the Proterozoic iron formation, Labrador Trough, Canada (Klein, 1978).
15, 16 Calcite from the Blackburn carbonatite dykes, near Ottawa, Ontario, Canada, carbonatitic calcite and xenolithic limestone calcite, respectively (Hogarth *et al.*, 1988).

Table 18. *Continued*

	17	18	19	20	21	22	23
CaO	53.02	51.42	48.41	54.90	49.37	50.32	43.88
MgO	0.48	0.09	1.20	0.06	2.58	0.04	0.41
FeO	2.21	—	—	0.12	3.38	5.21	—
MnO	0.49	2.59	4.57	—	0.65	1.06	10.21
		2.49 (PbO)	1.45 (PbO)				0.69 (BaO)
		0.11 (BaO)	1.18 (BaO)				3.46 (SrO)
		0.18 (oth.)	0.82 (oth.)				
CO_2[c]	43.78	42.57	42.64	43.22	44.03	43.37	42.88
Total	99.98	99.45	100.27	98.30	100.01	100.00	101.53
	Numbers of ions on the basis of six (O)						
Ca	1.900	1.895	1.780	1.993	1.760	1.821	1.606
Mg	0.024	0.005	0.061	0.003	0.128	0.001	0.021
Fe	0.062	—	—	0.003	0.094	0.147	—
Mn	0.014	0.075	0.133	—	0.018	0.030	0.295
		0.023 (Pb)	0.013 (Pb)				0.009 (Ba)
		0.001 (Ba)	0.016 (Ba)				0.060 (Sr)
C	2.000	2.000	2.000	2.000	2.000	2.000	2.000

17 Calcite from the differentiated tholeiitic Golden Mile dolerite sill, Kalgoorlie, Australia (Phillips and Brown, 1987).
18, 19 Calcite from manganese ores at Långan, Sweden (Sundius, 1965).
20 Calcite from contact-metamorphic lapis lazuli, the Italian Mountain deposits, Colorado, USA (Hogarth and Griffin, 1980).
21, 22 Calcite from metamorphosed iron formation, Gagnon region, Quebec (Bather, 1969).
23 Manganoan calcite from metamorphosed Mn deposits, India (Dasgupta *et al.*, 1990).

from the tropics to the poles, which is a direct consequence of temperature and sea-water saturation of carbonate ion (Mackenzie *et al.*, 1983). The range of Mg content and rate of change of Mg with temperature vary from organism to organism. Chave (1954a,b) showed that over a temperature range of 30 °C calcareous red algae skeletons have $MgCO_3$ contents which vary from 5 to 30 mol. per cent, whereas barnacles exhibit a variation of only a few per cent. Mackenzie *et al.* (1983) stated that calcification rates increase with increasing $(CO_3)^{2-}$ concentration, and that lower $(CO_3)^{2-}$ concentrations, both kinetically and thermodynamically, promote Mg-calcite precipitation of lower Mg content. Temperature and $(CO_3)^{2-}$ concentrations are also the factors which control the Mg content in marine calcite cements (James *et al.*, 1976; Schlager and James, 1978), which show decreasing Mg content with increasing latitude and with increasing depth. The occurrence of magnesian calcite in various geological environments was examined by Goldsmith *et al.* (1955), who found calcite solid solutions with $MgCO_3$ content up to 9 mol. per cent in association with dolomite. The variation of cell dimensions with composition for the magnesian calcite is shown in Fig. 33.

Trace amounts of boron in modern marine calcites are in the range 10–70 p.p.m. (Henning and Hansen, 1992); the B has been shown to reside in the crystal structure rather than in fluid inclusions, isolated BO_3^{3-} anions probably replacing CO_3^{2-} via a process of coupled substitution (Sen *et al.*, 1994).

The most important carbonate system for the experimental study of phase relationships is $CaCO_3$–$MgCO_3$. On the basis of experimental data produced by Harker and Tuttle (1955b), Graf and Goldsmith (1955, 1958), Goldsmith and Heard (1961), Goldsmith and Newton (1969), Irving and Wyllie (1975), Byrnes and Wyllie (1981) and Fanelli *et al.* (1983), Anovitz and Essene (1987) have constructed a calcite–dolomite solvus (Fig. 34), which is considered the best fit combining reversals from all available experimental data. In Fig. 34, the head of each arrow represents the final composition, and the direction of the arrow illustrates the

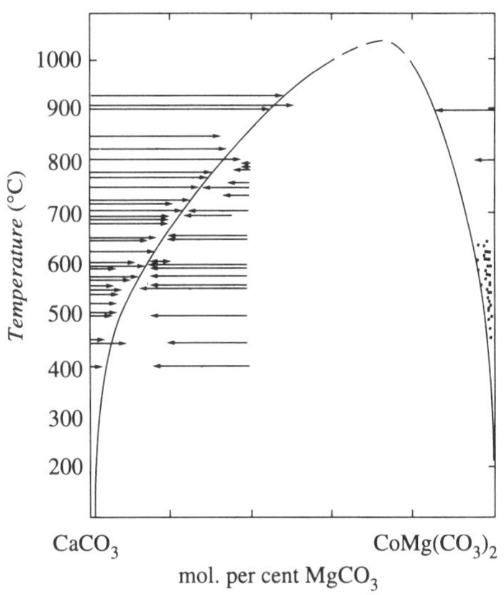

Fig. 34. **The calcite–dolomite solvus fitted to reversals compiled from published data (after Anovitz and Essene, 1987).**

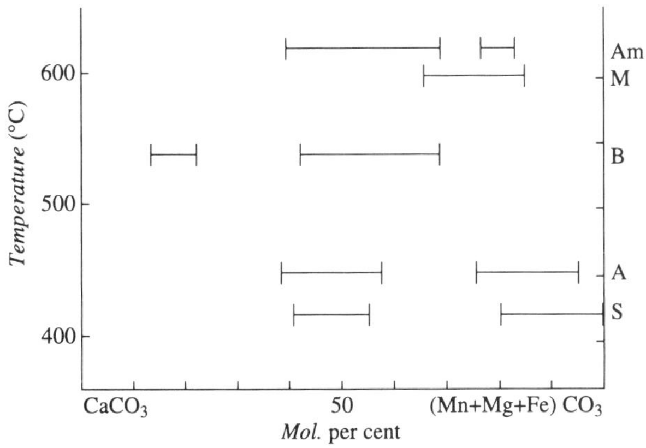

Fig. 35. Ranges of composition of manganese-rich carbonates in occurrences with different grades of metamorphism (after De Capitani and Peters, 1981). A, Alagna in northern Italy; S, Scerscen, Bernina; B, Buritirama, Paraña; M, Muretto Pass, Engadine; and Am, Serra do Navio, Amapa.

direction of the compositional shift. The two points on the calcite limb at 200 °C and below are compositions of natural calcite (Talantsev, 1976).

Manganese-bearing calcites were first described by Krieger (1930) to have a compositional range of 0–42 mol. per cent $MnCO_3$. Sundius (1965) reported Mn-calcites from the manganese ores at Långban, Sweden. In deep-sea sediments, Mn-bearing calcites have been found as precipitates on foraminifera tests (Boyle, 1983), and as authigenic carbonate in Panama Basin sediments (Pedersen and Price, 1982) with a composition of $(Mn_{0.48}Ca_{0.47}Mg_{0.05})CO_3$. Other manganese calcites have been recovered from shallow coastal sediments. Lynn and Bonatti (1965) observed a manganous calcite–calcian rhodochrosite association in an ash layer at depth in reducing sediments from the Guatemala Basin, off Guatemala coast, north Pacific ocean; an empirical formula, $(Mn_{0.5-0.8}Ca_{0.2-0.5})CO_3$, was assigned to it. Studies of Mn-calcite metamorphosed in greenschist facies (Peters *et al.*, 1978) showed the existence of a miscibility gap in the Mn-rich carbonates. Metamorphosed under low to medium amphibolite facies, however, Mn-carbonates in the calc-silicate marble have compositions ranging over the complete series (Peters *et al.*, 1977). As shown in Fig. 35, De Capitani and Peters (1981) demonstrated the effect of different degrees of metamorphism upon the chemical compositions of natural Mn-rich carbonates. Temperature has a greater impact than pressure, and the $FeCO_3$- and $MgCO_3$-contents are very minor. For the carbonates studied, the Muretto Pass (Engadine, Switzerland) occurrence was metamorphosed in the hornblende–hornfels facies, the Serra do Navio (Amapá, Brazil) deposit in the medium to upper amphibolite facies and the Alagna and Scerscen (northern Italy) occurrences in the greenschist facies. In addition, Winter *et al.* (1981) illustrated that there is a complete series without a break between rhodochrosite and kutnahorite composition for metamorphosed Mn-carbonates from Bald Knob, North Carolina, USA, suggesting temperatures >550 °C. The experimentally determined solvus (Goldsmith and Graf, 1957; De Capitani and Peters, 1981) is shown in Fig. 36. For synthetic samples of composition $(Ca_{0.5}Mn_{0.5})CO_3$, Goldsmith (1983) stated that reflections in X-ray powder diffraction data indicating ordering were not observed. Such reflections are generally weak, and cannot be identified even in some naturally occurring

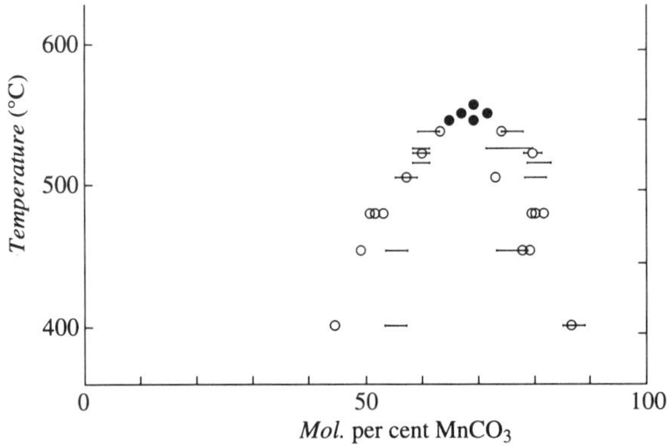

Fig. 36. Experimentally determined solvus for the system $CaCO_3$–$MnCO_3$ (after De Capitani and Peters, 1981). Bars and solid circles represent the spread of experimental data and homogeneous solid solutions, respectively; open circles are data from Goldsmith and Graf (1957).

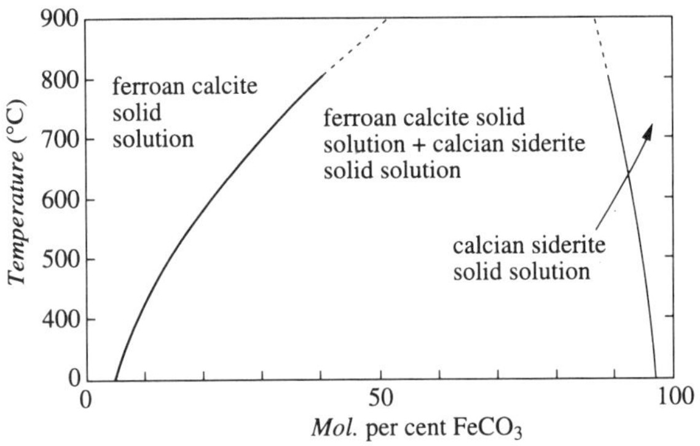

Fig. 37. Experimentally determined solvus for the system $CaCO_3$–$FeCO_3$ (after Goldsmith, 1983).

kutnahorite. Peacor *et al.* (1987) confirmed this observation. Thermal disordering in kutnahorite, as examined by single-crystal diffraction, starts at approximately 450 °C. A metastable solvus was calculated from temperature-dependent excess free energy functions by Capobianco and Navrotsky (1987). The immiscibility gap exists below 355 °C, with compositions ranging between 30 and 55 in mol. fraction of $MnCO_3$ at 300 °C to nearly 20–60 at 100 °C, as estimated from their phase diagram.

Calcite in its natural occurrences containing 2–3 per cent $FeCO_3$ is very common. It has been found, for example, in carbonate cements in Avalon Sand, Grand Banks of Newfoundland (Hutcheon *et al.*, 1985), in Golden Mill dolerite sill (Phillips and Brown, 1987), in the metamorphosed iron formation of the Gagnon Region, Quebec (Bather, 1969), in the Gunflint Formation, Minnesota–Ontario (Floran and Papike, 1978) and the Labrador Trough (Klein, 1974, 1978). In the last occurrence, $FeCO_3$ contents are up to a maximum of 12 per cent in calcite. Experimental studies of phase relationships in the system $CaCO_3$–$FeCO_3$ were made by Goldsmith *et al.* (1962) and Rosenberg (1963), and their results were used to plot the solvus shown in Fig. 37. Based upon these experimental data, a solvus

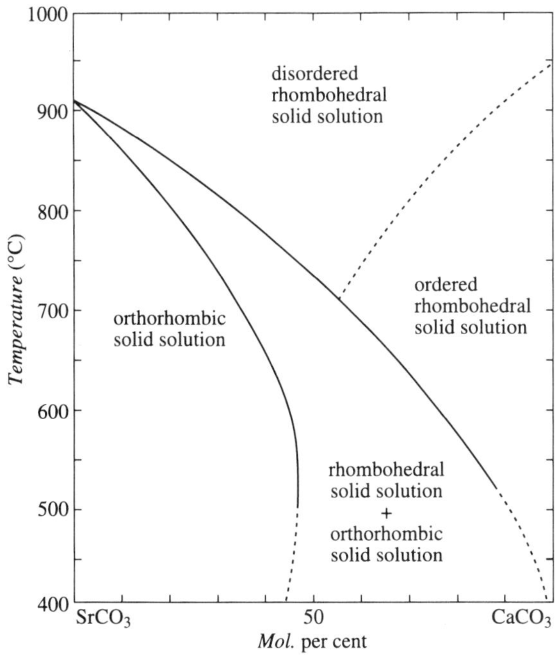

Fig. 38. Phase relationships in the system $CaCO_3$–$SrCO_3$ (after Chang, 1965).

fit was calculated by Anovitz and Essene (1987). They stated that the solvus is reasonably well constrained between 325 and 550 °C, but less well located at higher temperatures. The Fe analogue of dolomite has not been synthesized, nor is it known in nature.

Small amounts of strontium and barium may also substitute for calcium in calcite, but mostly in the p.p.m. range. Calcite cements from the Lower Cretaceous Avalon Sand of the Hibernia field, Newfoundland, have Sr and Ba concentrations ranging from 2.71 to 9.21 and 3102 to 4825 p.p.m., respectively (Hutcheon *et al.*, 1985). In the same field, some of the calcite fossils have a SrO content of 0.12 per cent (Table 18, anal. 2). The calcite in a carbonatite near the Loch Borralan intrustion, Assynt, Scotland, has up to 2 wt. per cent SrO (Young *et al.*, 1994). At Långban, Sweden, calcite from the manganese ores carries as high as 3.63 per cent BaO (Sundius, 1965). Calcites from carbonatites often contain relatively higher amounts of SrO and BaO, e.g. carbonates near Gatineau, Quebec (Hogarth *et al.*, 1985) (Table 18, anals. 12 and 13). Experimental studies of phase relationships in the systems $CaCO_3$–$SrCO_3$ and $CaCO_3$–$BaCO_3$ under low pressures in the stability region of calcite were made by Chang (1965). As shown in Fig. 38, the solubilities of $SrCO_3$ in calcite are 15.5 and 3.5 mol. per cent at 550 and 450 °C, respectively. At 350 °C no detectable amount was found. The dashed lines in the diagram represent the disordering temperature of $CaCO_3$ established by high-temperature X-ray diffraction. Fig. 39 illustrates the phase relationships in the system $CaCO_3$–$BaCO_3$. Below 600 °C, no significant amount of $BaCO_3$ was observed in $CaCO_3$. Among the polymorphic forms of $CaBa(CO_3)_2$, only barytocalcite is a stable phase in the system and it has a disordering temperature of 520 °C.

Direct solubility measurements for calcite by Plummer and Busenberg (1982) in CO_2–H_2O solutions between 0 and 90 °C indicate the following values of the

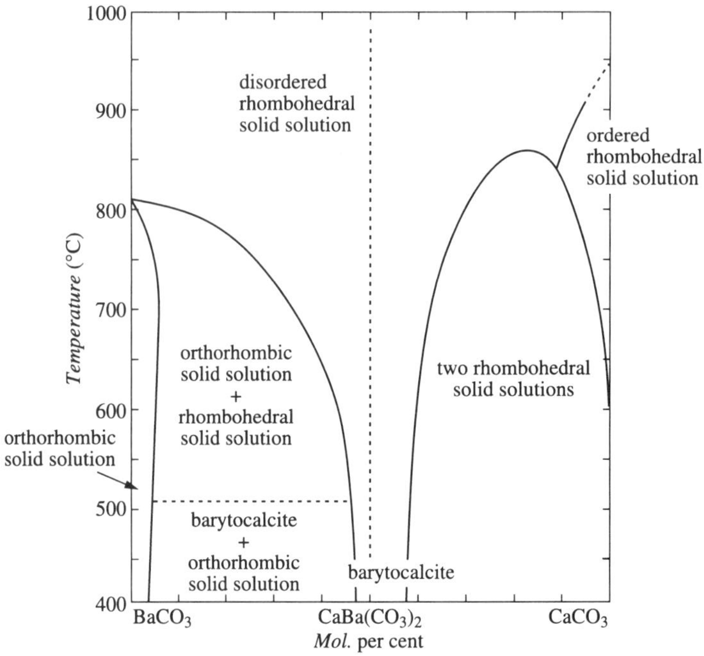

Fig. 39. Phase relationships in the system $CaCO_3$–$BaCO_3$ (after Chang, 1965).

equilibrium constant K for the reaction $CaCO_{3(s)} = Ca^{2+} + CO_3^{2-}$: $\log K_c = -171.9065 - 0.077993T + (2839.319/T) + 71.5951 \log T$, where T is in Kelvin. At 25 °C, the range of measured P_{CO_2} is between 0.9476 and 0.9497 atm, of total calcium is between 9.07 and 9.15 mmol./kg of solution, of calculated pH is between 6.015 and 6.018, and of $\log K_c$ is between -8.477 and -8.487. The solubility of calcite in sea-water of 35‰ at 25 °C and atmospheric pressure was measured by Morse *et al.* (1980); an apparent solubility constant of $4.36 \pm 0.20 \times 10^{-7}$ mol.^{2}/kg^2 was determined. This value is 22 per cent less than the solubility constant calculated from thermodynamic equilibrium constants and the total ion activity coefficient in sea-water. As explained by Morse *et al.* (1980), this may be due to the formation of a surface layer of different composition and lower solubility than the bulk solid. Chou *et al.* (1989) have observed that the rate of dissolution of calcite is proportional to the activity of H^+ at low pH and becomes constant in the neutral range. At high pH, the rate of dissolution decreases due to backward precipitation.

The precipitation reaction of calcite in sea-water solutions can be satisfactorily described by an empirical rate law of the form: $\log R = n \log(\Omega - 1) + \log k$, where n is the empirical reaction order, k is the rate constant, and Ω is the degree of supersaturation with respect to calcite (Zhang and Mucci, 1989). As shown in Fig. 40, calcite precipitation rates in sea-water solutions do not vary appreciably with salinity over the range 5–44‰. Zhang and Mucci further stated that the amount of Mg^{2+} incorporated in calcite overgrowth precipitated from sea-water solutions is independent of the precipitation rate. Wray and Daniels (1957) studied the formation of $CaCO_3$ polymorphs in highly supersaturated solutions of Ca^{2+} and CO_3^{2-} ions. They reported that aragonite was predominantly precipitated at 70 °C and vaterite was precipitated at 30 °C. Ogino *et al.* (1987) described the formation and transformation of calcium carbonate in water by the following

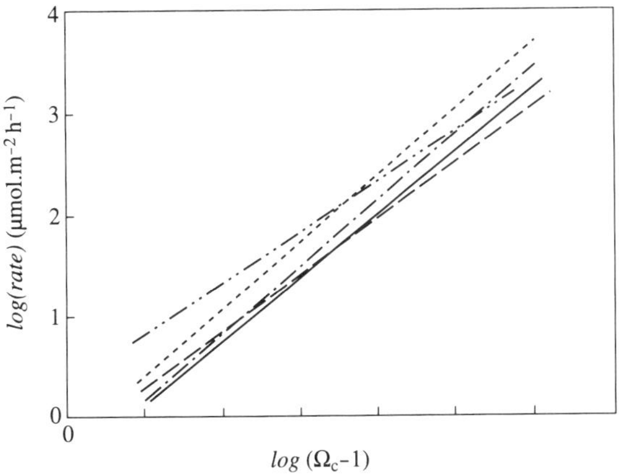

Fig. 40. Precipitation kinetics of calcite in artificial sea-water at various salinities (S in ‰). S=5, 15, 25, 35 and 44 are represented by dashed, bar-dot-bar, bar-dot-dot-bar, bar and solid lines, respectively. Ω_c is the degree of supersaturation with respect to calcite (after Zhang and Mucci, 1989).

sequence: amorphous calcium carbonate rapidly precipitates from highly supersaturated solutions of Ca^{2+} and CO_3^{2-}, and transforms to a mixture of several crystalline calcium carbonate polymorphs within several minutes. At temperatures between 14 and 30 °C, the transformed polymorphs are vaterite and calcite; at temperatures between 60 and 80 °C, calcite and aragonite are the products. At intermediate temperatures all three polymorphs are formed. Both vaterite and aragonite are metastable and are gradually transformed to the stable form, calcite. For the transformation of vaterite to calcite, it takes about 200 minutes at 25 °C and 370 minutes at 30 °C; 1000–1300 minutes are required for the transformation of aragonite to calcite in the temperature range 60–80 °C. The transformation of vaterite and aragonite to calcite takes place through dissolution of the metastable phase and growth of the stable phase, calcite. The rate-determining step of the transformation is the growth of calcite.

The melting of calcite in the presence of excess CO_2 was determined by Irving and Wyllie (1975) and Huang and Wyllie (1976). At 27 kbar, 1.5 wt. per cent CO_2 dissolved in molten $CaCO_3$, depressing the melting point of $CaCO_3$ from 1610 to 1505 °C. The dissociation curve terminates at an invariant point where melting begins at 40 bar and 1230 °C for CaO–CO_2. Wyllie and Tuttle (1960) and Wyllie and Boettcher (1969) determined the melting relationship of $CaCO_3$ in the system CaO–CO_2–H_2O. For the reaction $CaCO_3 + Ca(OH)_2$ = liquid, the univariant melting curve is almost linear and is very insensitive to pressure. Points measured are 655 °C at 4 kbar, 659 °C at 15.5 kbar, 662 °C at 25 kbar and 666 °C at 40.7 kbar. For the reaction $CaCO_3 + Ca(OH)_2$ + vapour = liquid, the melting point decreases with increasing pressure: 630 °C at 4 kbar, 565 °C at 15.5 kbar, 537 °C at 25 kbar and 524 °C at 40.8 kbar. In both reactions, with increasing pressure the $CaCO_3$ phase changes from calcite to aragonite. Phase relationships in the system CaO–CO_2–H_2O at 1 kbar determined by Wyllie and Tuttle (1960) and modified by Wyllie and Boettcher (1969) are shown in Fig. 41. The major features are the liquidus surface divided by field boundaries into three fields: L(portlandite), L(calcite) and L(lime). Thus, any mixture with a bulk composition lying within the triangle CaO–$CaCO_3$–$Ca(OH)_2$ can exist as a liquid, and after cooling and crystallization the final liquid has the composition E_2. Similarly, any mixture with

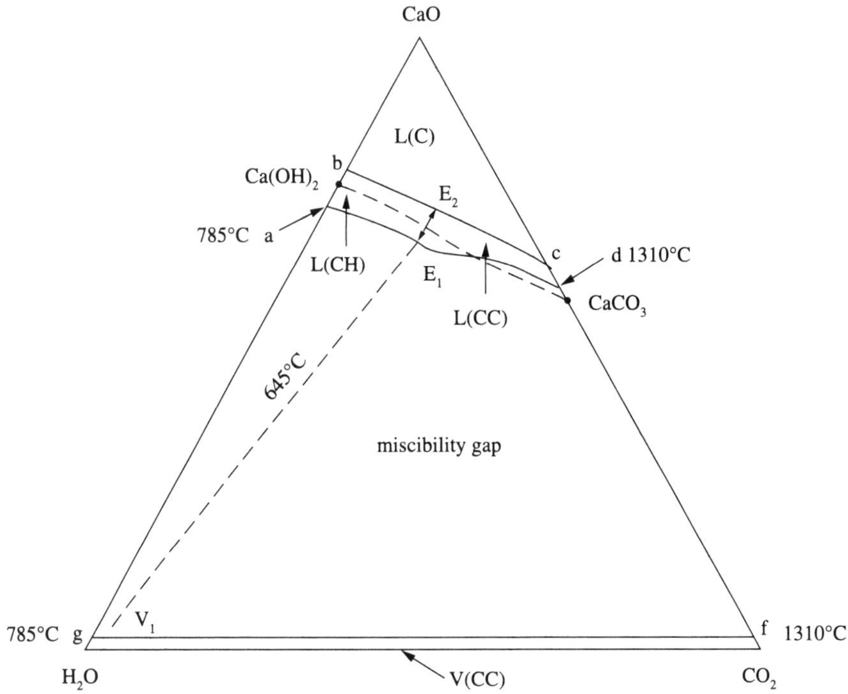

Fig. 41. Phase relationships in the system $CaO–CO_2–H_2O$ at 1 kbar determined by Wyllie and Tuttle (1960) and modified by Wyllie and Boettcher (1969). L(C), L(CH) and L(CC) are liquidus surfaces with CaO, $Ca(OH)_2$ and $CaCO_3$, respectively, as primary crystalline phases.

a composition within the triangle $CaCO_3$–$Ca(OH)_2$–H_2O can exist as a liquid, or as a mixture of liquid and vapour, and after cooling and crystallization the final liquid has the composition E_1. The development of liquid at moderate temperatures and pressures has important petrological implications.

The dissociation of calcite was determined by Harker and Tuttle (1955a) according to the reaction $CaCO_3 = CaO + CO_2$ between 7 and 35 bar. The univariant pressure (CO_2)–temperature curves for the dissociation pass through ~985 and 1100 °C. Due to its ease of solubility in slightly acid waters, the alteration of calcite is accomplished mainly by solution and replacement. It is commonly replaced by quartz, and less commonly by hematite, goethite and many other minerals.

Thermodynamic properties of calcite have been the subject of many measurements. Values of heat capacity and of derived thermodynamic properties at 298.15 K are as follows (Staveley and Linford, 1969): $C_p = 19.95$ cal/kmol.; $S^\circ_T - S^\circ_0 = 21.92$ cal/kmol.; $(H^\circ_T - H^\circ_0)/T = 11.61$ cal/kmol.; and $-(G^\circ - H^\circ_0)/T = 10.31$ cal/kmol. All properties listed were measured in the temperature range 10–300 K, and all have their values increased with increasing temperature.

Calcium carbonate is calcined to produce quicklime, CaO; further reaction with water produces slaked lime, $Ca(OH)_2$. As an industrial mineral, calcite is virtually the only important filler in the processing industry.

Optical and Physical Properties

Calcite is uniaxial and optically negative with extreme birefringence. As a consequence, interference colours in thin section are typically pale high-order colours, even if the optical axis is close to vertical. The structural dependence of birefringence of calcite and other carbonates has been the subject of many investigations. Lo (1973) computed the cation and anion polarizability for all carbonates and considered there to be no established trend between birefringence and cation/anion polarizations or with the cation species. He concluded that the polarizability of the polyatomic anion, such as the CO_3 group in crystals, is dependent on the nature of the constituent atoms as well as on the crystalline environment in which they are found. Isherwood and James (1976) suggested that the birefringence and oxygen bond length are influenced by the electronegativity of the cation. They proposed that the stronger interaction of the Ca^{2+} with the CO_3 in the plane of the CO_3 ion, in comparison with, for example, the Cd^{2+}–CO_3 bond in $CdCO_3$, results in the electron around the oxygen being distorted towards the Ca^{2+} ion and this reduces the overlap of the electron atomic orbitals of the C–O bond. The distortion will effectively increase the separation of the three oxygens in the CO_3 group, thus producing a reduction in the dipole–dipole coupling and consequently a decrease in birefringence. Weber (1988) developed a model to relate birefringence to anisotropic bond polarizabilities. A bond is assumed to have longitudinal and transverse polarizabilities, β^L and β^T, from which the average bond polarizability (β) and the anisotropic part of bond polarizability ($\Delta\beta$) can be derived. The average bond polarizability gives a better description of average refractive index than models using ionic polarizabilities, and the optical birefringence arises from $\Delta\beta$.

Indices of refraction of calcite vary in an approximately linear manner with $MgCO_3$, $MnCO_3$ and $FeCO_3$ contents, and are shown in Fig. 42 (Kennedy, 1947). The index $(n')_{\varepsilon(10\bar{1}1)}$ for a fragment lying on the cleavage surface is 1.566 for calcite. Barbaron (1949) measured the refractive indices in the temperature range between 18 and $-200\,^{\circ}C$. With decreasing temperature, there is a slight decrease in ω and a marked linear decrease in ε, resulting in an increase in birefringence. The ratio between the birefringence and the dispersion of birefringence has been suggested as a useful diagnostic feature (Winchell and Meek, 1947): for calcite this ratio has a value of 23.7, while for dolomite it is 27.3. Loupekine (1947) has devised a graphical method by which ε for trigonal carbonates may be derived rapidly from the values of ω and ε' from such cleavage fragments: ε' is related to ε by the equation:

$$\varepsilon' = \varepsilon\omega/(\omega^2 \sin^2 \phi + \varepsilon^2 \cos^2 \phi)^{1/2},$$

where ϕ is the angle between the z axis and the direction representing the optical axis of the microscope, which is normal to one of the rhombohedral cleavages: for calcite $\phi = 44^{\circ}\,37'$, and the angle varies for the other trigonal carbonates to $42^{\circ}\,57'$ for smithsonite.

Calcite may sometimes exhibit an abnormal biaxiality and show uneven extinction. Walker and Parsons (1925) attributed this to an inversion of high-temperature calcite of igneous origin to normal calcite during cooling. The biaxial calcite has $2V_\alpha$ close to 15°. Gillson (1927) related the formation of biaxial calcite (with $2V_\alpha$ from 4 to 14°) to deformational disturbance. Paulitsch (1956) noted an apparent relationship between the development of lamellar twinning and the biaxial calcite. Rath and Pohl (1971) pointed out that the value of the optic axial angle depends on the dimensions of the lamellae of similar orientation and on the twin angle. Nakissa and Paulitsch (1972) proposed that the degree of metamorphism

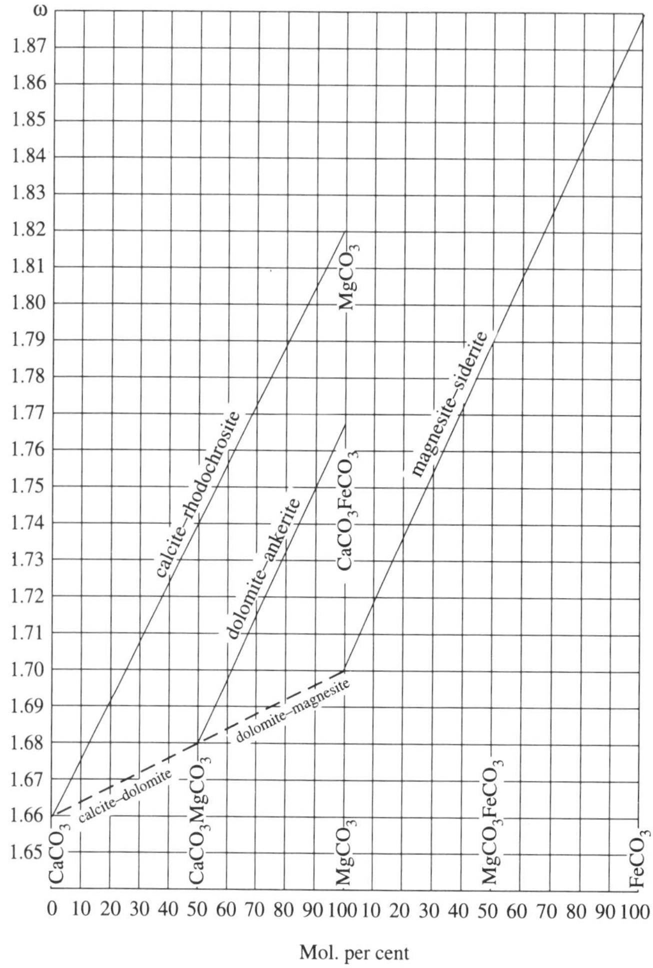

Fig. 42. Variation of ω with composition in the rhombohedral carbonates (after Kennedy, 1947).

of the carbonates can be deduced from the optic axial angle of biaxial calcite. Boettcher and Wyllie (1967) produced biaxial calcite from aragonite experimentally. Biaxiality in calcite with $2V_\alpha < 10°$ is widespread in metamorphic rocks of all grades and in carbonatites (Turner, 1975). In many cases, it is not directly related to visible $\{01\bar{1}2\}$ twinning.

Calcite when pure is colourless or white, but colour differs widely with compositional substitution or with mechanically included materials. The chemically pure and optically clear, colourless variety is known as Iceland spar. Dark bluish green calcite in weathered Deccan traps owes its colour to fine dusty inclusions of chloritic material (Kilpady and Adyalkar, 1954). Calcite containing appreciable amounts of manganese has pale purple to deep rose-red colours (Omori and Yamaoka, 1954). Red calcite has been described from cavities in limestones near thermal springs on the Danube and from Alpine crevices, Hohe Tauern, Austria (Haberlandt, 1944), where the colouring matter has been shown to be gallium- and iron-porphyrin, respectively. Vendl and Mándy (1958) considered that the blue and green colours of calcite are due to the ferric and ferrous iron content of the mineral. The pale-green calcite from a limestone quarry near Cairo, Egypt, contains both ferrous iron and manganese, but the very pale-pink shade due to the presence of manganese is masked by the colour of iron.

Many calcites possess fluorescent, thermoluminescent and cathodoluminescent characteristics. A comprehensive review of luminescence is given by Przibram (1956) and Barnes (1958). Handin *et al.* (1957) have reported that single crystals of calcite which had been subjected to γ radiation changed from amber to blue when the greatest principal stress applied during experimental deformation was parallel to the optic axis; they reported that no change occurred when the least principal stress was in this direction. Compression along z yields a new peak in the glow curves (at 280°) not present in undeformed irradiated calcite. Fonda (1940) synthesized calcite with 3.3 mol. per cent $MnCO_3$, which gave a strong pink fluorescence; with more Mn the fluorescence decreases. Haberlandt (1949) has reported investigations of 30 specimens of calcite and has tabulated their colour, crystal habit, paragenesis, thermoluminescence and fluorescence spectra for rare earth elements before and after ignition. He recorded a violet calcite which shows lines of Nd in the absorption spectrum. Medlin (1968a,b) described in detail the basic thermoluminescence in calcite, and Marfunin (1979) provided a literature review of Mn-activated cathodoluminescence in calcite. With ultraviolet radiation many calcites show fluorescence at various wavelengths. Gies (1975) has characterized calcites with three types of activated fluorescence: (a) white fluorescence activated by Zn content or due to lattice defects, (b) the Mn-activated red fluorescence with Pb as a sensitizer, and (c) the *REE*-activated flesh-coloured fluorescence. Carlson (1955), using synthetic calcite, has shown that the addition of the stannous ion activates a bright white fluorescence. The additional presence of Bi, La and Nd, Mn, and Mg alters the Sn^{2+} fluorescence to green, pale yellow, deep orange and bluish green, respectively.

An investigation of the ultraviolet luminescence of Iceland spar from the Siberian platform revealed the presence of two types of luminescence centre related to Ce^{3+} impurity ions in natural calcite crystals (Kas'yanenko *et al.*, 1993). The predominance of one or other type depends on the oxidation–reduction potential of the mineralization environment and reflects its variability; centres of the type typical for crystals formed under oxidizing conditions are related to ionized oxygen–Ce^{3+} pairs forming in the lattice of the calcite crystals.

On heating, calcite has both natural and artificially induced thermoluminescence curves. Generally, the inducement is made by irradiating the sample in a ^{60}Co γ-cell. Lewis (1956) demonstrated that the peak heights vary regularly with composition in the calcite–dolomite range. Nambi and Mitra (1978) examined the thermoluminescence characteristics of calcites from three different environments, hydrothermal, sedimentary and metamorphic, with a Mn content of 27, 2900 and 1200 p.p.m., respectively. Features of their natural and artificially induced thermoluminescence glow curves relate approximately to relative temperatures of formation, decreasing from metamorphic to hydrothermal to sedimentary calcite. The intensity of the low-temperature peak (110 °C) in relation to other peaks present in the thermoluminescence curves is related directly to the Mn content (Siegel, 1963). The sedimentary calcite, with the highest Mn content among the three samples, yields the lowest value for the ratio of 330/110° peaks. Pacey (1986) measured the thermoluminescence characteristics of calcite in English chalks and fossils, which show systematic changes in peak intensities and intensity ratios. Pacey suggested that these characteristics are related to diagenetic history: partial recrystallization of the chalk due to pressure solution and total replacement of fossil shells by secondary calcite.

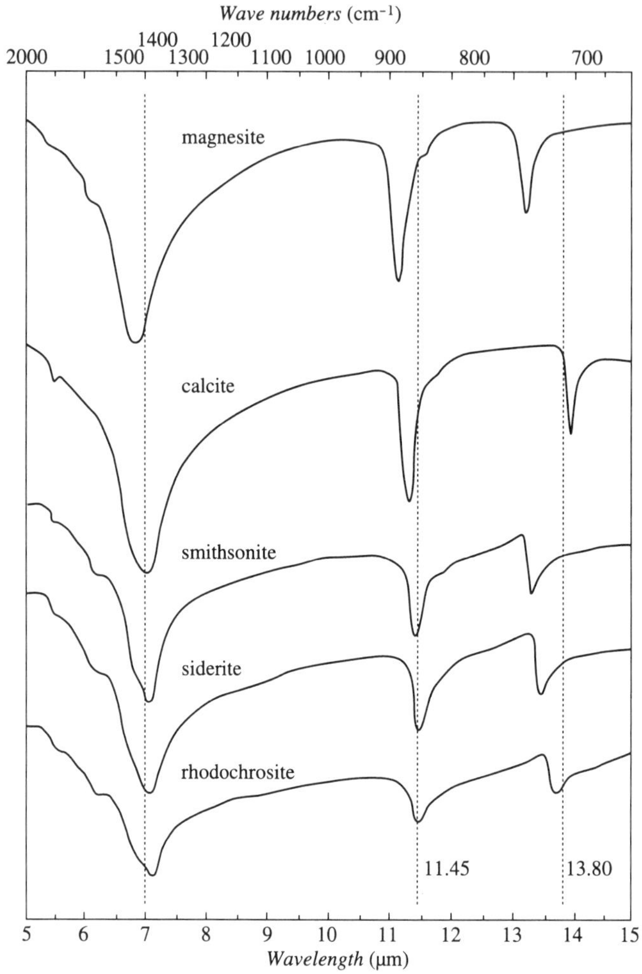

Fig. 43. Infrared spectra of calcite-type carbonates (after Adler and Kerr, 1963).

Under electron beam excitation calcite produces cathodoluminescence, which was used by Wackden and Berry (1984) to study the crystal growth of calcite during diagenesis. Four types of syntaxial overgrowth were recognized: spiral, needles, contouring and idiomorphic. Dorobek and Watkinson (1988), in their analysis of strain microfabrics in deformed limestones, revealed the varied and heterogeneous nature of microstructures within calcite visible only by cathodoluminescence.

Infrared spectra of calcite are well established (Adler and Kerr, 1963; Chester and Elderfield, 1967; White, 1974). As shown in Fig. 43, predominant absorption bands above 600 cm^{-1} are attributed to the fundamental vibration of the planar CO_3 group, which occurs in all carbonates. White (1974) discussed the effect of cation substitution on the infrared spectra of carbonates and stated that the presence of larger and heavier cations in place of smaller and lighter cations produces frequency decrease. Raman spectra of calcite have also been measured by many investigators (see White, 1974). The doubly degenerate asymmetric stretching, ν_3, is the strongest band in the infrared spectrum, but it is very weak in the Raman spectrum. The symmetric stretching frequency, ν_1, is the strongest band in the Raman spectrum. Bischoff *et al.* (1985) have used Raman spectra to show that positional disordering of the carbonate ions is greater for biogenic magnesian calcite than for synthetic magnesian calcite of similar composition. Raman spectra of calcite II and calcite III were examined under hydrostatic (Liu and Mernagh, 1990) and non-hydrostatic (Fong and Nicol, 1971) pressure environments. The Raman spectrum of calcite II is in general similar to that of calcite I, while two distinct types of Raman spectrum of calcite III were observed by Liu and Mernagh (1990).

Physically, calcite is one of the most beautiful crystalline minerals known. Its crystals are common, and sometimes very large. A large volume of data are available on the physical properties of calcite. Earlier data were summarized by Birch *et al.* (1942), and discussed in a review by Graf and Lamar (1955).

Calcite crystals are extremely varied in habit, often highly complex, and over 300 well-established forms have been identified. Two major types of habit are: (a) prismatic–scalenohedral, in which blunt to sharper pointed crystals have prism faces such as $\{10\bar{1}0\}$ and scalenohedral faces such as $\{21\bar{3}1\}$ and (b) rhombohedral, in which rhombohedral faces are prominent and furnish blocky crystals. The specific gravity of calcite varies with chemical composition. It increases to 2.793 from 2.711 with 6 mol. per cent FeO in solid solution (Túcan, 1937) and to 2.775 with 7 mol. per cent MnO (Rost and Tuček, 1944). Calcite with PbO and BaO (Table 18, anals. 18 and 19) has a specific gravity of 2.95. The hardness on the other hand, is controlled by crystallographic directions, varying from $2\frac{1}{2}$ on $\{0001\}$ to $3\frac{1}{4}$ on the surface parallel to the z axis (Tertsch, 1934). In correlating Mohs scale with Vickers' hardness numbers, Taylor (1949) reported that for a surface perpendicular to z the Vickers' number was 105, parallel to z it was 145, while on a cleavage surface it was 136.

Calcite is thermally anisotropic. The directions in which the linear expansion is zero lie around the trigonal axis in a cone with a semiangle of 64° 26′ (Graf and Lamar, 1955). Coefficients of thermal expansion of calcite measured by a high-temperature X-ray method (Rao *et al.*, 1968) are $\bar{\alpha}_a$ $-3.7\times10^{-6}/°C$ and $\bar{\alpha}_c$ $25.6\times10^{-6}/°C$. However, results from structure refinement at high temperatures by Markgraf and Reeder gave $\bar{\alpha}_a$ $-2.8\times10^{-6}/°C$ and $\bar{\alpha}_c$ $32.3\times10^{-6}/°C$. Markgraf and Reeder (1985) ran their measurements to 900 °C, which is 376 °C above the upper temperature limit of Rao *et al.*'s experimentation. The negative thermal expansion of '*a*' becomes less negative at these higher temperatures. The calcite structure opens slightly with increased temperatures, as illustrated by the change of the *M*–O–*M* bond angle: ~0.1°/100 °C (Markgraf and Reeder, 1985). The mean thermal expansion coefficient for the octahedral volume is $47.2\times10^{-6}/°C$.

Bridgman (1925) first determined the linear compressibility of calcite and observed that the compressibility of calcite III is abnormal in that it is higher than that of calcite, despite the fact that calcite III is the denser form. The fractional changes in volume at 5 and 10 kbar are 0.0066 and 0.0130, respectively. The deformation of calcite single crystals, when subjected dry at room temperature to a constant confining pressure of 2000 or 2750 bar at a strain rate of 1 per cent per minute, was investigated by Handin *et al.* (1957). They demonstrated that the principal deformation mechanisms were twin gliding on $\{01\bar{1}2\}$ and translation gliding on $\{10\bar{1}1\}$. Vaidya *et al.* (1973) obtained compressibility data for calcite very similar to those of Bridgman (1925, 1939). They were not able to measure the compressibility of calcite II because it exists over only a small pressure range. By using a piston-cylinder apparatus, Singh and Kennedy (1974) measured the compressibility of calcite to 40 kbar. Calcite I was found to transform to calcite II at 14.5 kbar with a volume change of 0.0048 cm^3/g. The volume compressibility data for three calcite polymorphic forms can be described by the following quadratic relationships:

$$-(\Delta V/V_0)=(-14.07\times10^{-4}P)+(5.107\times10^{-6}P^2) \text{ for calcite I,}$$

$$-(\Delta V/V_0)=0.0382-(67.80\times10^{-4}P)+(1.291\times10^{-4}P^2) \text{ for calcite II,}$$

and

$$-(\Delta V/V_0)=-0.0476-(19.35\times10^{4}P)+(17.34\times10^{-6}P^2) \text{ for calcite III,}$$

where P is pressure in kbar.

The variations of the dielectric constant of calcite with frequency and temperature were measured by Rao and Rao (1968) in two crystal orientations with the electric field parallel and perpendicular to the optic axis. The dielectric constant at 25 °C in both orientations is frequency-independent. As a function of temperature, the dielectric constant exhibits two distinct regions: (a) frequency-independent between −193 and 50 °C and (b) frequency-dependent between 50 and 450 °C. The electric conductivity of calcite cleavage fragments was measured by Mirwald (1979b) using direct current between 300 and 1200 °C. He identified five temperature intervals of different conductivity (in Ω^{-1} cm^{-1}): 7×10^{-7} between 300 and 1500 °C, 7×10^{-2} between 400 and 725 °C, 6×10^{-5} between 725 and 800 °C, 1×10 between 800 and 935 °C, and 3×10^3 between 935 and 1200 °C. Three of the four discontinuities in conductivity appear to be clearly related to the high-temperature phase transitions in $CaCO_3$. Measurements in the low-temperature region between 300 and 500 °C produced data with large variations and no systematic trend.

Twinning is common in calcite, and four well-defined laws are distinguished. Twin and composition plane on $\{01\bar{1}2\}$ forms a pair of individual crystals diverging from each other at an angle close to 120°. It is this type of twinning that gives the main lamellar twinning in calcite, which produces striae that lie parallel to the edge and to the long diagonal of the cleavage rhomb. Twin and composition plane on $\{0001\}$ introduces an equatorial plane of symmetry indicated by re-entrant angles around the middle of the crystal. Twin and composition plane on $\{10\bar{1}1\}$ produces heart-shaped twins, and on $\{02\bar{2}1\}$ produces crystals diverging from each other at an angle of about 45°. The most common deformation twinning is on $\{01\bar{1}2\}$, for which the shear displacement is parallel to the intersection of the twin plane with the cleavage plane $\{10\bar{1}1\}$ (Turner *et al.*, 1954). Interpenetrating twins are more likely to occur in calcite which is not so extensively twinned.

Distinguishing Features

Calcite is characterized by perfect rhombohedral cleavages $\{10\bar{1}1\}$, extremely high birefringence, variable relief and effervescence with dilute HCl. The refractive indices of calcite are lower than those for other carbonates, and ε is appreciably below the index of the standard mounting material. Calcite commonly shows lamellar glide twinning. In a cleavage rhombohedron the twin lamellae lie parallel to the long diagonal of the rhomb, and not to the short diagonal as in dolomite and ankerite. Twinned crystals of calcite may also be distinguished from dolomite by

Table 19. Etching and Staining Characteristics of Calcite and Dolomite (from Adams *et al.*, 1984)

Mineral	Effect of etching	Stain colour with alizarin red S	Stain colour with potassium ferricyanide	Combined result
Calcite (non-ferroan)	Considerable (relief reduced)	Pink to red-brown	None	Pink to red-brown
Calcite (ferroan)	Considerable (relief reduced)	Pink to red-brown	Pale to deep blue, depending on iron content	Mauve to blue
Dolomite (non-ferroan)	Negligible (relief maintained)	None	None	Colourless
Dolomite (ferroan)	Negligible (relief maintained)	None	Very pale blue	Very pale blue (appears turquoise or greenish in thin section)

their extinction angle, measured between the fast direction ε' and the trace of the twin lamellae (Williams *et al.*, 1982). It is $>55°$ in calcite and usually between 20 and 40° in dolomite. In a crystal which shows two sharply defined sets of twin lamellae, ε' lies in the obtuse angle of intersection in calcite, but in the acute angle in dolomite. Phillips (1971) described a universal stage method to distinguish calcite, dolomite and magnesite. Stevens and Carron (1948) have recommended an abrasion pH, which has a value of 8 for calcite, 9–10 for dolomite and 10–11 for magnesite. A gasometric procedure was proposed by Wolfe and Bartlett (1958) in which calcite on treatment with 0.5 N acetic acid releases all its CO_2 in about 15 minutes, whereas dolomite requires >3 hours for complete reaction.

Differential staining is widely used in the distinction of calcite from other carbonates. The method in general is based upon the fact that organic dyes stain calcite in acid solutions, and dolomite and magnesite in basic solutions. Extensive reviews of this method have been given by Rodgers (1940), Hügi (1945), Friedman (1959), Warne (1962) and Dickson (1966). For routine application, both Friedman and Warne recommend a combination of two different stains, alizarin red S and Feigl's solution, to differentiate calcite, dolomite, aragonite, high-magnesian calcite, gypsum and anhydrite. Adams *et al.* (1984) have summarized the characteristics of staining procedures, which are shown in Table 19. Rodgers (1940) recommends copper nitrate, $Cu(NO_3)_2$, with ammonia, which gives calcite a bluish green stain and will not affect dolomite. Other organic stains specific for calcite are Harris' hematoxylin and Trypan blue. Dickson (1966) used potassium ferricyanide to distinguish ferroan calcite.

Where both calcite and dolomite are present in a rock, their relative proportions may be determined by X-ray powder diffraction (Tennant and Berger, 1957). It involves a comparison of either integrated intensities or ratios of intensities of selected reflections from the two minerals in the unknown mixture with standard calibration curves. The analysis is rapid and can be precise, but a number of factors may contribute to erroneous results. Evaluations of their methods have been given by Runnells (1970) and Royse *et al.* (1971). Gensmer and Weiss (1980) suggested procedures to improve the accuracy.

Paragenesis

Calcite is a very common and widely distributed mineral in the Earth's crust. It is an important rock-forming mineral in sedimentary rocks, and also occurs in metamorphic and basic igneous rocks. It is a common mineral in the late stages of hydrothermal deposition and is found in vughs or fissures.

Sedimentary rocks. In sedimentary rocks, calcite occurs in the form of shells and skeletons of marine organisms and as a primary precipitate from solution. It is the principal constituent of limestone, chalk, travertine, oolite and other carbonate sedimentary rocks. $CaCO_3$ precipitated in marine waters or freshwater lakes and the shells of many marine organisms is in the form of aragonite, but aragonite transforms in time to calcite, the stable form at low temperature. Calcite is a common cementing material in detrital sediments. Fine-grained calcite occurs with clay minerals in marl and calcareous shale. It is also common in cavities, sometimes as geodes, and in cave deposits, often as stalactites and stalagmites. Hot springs and streams may precipitate calcite tufa.

The crystallography of calcite in echinoids has been examined by Raup (1959, 1962, 1966). Varying from species to species, the z axis of calcite has either one of two distinct orientations: normal to the plate and tangential meridional. In brachiopods, sections through valves show a two-layered wall (Williams, 1966). The primary layer is made of fine fibres of low-magnesian calcite, elongated perpendicular to the shell surface; the secondary layer, also of low-magnesian calcite and lying under the primary layer at low angles to its inner surface, is made of long and slender fibres. The optic axis of the former is parallel to the elongation of the fibres; that of the latter is not (Brunton, 1969). Molluscan shells consist of complicated aggregates of layers (Oberling, 1964). Each layer differs from other layers in structure, orientation of structure and mineralogy. In any one altered species, the layers are either all aragonite or interlayered aragonite and calcite which may be of a high- or low-magnesian type. The Recent calcareous algae that are important limestone builders consist of both aragonite and calcite, whereas fossil calcareous algae are all calcite (Johnson, 1961).

The separation of the foraminifera into suborders is based upon mineralogy and crystal fabric of the test wall. The crystal fabric of calcite is described as microgranular, perforate or spiculate (Towe and Cifelli, 1967). Foraminifera and pelagic plants such as coccolithophores dominate the formation of deep-sea $CaCO_3$ in the present-day ocean. Both contain low-magnesian calcite. For shoal-water organisms, intermediate- to high-magnesian calcite and aragonite are the principal constituents, as summarized in Table 20.

The precipitation of $CaCO_3$ from aqueous solutions is controlled by the equilibrium:

$$CaCO_3 + CO_2 + H_2O \rightleftharpoons Ca^{2+} + 2HCO_3^-.$$

This reaction is very sensitive to changes in the amount of dissolved CO_2, and CO_2 concentration in solution is dependent on a variety of influences. Much surface water and groundwater are practically saturated with $CaCO_3$, and any slight alteration in external conditions will cause precipitation or dissolution in such waters. Precipitation is favoured where the water is warm and where CO_2 is being lost through evaporation or photosynthesis, as on the shallow banks off Florida and the Bahamas.

Calcium carbonate also occurs in sediments as subspherical grains, generally referred to as ooliths. Sediments that are dominantly composed of such material

Table 20. Mineralogy and Chemistry of Major Shoal-water Organism Groups (from Morse and Mackenzie, 1990)

Type of organism	Most common mineralogy	Range in chemical composition: $CaCO_3$ (mol. per cent)	$MgCO_3$ (mol. per cent)	$SrCO_3$ (mean (mol. per cent)
Foraminifera	Calcite	77–99	< 1.2–19	0.249
Calcareous sponges	Calcite	71–85	0.08–0.7	0.079
Scleractinian corals	Aragonite	98–99	5.9–16.7	0.911
Octacorals	Calcite	73–99	0.4–19	0.264
Echinoids	Calcite	78–92	4.8–19	0.229
Crinoids	Calcite	83–92	8.3–19	0.167
Asteroids	Calcite	84–91	10.7–19	0.154
Ophiuroids	Calcite	83–91	0.2–13.1	0.262
Bryozoa	Calcite, aragonite	63–97	0.2–13.1	0.262
Calcareous brachiopods	Calcite	89–99	0.6–10.7	0.130
Phosphatic brachiopods	Chitinophosphatic	7–8	2.4–8.3	—
Annelid worms	Calcite, aragonite	83–94	7.1–20.2	0.484
Pelecypods	Calcite, aragonite	98.6–99.8	0–3.6	0.177
Gastropods	Calcite, aragonite	96.6–99.9	0–2.4	0.160
Cephalopods	Aragonite	93.8–99.5	Trace–0.4	0.337
Crustaceans	Calcite, calcium phosphate	29–83	1.2–19	0.392
Calcareous algae	Calcite, aragonite	65–88	8.3–35	0.221

are called oolites, which are common constituents of modern shoal-water tropical and subtropical carbonate-rich sediments. In ooliths, calcite, high-Mg-calcite and aragonite are all present (Land *et al.*, 1979). Three hypotheses exist for the formation of ooliths: the bacterial–mechanical formation hypothesis (Oppenheimer, 1961; Cloud, 1962; Wilson, 1967; Folk, 1973), the algal formation hypothesis (Fabricius, 1977; Bathurst, 1976) and the direct chemical formation hypothesis (Berner, 1971; Land *et al.*, 1979).

Calcite is also formed in sedimentary rocks as a cementing medium by diagenetic processes. An idealized scheme has been proposed by Land *et al.* (1967) from their study of Bermudan limestones. They divided the development into five stages. The initial stage is marked by the unconsolidated lime sand consisting of aragonite (molluscan, coral and hallimed debris), high-magnesian calcite (foraminiferids, coralline algae and echinodermal fragments) and low-magnesian calcite (lithoclasts). The second stage is a stage of cementation, which is marked by the appearance of low-magnesian calcite cement on the surface of the sand grains. The third and fourth stages are marked, respectively, by the disappearance of the high-magnesian calcite and the dissolution of aragonite. The final stage is marked by a limestone consisting of low-magnesian calcite only. In a study of the Pleistocene limestones of Barbados, Pingitore (1970) found that 80 per cent of the volume of the sparry calcite could be attributed to the dissolution of the coral skeleton and reprecipitation. A particular form of calcite as cementing medium in sandstones may occur in which large crystals of calcite contain many thousands of sand grains, a famous example being the sand calcite of the Fontainbleau district in the Paris basin. In other cases, the calcite may be in crystals of up to 1 cm across, which cause the rock to break preferentially along the calcite cleavage planes giving rise to a phenomenon known as lustre mottling, as each crystal unit gives a more or less brilliant reflection when viewed from an appropriate angle.

The formation of calcite by the devitrification of volcanic glass in sea-water was reported by Zen (1959), who studied the possible equilibrium relationship between calcite and kaolinite. Lublinite, with type locality in Lublin, Poland (Iwanoff, 1906), is a special variety of calcite. It consists of colourless needles and fibres displaying an extinction angle of 40–50°. Stoops (1976) reported a lublinite with en échelon stacking of equidimensional calcite crystals with *z* axes mutually parallel, so that all parts of the aggregate extinguish simultaneously. The $CaCO_3$ in pearls is generally in the form of aragonite, but Alexander (1941) reported pearls of calcite. Veins of fibrous calcite, sometimes known as 'beef', occur in some shales, and have the fibres parallel to *z*. The veins may show a cone-in-cone structure due to the sensitivity of the rhombohedral cleavage of the fibres to shear.

Metamorphic rocks. When sedimentary calcite in limestones undergoes metamorphism, a relatively small amount of overburden is sufficient to prevent its breakdown and the escape of CO_2. In the normal course of events the calcite (or aragonite) merely recrystallizes to form a marble. Calcite is thus a relatively common mineral in calcareous sediments which have been thermally or regionally metamorphosed. When the original sediments contained other materials in addition to $CaCO_3$, or when the metasomatic introduction of further elements occurs during metamorphism, the calcite may react and give rise to various mineral assemblages. When SiO_2 is present, the reaction $CaCO_3+SiO_2 \rightleftharpoons CaSiO_3+CO_2$ takes place with the formation of wollastonite. This reaction has been investigated extensively by Harker and Tuttle (1956), Greenwood (1967), Gordon (1971) and Kridelbaugh (1973) because of its geological interest. Isobaric changes of equilibrium temperature with change of composition of the co-existing gas phase (H_2O+CO_2) are shown in Fig. 44. With an increase in CO_2 concentration (or P_{CO_2}), this reaction

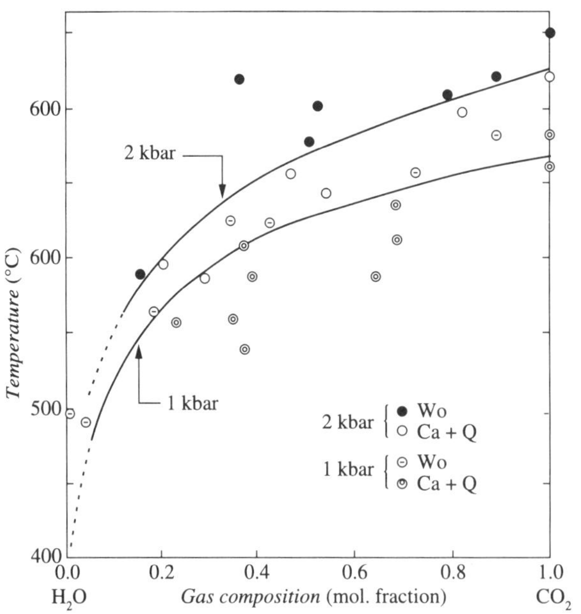

Fig. 44. Stability of calcite, quartz and wollastonite in supercritical mixtures of H_2O and CO_2 (after Greenwood, 1967). Curves drawn through experimental data points were calculated assuming ideal mixing of H_2O and CO_2.

temperature will increase. If, on the other hand, there is a reduction of P_{CO_2}, either by the presence of another volatile such as H_2O or by the escape of CO_2 through fissures, the calcite may react at low temperature. In the study of kinetics of the reaction calcite + quartz $\rightleftharpoons$ wollastonite + CO_2, Kridelbaugh (1973) established the activation energy to be in the range 25.5–29.0 kcal/mol. When the fluid phase is CO_2, granular wollastonite is formed with nucleation occurring on the quartz grain. If H_2O is present in the fluid phase, the wollastonite has its characteristic fibrous habit, and it nucleates on the calcite. Calcite may further react with wollastonite at higher temperatures to produce the rare minerals spurrite and tilleyite (Tuttle and Harker, 1957; Zharikov and Shmulovich, 1969).

The common presence of hydrous silicates indicates that water vapour is often active during the metamorphism of carbonate rocks in addition to the CO_2 produced by decarbonation reaction. Metamorphism of a siliceous carbonate rock consisting of both calcite and dolomite causes the formation of hydrous phases such as talc and tremolite (Gordon and Greenwood, 1970; Skippen, 1971):

$$\underset{\text{dolomite}}{3CaMg(CO_3)_2} + \underset{\text{quartz}}{4SiO_2} + H_2O \rightarrow \underset{\text{talc}}{Mg_3Si_6O_{10}(OH)_2} + \underset{\text{calcite}}{3CaCO_3} + 3CO_2$$

and

$$\underset{\text{talc}}{5Mg_3Si_4O_{10}(OH)_2} + \underset{\text{calcite}}{6CaCO_3} + \underset{\text{quartz}}{4SiO_2} \rightarrow \underset{\text{tremolite}}{3Ca_2Mg_5Si_8O_{22}(OH)_2} + 6CO_2 + 2H_2O.$$

Extensive metasomatic zones may occur around some intrusions. At Crestmore, California, USA, where brucite marbles were intruded by an epizonal plug of quartz monzonite porphyry (Burnham, 1959), four skarn mineral zones were developed with increasing concentration of silica from the marble inward to the porphyry in the following order: forsterite zone, monticellite zone, idocrase zone and garnet zone. Calcite and brucite react with silica in the hydrothermal solution to form monticellite, which marks the transition from the forsterite to the monticellite zone. Further reactions produce spurrite and merwinite, and with the introduction of Al melilite is formed. Reactions of monticellite, spurrite, merwinite and melilite with silica produce vesuvianite, which in turn reacts with silica to form garnet, diopside and wollastonite. This series of reactions indicates progressive silicification and decarbonation of the country rocks.

Prehnite and pumpellyite form from the Ca-zeolites in the presence of calcite:

$$\underset{\text{laumontite}}{CaAl_2Si_4O_{12}\cdot 4H_2O} + \underset{\text{calcite}}{CaCO_3} \rightarrow \underset{\text{prehnite}}{Ca_2Al_2Si_3O_{10}(OH)_2} + \underset{\text{quartz}}{SiO_2} + 3H_2O + CO_2.$$

This dehydration–decarbonation marks the transition from the zeolite to the prehnite-pumpellyite facies (Boles and Coombs, 1977).

In the subsea-floor hydrothermal system, initial sea-water has relatively high concentrations of O^{2-}, Cl^-, Br^-, HCO_3^-, CO_3^{2-} and SO_4^{2-}. During penetrative convection into magmatic rocks, reactions between Ca-silicates and carbonate anions produce calcite (Spooner and Fyfe, 1973).

Igneous rocks. Calcite occurs in alkaline igneous rocks, notably carbonatites, kimberlites (Meyer, 1979), lamprophyres (Rock, 1977) and nepheline–clinopyroxene rocks (Bailey, 1974). Sundius (1958) described a calcite-bearing pegmatite. Carbonatites are fine- to coarse-grained igneous rocks which characteristically contain >50 per cent by volume carbonate minerals, especially calcite. The igneous origin of carbonatites had long been debated, but the observations by Dawson (1962, 1966) of active carbonatite volcanism in the crater of Oldoinyo Lengai in

East Africa ended the discussion. It is now accepted that carbonatites can be intrusive and extrusive. Experimental studies in the system $CaO–CO_2–H_2O$ by Wyllie and Tuttle (1960) illustrated that the addition of water at 1 kbar lowers the solidus temperature in the system $CaO—CO_2$ from >1200 to about 740 °C, i.e. into the same temperature range as for granitic liquids. Cooper *et al.* (1975) studied the phase relationships in the system $Na_2CO_3–K_2CO_3–CaCO_3$ and compared their results with the mineralogy of the Oldoinyo Lengai lavas. The lava (erupted 1960) contained small phenocrysts of two-carbonate solid solutions, $Na_2CO_3+K_2CO_3$, with and without $CaCO_3$ in the groundmass of the same phases. In other carbonatites, the alkalis are in the silicates rather than the carbonates. Cooper *et al.* (1975) concluded that carbonatite magmas probably contained higher proportions of Na and K, but they were lost with escaping water with decreasing pressure. By using Oldoinyo Lengai lavas as starting materials in their experiments, Freestone and Hamilton (1980) found that with increasing P_{CO_2} and decreasing temperature, the immiscibility gap between silicate and carbonate liquids becomes wider, and calcium becomes enriched in the carbonate liquid at the expense of alkalis.

Most carbonatites occur in composite plutons with coeval silicate rocks, which are silica-undersaturated and usually mafic (Le Bas, 1987). In the carbonatite-bearing pluton of the Homa Bay area, western Kenya, the earliest carbonatites are calcitic, and were followed by cone sheets and dykes of ankeritic carbonatites. The final stage was marked by veins of baryte, calcite, fluorite and quartz. The Jacupiranga carbonatite plug, Brazil (Gaspar and Wyllie, 1987), consists of five intrusions: the first, third and fourth are calcitic, the second contains both calcite and dolomite, and the fifth is dolomitic. A decreasing trend of Mg/Fe is observed in associated magnetite in the carbonatite emplaced during the second to fifth intrusive events. The most widely publicized occurrence of carbonate-rich rock in kimberlite is that of late dykes cutting kimberlite in the Premier Mine, South Africa. These dykes contain about 50 per cent calcite, with serpentine and magnetite (both pseudomorphous after olivine), and traces of apatite and phlogopite. Robinson (1975) interpreted them as products of residual liquid from the crystallization of kimberlite magma.

Carbonatite in small hypabyssal intrusions is exemplified by dykes of the Maimecha–Kotui region, north Siberia (Zhabin and Cherepivskaya, 1965). These are vertical dykes with widths up to 0.5 m. They consist of 60–80 per cent euhedral tabular calcite phenocrysts set in a groundmass of apatite, calcite and dolomite. A plug of carbonatite containing calcite, magnetite, monticellite and periclase is exposed at the centre of the summit crater of Kerimasi volcano (Marino and Roeder, 1983).

Calcite frequently crystallizes as a gangue mineral in many kinds of hydrothermal vein and as a secondary mineral with quartz and zeolite in amygdales in basic igneous rocks. One of the best known localities for the large crystals of calcite is from within basalt at Helgustadir, on Eskifjord, Iceland, the Iceland spar. Calcite is also found as a deuteric alteration product illustrated by the reaction:

$$\underset{\text{plagioclase}}{2CaAl_2Si_2O_8}+4H_2O+2CO_2\rightleftharpoons\underset{\text{kaolinite}}{Al_4Si_4O_{10}(OH)_8}+2CaCO_3.$$

In the severely shocked areas of the Houghton impact crater, Canada (where peak shock reached 50–60 Gpa), carbonates form around 10 vol. per cent of the sample. They are mainly calcite with $\delta^{13}C\leqslant+9‰$; the $\delta^{18}O$ values range from +15 to +20‰ and are ~5‰ lower than in unshocked reference material (Martinez *et al.*, 1994).

References

Adams, A. E., MacKenzie, W. S. and **Guilford, C.**, 1984. *Atlas of Sedimentary Rocks under the Microscope*: London (Longman) and New York (John Wiley & Sons), 104 pp.

Adler, H. H. and **Kerr, P. F.**, 1963. Infrared absorption frequency trends for anhydrous normal carbonates. *Amer. Min.*, **48**, 124–37.

Alexander, A. E., 1941. The mineralogy of five unusual natural pearls. *Gemmologist*, **10**, 93–4.

Althoff, P. L., 1977. Structural refinements of dolomite and a magnesian calcite and implications for dolomite formation in the marine environment. *Amer. Min.*, **62**, 772–83.

Anovitz, L. M. and **Essene, E. J.**, 1987. Phase equilibria in the system $CaCO_3$–$MgCO_3$–$FeCO_3$. *J. Petr.*, **28**, 389–414.

Bailey, D. K., 1974. Nephelinites and ijolites. In *The Alkaline Rocks*: Ed. H. Sørensen, New York (John Wiley & Sons), 622 pp.

Barbaron, M., 1949. Constantes optique de la calcite à basse température. *C. R. Avad. Sci. Paris*, **229**, 875–6.

Barnes, D. F., 1958. Infrared luminescence of minerals. *U.S. Geol. Surv., Bull.*, **1052C**, 157 pp.

Bather, P., 1969. Mineral compositions and equilibria in the metamorphosed iron-formation of the Gagnon region, Quebec. *J. Petr.*, **10**, 56–101.

Bathurst, R. G. C., 1976. *Carbonate Sediments and their Diagenesis, Developments in Sedimentology*, **12** (2nd edn.), Amsterdam (Elsevier Publishing Co.), 658 pp.

Bentor, Y. K., Gross, S. and **Heller, L.**, 1963. Some unusual minerals from the 'Mottled zone' complex, Israel. *Amer. Min.*, **48**, 924–30.

Berner, R. A., 1971. *Principles of Chemical Sedimentology*. New York (McGraw-Hill), 240 pp.

Birch, F., Schairer, J. F. and **Spicer, H. C.**, 1942. Handbook of Physical Constants. *Geol. Soc. Amer., Spec. Paper*, **36**, 325 pp.

Bischoff, J. L., 1968. Catalysis, inhibition and the calcite–aragonite problem. II. The vaterite–aragonite transformation. *Amer. J. Sci.*, **266**, 80–90.

Bischoff, W. D., Bishop, F. C. and **MacKenzie, F. T.**, 1983. Biogenically produced magnesian calcite: inhomogeneities in chemical and physical properties; comparison with synthetic phases. *Amer. Min.*, **68**, 1183–8.

Bischoff, W. D., Sharma, S. K. and **MacKenzie, F. T.**, 1985. Carbonate ion disorder in synthetic and biogenic magnesian calcites: a Raman spectral study. *Amer. Min.*, **70**, 581–9.

Boeke, H. E., 1912. Die Schmelzerscheinungen und die Umkehrbare Umwandlung des Calcium Carbonates. *Neues Jahrb. Min. Geol.*, **1**, 91–121.

Boettcher, A. L. and **Wyllie, P. J.**, 1967. Biaxial calcite inverted from aragonite. *Amer. Min.*, **52**, 1527–9.

Boles, D. R. and **Coombs, D. S.**, 1977. Zeolite facies alteration of sandstones in the Southland syncline, New Zealand. *Amer. J. Sci.*, **277**, 982–1012.

Boyle, E. A., 1983. Manganese carbonate overgrowth on foraminifera tests. *Geochim. Cosmochim. Acta*, **47**, 1815–9.

Bragg, W. L., 1914. The analysis of crystals by the X-ray spectrometer. *Proc. Roy. Soc. London*, **A89**, 468–89.

Bridgman, P. W., 1925. Linear compressibility of fourteen natural crystals. *Amer. J. Sci.*, **210**, 483–98.

Bridgman, P. W., 1939. The high-pressure behavior of miscellaneous minerals. *Amer. J. Sci.*, **237**, 7–18.

Brunton, C. N., 1969. Electron microscopic studies of growth margins of articulate brachiopods. *Z. Zellforsch.*, **100**, 189–200.

Burnham, C. W., 1959. Contact metamorphism of magnesian limestone at Crestmore, California. *Geol. Soc. Amer. Bull.*, **70**, 879–920.

Burns, S. J. and **Baker, P. A.**, 1986. A geochemical study of dolomite in the Monterey formation, California. *J. Sed. Petr.*, **57**, 128–39.

Byrnes, A. P. and **Wyllie, P. J.**, 1981. Subsolidus and melting relations for the join $CaCO_3$–$MgCO_3$ at 10 kbar. *Geochim. Cosmochim. Acta*, **45**, 321–8.

Capobianco, C. and **Navrotsky, A.**, 1987. Solid solution thermodynamics in $CaCO_3$–$MnCO_3$. *Amer. Min.*, **72**, 312–8.

Carlson, S. J., 1955. Activation of photoluminescence in artificial calcite by stannous ion. *Amer. Min.*, **40**, 540–2.

Carlson, W. D., 1983. The polymorphism of $CaCO_3$ and the aragonite–calcite transformation. In *Carbonates: Mineralogy and Chemistry*: Ed. R. J. Reeder, *Reviews in Mineralogy*, Mineral. Soc. Amer., Washington, DC, **11**, 191–225.

Chang, L. L. Y., 1965. Subsolidus phase relations in the systems $BaCO_3$–$SrCO_3$, $SrCO_3$–$CaCO_3$, and $BaCO_3$–$CaCO_3$. *J. Geol.*, **73**, 346–68.

Chave, K. E., 1954a. Aspects of the biogeochemistry of magnesium. 1. Calcareous and marine organisms. *J. Geol.*, **62**, 266–83.

Chave, K. E., 1954b. Aspects of the biogeochemistry of magnesium. 2. Calcareous sediments and rocks. *J. Geol.*, **62**, 587–99.

Chave, K. E., 1981. Summary of background information on Mg-calcites. In *Some Aspects of the Role of the Shallow Ocean in Global Carbon Dioxide Uptake*: Eds. R. M. Garrels and F. T. MacKenzie, U.S.D.O.E. Conf. 8003115, A4–A9, Nat'l Tech. Information Service, Springfield, Virginia, USA.

Chester, R. and **Elderfield, H.**, 1967. The application of infrared absorption spectroscopy to carbonate mineralogy. *Sedimentology*, **9**, 5–22.

Chou, L., Garrels, R. M. and **Wollast, R.**, 1989. Comparative study of the kinetics and mechanisms of dissolution of carbonate minerals. *Chem. Geol.*, **78**, 269–82.

Cloud, P. E., 1962. Environment of calcium carbonate deposition, west of Andros Island. Bahamas. *U.S. Geol. Surv. Prof. Papers*, **350**, 1–138.

Cohen, C. H. and **Klement, W.**, 1973. Determination of high-temperature transition in calcite to 5 kbar by differential thermal analysis in hydrostatic apparatus. *J. Geol.*, **81**, 724–7.

Cooper, A. F., Gittins, J. and **Tuttle, O. F.**, 1975. The system Na_2CO_3–K_2CO_3–$CaCO_3$ at 1 kbar and its significance in carbonatite petrogenesis. *Amer. J. Sci.*, **275**, 534–60.

Crawford, W. A. and **Hoersch, A. C.**, 1972. Calcite–aragonite equilibrium from 50 to 150 °C. *Amer. Min.*, **57**, 995–8.

Dasgupta, S., Chakraborti, S., Fukuoka, M., Sengupta, P. and **Roy, S.**, 1990. Ca–Ba–Sr carbonates from metamorphosed manganese deposits of the Sausar group, India, and their petrological significance. *Min. Mag.*, **54**, 511–3.

Davis, B. L., 1964. X-ray diffraction data on two high-pressure phases of calcium carbonates. *Science*, **145**, 489–91.

Dawson, J. B., 1962. The geology of Oldoinyo Lengai. *Bull. Volcanol.*, **24**, 349–87.

Dawson, J. B., 1966. Oldoinyo Lengai – an active volcano with sodium carbonatite lava flows. In *Carbonatites*: Eds. O. F. Tuttle and J. Gittins, New York (Interscience), 591 pp.

De Capitani, C. and **Peters, T.**, 1981. The solvus in the system $MnCO_3$–$CaCO_3$. *Contrib. Min. Petr.*, **76**, 394–400.

DeKeyser, W. L. and **Degueldre, L.**, 1950. Contribution a l'étude de la formation de la calcite, aragonite et vaterite. *Bull. Soc. Chim. Belg.*, **59**, 40–71.

Dickson, J. A. D., 1966. Carbonate identification and genesis as revealed by staining. *J. Sed. Petr.*, **36**, 491–505.

Dorobek, S. L. and **Watkinson, A. J.**, 1988. Cathodoluminescence petrography of micro-structural fabrics in deformed limestones. *J. Struct. Geol.*, **10**, 305–9.

Effenberger, H., Mereiter, K. and **Zemann, J.**, 1981. Crystal structure refinements of magnesite, calcite, rhodochrosite, siderite, smithsonite and dolomite with discussion of some aspects of the stereochemistry of calcite-type carbonates. *Z. Krist.*, **156**, 233–43.

Fabricius, F. H., 1977. *Origin of Marine Oöids. Contr. Sedimentol.*, **7**, Stuttgart (E. Schweizerbartshe Verlagsbuchhandlung), 113 pp.

Fanelli, M. F., Cava, N. and **Wyllie, P. J.**, 1983. Calcite and dolomite with portlandite at a new eutectic in CaO–MgO–CO_2–H_2O, with application to carbonatites. *Proc. 13th Gen. Meeting, IMA*, Varna, Bulgaria, 1982, p. 271.

Figuet, G., Guyot, F. and **Itié, J.**, 1994. High-pressure X-ray diffraction study of carbonates: $MgCO_3$, $CaMg(CO_2)_3$ and $CaCO_3$. *Amer. Min.*, **79**, 15–23.

Floran, R. J. and **Papike, J. J.**, 1978. Mineralogy and petrology of the Gunflint iron formation, Minnesota–Ontario: correlation of compositional and assemblage variations at low to moderate grade. *J. Petr.*, **19**, 215–88.

Folk, R. L., 1973. Carbonate petrography in the post-Sorbian age. In *Evolving Concepts in Sedimentology*: Ed. R. N. Ginsberg, Baltimore (John Hopkins University Press), pp. 118–158.

Fonda, G. R., 1940. The preparation of fluorescent calcite. *J. Phys. Chem.*, **44**, 435–9.

Fong, M. Y. and **Nicol, M.**, 1971. Raman spectrum of calcium carbonate at high pressure. *J. Chem. Phys.*, **54**, 579–85.

Freestone, I. C. and **Hamilton, D. L.**, 1980. The role of liquid immiscibility in the genesis of carbonatites – an experimental study. *Contrib. Min. Petr.*, **73**, 105–17.

Friedman, G. M., 1959. Identification of carbonate minerals by staining methods. *J. Sed. Petr.*, **29**, 87–91.

Friedman, G. M. and **Schultz, D. J.**, 1994. Precipitation of vaterite ($CaCO_3$) during oil field drilling. *Min. Mag.*, **58**, 399–406.

Gaspar, J. L. and **Wyllie, P. J.**, 1987. The phlogopites from the Jacupiranga carbonatite intrusions. *Min. Petr.*, **36**, 121–34.

Gensmer, R. P. and **Weiss, M. P.**, 1980. Accuracy of calcite/dolomite ratios by X-ray diffraction and comparison with results from staining techniques. *J. Sed. Petr.*, **50**, 626–9.

Gies, H., 1975. Activation possibilities and geochemical correlation of photoluminescing carbonates, particularly calcite. *Min. Deposita*, **10**, 216–27.

Gillson, J. L., 1927. Biaxial calcite. *Amer. Min.*, **12**, 357–60.

Goldsmith, J. R., 1983. Phase relations of rhombohedral carbonates. In *Carbonates: Mineralogy and Chemistry*: Ed. R. J. Reeder, *Reviews in Mineralogy*, **11**, 49–76, Mineralogical Society of America, Washington, DC.

Goldsmith, J. R. and **Graf, D. L.**, 1957. The system CaO–MnO–CO_2: solid solution and decomposition relations. *Geochim. Cosmochim. Acta*, **11**, 310–34.

Goldsmith, J. R. and **Heard, H. C.**, 1961. Subsolidus phase relations in the system $CaCO_3$–$MgCO_3$. *J. Geol.*, **69**, 45–74.

Goldsmith, J. R. and **Newton, R. C.**, 1969. *P–T–X* relations in the system $CaCO_3$–$MgCO_3$ at high temperatures and pressures. *Amer. J. Sci.*, **267A**, 160–90.

Goldsmith, J. R., Graf, D. L. and **Joensuu, O. I.**, 1955. The occurrence of magnesian calcites in nature. *Geochim. Cosmochim. Acta*, **7**, 212–30.

Goldsmith, J. R., Graf, D. L. and **Heard, H. C.**, 1961. Lattice constants of the calcium–magnesium carbonates. *Amer. Min.*, **46**, 453–7.

Goldsmith, J. R., Graf, D. L., Witters, J. and **Northrop, D. A.**, 1962. Studies in the system $CaCO_3$–$MgCO_3$–$FeCO_3$. 1. Phase relations; 2. A method for major element spectrochemical analysis; 3. Compositions of some ferroan dolomites. *J. Geol.*, **70**, 659–88.

Gordon, T. M., 1971. Some observations on the formation of wollastonite from calcite and quartz. *Can. J. Earth Sci.*, **8**, 845–51.

Gordon, T. M. and **Greenwood, H. J.**, 1970. The reaction: dolomite + quartz + water = talc + calcite + carbon dioxide. *Amer. J. Sci.*, **268**, 225–42.

Graf, D. L. and **Goldsmith, J. R.**, 1955. Dolomite–magnesian calcite relations at elevated temperatures and CO_2 pressures. *Geochim. Cosmochim. Acta*, **7**, 109–28.

Graf, D. L. and **Goldsmith, J. R.**, 1958. The solid solubility of $MgCO_3$ in $CaCO_3$: a revision. *Geochim. Cosmochim. Acta*, **13**, 218–9.

Graf, D. L. and **Lamar, J. E.**, 1955. Properties of calcium and magnesium carbonates and their bearing on some uses of carbonate rocks. *Econ. Geol.*, **50th Anniv. Vol.**, 639–713.

Greenwood, H. J., 1967. Wollastonite: stability in H_2O–CO_2 mixture and occurrence in a contact metamorphic aureole near Salmo, British Columbia, Canada. *Amer. Min.*, **52**, 1669–80.

Haberlandt, H., 1944. Porphyrinkomplexverbindungen als Farbende Einlagerungen in hydrothermalen Kalkspatkristallbildungen. *Weiner Chem. Z.*, **47**, 80–103.

Haberlandt, H., 1949. Neue Lumineszuntersuchungen an Fluoriten und anderen Mineralien IV. *Sitz.-Ber. Östereich. Akad. Wiss., Math.-Nat. Kl.*, Wien, Abt. I., **158**, 609–46.

Hall, A. and **Taylor, J. D.**, 1971. The occurrence of vaterite in gastropod egg-shells. *Min. Mag.*, **38**, 521–5.

Handin, J., Higgs, D. V., Lewis, D. R. and **Weyl, P. K.**, 1957. Effects of gamma radiation on the experimental deformation of calcite and certain rocks. *Bull. Geol. Soc. Amer.*, **68**, 1203–24.

Harker, R. I. and **Tuttle, O. F.**, 1955a. Studies in the system CaO–MgO–CO_2. Part 1. The thermal dissociation of calcite, dolomite and magnesite. *Amer. J. Sci.*, **253**, 209–24.

Harker, R. I. and **Tuttle, O. F.**, 1955b. Studies in the system CaO–MgO–CO_2. II. Limits of solid solutions along the binary join $CaCO_3$–$MgCO_3$. *Amer. J. Sci.*, **253**, 274–82.

Harker, R. I. and **Tuttle, O. F.**, 1956. Experimental data on the P_{CO_2}–T curve for the reaction: calcite + quartz ⇌ wollastonite + carbon dioxide. *Amer. J. Sci.*, **254**, 239–56.

Hemming, N. G. and Hansen, G. N., 1992. Boron isotopic composition and concentration in modern marine carbonates. *Geochim. Cosmochin. Acta*, **56**, 537–43.

Hogarth, D. D. and **Griffin, W. L.**, 1980. Contact-metamorphic lapis lazuli: the Italian Mountain deposits, Colorado. *Can. Min.*, **18**, 59–70.

Hogarth, D. D., Hartree, R. and **Loop, J.**, 1985. Rare-earth element minerals in four carbonatites near Gatineau, Quebec. *Amer. Min.*, **70**, 1135–42.

Hogarth, D. D., Rushforth, P. and **McCorkell, R. H.**, 1988. The Blackburn carbonatites, near Ottawa, Ontario: dykes with fluidized emplacement. *Can. Min.*, **26**, 377–90.

Huang, W. L. and **Wyllie, P. J.**, 1976. Melting relationships in the system CaO–CO_2 and MgO–CO_2 to 33 kilobars. *Geochim. Cosmochim. Acta*, **40**, 129–32.

Hügi, Th., 1945. Gesteinsbildend wichtige Karbonate und deren Nachweis mittels Färbmethoden. *Schweiz, Min. Petr. Mitt.*, **25**, 114–40.

Hutcheon, I., Nahnybida, C. and **Krouse, H. R.**, 1985. The geochemistry of carbonate cements in the Avalon Sand, Grand Banks of Newfoundland. *Min. Mag.*, **49**, 457–67.

Irving, A. J. and **Wyllie, P. J.**, 1975. Subsolidus and melting relationships for calcite, magnesite and the join $CaCO_3$–$MgCO_3$ to 36 kb. *Geochim. Cosmochim. Acta*, **39**, 35–53.

Isherwood, B. J. and **James, J. A.**, 1976. Structural dependence of the optical birefringence of crystals with calcite- and aragonite-type structures. *Acta Cryst.*, **A32**, 340–1.

Iwanoff, L. L., 1906. Ein Wasserhaltiges Calciumcarbonat aus den Umgebungen von Nowo-Alexandria. *Annu. Geöl. Minéral. Russie*, **8**, 23–5.

James, N. P., Ginsberg, R. N., Marszalek, D. S. and **Choquette, P. W.**, 1976. Facies and fabric specificity of early subsea cements in shallow Belize (British Honduras) reefs. *J. Sed. Petr.*, **46**, 523–44.

Jamieson, J. C., 1957. Introductory studies of high pressure polymorphism to 24 000 bars by X-ray diffraction with some comments on calcite II. *J. Geol.,* **65**, 334–43.

Johannes, W. and **Puhan, D.,** 1971. The calcite–aragonite transition, reinvestigated. *Contrib. Min. Petr.,* **31**, 28–38.

Johnson, J. H., 1961. *Limestone-Building Algae and Algal Limestones.* Colorado School of Mines, Boulder, Colorado, USA, 297 pp.

Kamhi, S. R., 1963. On the structure of vaterite, $CaCO_3$. *Acta Cryst.,* **16**, 770–2.

Kas'yanenko, E. V., Kukuy, A. L. and **Matveeva, O. P.,** 1993. [About the influence of natural Iceland spar formation conditions upon its luminescence.] *Proc. Russian Min. Soc.,* **122(5)**, 39–43 (in Russian with English abstract).

Kennedy, G. C., 1947. Charts for correlation of optical properties with chemical composition of some rock-forming minerals. *Amer. Min.,* **32**, 561–73.

Kilpady, S. and **Adyalkar, P. G.,** 1954. A note on a green calcite from the Deccan traps. *J. Univ. Geol. Soc. Nagpur,* **1**, 11–3.

Klein, C., *Jr.,* 1974. Greenalite, stipnomelane, minnesotaite, crocidolite and carbonates in a very low-grade metamorphic Precambrian iron formation. *Can. Min.,* **12**, 475–98.

Klein, C., *Jr.,* 1978. Regional metamorphism of Proterozoic iron-formation, Labrador Trough, Canada. *Amer. Min.,* **63**, 898–912.

Kondo, S., Suto, K. and **Matsushima, S.,** 1972. Ultrasonic observation of calcite I–II inversion to 700 °C. *J. Phys. Earth,* **20**, 245–50.

Kridelbaugh, S. J., 1973. The kinetics of the reaction: calcite + quartz = wollastonite + carbon dioxide at elevated temperatures and pressures. *Amer. J. Sci.,* **273**, 757–77.

Krieger, P., 1930. Notes on an X-ray diffraction study of the series calcite–rhodochrosite. *Amer. Min.,* **15**, 23–9.

Land, L. S., Mackenzie, F. T. and **Gould, S. J.,** 1967. Pleistocene history of Bermuda. *Geol. Soc. Amer. Bull.,* **78**, 993–1006.

Land, L. S., Behrens, E. W. and **Frishman, S. A.,** 1979. The oöids of Baffin Bay, Texas. *J. Sed. Petr.,* **49**, 1269–78.

Le Bas, M. J., 1987. Nephelinites and carbonatites. In *Alkaline Igneous Rocks*: Eds. J. G. Fitton and B. G. J. Upton, *Geol. Soc. London, Spec. Publ.,* **30**, 53–83.

Lewis, D. R., 1956. The thermoluminescence of dolomite and calcite. *J. Phys. Chem.,* **60**, 698–701.

Lindsay, C. G. and **Jackson, M. D.,** 1994. Modified electron gas modelings of calcite and aragonite; comparison of polarizable anion and fully ionic methods. *Amer. Min.,* **79**, 215–20.

Lippman, F., 1973. *Sedimentary Carbonate Minerals.* New York (Springer-Verlag), 228 pp.

Liu, L. G. and **Mernagh, T. P.,** 1990. Phase transitions and Raman spectra of calcite at high pressures and room temperature. *Amer. Min.,* **75**, 801–6.

Lo, B. W. N., 1973. The polarizabilities of planar nitrate, carbonate and borate anions in crystals. *J. Phys. Chem. Solids,* **34**, 513–20.

Loupekine, I. S., 1947. Graphical derivation of refractive index ε for the trigonal carbonates. *Amer. Min.,* **32**, 502–7.

Lynn, D. C. and **Bonatti, E.,** 1965. Mobility of manganese in diagenesis of deep sea sediments. *Marine Geol.,* **3**, 357–74.

McConnell, J. D. C., 1959. Vaterite from Ballycraigy, Larne, Northern Ireland. *Min. Mag.,* **32**, 535–45.

McDowell, S. D. and **Paces, J.,** 1985. Carbonate alteration minerals in the Salton Sea geothermal system, California, U.S.A. *Min. Mag.,* **49**, 469–79.

Mackenzie, F. T., Bischoff, W. D., Bishop, F. C., Loijens, M., Schoonmaker, J. and **Wollast, R.,** 1983. Magnesian calcites: low-temperature occurrence, solubility and solid solution behavior. In *Carbonates: Mineralogy and Chemistry*: Ed. R. J. Reeder, *Reviews in Mineralogy,* **11**, 97–144, Mineralogical Society of America, Washington, DC.

Marfunin, A. S., 1979. *Spectroscopy, Luminescence and Radiation Centers in Minerals.* Berlin (Springer-Verlag), 352 pp.

Marino, A. N. and **Roeder, P. L.,** 1983. Kerimasi: a neglected carbonatites volcano. *J. Geol.,* **91**, 449–55.

Markgraf, S. A. and **Reeder, R. J.,** 1985. High-temperature structure refinements of calcite and magnesite. *Amer. Min.,* **70**, 590–600.

Martinez, I., Agrinier, P., Schärer, U., and **Jaroy, M.,** 1994. A SEM-ATEM and stable isotope study of carbonates from the Haughton impact crater, Canada. *Earth & Planet. Sci. Letters,* **121**, 559–74.

Mauguin, C., 1923. Reflexions des rayons de Rontgen sur certains plans reticularies remarquables de la calcite. *C. R. Acad. Sci. Paris,* **170**, 1331–4.

Medlin, W. L., 1968a. Thermoluminescence growth curves in calcite. In *Thermoluminescence of Geological Material*: Ed. D. J. McDougall, London (Academic Press), pp. 91–101.

Medlin, W. L., 1968b. The nature of traps and emission centers in thermoluminescent rock materials. In *Thermoluminescence of Geological Materials*: Ed. D. J. McDougall, London (Academic Press), pp. 193–223.

Merrill, L. and **Bassett, W. A.**, 1975. The crystal structure of $CaCO_3$(II), a high-pressure metastable phase of calcium carbonate. *Acta Cryst.*, **B31**, 343–9.

Meyer, H. J. Von, 1965. Bildung und morphologie des Vaterites. *Z. Krist.*, **121**, 220–42.

Meyer, H. O. A., 1979. Kimberlites and the mantle. *Rev. Geophys. Space Phys.*, **17**, 779–88.

Mirwald, P. W., 1976. A differential thermal analysis of the high-temperature polymorphism of calcite at high pressure. *Contrib. Min. Petr.*, **59**, 33–44.

Mirwald, P. W., 1979a. Determination of a high-temperature transition of calcite at 800 °C and one bar CO_2 pressure. *Neues Jahrb. Min., Mh.*, 309–15.

Mirwald, P. W., 1979b. The electrical conductivity of calcite between 300° and 1200 °C at a CO_2 pressure of 40 bars. *Phys. Chem. Min.*, **4**, 291–3.

Morse, J. W. and **Mackenzie, F. T.**, 1990. *Geochemistry of Sedimentary Carbonates, Developments in Sedimentology*, **48**, Oxfoıd and New York (Elsevier), 707 pp.

Morse, J. W., **Mucci, A.** and **Millero, F. J.**, 1980. The solubility of calcite and aragonite in seawater of 35‰ salinity at 25 °C and atmospheric pressure. *Geochim. Cosmochim. Acta*, **44**, 85–94.

Nakissa, M. and **Paulitsch, P.**, 1972. Zweiachsig Dolomit-Labellen. *Contrib. Min. Petr.*, **34**, 224–8.

Nambi, K. S. V. and **Mitra, S.**, 1978. Thermoluminescence investigations of natural calcite crystals of differing genesis. *Thermochim. Acta*, **27**, 61–7.

Oberling, J. J., 1964. Observations on some structural features of the pelecypod shell. *Mitt. Naturforsch. Ges. Bern*, **20**, 1–60.

Ogino, T., **Suguki, T.** and **Sawada, K.**, 1987. The formation and transformation mechanism of calcium carbonate in water. *Geochim. Cosmochim. Acta*, **51**, 2757–67.

Omori, K. and **Yamaoka, K.**, 1954. Towery mangano–calcite from Hosokara mine, Miyagi Prefecture. *J. Japan Assoc. Min. Petr. Econ. Geol.*, **38**, 81–6.

Oppenheimer, C. H., 1961. Note on the formation of spherical aragonitic bodies in the presence of bacteria from the Bahama Bank. *Geochim. Cosmochim. Acta*, **23**, 295–9.

Pacey, N. R., 1986. Luminescence properties of calcite in the English chalk. *Modern Geol.*, **10**, 29–38.

Paquette, J. and **Reeder, R. J.**, 1990. Single-crystal X-ray structure refinements of two biogenic magnesian calcite crystals. *Amer. Min.*, **75**, 1151–8.

Paulitsch, P., 1956. Die optische Achsenebene als Ebene der Deformation im Einkorn und Gefüge von Kaltzit. *Carinthia II*, **20** Sonderheft (Angel-Bd.), 123–32.

Peacor, D. R., **Essene, E. J.** and **Gaines, A. M.**, 1987. Petrologic and crystal–chemical implications of cation order–disorder in kutnahorite [$CaMn(CO_3)_2$]. *Amer. Min.*, **72**, 319–28.

Pedersen, T. F. and **Price, N. B.**, 1982. The geochemistry of manganese carbonate in Panama Basin sediments. *Geochim. Cosmochim. Acta*, **46**, 59–68.

Peters, Tj., **Valarelli, J. V.**, **Continho, J. M.**, **Sommerauer, J.** and **Raumer, J. von,** 1977. The manganese deposits of Buritirama (Para, Brazil). *Schweiz. Min. Petr. Mitt.*, **57**, 313–27.

Peters, Tj., **Trommsdorff, V.** and **Sommerauer, J.**, 1978. Manganese pyroxenoids and carbonates: critical phase relations in metamorphic assemblages from the Alps. *Contrib. Min. Petr.*, **66**, 383–8.

Phillips, G. N. and **Brown, I. J.**, 1987. Host rock and fluid control on carbonate assemblages in three Golden Mile dolerites, Kalgoorlie gold deposit, Australia. *Can. Min.*, **25**, 265–73.

Phillips, W. R., 1971. *Mineral Optics: Principles and Techniques.* San Francisco (W. H. Freeman), 249 pp.

Pingitore, N. E., 1970. Diagenesis and porosity modification in Acropora Palmata, Pleistocene of Barbados, West Indies. *J. Sed. Petr.*, **40**, 712–21.

Plummer, L. N. and **Busenberg, E.**, 1982. The solubilities of calcite, aragonite and vaterite in CO_2–H_2O solutions between 0° and 90 °C and an evaluation of the aqueous model for the system $CaCO_3$–CO_2–H_2O. *Geochim. Cosmochim. Acta*, **46**, 1011–40.

Przibram, K., 1956. *Irradiation Colours and Luminescence.* London (Pergamon Press), 332 pp.

Rao, K. S. and **Rao, K. V.**, 1968. Dielectric dispersion and its temperature variation in calcite single crystals. *Z. Phys.*, **216**, 300–6.

Rao, K. V. K., **Naidu, N.** and **Murthy, K. S.**, 1968. Precision lattice parameters and thermal expansion of calcite. *J. Phys. Chem. Solids*, **29**, 245–8.

Rath, R. and **Pohl, D.**, 1971. Zweiachsig erscheinende Kalzite und Dolomite. *Contrib. Min. Petr.*, **32**, 74–8.

Raup, D. M., 1959. Crystallography of echinoid calcite. *J. Geol.*, **67**, 661–74.

Raup, D. M., 1962. The phylogeny of calcite crystallography in echinoids. *J. Paleont.*, **36**, 793–810.

Raup, D. M., 1966. Crystallographic data for echinoid coronal plates. *J. Paleont.*, **40**, 555–68.

Reeder, R. J., 1983. Crystal chemistry of the rhombohedral carbonates. In *Carbonates: Mineralogy and Chemistry*: Ed. R. J. Reeder, *Reviews in Mineralogy*, **11**, 1–47, Mineralogical Society of America, Washington, DC.

Reeder, R. J. and **Markgraf, S. A.**, 1986. High-temperature crystal chemistry of dolomite. *Amer. Min.*, **71**, 795–804.

Robinson, D. N., 1975. Magnetite–serpentine–calcite dykes at Premier Mine and aspects of their relationship to kimberlite and to carbonatite of alkalic carbonatite complexes. *Phys. Chem. Earth*, **9**, 61–70.

Rock, N. M. S., 1977. The nature and origin of lamprophyres: some definitions, distinctions and derivations. *Earth Sci. Rev.*, **13**, 123–69.

Rodgers, J., 1940. Distinction between calcite and dolomite on polished surface. *Amer. J. Sci.*, **238**, 788–98.

Rosenberg, P. E., 1963. Subsolidus relations in the system $CaCO_3$–$FeCO_3$. *Amer. J. Sci.*, **261**, 683–90.

Rost, R. and **Tuček, K.**, 1944. Manganocalcite from Pribram. *Rozpravy. Ceskê. Akad.*, **54**, 17 pp.

Rowland, D. L. G. and **Webster, R. K.**, 1971. Precipitation of vaterite in lake water. *Nature (Phys. Sci.)*, **229**, 158.

Royse, C. F., *Jr.*, **Wadell, J. S.** and **Petersen, L. E.**, 1971. X-ray determination of calcite–dolomite: an evaluation. *J. Sed. Petr.*, **41**, 483–8.

Runnells, D. D., 1970. Errors in X-ray analysis of carbonates due to solid solution variation in composition of component minerals. *J. Sed. Petr.*, **40**, 1156–66.

Schlager, W. and **James, N. P.**, 1978. Low-magnesian calcite limestones forming at the deep-sea floor, Tongue of the Ocean, Bahamas. *Sedimentology*, **25**, 675–702.

Sen, S., Stebbins, J. F., Hemming, N. G., and Ghosh, B., 1994. Coordination environments of B impurities in calcite and aragonite: a B MASNMR study. *Amer. Min.*, **79**, 819–25.

Siegel, F. R., 1963. Artificially induced thermoluminescence of sedimentary dolomites. *J. Sed. Petr.*, **33**, 64–72.

Singh, A. K. and **Kennedy, G. C.**, 1974. Compression of calcite to 40 kbars. *J. Geophys. Res.*, **79**, 2615–22.

Skinner, A. J., La Femina, J. P. and **Jansen, H. J. F.**, 1994. Structure and bonding of calcite: a theoretical study. *Amer. Min.*, **79**, 205–14.

Skippen, G. B., 1971. Experimental data for reactions in siliceous marbles. *J. Geol.*, **79**, 457–81.

Spooner, E. T. C. and **Fyfe, W. S.**, 1973. Sub-sea-floor metamorphism, heat and mass transfer. *Contrib. Min. Petr.*, **42**, 287–304.

Staveley, L. A. K. and **Linford, R. G.**, 1969. The heat capacity and entropy of calcite and aragonite, and their interpretation. *J. Chem. Thermodyn.*, **1**, 1–11.

Stevens, R. G. and **Carron, M. K.**, 1948. Simple field test for distinguishing minerals by abrasion pH. *Amer. Min.*, **33**, 31–49.

Stoops, G. J., 1976. On the nature of 'lublinite' from Hollanto (Turkey). *Amer. Min.*, **61**, 172.

Sundius, N., 1958. Alkaline rocks and carbonates of alkalis, calcium and magnesium. *Arkiv. Min. Geol.*, **2**, 319–31.

Sundius, N., 1965. Carbonates in the manganese ores at Långban. *Arkiv. Min. Geol.*, **4**, 279–85.

Talantsev, A. S., 1976. Dolomite–calcite geothermobarometer. *Dokl. Acad. Sci. USSR, Earth Sci. Sect.*, **228**, 166–8.

Taylor, E. W., 1949. Correlation of the Moh's scale of hardness with the Vickers's hardness numbers. *Min. Mag.*, **28**, 718–21.

Tennant, C. B. and **Berger, R. W.**, 1957. X-ray determination of dolomite–calcite ratio of a carbonate rock. *Amer. Min.*, **42**, 23–9.

Tertsch, H., 1934. Über Sehleifhärtenanisotropie. *Z. Krist.*, **89**, 541–2.

Towe, K. M. and **Cifelli, R.**, 1967. Wall ultrastructure in the calcareous foraminifera: crystallographic aspects and a model for calcification. *J. Paleont.*, **41**, 742–62.

Túcan, F., 1937. Onyx–marble and aragonite in the neighborhood of Tetovo (S. Serbia). *Z. Prakt. Geol.*, **45**, 203–10.

Turnbull, A. G., 1973. A thermochemical study of vaterite. *Geochim. Cosmochim. Acta*, **37**, 1593–601.

Turner, F. J., 1975. Biaxial calcite: occurrence, optics, and associated minor strain phenomena. *Contrib. Min. Petr.*, **50**, 247–55.

Turner, F. J., Griggs, D. T. and **Heard, H. C.**, 1954. Experimental deformation on calcite crystals. *Bull. Geol. Soc. Amer.*, **65**, 883–934.

Tuttle, O. F. and **Harker, R. I.**, 1957. Synthesis of spurrite and the reaction wollastonite + calcite = spurrite + carbon dioxide. *Amer. J. Sci.*, **255**, 226–34.

Vaidya, S. N., Bailey, S., Pasternack, T. and **Kennedy, G. C.**, 1973. Compressibility of fifteen minerals to 45 kilobars. *J. Geophys. Res.*, **78**, 6893–8.

Vendl, A. and **Mándy, T.**, 1958. Über die blaue Farbe einiger Mineralien. *Acta Min. Petr., Univ. Szeged.*, **11**, 61–7.

Wackden, G. M. and **Berry, J. R.**, 1984. Natural calcite in cathodoluminescence: crystal growth during diagenesis. *Nature*, **308**, 525–7.

Walker, T. L. and **Parsons, A. L.**, 1925. The characteristics of primary calcite. *Univ. Toronto Studies, Geol. Ser.*, **20**,, 14–7.

Wang, C., 1968. Ultrasonic study of phase transition in calcite to 20 kilobars and 180 °C. *J. Geophys. Res.*, **73**, 3937–44.

Warne, S. St J., 1962. A quick field or laboratory staining scheme for the differentiation of the major carbonate minerals. *J. Sed. Petr.*, **32**, 29–38.

Weber, H. J., 1988. Anisotropic bond polarizabilities in birefringent crystals. *Acta Cryst.*, **A44**, 320–6.

White, W. B., 1974. The carbonate minerals. In *The Infrared Spectra of Minerals*: Ed. V. C. Farmer, London (Min. Soc. Monograph **4**), pp. 227–84.

Williams, A., 1966. Growth and structure of the shell of living articulate brachiopods. *Nature*, **211**, 1146–8.

Williams, H., Turner, F. J. and **Gilbert, C. W.,** 1982. *Petrography*, 2nd edn. San Francisco (Freeman and Co.), 626 pp.

Wilson, R. C. L., 1967. Particle nomenclature in carbonate sediments. *Neues Jahrb. Geol. Paleont., Mh.*, 498–510.

Winchell, A. N. and **Meek, W. B.,** 1947. Birefringence-dispersion ratio as a diagnostic. *Amer. Min.*, **32**, 336–43.

Winter, G. A., Essene, E. J. and **Peacor, D. R.,** 1981. Carbonates and pyroxenoids from the manganese deposit near Bald Knob, North Carolina. *Amer. Min.*, **66**, 278–89.

Wolfe, J. A. and **Bartlett, V. C.,** 1958. Gasometric determination of calcite and dolomite. *Bull. Geol. Soc. Amer.*, **69**, 1664–5.

Wray, J. L. and **Daniels, F.,** 1957. Precipitation of calcite and aragonite. *J. Amer. Chem. Soc.*, **79**, 2031–4.

Wyllie, P. J. and **Boettcher, A. J.,** 1969. Liquidus phase relationships in the system $CaCO_3$–CO_2–H_2O to 40 kbars pressure with petrological applications. *Amer. J. Sci.*, **267A**, 489–508.

Wyllie, P. J. and **Tuttle, O. F.,** 1960. The system CaO–CO_2–H_2O and the origin of carbonatites. *J. Petr.*, **1**, 1–46.

Young, B. N., Parsons, I. and **Threadgold, R.,** 1994. Carbonatite near the Loch Borralan intrusion, Assynt. *J. Geol. Soc.*, **151**, 945–54.

Zen, E.-An, 1959. Clay mineral–carbonate relations in sedimentary rocks. *Amer. J. Sci.*, **257**, 29–53.

Zhabin, A. G. and **Cherepivskaya, G. Y.,** 1965. Carbonatite dikes as related to ultrabasic alkali extrusive igneous activity. *Dokl. Akad. Nauk SSSR, Earth Sci. Sect.*, **160**, 135–8.

Zhang, S. and **Mucci, A.,** 1989. Calcite and aragonite precipitation from seawater solutions of various salinities: precipitation rates and overgrowth compositions. *Chem. Geol.*, **78**, 283–99.

Zharikov, V. A. and **Shmulovich, K. I.,** 1969. High temperature mineral equilibria in the system CaO–SiO_2–CO_2. *Geochem. Internat.*, **6**, 853–69.

Magnesite $MgCO_3$

Trigonal (−)

ε	1.509
ω	1.700
δ	0.191
Dispersion:	Very strong.
D	3.00
H	4
Cleavage[1]:	$\{10\bar{1}1\}$ perfect.
Twinning[1]:	Rare, translational gliding on {0001}.
Colour:	White or colourless; colourless in thin section.
Unit cell:	a_{hex} 4.632 Å, c_{hex} 15.012 Å, $Z_{hex}=6$. a_{rh} 5.675 Å, α 48° 10′, $Z_{rh}=2$. Space group $R\bar{3}c$

Slightly soluble in HCl with effervescence in warm HCl.

Magnesite usually occurs in fine-grained white masses as an alteration product of magnesium-rich metamorphic and igneous rocks. It is also found in sedimentary rocks, from direct precipitations or as a replacement, and as a primary mineral in igneous rocks. Magnesite rarely occurs as well-developed crystals. The name magnesite is in allusion to its composition.

Structure

The crystal structure of magnesite was determined by Oh *et al.* (1973) from a single crystal prepared by a hydrothermal method. The calcite-type structure was confirmed with *a* 4.637, *c* 15.023 Å. Using a magnesite crystal from Oberdorf, Styria, Austria, Effenberger *et al.* (1981) refined the unit cell to give *a* 4.6328(2), *c* 15.0129(5) Å. The interatomic distances are: C–O, 1.2851(4) Å (1.2873 Å with a correction for riding motion); Mg–O, 2.1018 Å; d_1 (edge of the MgO_6 octahedron inclined to (0001)), 3.0188(5) Å; d_2 (edge of the MgO_6 octahedron parallel to (0001), 2.9252(7) Å; and d_3 (shortest O–O distance outside the coordination figure), 2.8485 Å. The distortion of the MgO_6 octahedron is less than that of CaO_6 in calcite, and Oh *et al.* (1973) attribute this to the fact that Mg^{2+} has a more suitable size for octahedral coordination by oxygen ions. The thermal ellipsoid of oxygen in magnesite is an oblate spheroid, as in orthorhombic carbonates (Dal Negro and Ungaretti, 1971; DeVilliers, 1971), with the length of the three principal axes being 0.063, 0.084 and 0.099 Å. High-temperature structure refinement of magnesite by Markgraf and Reeder (1985) shows that both *a* and *c* increase with increasing temperature (Fig. 45). Both C–O and $(O\text{–}O)_{CO_3}$ show a decrease in the uncorrected

[1]Cleavage, twinning and morphology are indexed using the hexagonal morphological cell with quadrupled *a* parameter (see Table 16, p. 103).

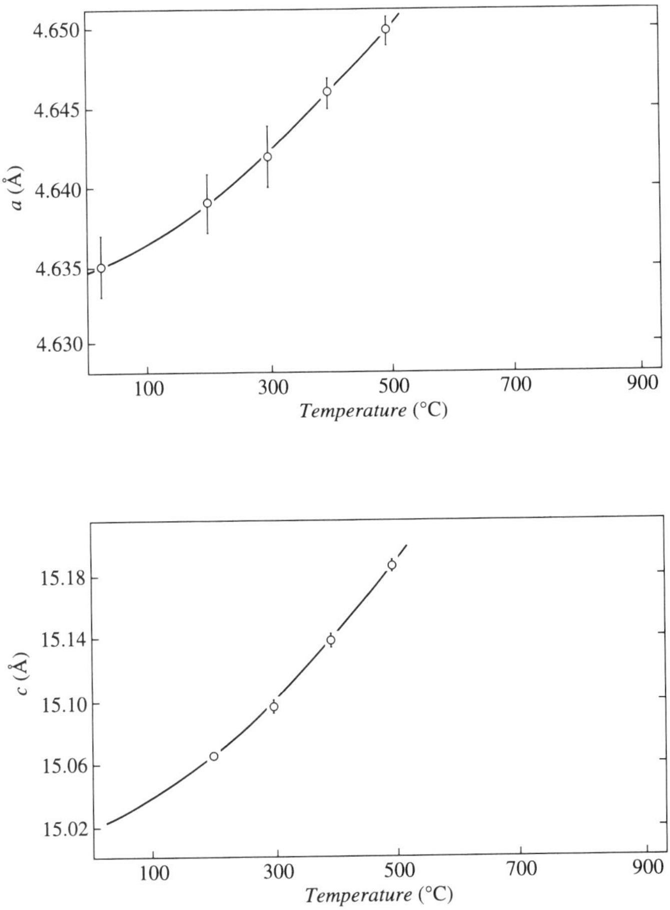

Fig. 45. Variation of cell dimensions with increasing temperature for magnesite. Error bars represent two standard deviations (after Markgraf and Reeder, 1985).

bond lengths, which indicates an increased thermal motion, specifically libration. In comparison with calcite, the decrease is small. With increasing temperature from 24 to 500 °C, the bond angle of O(1)–Mg–O(2) decreases, whereas those of both O(1)–Mg–O(6) and Mg–O–Mg increase. MgO_6 octahedra are also distorted to a greater degree (Markgraf and Reeder, 1985). High-pressure X-ray powder diffraction data for magnesite were obtained by Figuet *et al.* (1994). Cell dimensions measured at room pressure are *a* 4.6348, *c* 15.034 Å, which compare well with those given by Markgraf and Reeder. As shown in Fig. 46, both *a* and *c* decrease with increasing pressure, and anomalies occur at pressures near 250 kbar. It remains unclear whether they represent phase transitions in magnesite.

Using the (0006) reflection, Shaikh and Brusewitz (1975) examined the relationship between the FeO content and *c* dimension of magnesite occurring as crystalline lenses in basic rocks, and concluded that there was no variation of the cell dimension, at least to 8.55 per cent FeO.

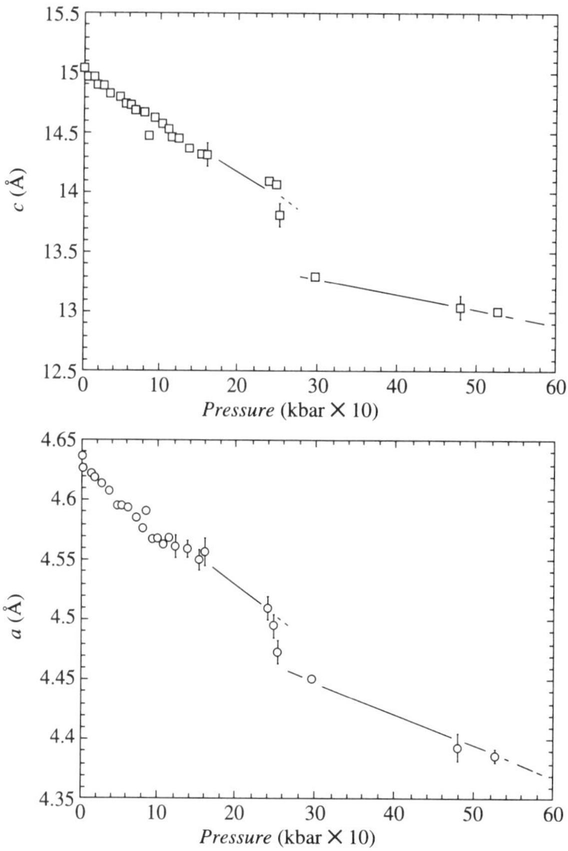

Fig. 46. Unit cell parameters of magnesite as a function of pressure (after Figuet *et al.*, 1994).

Chemistry

Pure magnesite has the composition $MgCO_3$, but there appears to be a complete solid solution series between magnesite and siderite ($FeCO_3$). The composition range between 5 and 50 mol. per cent $FeCO_3$ has been known as breunnerite. The substitution of either Ca or Mn for Mg also occurs, but to a much restricted extent. Other naturally occurring replacements are Ni and Co. Magnesite analyses are given in Table 21. Hydromagnesite $Mg_4(CO_3)_3(OH)_2{\cdot}3H_2O$ is a hydrated basic carbonate. Other hydrated magnesium carbonates are lansfordite ($MgCO_3{\cdot}5H_2O$) and nesquehonite ($MgCO_3{\cdot}3H_2O$).

Phase relationships in the system $MgCO_3$–$FeCO_3$ with and without $CaCO_3$ were studied by Goldsmith *et al.* (1962) and Rosenberg (1963, 1967). The complete solid solution series $MgCO_3$–$FeCO_3$ has been well established. In the presence of $CaCO_3$, the series forms an equilibrium assemblage with the dolomite–ankerite series and with calcite. In both of these equilibrium assemblages, the $MgCO_3/FeCO_3$ ratio of the magnesite–siderite series increases with increasing temperature.

Table 21. Magnesite Analyses

	1	2	3	4	5	6
MgO	47.1	47.09	46.35	45.34	47.32	46.92
FeO	0.6	0.60	1.22	2.93	0.43	0.71
CaO	0.1	0.07	0.03	0.31	0.07	0.39
MnO	0.1	0.67	0.62	0.12	0.13	—
CO_2	51.93(c)	51.74(c)	51.77	51.47	52.06	51.97
Total	99.83	100.17	99.99	100.17	100.01	99.99
	Numbers of ions on the basis of six (O)					
Mg	1.980	1.968	1.967	1.928	1.982	1.990
Fe	0.014	0.014	0.029	0.070	0.010	0.017
Ca	0.003	0.002	0.001	0.004	0.002	0.012
Mn	0.003	0.016	0.015	0.003	0.003	—
C	2.000	2.000	2.012	2.005	1.997	2.019

(c) Calculated from stoichiometry.

1 Sedimentary magnesite from Oberdorf, Styria, Austria (Effenberger *et al.*, 1981).
2 Magnesite from Brumado magnesite mine, Brazil (Markgraf and Reeder, 1985).
3–5 Magnesite from Main Creek magnesite deposits, Savage River, Tasmania (Frost, 1982).
6 Sedimentary magnesite from Mundallio Creek, South Australia (Frost, 1982).

Table 21. *Continued*

	7	8	9	10	11
MgO	19.37	47.60	46.15	22.89	17.3
FeO	34.64	0.04	1.81	0.70	5.7
CaO	0.81	0.20	0.19	1.89	—
MnO	0.79	—	0.12	0.03	—
				29.64 (NiO)	35.0 (NiO)
				0.08 (CoO)	
				5.10 (oth.)	
CO_2	43.48(c)	52.15	51.72	40.69	43.0(c)
Total	99.09	99.99	99.99	101.02	101.0
	Numbers of ions on the basis of six (O)				
Mg	0.973	2.024	1.997	1.190	0.88
Fe	0.976	0.001	0.044	0.020	0.16
Ca	0.029	0.006	0.006	0.071	—
Mn	0.022	—	0.003	0.001	—
				0.831 (Ni)	0.96 (Ni)
				0.002 (Co)	
C	2.000	2.031	2.050	1.938	2.000

7 Magnesian siderite from carbonatite complex, Newania, India (Buckley and Woolley, 1990).
8 Magnesite, an alteration from ultrabasic igneous rock, Main Creek, Savage River, Tasmania (Frost, 1982).
9 Magnesite of metasomatic origin from Balcanoona Station, South Australia (Frost, 1982).
10 A Ni–Mg carbonate from the oxidation zone of a nickeliferous copper sulphide ore, west of Yellow River, China (Yue *et al.*, 1964).
11 Magnesian gaspéite from dolomitic rock, Gaspé Peninsula, Quebec, Canada (Kohls and Rodda, 1966); ε 1.61, ω 1.83, D 3.71.

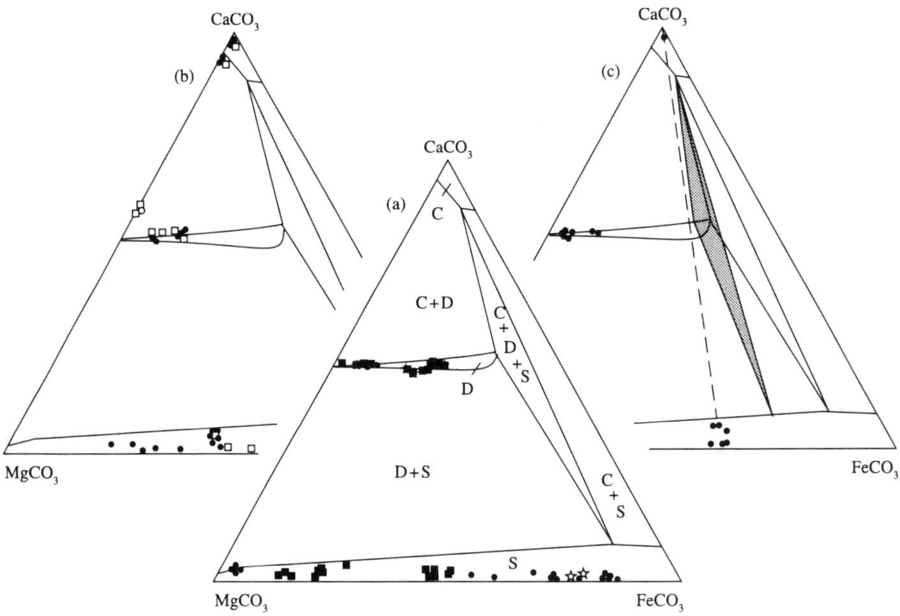

Fig. 47. Compositions of the magnesite–siderite series (after Buckley and Woolley, 1990). The field boundaries are for the subsolidus relations at 450 °C with the three-phase field (shaded) at 400 °C. (a) Carbonatites from the Lueshe carbonatite, Zaire (squares), and Newmania, India (circles); the stars represent magnesite–siderite solid solutions from Russia and the USA. (b) Carbonatites from Kangankunde, Malawi (circles and open squares). (c) Carbonatites from Chapman Lake, Canada. In these diagrams the minerals are represented by C (calcite), D (dolomite–ankerite) and S (siderite).

A complete magnesite–siderite series was found by Buckley and Woolley (1990) in their examination of four carbonatite complexes; their analytical data are given in Fig. 47 (see also Table 21, anal. 7). Also shown in the figure are field boundaries between the magnesite–siderite series, the dolomite–ankerite series and the calcite solid solution at 450 °C taken from Rosenberg (1967). The three-phase field at 400 °C (shaded) is also shown in Fig. 47. In nearly all the carbonates examined, the magnesite–siderite series occurs in association with the dolomite–ankerite series. In one sample, magnesian siderite was found within calcite grains (dashed line in Fig. 47).

A complete series of solid solution exists in the system $MgCO_3$–$MnCO_3$ (Goldsmith and Graf, 1960), but magnesite containing a significant amount of manganese has not been reported. Phase relationships between $MgCO_3$ and $CaCO_3$ will be treated in connection with dolomite (p. 199), and the Ca-rich portion of the system has been described in the section on calcite (p. 111).

Yue *et al.* (1964) described a series of magnesium–nickel carbonates from China with a NiO content varying from 0.77 to 29.64 per cent. The name hoshiite was given to the composition $NiMg(CO_3)_2$. Its cell dimensions are *a* 4.637, *c* 15.004 Å. Magnesian gaspéite described by Kohls and Rodda (1966) has a composition of $(Ni_{0.98}Mg_{0.86}Fe_{0.16})(CO_3)_2$ (Table 21, anal. 11) and cell dimensions of *a* 4.621, *c* 14.93 Å. Although Kohls and Rodda (1966) did not relate their magnesian gaspéite to Yue *et al.*'s hoshiite it appears from a comparison of chemical composition and cell dimension that no distinction exists between these two minerals. Gaspéite of relatively pure $NiCO_3$ composition has been reported by Mariano *et al.* (1969) from Pafuri, Transvaal, South Africa, and by Nickel (1973) from Nullagine, Western Australia. The cell dimensions of the Australian gaspéite are given in Table 13 (p. 97). A miscibility gap exists in the system $MgCO_3$–$NiCO_3$, and the system has solid solutions with restricted ranges even at a temperature as high as 600 °C (Goldsmith and Northrop, 1965; Gaines and Goldsmith, 1971). Results indicate that magnesian gaspéite or hoshiite is thermodynamically unstable with respect to the end-members in the system at temperatures >100 °C.

The melting of magnesite was studied by Irving and Wyllie (1975), Huang and Wyllie (1976) and Byrnes and Wyllie (1981). At 10 kbar, magnesite solid solution dissociates at 1090 °C to produce dolomite + periclase + CO_2. At 27 kbar, 6.5 per cent CO_2 dissolves in molten $MgCO_3$, which depresses the melting temperature from 1590 to 1510 °C. The dissociation curve for $MgCO_3$ terminates at a univariant point where melting begins at 23 kbar and 1550 °C. From 23 to 24.5 kbar, $MgCO_3$ melts incongruently to produce periclase and liquid, while above 24.5 kbar it has a congruent melting point as shown schematically in Fig. 48.

The thermal dissociation of magnesite was determined by Harker and Tuttle (1955). The dissociation temperature for the reaction $MgCO_3 = MgO + CO_2$ increases from 650 to 875 °C in the pressure (CO_2) range 62 to 2760 bar. Harker (1958) stated that the dissociation temperature of magnesite at various pressures of CO_2 + inert gas (He, N or Ar) was largely dependent on the proportion of CO_2 at a given total pressure. Results of differential thermal analysis reported by Smykatz-Kloss (1964) show endothermal peaks at 625 and 643 °C for dense and sparry magnesite, respectively. For 'breunnerite', the iron-rich magnesite, an endothermal peak occurs at 654 °C and an exothermal peak at 677 °C. The latter

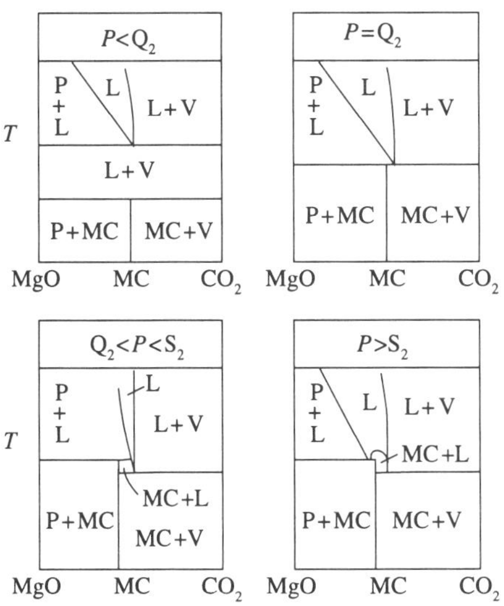

Fig. 48. Schematic isobaric (*T*–*X*) sections for the system MgO–CO_2. 'Q' at 23 kbar is an invariant point at which the decomposition of magnesite ends and melting begins. 'S_2' is a singular point lying above 24.5 kbar (after Huang and Wyllie, 1976). Symbols are V (vapour), L (liquid), P (periclase) and MC (magnesite).

may be due to the oxidation of $FeCO_3$ in the solid solution. Experiments on the assemblage magnesite + diopside are referred to on p. 146.

Stability relationships of magnesite in the system $MgO–CO_2–H_2O$ have been examined by Langmuir (1965), Christ and Hostetler (1970) and others. Using $10^{-5.1}$ as the 25 °C value for the activity product constant of magnesite, Langmuir calculated the stability relationships between nesquehonite ($MgCO_3{\cdot}3H_2O$), lansfordite ($MgCO_3{\cdot}5H_2O$), hydromagnesite ($Mg_4(CO_3)_3(OH)_2{\cdot}3H_2O$) and magnesite. According to his results, magnesite is not stable in the presence of water, at unit activity of water, below 60 °C. Between 60 and 10 °C, nesquehonite and hydromagnesite are stable, depending on P_{CO_2}; at 10 °C nesquehonite transforms to lansfordite. At surface temperatures, nesquehonite is stable for $P_{CO_2} > 10^{-2}$ bar and hydromagnesite is stable for $P_{CO_2} < 10^{-2}$ bar. The solubility studies indicate that for a P_{CO_2} of approximately 1 bar, the saturated solution is in equilibrium with lansfordite at temperatures up to about 10 °C, with nesquehonite up to 55 °C, and with hydromagnesite above 55 °C. For an activity product constant of $10^{-8.1}$, the only stable mineral under surface or near-surface geological conditions is magnesite (Langmuir, 1965). The experimental study of Christ and Hostetler (1970) confirmed that the activity-product constant of magnesite at 25 °C is $10^{-8.1}$. They pointed out that although the experimental solubility results discussed are reproducible, these results represent metastable relationships with respect to magnesite plus solution. The stability of magnesite in conditions of high temperature and pressure has been studied (Gillet, 1993) using Raman spectroscopy; the results confirm that magnesite can act as a host for the storage of carbon down to at least 1000 km, and can also be a carrier for carbon in subducting plates.

The crystallization kinetics of magnesite in aqueous solutions were studied utilizing the reaction:

$$\underset{\text{hydromagnesite}}{(5MgO + 4CO_2 + 5H_2O)} + xCO_2 = (4 + x)MgCO_3 + (4 + x)H_2O + (1 - x)Mg(OH)_2$$

(Sayles and Fyfe, 1973). Results illustrate inhibitions of magnesite crystallization by a species of Mg in solution and strong positive catalysis by ionic strength and P_{CO_2}. The Mg inhibition has been demonstrated to be limited by increasing concentration, but no such limit was observed by Sayles and Fyfe for positive catalysis by ionic strength. Therefore, highly saline conditions offer a favourable environment for magnesite crystallization.

The dissolution of magnesite at 25 °C and over a pH range of 2–8 has been studied by Chou *et al.* (1989) and is shown in Fig. 49. The curves were calculated according to the equation:

$$R_f = k_1 a_{(H^+)} + k_2 a_{(H_2CO_3^*)} + k_3 a_{(H_2O)},$$

where R_f is the forward rate of dissolution; k_1, k_2 and k_3 are rate constants which have experimentally determined values of 2.5×10^{-9}, 6.0×10^{-10} and 4.5×10^{-14}, respectively, $a_{(i)}$ denotes the activity of the subscript species, and $H_2CO_3^*$ represents by convention $H_2CO_3^\circ + CO_{2(aq)}$. The rate of dissolution of magnesite is very slow, and no backward reaction was observed by Chou *et al.* At pH values <5, the rate is a linear function of the activity of H^+, and becomes constant in the neutral pH range. The influence of the partial pressure of CO_2 appears to be more pronounced in the dissolution of magnesite than for calcite and aragonite. Its effect on the rate is shown in Fig. 49(b).

The thermodynamic properties of magnesite have been the subject of many measurements. From Hemingway *et al.* (1977), values (in J/kmol.) at 298.15 K of

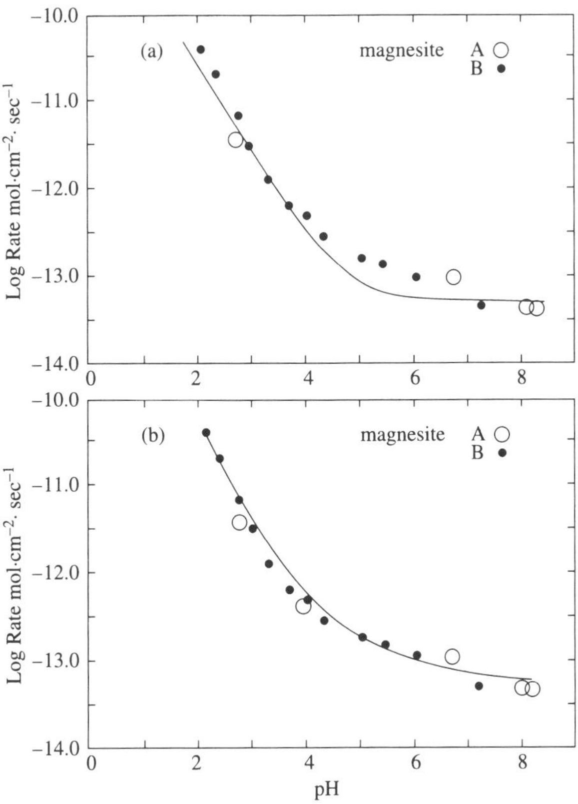

Fig. 49. Dissolution of magnesite (after Chou *et al.*, 1989). Magnesite A has a composition $(Mg_{0.965}Ca_{0.020}Fe_{0.015})CO_3$ and magnesite B is $MgCO_3$. The curves were calculated (a) without considering the effect of P_{CO_2} and (b) with consideration of the effect of P_{CO_2}.

C_p, $S^\circ_T - S^\circ_o$, $(H^\circ_T - H^\circ_o)/T$, and $-(G^\circ_T - H^\circ_o)/T$ are, respectively, 76.09, 65.09, 39.01 and 26.09. With increasing temperature between 13 and 380 K, all properties decrease in value.

Magnesite is an important raw material for basic refractories: after heating and sintering, the product known as 'dead-burned magnesite' is produced which has the composition MgO.

Optical and Physical Properties

Like calcite, magnesite is uniaxial and optically negative with extreme birefringence ($\omega - \varepsilon = 0.191$). Interference colours seen in thin section or grain mount are typically pale high-order colours, even if the optic axis is close to vertical. The structural dependence of the birefringence of carbonates has been examined by Lo (1973), Isherwood and James (1976) and Weber (1988). As shown in Fig. 42 (p. 118) the refractive indices of magnesite increase linearly with the substitution of Fe^{2+} for Mg (Kennedy, 1947). Data from Yue *et al.* (1964) show an increase in both ω and ε with increasing NiO content. For 0.05, 8.0, 10.0, 13.0 and 18.0 mol. per cent

NiO, ω and ε are 1.691 and 1.510, 1.711 and 1.519, 1.715 and 1.527, 1.720 and 1.531, and 1.728 and 1.534, respectively.

Coefficients of thermal expansion for magnesite in both the *a* and *c* dimensions are positive and have the values of 6.75×10^{-6}/°C and 22.9×10^{-6}/°C, respectively (Markgraf and Reeder, 1985). Axial compressibilities measured by high-pressure X-ray diffraction (Figuet *et al.*, 1994) are 1.08×10^{-4}/kbar for *a* and 2.66×10^{-4}/kbar for *c*.

Magnesite is usually white, but it may be light to dark brown if iron-bearing. With an appreciable content of nickel, its colour appears to be pale to emerald green (Yue *et al.*, 1964; Kohls and Rodda, 1966). Magnesite is not one of the strong luminescent carbonates (Gies, 1975).

Magnesite usually occurs as compact granular aggregates. Crystals are rare. The finest specimens are from deposits in the Serra Das Equas, Bahia, Brazil, near the town of Brumado (De Aranjo, 1927). The specific gravity of magnesite is increased by the introduction of iron, rising to 3.50 for 'breunnerite' with 50 per cent $FeCO_3$. With 0.77, 13.55, 18.13, 22.18 and 29.13 per cent NiO, the specific gravities are 2.91, 3.15, 3.25, 3.26 and 3.36, respectively (Yue *et al.*, 1964). The hardness of magnesite is normally $3\frac{1}{2}$–$4\frac{1}{2}$, but a porcellanean variety from Tanzania has been reported to have a hardness of $7\frac{1}{2}$ (Bassett, 1951).

The infrared spectra of magnesite, like those of calcite, are characterized by three prominent absorption maxima, as shown in Fig. 43 (p. 120). The shift in the frequencies of carbonate-ion vibrations is related to the difference in the radius of calcium and magnesium in the structure (Huang and Kerr, 1960; White, 1974). For the magnesite–siderite series, the infrared absorption frequencies shift towards lower wavenumbers with increasing Fe content (Peng *et al.*, 1985). A linear equation has been established from their experimental data as $Y = 880.7–0.1450X$, where Y is the absorption frequency ν_2 (867–880 cm^{-1}) with $X = Fe/(Fe + Mg)$ per cent.

The reflectivity of carbonates in the magnesite–siderite series varies linearly with composition (Beran, 1978). An examination of the series by Fourier transform infrared spectroscopy was made by Dubrawski *et al.* (1989). The shifts in two of the carbonate vibrational bands are very pronounced, showing that Fourier transform infrared spectroscopy can be used to distinguish between members of this series.

Distinguishing Features

Magnesite resembles dolomite in being only slightly soluble in cold dilute HCl, but it dissolves with effervescence in warm acid. Its refractive indices are higher than calcite and dolomite and lower than siderite and rhodochrosite. Crystals or grains of magnesite are usually anhedral and lack lamellar twinning.

Magnesite can be differentiated from calcite and dolomite by staining with either alizarin red S or Harris' hematoxylin (Friedman, 1959; Warne, 1962). Warne (1962) proposed a staining method using tropaeolin OO to differentiate magnesite from smithsonite. Stevens and Carron (1948) suggest the use of the 'abrasion pH' values with a series of indicator papers to distinguish between calcite which has an abrasion pH of 8, dolomite 9–10 and magnesite 10–11. Magnesite may also be distinguished from other trigonal carbonates by its differential thermal analysis curve; its main endothermal peak occurs at about 100–150 °C lower than those of dolomite and calcite, and about 100–150 °C higher than those of siderite and smithsonite.

Paragenesis

Magnesite usually occurs in veins and masses associated with serpentine and other magnesium-rich igneous and metamorphic rocks from which it has been formed by alteration. For example, magnesite deposits in Serbia occur in interrupted veining along lines of fissure in a serpentine which has been derived from peridotite (Ignatieff, 1930). In Silesia, magnesite veins occupy a 100 m zone in altered peridotite (Spangenberg, 1949). Barnes *et al.* (1973) examined wall-rock alteration in small mercury deposits of the California Coast Range, USA. They found that the acidic CO_2-rich waters reacted with serpentine to form silica (quartz + chalcedony or opal) and carbonates (magnesite + dolomite). The waters are of a metamorphic origin and in part are locally derived meteoric waters. The CO_2 is entirely derived from metamorphic reactions at depth. If the CO_2-rich fluids react directly with peridotite or dunite, massive magnesite deposits may form. The Deloro magnesite deposits near Timmins, Ontario, Canada, are characterized by four distinct mineral assemblages which represent various stages in the progressive carbonate alteration of a serpentinized dunite (Griffin, 1972). In order of increasing intensity of alteration, they are (a) serpentinite with magnetite and minor amounts of magnesite, (b) fine-grained magnesite, talc, serpentine and accessory iron minerals, (c) fine-grained magnesite and quartz with minor amounts of talc and hematite, and (d) medium-grained magnesite, talc, quartz and accessory iron-oxide minerals. Chemical and modal analyses indicate that no significant amounts of magnesite were added or removed from the system during the alteration. Similarly, very little silica was added during the talc–carbonate alteration. Experimental data in the system $MgO–SiO_2–H_2O–CO_2$ (Greenwood, 1967) suggest that the early stage of talc–carbonate alteration of the Deloro rocks may have been characterized by temperatures in the range 150–300 °C. A decrease in CO_2 in the fluids and an increase in temperature may have caused the late stage of decarbonization and talc crystallization.

Carbonate beds up to 700 m thick in the Rum Jungle uranium field, Northern Territory, Australia, consist of coarsely crystalline magnesite and dolomite (Aharon, 1988). The temperature of fluid inclusion homogenization and oxygen isotopic data indicate that magnesites were formed in the temperature range 100–200 °C from fluids that became progressively enriched in ^{18}O. Aharon suggests an epigenetic–metasomatic model for the replacement of dolomite by Mg-rich fluids and the formation of magnesite. The large magnesite deposit in northwestern Tasmania is the product of similar processes, but at a temperature as high as 400 °C and a total pressure of 2 kbar (Frost, 1982). The extensive banded magnesite deposits in the Proterozoic dolomites and associated rocks of Bahia State, Brazil, illustrate a complete gradation from gneissic and pegmatitic rocks through intermediary dolomite into magnesite (Grubb, 1981). Grubb concluded that Mg-metamorphism assisted by structural movement was the essential factor in the formation of the magnesite deposits. The magnesite deposits in Chalkidiki, Greece, are also the products of structural deformation in the host ultramafic rocks (Chatzidimitriadis *et al.*, 1981).

In Kamataka, India, magnesite mineralization in the Copper Mountain area is in Archaean grunerite schist; the magnesite occurs as a series of parallel to subparallel veins 1–50 cm thick. It is cryptocrystalline and associated with minor dolomite and/or magnesian calcite. Field relationships, rock and mineral chemistry, and stable isotope data indicate that the magnesite was precipitated at ambient temperature from descending meteoric water enriched in biogenic and atmospheric CO_2 and Mg^{2+}, the latter being derived from the upper, weathered part of the grunerite schist (Krishna Rao *et al.*, 1994).

From an investigation of physical and chemical conditions leading to the formation of magnesite, Johannes (1970) concluded that all magnesite occurrences can be classified into two types: (a) those formed by CO_2 metamorphism and (b) those formed by Mg metamorphism. In the CO_2 metamorphism, the assemblages of quartz + magnesite and talc + magnesite are stable in contact with both CO_2-rich and CO_2-poor fluid phases, whereas the assemblages of enstatite + magnesite and serpentine + magnesite indicate a low CO_2 content in the fluid phase. In the Mg metamorphism, magnesite is formed by the action of salt-bearing solutions. In equilibrium with chloride solutions, the formation of calcite + dolomite or dolomite + magnesite is a function of temperature, pressure, CO_2 content, NaCl content and the concentration of $CaCl_2 + MgCl_2$ in the fluid.

Magnesite is relatively rare in carbonatites, but recent studies by Buckley and Woolley (1990) have described its occurrence in detail in four complexes: Lueshe, Zaïre; Newania, India; Kangankunde, Malawi; and Chipman Lake, Ontario, Canada. In each location, fine-grained ferrous magnesite exists in the coarse-grained dolomite–ankerite. Textural evidence in the Newania carbonatites indicates that the magnesian siderite (Table 21, anal. 7) and ankerite are primary liquidus phases evolving directly from the Ca-poor, Mg-rich carbonatitic liquid generated by partial melting of phlogopite–carbonate peridotite in the mantle at pressures >32 kbar. The occurrence of magnesite with enstatite and talc in the rock-type sagvandite was considered to be of primary igneous origin, but Johannes (1970) suggested that unless originating in the Earth's mantle, sagvandites are not magmatic rocks but are formed by CO_2 metasomatism. Nodules and veins of compact magnesite occur with other carbonates and quartz in vitreous volcanic tuffs and associated dacitic and andesitic lava flows in Nevada, USA, where the magnesite is considered to be due to hypogene alteration by solution of $MgH_2(CO_3)_2$ (Faust and Callaghan, 1948).

Zhang and Liou (1994) have reported the occurrence of magnesite with 0.13 wt. per cent CaO and Mg/(Mg + Fe) = 0.94, together with diopside, in garnet websterite in the Dabie ultra-high-pressure metamorphic terrain of central China, which may have formed at about 56 kbar and 800–900 °C. In the same region, magnesite ($Mg_{0.75}Fe_{0.25}CO_3$) occurs in coesite-bearing eclogite and is estimated to have formed at a pressure >26 kbar. These and other reported occurrences (e.g. Katsura *et al.*, 1991; Yang *et al.*, 1993) show clearly that magnesite is a stable mineral of the mantle. Experimental work, e.g. by Katsura *et al.* (1991) and Brey *et al.* (1983), has also indicated that the assemblage diopside + magnesite is stable at mantle depths.

Experimental work at high pressures has shown that dolomite breaks down to a $CaCO_3 + MgCO_3$ assemblage at 20–50 GPa (Biellmann *et al.*, 1993); equimolar mixed powders of dolomite + enstatite were transformed into magnesite + calcite perovskite. These results appear to indicate that magnesite is the stable carbonate in the presence of silicates in the Earth's lower mantle down to at least 1500 km. Biellmann *et al.* also noted the eutectic intergrowth of magnesite and magnesiowüstite, suggesting a possible mutual solubility between these two phases at high P and T.

Magnesite may form in sedimentary environments as either primary precipitates or secondary recrystallized products. In the modern sediments of Glacial Lake, Bonneville, Utah, and of the Coorong, South Australia, Graf *et al.* (1961) and Borch (1965), respectively, reported the formation of magnesite in the modern sediments. Magnesite derived from hydromagnesite was described by Alderman (1965). The Trucial Coast of the Arabian Gulf is an area of extensive subtidal or intertidal carbonate sedimentation, and supratidal dolomite and magnesite precipitation (Bush, 1973; Purser and Evans, 1973). Magnesite occurs within the sediments of the mid-sabkha above the buried microbial mats. There is no aragonite in the carbonate sediment. According to Bush, the magnesite probably forms after dolomitization has removed all aragonite.

In the British Permian evaporites, magnesite is the normal carbonate of the potash zones of the Upper and Middle evaporites, and the only carbonate in the zones of halides in the Lower evaporite (Stewart, 1965).

Occurrences of detrital magnesite have been described in Yugoslavia in the Janok and Badanj deposits, and in Greece in the Euboea deposits; their most outstanding feature is that they are typical clastic rocks of magnesite detritus bound by various kinds of cement (Ilić, 1981).

References

Aharon, P., 1988. A stable isotope study of magnesite from the Rum Jungle uranium field, Australia: implications for the origin of strata-bound massive magnesites. *Chem. Geol.*, **69**, 127–45.

Alderman, A. R., 1965. Dolomitic sediments and their environment in the south-east of South Australia. *Geochim. Cosmochim. Acta*, **29**, 1355–65.

Barnes, I., O'Neil, J. R., Rapp, J. B. and **White, D. E.**, 1973. Silica–carbonate alteration of serpentine: wall-rock alteration in mercury deposits of the California coast ranges. *Econ. Geol.*, **68**, 388–98.

Bassett, H., 1951. New and old minerals in Tanganyika. *Rec. Geol. Surv. Tanganyika*, **1**, 60 pp.

Beran, A., 1978. Die Bestimmung der Zusammensetzung von Dolomit–Ankerit und Magnesit–Siderit–Mischkristallen mit Hilfe von Reflexionsmessungen. *Neues Jahrb. Min., Mh.*, 559–65.

Biellmann, C., Gillet, P., Guyot, F., Peyronneau, J. and **Reynard, B.**, 1993. Experimental evidence for carbonate stability in the Earth's lower mantle. *Earth Planet. Sci. Lett.*, **118**, 31–41.

Borch, C. Von Der, 1965. The distribution and preliminary geochemistry of modern carbonate sediments of the Coorong area, South Australia. *Geochim. Cosmochim. Acta*, **29**, 781–99.

Brey, G., Brice, W. R., Ellis, D. J., Green, D. H., Harris, K. L. and **Ryabchikov, I. D.**, 1983. Pyroxene–carbonate reactions in the upper mantle. *Earth Planet. Sci. Lett.*, **62**, 63–74.

Buckley, H. A. and **Woolley, A. R.**, 1990. Carbonates of the magnesite–siderite series from four carbonatite complexes. *Min. Mag.*, **54**, 413–8.

Bush, P., 1973. Some aspects of the diagenetic history of the Sabkha in Abu Dhabi, Persian Gulf. In *The Persian Gulf*: Ed. B. H. Purser, Berlin (Springer-Verlag), pp. 395–407.

Byrnes, A. P. and **Wyllie, P. J.**, 1981. Subsolidus and melting relations for the join $CaCO_3$–$MgCO_3$ at 10 kbars. *Geochim. Cosmochim. Acta*, **45**, 321–8.

Chatzidimitriadis, E., Kelepertsis, A. B. and **Christaras, V. B.**, 1981. Magnesite occurrences at the areas of 'Basilika' and 'Galavinos' of the Chalkidiki province (Greece). In *An International Symposium on Metallogeny of Mafic and Ultramafic Complexes: the Eastern Mediterranean–Western Asia area, and its Comparison with Similar Metallogenic Environments in the World*: Ed. S. S. Augustithis, **3**, 246–68.

Chou, L., Garrels, R. M. and **Wollast, R.**, 1989. Comparative study of the kinetics and mechanisms of dissolution of carbonate minerals. *Chem. Geol.*, **78**, 269–82.

Christ, C. L. and **Hostetler, P. B.**, 1970. Studies in the system MgO–SiO_2–CO_2–H_2O. II: The activity-product constant of magnesite. *Amer. J. Sci.*, **268**, 439–53.

Dal Negro, A. and **Ungaretti, L.**, 1971. Refinement of the crystal structure of aragonite. *Amer. Min.*, **56**, 768–72.

De Aranjo, J. P., 1927. Notes on the mineralogy of Brazil. *Bull. Soc. Franc. Min. Crist.*, **50**, 6–41.

DeVilliers, J. P. R., 1971. Crystal structures of aragonite, strontianite, and witherite. *Amer. Min.*, **56**, 758–67.

Dubrawski, J. V., Channon, A. L. and **Warne, S. St J.**, 1989. Examination of the siderite–magnesite mineral series by Fourier transform infrared spectroscopy. *Amer. Min.*, **74**, 187–90.

Effenberger, H., Mereiter, K. and **Zemann, J.**, 1981. Crystal structure refinements of magnesite, calcite, rhodochrosite, siderite, smithsonite and dolomite, with discussion of some aspects of the stereochemistry of calcite-type carbonates. *Z. Krist.*, **156**, 233–43.

Faust, G. T. and **Gallaghan, E.**, 1948. Mineralogy and petrology of the Currant Creek magnesite deposits and associated rocks of Nevada. *Bull. Geol. Soc. Amer.*, **59**, 11–74.

Figuet, G., Guyot, F. and **Itié, J.**, 1994. High-pressure X-ray diffraction study of carbonates: $MgCO_3$, $CaMg(CO_3)_2$, and $CaCO_3$. *Amer. Min.*, **79**, 15–23.

Friedman, G. M., 1959. Identification of carbonate minerals by staining methods. *J. Sed. Petr.*, **29**, 87–97.

Frost, M. T., 1982. The magnesite deposit at Main Creek, Savage River, Tasmania. *Econ. Geol.*, **77**, 1901–11.

Gaines, A. M. and **Goldsmith, J. R.**, 1971. Crystal chemistry and stability relations in the system $MgCO_3$–$NiCO_3$. *Z. Krist.*, **133**, 432–44.

Gies, H., 1975. Activation possibilities and geochemical correlation of photoluminescing carbonates, particularly calcites. *Min. Deposita*, **10**, 216–27.

Gillet, P., 1993. Stability of magnesite ($MgCO_3$) at mantle pressure and temperature conditions: a Raman spectroscopy study. *Amer. Min.*, **78**, 1328–31.

Goldsmith, J. R. and **Graf, D. L.**, 1960. Subsolidus phase relations in the system $CaCO_3$–$MgCO_3$–$MnCO_3$. *J. Geol.*, **68**, 324–35.

Goldsmith, J. R. and **Northrop, D. A.**, 1965. Subsolidus phase relations in the systems $CaCO_3$–$MgCO_3$–$CaCO_3$ and $CaCO_3$–$MgCO_3$–$NiCO_3$. *J. Geol.*, **73**, 817–29.

Goldsmith, J. R., Graf, D. L., Witters, J. and **Northrop, D. A.**, 1962. Studies in the system $CaCO_3$–$MgCO_3$–$FeCO_3$: 1. Phase relations; 2. A method for major element spectrochemical analysis; 3. Compositions of some ferroan dolomites. *J. Geol.*, **70**, 659–88.

Graf, D. L., Earley, A. J. and **Shimp, N. F.**, 1961. A preliminary report on magnesium carbonate formation in Glacial Lake, Bonneville. *J. Geol.*, **69**, 219–23.

Greenwood, H. J., 1967. Mineral equilibria in the system MgO–SiO_2–CO_2–H_2O. In *Researches in Geochemistry*: Ed. P. H. Abelson, **2**, 542–67.

Griffin, R., 1972. Genesis of a magnesite deposit, Deloro Twp., Ontario. *Econ. Geol.*, **67**, 63–71.

Grubb, P. L. C., 1981. Genesis of an unusual magnesite deposit in Bahia State, Brazil. In *An International Symposium on Metallogeny of Mafic and Ultramafic Complexes: the Eastern Mediterranean–Western Asia Area, and its Comparison with Similar Metallogenic Environments in the World*: Ed. S. S. Augustithis, **3**, 361–74.

Harker, R. I., 1958. The system MgO–CO_2–A, and the effect of inert pressure on certain types of hydrothermal reaction. *Amer. J. Sci.*, **256**, 128–38.

Harker, R. I. and **Tuttle, O. F.**, 1955. Studies in the system CaO–MgO–CO_2. Part I. The thermal dissociation of calcite, dolomite and magnesite. *Amer. J. Sci.*, **253**, 209–24.

Hemingway, B. S., Robie, R. A., Fisher, J. R. and **Wilson, W. H.**, 1977. Heat capacities of gibbsite, $Al(OH)_3$, between 13 and 480K magnesite, $MgCO_3$, between 13 and 380K and their standard entropies at 298.15K, and the heat capacities of calorimetry conference benzoic acid between 12 and 316K. *J. Res. U. S. Geol. Surv.*, **5**, 797–806.

Huang, C. K. and **Kerr, P. F.**, 1960. Infrared study of the carbonate minerals. *Amer. Min.*, **45**, 311–24.

Huang, W. L. and **Wyllie, P. J.**, 1976. Melting relationships in the system CaO–CO_2 and MgO–CO_2 to 33 kbars. *Geochim. Cosmochim. Acta*, **40**, 129–32.

Ignatieff, A., 1930. Magnesite deposits in Serbia. *Mining Mag.*, **42**, 217–21.

Ilić, M., 1981. A brief survey of magnesite deposits of the detrital type. In *An International Symposium on Metallogeny of Mafic and Ultramafic Complexes: the Eastern Mediterranean–Western Asia Area, and its Comparison with Similar Metallogenic Environments in the World*: Ed. S. S. Augustithis, **3**, 284–92.

Irving, A. J. and **Wyllie, P. J.**, 1975. Subsolidus and melting relationships for calcite, magnesite and the join $CaCO_3$–$MgCO_3$ to 36 kbars. *Geochim. Cosmochim. Acta*, **39**, 35–53.

Isherwood, B. J. and **James, J. A.**, 1976. Structural dependence of the optical birefringence of crystals with calcite- and aragonite-type structures. *Acta Cryst.*, **A32**, 340–1.

Johannes, W., 1970. Zur Entstehung von Magnesitvorkommen. *Neues Jahrb. Min., Abh.*, **113**, 274–325.

Katsura, T., Tsuchida, Y., Ito, E., Yagi, T., Utsumi, W. and **Akimoto, S.**, 1991. Stability of magnesite under the lower mantle conditions. *Proc. Japan Acad.*, **67**, 57–60.

Kennedy, G. C., 1947. Charts for correlation of optical properties with chemical composition of some rock-forming minerals. *Amer. Min.*, **32**, 561–73.

Kohls, J. W. and **Rodda, J. L.**, 1966. Gaspéite $(Ni,Mg,Fe)CO_3$, a new carbonate from the Gaspé Peninsula, Quebec. *Amer. Min.*, **51**, 677–84.

Krishna Rao, B., Prasad, M. H., Vasudev, V. N. and **Sethumadluv, M. S.**, 1994. Origin of grunerite schist-hosted magnesite mineralization in Copper Mountain area, Sandur schist belt, Kamataka. *J. Geol. Soc. India*, **44**, 267–74.

Kushiro, I., Satake, H. and **Akimoto, S.**, 1975. Carbonate–silicate reactions at high pressures and possible presence of dolomite and magnesite in the upper mantle. *Earth Planet. Sci. Lett.*, **28**, 116–20.

Langmuir, D., 1965. Stability of carbonates in the system MgO–CO_2–H_2O. *J. Geol.*, **73**, 730–54.

Lo, B. W. M., 1973. The polarizabilities of planar nitrate, carbonate and borate anions in crystals. *J. Phys. Chem. Solids*, **34**, 514–520.

Mariano, A. N., Pojasek, W. J. and **Bender, S. L.**, 1969. Nickel carbonate from northeast Transvaal. *Can. Min.*, **10**, 140 (abstr.).

Kushiro, I., Satake, H. and **Akimoto, S.**, 1975. Carbonate–silicate reactions at high pressures and possible presence of dolomite and magnesite in the upper mantle. *Earth Planet. Sci. Lett.*, **28**, 116–20.

Langmuir, D., 1965. Stability of carbonates in the system $MgO–CO_2–H_2O$. *J. Geol.*, **73**, 730–54.

Lo, B. W. M., 1973. The polarizabilities of planar nitrate, carbonate and borate anions in crystals. *J. Phys. Chem. Solids*, **34**, 514–520.

Mariano, A. N., Pojasek, W. J. and **Bender, S. L.**, 1969. Nickel carbonate from Northeast Transvaal. *Can. Min.*, **10**, 140 (abstr.).

Markgraf, S. A. and **Reeder, R. J.**, 1985. High-temperature structure refinements of calcite and magnesite. *Amer. Min.*, **70**, 590–600.

Nickel, E. H., 1973. An occurrence of gaspéite and pecoraite in the Nullagine region of Western Australia. *Min. Mag.*, **39**, 113–5.

Oh, K. D., Morikawa, H., Iwai, S. and **Aoki, H.**, 1973. The crystal structure of magnesite. *Amer. Min.*, **58**, 1029–33.

Peng, W., Liu, G. and **Ke, L.**, 1985. Infrared spectra study of magnesite–siderite series. *Acta Min. Sinica*, **5**, 229–33.

Purser, B. H. and **Evans, G.**, 1973. Regional sedimentation along the Trucial Coast, S. E. Persian Gulf. In *The Persian Gulf*: Ed. B. H. Purser, Berlin (Springer-Verlag), pp. 211–32.

Rosenberg, P. E., 1963. Synthetic solid solutions in the systems $MgCO_3–FeCO_3$ and $MnCO_3–FeCO_3$. *Amer. Min.*, **48**, 1396–400.

Rosenberg, P. E., 1967. Subsolidus relations in the system $CaCO_3–MgCO_3–FeCO_3$ between 350 and 550 °C. *Amer. Min.*, **52**, 787–96.

Sayles, F. L. and **Fyfe, W. S.**, 1973. The crystallization of magnesite from aqueous solution. *Geochim. Cosmochim. Acta*, **37**, 87–99.

Shaikh, N. A. and **Brusewitz, A. M.**, 1975. FeO content and cell parameter of magnesite from the Rakas and Äpartjåkka deposits, Sweden. *Geol. Fören. Stockholm. Förh.*, **97**, 195–7.

Smykatz-Kloss, W., 1964. Differential-Thermal-Analyse von einigen Karbonat-Mineralen. *Beit. Min. Petrogr.*, **9**, 481–502.

Spangenberg, K., 1949. Zur Genesis der Magnesitlagerstätte vom Galgenberg bei Zobten (Schlesien). *Neues Jahrb. Min., Mh. A.*, 117–190.

Stevens, R. G. and **Carron, M. K.**, 1948. Simple field test for distinguishing minerals by abrasion pH. *Amer. Min.*, **33**, 31–49.

Stewart, F. H., 1965. The mineralogy of the British Permian evaporites. *Min. Mag.*, **34**, 460–70.

Warne, S. St J., 1962. A quick field or laboratory staining scheme for the differentiation of the major carbonate minerals. *J. Sed. Petr.*, **32**, 29–38.

Weber, H. J., 1988. Anisotropic bond polarizabilities in birefringent crystals. *Acta Cryst.*, **A44**, 320–6.

White, W. B., 1974. The carbonate minerals. In *The Infrared Spectra of Minerals*: Ed. V. C. Farmer, *Min. Soc. Monograph*, **4**, 227–84.

Yang, J., Godard, G., Kienast, J. R., Lu, Y. and **Sun, J.**, 1993. Ultrahigh-pressure (60 kbar) magnesite-bearing garnet peridotites from north-eastern Jiangsu, China. *J. Geol.*, **101**, 541–54.

Yue, C. S., Fuo, K. F. and **Shih, C. I.**, 1964. Hoshiite, $NiMg(CO_3)_2$, a new mineral. *Acta Geol. Sinica*, **44**, 213–8.

Zhang, R. Y. and **Liou, J. G.**, 1994. Significance of magnesite paragenesis in ultrahigh-pressure metamorphic rocks. *Amer. Min.*, **79**, 397–400.

Rhodochrosite $MnCO_3$

Trigonal (−)

ε	1.597
ω	1.816
δ	0.219
Dispersion:	Strong.
D	3.70
H	$3\frac{1}{2}$–4
Cleavage[1]:	$\{10\bar{1}1\}$ perfect.
Twinning[1]:	Rare lamellar twinning on $\{01\bar{1}2\}$.
Colour:	Rose-pink, pink, red or brown; colourless to pale pink in thin section.
Pleochroism:	Red varieties may show pleochroism with absorption $\omega > \varepsilon$.
Unit cell:	a_{hex} 4.768 Å, c_{hex} 15.635 Å, $Z_{hex}=6$. a_{rh} 5.915 Å, α 47° 43′, $Z_{rh}=2$. Space group $R\bar{3}c$

Dissolves with effervescence in warm dilute acids. On exposure to air, may develop a brown or black surface alteration product.

Rhodochrosite may occur in distinct crystals, but it is usually found in a massive form or as a granular aggregate. Its common occurrences are in hydrothermal veins with ores of silver, lead and copper, and in contact metasomatic deposits. It is also found in Mn-rich sediments in association with siderite. The name is from the Greek *rhodo*, for *rose*, and *chrosis*, for a colouring.

Structure

The structure of rhodochrosite was investigated by Wyckoff (1920) and shown to be isostructural with calcite. Refinements by Effenberger *et al.* (1981) assign a space group of $R\bar{3}c$ to rhodochrosite with a 4.7682, c 15.6354 Å and Z = 6. The interatomic distances are: C–O, 1.2867(5) Å (1.2898 Å with a correction for riding motion); Mn–O, 2.1900(5) Å; d_1 (edge of the MnO_6 octahedron inclined to (0001)), 3.1447(6) Å; d_2 (edge of the MnO_6 octahedron parallel to (0001)), 3.0489(8) Å; and d_3 (shortest O–O distance outside the coordination figure), 2.9531(5) Å. The cell dimensions of rhodochrosite were measured by high-temperature X-ray diffraction (Rao and Murthy, 1970). Both a and c increase with increasing temperature from 30 to 265 °C, as shown in Fig. 50. The cell dimensions measured by Rao and Murthy at room temperature are 4.7719 and 15.6443 Å.

Goldsmith and Graf (1957) found that the variations of $d_{10\bar{1}4}$ and $d_{31\bar{4}8}$ (for the

[1]Cleavage, twinning and morphology are indexed using the hexagonal morphological cell with quadrupled a parameter (see Table 16, p. 103).

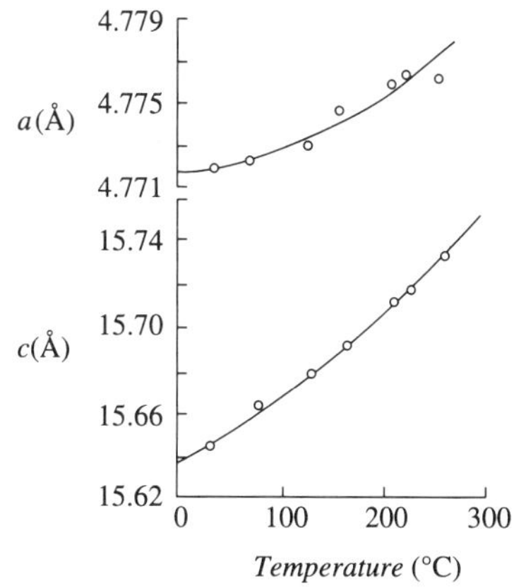

Fig. 50. Variation of cell dimensions of $MnCO_3$ with temperature (after Rao and Murthy, 1970).

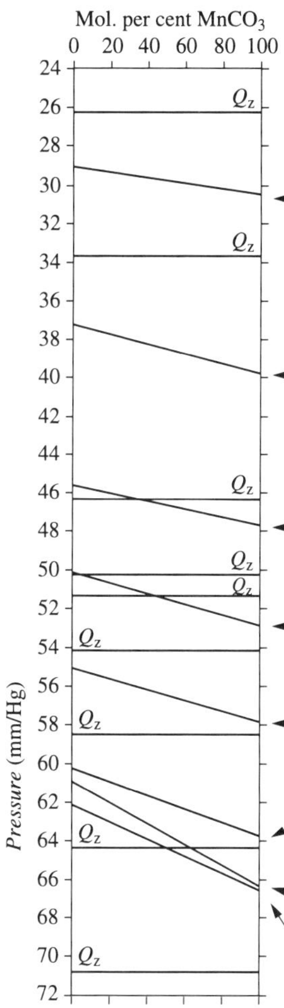

h k l	Mol. per cent $MnCO_3$	uncert. Mol. per cent
102	$\frac{x - 29.03}{1.52} \times 100$	3.3
104	$\frac{x - 37.16}{2.53} \times 100$	2.0
110	$\frac{x - 45.65}{1.95} \times 100$	2.6
113	$\frac{x - 50.12}{2.61} \times 100$	1.9
202	$\frac{x - 55.02}{2.66} \times 100$	1.9
204	$\frac{x - 60.25}{3.42} \times 100$	1.5
208	$\frac{x - 60.80}{5.34} \times 100$	1.0
116	$\frac{x - 62.12}{4.29} \times 100$	1.2

Fig. 51. X-ray calibration curves for eight reflections and the formulae to determine the composition of $(Mn,Ca)CO_3$ solid solutions (after De Capitani and Peters, 1981).

hexagonal structural cell) compared with those of calcite were linear for 0–100 mol. per cent $MnCO_3$, and used such variations to determine compositions within this series. X-ray calibration curves for the eight strongest reflections of $(Ca,Mn)CO_3$ solid solutions have been established by De Capitani and Peters (1981). As shown in Fig. 51, linear relationships exist for all reflections, and formulae to determine the compositions of $(Ca,Mn)CO_3$ solid solutions are also listed. Rosenberg (1963) found a linear relationship between $d_{21\bar{3}1}$ and composition in the system $MnCO_3$–$FeCO_3$.

Chemistry

Pure rhodochrosite, $MnCO_3$, is rare because of the existence of solid solutions between $CaCO_3$, $MgCO_3$, $FeCO_3$ and $MnCO_3$. A variety of rhodochrosite containing up to 20 per cent $FeCO_3$ has been called ponite, the term manganosiderite being used where Mn ~ Fe. A ferroan zincian variety, capillitite, was described by Galloni (1950). Among the purest recorded, the rhodochrosite from N'chwaning mine, Kalahari manganese field, has only 0.1 per cent CaO and 0.2 per cent FeO (Wilson and Dunn, 1978). A list of selected compositions of rhodochrosite is shown in Table 22.

Subsolidus phase relationships in the system $MnCO_3$–$CaCO_3$ were determined by Goldsmith and Graf (1957) and De Capitani and Peters (1981), showing a miscibility gap stable at temperatures <600 °C (Fig. 36, p. 112). Kutnahorite ($CaMn(CO_3)_2$) was not found in the system. However, as observed by Goldsmith (1983) and Peacor *et al.* (1987), the ordering reflections of kutnahorite are generally weak, and they cannot be identified even in some naturally occurring kutnahorite. Single-crystal examination is necessary. Complete series of solid solutions exist in the system $MnCO_3$–$MgCO_3$ above 500 °C (Goldsmith and Graf, 1960) and the

Table 22. Rhodochrosite Analyses

	1	2	3	4	5	6
MnO	29.90	52.03	59.07	58.96	31.91	55.77
FeO	—	0.19	0.49	1.32	—	2.50
MgO	2.11	1.29	0.96	0.66	2.74	1.31
CaO	25.30	7.28	1.90	0.71	23.85	1.51
CO_2	40.72	39.51(c)	39.40	38.66(c)	41.50(c)	38.35
Total	98.03	100.30	101.82	100.31	100.00	99.44
	Numbers of ions on the basis of six (O)					
Mn	0.911	1.642	1.859	1.892	0.954	1.817
Fe	—	0.006	0.015	0.042	—	0.080
Mg	0.113	0.072	0.053	0.037	0.144	0.075
Ca	0.975	0.289	0.075	0.029	0.902	0.062
C	2.001	2.000	1.998	2.000	2.000	2.013

(c) Calculated from stoichiometry.

1 Manganese carbonate in Panama Basin sediments (Pedersen and Price, 1982).
2 Rhodochrosite from deep-sea sediments, northwestern Pacific Ocean (Logvinenko *et al.*, 1972).
3, 4 Manganese carbonate nodules in mudstone of the Cretaceous Susaki Formation, Kochi, Shikoku, Japan (Kashima and Motomura, 1977).
5 Average composition of manganese carbonate concretions in Recent sediments of Loch Fyne, Argyllshire, Scotland (Calvert and Price, 1970).
6 Rhodochrosite from Ryujima mine, Japan (Tsusue, 1967); ε 1.592, ω 1.812; a 4.7693, C 15.654 Å.

Table 22. *Continued*

	7	8	9	10	11	12
MnO	31.00	24.0	55.82	39.92	50.61	57.80
FeO	29.2	32.4	2.40	0.08	4.60	0.82
MgO	0.2	0.4	2.05	2.01	0.33	1.50
CaO	0.4	0.8	—	12.41	0.38	1.29
	1.0 (ZnO)	4.1 (ZnO)		5.37 (BaO) 0.11 (PbO) 3.48 (oth.)	9.14 (oth.)	
CO_2	38.3	38.0	38.35	36.47	35.04	38.00
Total	100.1	99.7	98.62	99.85	100.00	99.41
	Numbers of ions on the basis of six (O)					
Mn	1.003	0.78	1.841	1.335	1.792	1.871
Fe	0.946	1.04	0.078	0.003	0.161	0.026
Mg	0.011	0.023	0.003	0.118	0.021	0.058
Ca	0.016	0.033	—	0.525	0.017	0.053
	0.028 (Zn)	0.12 (Zn)		0.083 (Ba) 0.001 (Pb)		
C	1.997	2.000	2.038	1.961	2.004	1.982

7, 8 Rhodochrosite–siderite carbonate coexisting with late-stage sphalerite from Au–Ag vein deposits, Japan (Shikazono, 1977).
9 Secondary rhodochrosite in sulphide vein, Catamarca, Argentina (Della Sala *et al.*, 1973).
10 Rhodochrosite from the Långban mine, Sweden (Sundius, 1962).
11 Vein manganese carbonate from the Chvaletice deposit, Bohemia (Žák and Povondra, 1980).
12 Rhodochrosite from metamorphosed manganese deposits at Koduru and Garbham, India (Sivaprakash, 1980).

Table 22. *Continued*

	13	14	15	16	17
MnO	56.61	48.60	26.35	55.85	32.58
FeO	0.72	0.18	1.68	—	—
MgO	0.34	0.25	1.71	—	0.38
CaO	3.26	11.42	27.75	4.80	26.16
CO_2(c)	38.45	39.47	41.01	38.42	41.15
Total	99.38	99.92	98.5	99.07	100.27
	Numbers of ions on the basis of six (O)				
Mn	1.826	1.527	0.797	1.803	0.982
Fe	0.023	0.006	0.050	—	—
Mg	0.019	0.014	0.091	—	0.020
Ca	0.133	0.454	1.062	0.196	0.997
C	2.000	2.000	2.000	2.000	2.000

13–15 Rhodochrosite–kutnahorite series from the metamorphosed manganese deposit at Bald Knob, North California, USA (Winter *et al.*, 1981).
16, 17 Manganese carbonates in metamorphic assemblages from the Alps (Peters *et al.*, 1978).

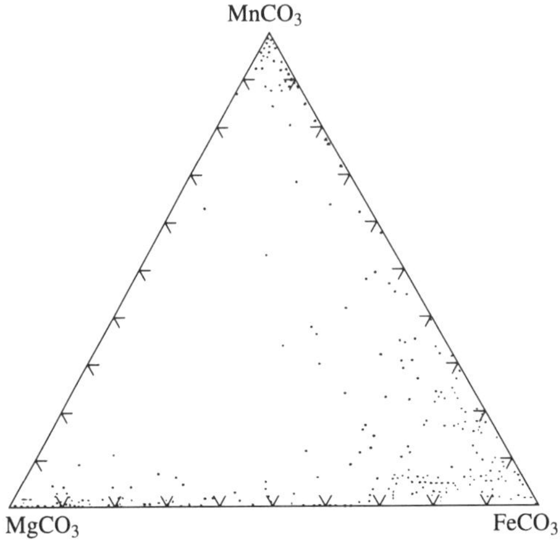

Fig. 52. Reported chemical compositions (in mol. per cent) of carbonates in the system $MnCO_3$–$MgCO_3$–$FeCO_3$ (after Essene, 1983).

system $MnCO_3$–$FeCO_3$ above 450 °C (Rosenberg, 1963). Both temperatures represent the lowest temperatures to which experimental studies have been conducted. A compositional diagram has been constructed by Essene (1983) which includes the majority of published analyses of carbonates in the system $MnCO_3$–$MgCO_3$–$FeCO_3$ from 1911 to 1982 (Fig. 52). The lack of known carbonate composition between rhodochrosite and magnesite, as highlighted by Essene, may be due to the presence of a miscibility gap between $MgCO_3$ and $MnCO_3$ at temperatures <500 °C or to petrological phenomena such as the partitioning of cations between carbonates and other phases.

Zinc is also a common substituent for Mn in rhodochrosite (Table 22, anal. 8). In the oxidized zone at Broken Hill, New South Wales, Australia, Birch (1986) reported a suite of rhodochrosite samples with $ZnCO_3$ contents up to 47 per cent. For the Mn-rich members of this suite, $PbCO_3$ was detected in solid solution with a maximum of 3 per cent. Birch (1986) found that $d_{10\bar{1}4}$ varies linearly with the ratio of $MnCO_3/(MnCO_3+ZnCO_3)$, and was greater than $d_{10\bar{1}4}$ for the corresponding members of the pure $ZnCO_3$–$MnCO_3$ series. He attributed this difference to the Pb and Ca content of the carbonates of the Broken Hill suite.

The dissociation of $MnCO_3$ was detected at 125 °C in air and at 325 °C in nitrogen (Seguin, 1972). A first-order equation is established for the reaction, and the activation energies calculated and extrapolated to 120 °C are as follows: for 325 mesh samples 18.1 kcal/mol. in nitrogen and 19.9 kcal/mol. in air; for 100–200 mesh samples, 26.8 kcal/mol. in nitrogen and 27.2 kcal/mol. in air. The intermediate and final products of dissociation are, respectively, manganosite (MnO) and hausmannite (Mn_3O_4); and hausmannite, γ-Mn_2O_3 and bixbyite (α-Mn_2O_3). No pyrolusite was observed. Differential thermal analysis curves obtained by Seguin show a double endothermal peak at 612 and 643 °C, due to the decomposition of manganese carbonate, and an exothermal peak at 736 °C, due to the oxidation of manganosite to hausmannite. Under pressure and as a function of f_{O_2}, the stability of rhodochrosite was determined by Huebner (1969). His results are shown in Fig. 53. The equilibria rhodochrosite–manganosite–gas and rhodochrosite–

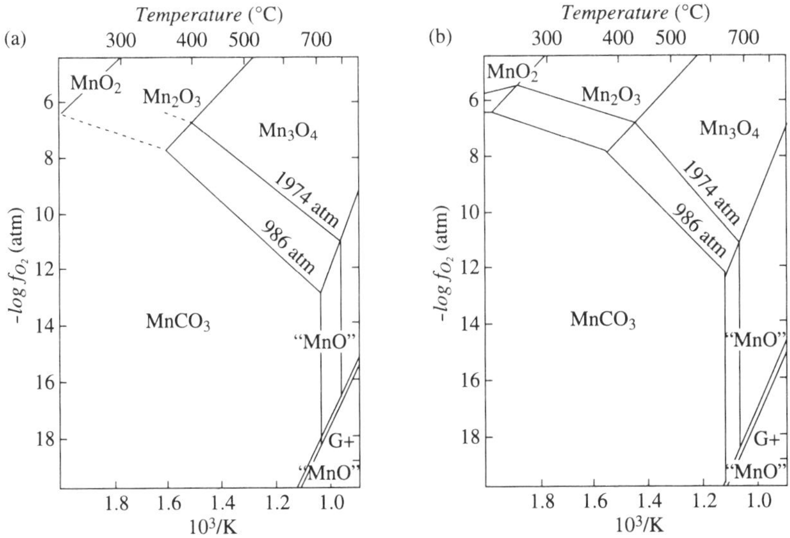

Fig. 53. Stability field of rhodochrosite (a) experimentally determined and (b) calculated (after Huebner, 1969); G represents a gas phase.

hausmannite–gas were calculated from equations:

$$\log f_{CO_2}(\text{atm}) = 8.662 - 5556/T + 0.0944(P/T)$$

and

$$\log K = 40.39 - 9429/T(°\text{K}) + 0.486(P/T).$$

A comparison between the experimentally determined and calculated relationships shows a best correlation for the reaction rhodochrosite to bixbyite + gas at 960 atm, and a least correlation for the reaction rhodochrosite to manganosite + gas. The reaction $MnCO_3 + SiO_2 = MnSiO_3 + CO_2$ was investigated by Abrecht (1988). Both experimentally determined and calculated phase relationships are shown in Fig. 54. The experimental data presented by Abrecht (1988) require non-ideal mixing behaviour of the CO_2–H_2O fluid under the given P–T conditions. In comparison with the reaction $CaCO_3 + SiO_2 = CaSiO_3 + CO_2$ (Greenwood, 1967), the increase in non-ideality appears to be caused by the reaction with $MnCO_3$ at lower temperatures (Abrecht, 1988). Peters *et al.* (1973) have examined the manganese carbonate–silicate assemblages experimentally in the system MnO–SiO_2–CO_2–H_2O at 2 kbar. Results are expressed in the following equations:

rhodochrosite + quartz = pyroxmangite + CO_2

with $\log_{10} K = -11.765/T + 18.618$, and

rhodochrosite + pyroxmangite = tephroite + CO_2

with $\log_{10} K = -7.083/T + 11.870$.

The reaction, 2 rhodochrosite + quartz = tephroite + $2CO_2$, was not determined experimentally, but a calculation from the two reactions above gave an equilibrium relationship $\log_{10} K = -18.848/T + 30.488$. For the fourth reaction involved, tephroite + quartz = 2 pyroxmangite, experimental data illustrate that starting

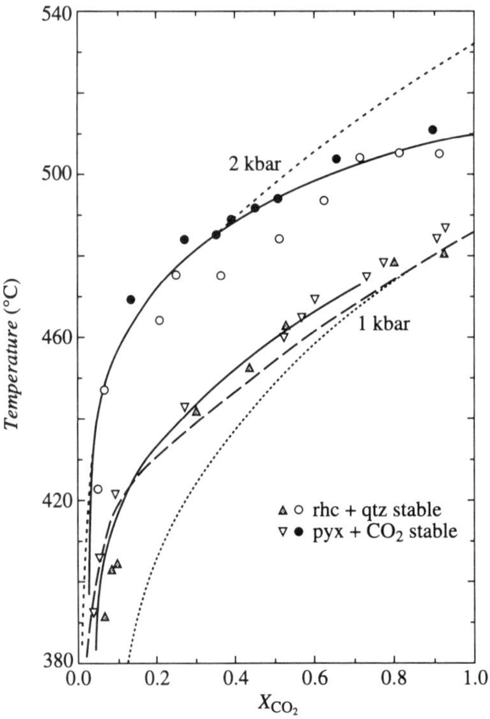

Fig. 54. Temperature–composition relationships for the reaction: $MnCO_3 + SiO_2 = MnSiO_3 + CO_2$ at 1 kbar (after Abrecht, 1988). Calculated relationships for non-ideal fluid mixing and ideal fluid mixing are represented by dashed and dotted lines, respectively; rhc, pyx and qtz denote rhododochrosite, pyroxmangite and quartz, respectively.

from rhodo-chrosite and quartz at very low X_{CO_2}, tephroite is obtained below 420 °C and pyroxmangite is formed above 430 °C.

The heat capacity of rhodochrosite having the composition $(Mn_{0.994}Fe_{0.005}Mg_{0.001})CO_3$ was measured between 5 and 600 K by Robie *et al.* (1984). At 298.15 K, the heat capacity (C_p), entropy ($S^\circ_T - S^\circ_0$), enthalpy $(H^\circ_T - H^\circ_0)/T$, and Gibbs energy $(-(G^\circ_T - H^\circ_0)/T)$ of rhodochrosite are (in J/mol.K) 80.62 ± 0.10, 98.03 ± 0.10, 47.37 and 50.66, respectively. With temperature increasing from 5 to 370 K, the values of all thermodynamic functions increase. The heat capacity between 350 and 600 K can be represented by the equation

$$C^\circ_p = 149.70 + 0.0187T - 1314.2T^{-0.5} + (1.417 \times 10^5 T^{-2}).$$

The stoichiometric solubility products of rhodochrosite in sea-water of 34.27‰ salinity at 25.0 and 3.3 °C are $(3.24 \pm 0.23) \times 10^{-9}$ and $(2.28 \pm 0.24) \times 10^{-9}$, respectively (Johnson, 1982). The solubility product extrapolated to infinite dilution at 25 °C is $(2.60 \pm 0.07) \times 10^{-11}$. This value agrees well with the value (2.9×10^{-11}) calculated from the free energy data tabulated by Robie *et al.* (1978), but differs from solubility products measured by Morgan (1967) (3.9×10^{-11}) and Garrels *et al.* (1960) (5.0×10^{-11}). Johnson attributed the differences to the effects of adsorption.

Rhodochrosite alters rather readily, developing a superficial brown or black layer of complex manganese oxides. It may eventually alter to manganite (MnO·OH) or pyrolusite (MnO_2).

As an industrial mineral, rhodochrosite is used in the manufacture of other Mn^{2+} salts, ferrites and welding rods.

Optical and Physical Properties

Rhodochrosite is uniaxial and optically negative. Basal sections show a distinct uniaxial cross with numerous isochromes, and isogyres that fan out strongly towards the periphery. Birefringence is extreme ($\omega - \varepsilon = 0.219$) and produces pale high-order colours in thin sections and grain mounts, even for orientations near the optic axis. The structural dependence of the birefringence of carbonates has been examined by Lo (1973), Isherwood and James (1976) and Weber (1988). The refractive indices of rhodochrosite increase linearly towards those of other members when Mn is substituted by Fe^{2+} and Zn^{2+}, but decreases with Ca^{2+} (Winchell and Meek, 1947).

Generally, rhodochrosite is pink to rose-red, and becomes brownish or yellowish when appreciable amounts of Fe^{2+} are present. The rhodochrosite from Långban, Sweden, with 12.41 per cent CaO and 5.37 per cent BaO (Table 22, anal. 10) is in most cases very pale and sometimes pure white. The finest crystals of rhodochrosite occur in the ore vein of the St Elmo and May Murphy mines, Chaffee County; Moose mine, Gilpin County; John Reed mine, Lake County; and Eagle and Rawley mines, Saguache County, all in Colorado, USA; all specimens have the rose-red colour (Eckel, 1961). From the White Raven mine, Boulder County, Colorado, Eckel reported a ferroan rhodochrosite of light greenish colour. Manganese oxide alteration products may cause rhodochrosite to become brown along cleavages or superficially brown or black.

The specific gravity of rhodochrosite obtained by Wayland (1942) is 3.70, but its variations with composition have not yet been determined, especially towards the end-members that form extensive solid solutions with $MnCO_3$. It is expected that the specific gravity should be increased by the substitution of Fe^{2+} or Zn, and decreased by the substitution of Ca or Mg for Mn. The hardness of rhodochrosite ranges from $3\frac{1}{2}$ to 4, but the rhodochrosite from Argentina (Table 22, anal. 9), with 2.40 per cent FeO and 2.05 per cent MgO, has a hardness of $4\frac{1}{2}$ (Della Sala *et al.*, 1973).

The infrared spectra of rhodochrosite, like those of calcite, are characterized by three prominent absorption maxima, as shown in Fig. 43 (p. 120). In the calcite–rhodochrosite series, significant frequency shifts were observed by Böttcher *et al.* (1991) in the ν_2 and ν_4 vibrational bands, reflecting the different ionic radii and masses of the end-members.

Coefficients of thermal expansion of $MnCO_3$ from 30 to 265 °C were determined by Rao and Murthy (1970). The temperature dependence of the coefficients '$\alpha_{\parallel}$' along the z axis and '$\alpha_{\perp}$' perpendicular to the z axis is represented by the following equations:

$$\alpha_{\parallel} = 22.942 \times 10^{-6} - 5.55 \times 10^{-8} T + 3.361 \times 10^{-10} T^2$$

and

$$\alpha_{\perp} = 0.740 \times 10^{-6} + 2.812 \times 10^{-8} T - 6.722 \times 10^{-12} T^2.$$

The thermal behaviour of rhodochrosite is similar to that of calcite and magnesite in having a relatively large coefficient of expansion along the *z* axis and a small coefficient of expansion in the perpendicular direction. The *z* direction is normal to the CO_3 layers in carbonates.

From alternating-current susceptibility measurement on a 250 μm-sized synthetic crystal of rhodochrosite, Maartense (1969, 1971) obtained a value for T'_N of 34.5 ± 0.1 K. Meijer *et al.* (1970) gave $T'_N = 32.55$ and 32.69 K by static (direct-current) susceptibility measurement on rhodochrosite from Sonora, Mexico, which has appreciably less manganese than required by $MnCO_3$. In $MnCO_3$, the magnetic moments of Mn^{2+} ions are aligned, antiferromagnetically, normal to $[11\bar{2}1]$ (Alikhanov, 1959). Dzyaloshinsky (1958) has shown that because of this alignment in $MnCO_3$, a weak ferromagnetism can be produced in the $(11\bar{2}1)$ plane.

Distinguishing Features

Rhodochrosite is distinguished from other calcite-type carbonate minerals by its pink colour and by its common alteration to a brown or black oxide of manganese. In association, rhodochrosite is commonly found with other manganese minerals. X-ray powder diffraction was used for the determination of members in the systems $CaCO_3$–$MnCO_3$ (Goldsmith and Graf, 1957), $FeCO_3$–$MnCO_3$ (Rosenberg, 1963) and $MgCO_3$–$MnCO_3$ (Goldsmith and Graf, 1960), and for natural samples in the series $ZnCO_3$–$MnCO_3$ by Birch (1986). Rhodochrosite may be distinguished from a member of the siderite–magnesite series by its ω refractive index and its specific gravity, because for a given value of ω its specific gravity is about 0.1 higher than that for a magnesian siderite (Wayland, 1942).

Like siderite and magnesite, rhodochrosite is not stained by alizarin red S, but Dickson (1966) has suggested that using potassium ferricyanide on rhodochrosite, a very pale brownish stain was produced which was not evident on other carbonate minerals. Ferroan calcite and ferroan dolomite do react with potassium ferricyanide, but the result is a pale to deep turquoise stain. Warne (1962) recommended the use of benzidine, which gives rhodochrosite a blue stain. Rhodochrosite does not show luminescence as strongly as calcite and smithsonite (Gies, 1975).

Paragenesis

Rhodochrosite is commonly found as a primary mineral in hydrothermal vein deposits. In the well-known Butte deposits of Montana, USA, rhodochrosite is abundant in the peripheral epithermal zone, which is characterized by Ag, Pb and Zn mineralization; the rhodochrosite is in association with rhodonite, hübnerite, alabandite, helvite and quartz. The central mesothermal zone and the intermediate zone of the deposit are barren or contain minor amounts of Zn, Pb and Mn, but these zones have abundant Fe- and Cu-sulphide ores. The Cu and Fe mineralization in the epithermal zone is minor (Allsman, 1956). In the Philipsburg mining district, Montana (Prinz, 1967), rhodochrosite (87 per cent $MnCO_3$) and manganoan dolomite (30 per cent $MnCO_3$) are present as important gangue minerals associated with Pb–Zn–Ag sulphide ores. In the oxidized part of the hydrothermal veins, there is the fairly widespread development of supergene Mn minerals, including cryptomelane, pyrolusite, chalcophanite and coronadite. Rhodochrosite and calcite are the dominant gangue minerals in the Taebaek Pb–Zn–Ag deposit, Yeonhwa, Korea (So *et al.*, 1993).

Casadevall and Ohmoto (1977) examined the Sunnyside deposits in Eureka, Colorado, USA, and established six periods of mineralization: (1) pyrite–quartz ores, (2) banded quartz–sulphide ores, (3) massive Pb–Zn–Fe–Cu ores, (4) gold telluride–quartz ores, (5) Mn ores, and (6) quartz–fluorite–carbonate–sulphate ores. The Mn ores are mainly pyroxmangite and rhodochrosite, with minor amounts of alabandite, tephroite, hübnerite and helvite. The temperature of mineralization during periods (1)–(5) was determined to be between 250 and 320 °C.

Hydrothermal Mn vein deposits have been described from Precambrian schists, Palaeozoic dolomites and limestones, Mesozoic andesite, granite and sedimentary rocks, and Cainozoic rhyolite of Korea (Kim, 1972). Manganese-bearing veins of rhodochrosite and magnesian kutnahorite transect a slate–chert–sandstone sequence in the Ryujima mine, Nagano, Japan, and were found to be the result of a single period of mineralization from a hydrothermal solution related to volcanism (Tsusue, 1967). Vein rhodochrosite in mineralized limestone and calcareous sandstone at Chvaletice, Czechoslovakia, was thought to have originated from a sedimentary Proterozoic Fe-rhodochrosite by Alpine paragenesis (Žák and Povondra, 1980).

Volcanogenic sedimentary manganese deposits of Precambrian to Recent age are widely distributed in the world. The greenstone formation of the Olympic Peninsula, Washington, USA, contains rhodochrosite and manganoan calcite in association with Mn-silicates and oxides. The spilite, diabase, tuff, greywacke, limestone and jasper are in a geosynclinal sequence, overlain by a volcanic–sedimentary sequence (Park, 1946). Boström (1973) considered that the replacement of whole pillows of spilites by Mn minerals is an indication of a later influx of hot manganiferous fluid.

Rhodochrosite has also been recorded from pegmatites, as at Berry, Maine, USA, where it occurs with manganese phosphate minerals on amblygonite lining pockets in microcline (Berman and Gonyer, 1930).

Thin layers and lenses of manganese carbonate (rhodochrosite)–silicate (tephroite, pyroxmangite and quartz) rocks of the jasperoid formation occur in meta chert, pelitic schists and metavolcanic rocks in the Swiss Alps. Peters *et al.* (1973) examined this mineral assemblage experimentally at 2 kbar. Results, as discussed above (p. 155), illustrate that starting with rhodochrosite and quartz at very low X_{CO_2}, tephroite was obtained below 420 °C and pyroxmangite was formed above 430 °C. The manganese carbonate–silicate assemblage in the Koduru deposits, Andhra Pradesh, India, consists of rhodochrosite–pyroxmangite–tephroite–rhodonite (Sivaprakash, 1980). Quartz was not observed in the assemblage, and has apparently been consumed in a reaction with rhodochrosite to produce pyroxmangite. The temperatures of the reaction, rhodochrosite + pyroxmangite = tephroite + CO_2, may have been lower in the Koduru deposit in comparison with the equilibrium temperatures established by Peters *et al.* (1973), because P_{CO_2} was less than P_{total}. However, because total pressures are much higher than in the experiments the reaction, rhodochrosite + pyroxmangite = tephroite + CO_2, may have occurred at a temperature higher than 581 °C in the Koduru deposit.

Rhodochrosites of both hydrothermal and sedimentary types have been reported from Pay Khoy in the central part of the Yugorskiy Peninsula, NE Russia (Yudovich and Ketris, 1982). The hydrothermal rhodochrosite occurs with quartz in thin, closely spaced, parallel veins and is covered by black crusts and sintered deposits of iron and manganese hydroxides. The sedimentary rhodochrosite was deposited as a result of the accumulation of manganese and iron in the form of concretions.

Manganese deposits in Nikopal and Bol'shoi Tokmak in south Ukraine are of sedimentary origin and consist of three distinct zones: the oxide (pyrolusite, psilomelane), the oxide–carbonate (psilomelane, manganite, manganoan calcite, rhodochrosite) and the carbonate (calcian rhodochrosite, manganoan calcite) (Hewitt, 1966). In the Molango District, Mexico, Okita (1992) reported the largest Mn deposit in North America, which consists predominantly of rhodochrosite. It was formed by the early diagenetic reduction of manganese oxides through the oxidation of organic matter and iron sulphides. There is a mineralogical change in the deposit upwards, from rhodochrosite in the ore zone, to Mn-calcite + kutnahorite ± rhodochrosite, and to Mn-calcite. Manganese carbonate ores hosted in the Black Shale Series have been described by Fan *et al.* (1992). The ores include rhodochrosite, calcian rhodochrosite, kutnahorite, manganocalcite, Mn-containing calcite and Mn-containing dolomite. Processes proposed for the formation of the carbonate ores are concentration in sea-water, precipitation from sea-water and enrichment in the sediments.

Manganese deposits in present-day basins occur mainly as nodules, and to a lesser extent as crusts and coatings on volcanic rock fragments, detrital sediments and dead or living organisms. The most common manganese minerals formed in these deposits are todorokite, birnessite and pyrolusite. Rhodochrosite and other Mn-rich carbonates are formed through diagenetic processes in freshwater lakes (Damiani *et al.*, 1973) and in a marine environment (Logvinenko *et al.*, 1972; Rozhanov *et al.*, 1973). Compositions of manganese carbonates formed in Recent sediments range from $(Mn_{70-60}Ca_{30-32}Mg_{0-8})CO_3$ in the Baltic Sea (Manheim, 1961), to $(Mn_{50-80}Ca_{50-20})CO_3$ in the eastern marginal Pacific (Lynn and Bonatti, 1965), and to $(Mn_{47.7}Ca_{45.1}Mg_{7.2})CO_3$ in Loch Fyne, Argyllshire, Scotland (Calvert and Price, 1970). The rhodochrosite from the deep-sea sediments of the Pacific Ocean has an average composition of $(Mn_{0.82}Ca_{0.14}Mg_{0.036}Fe_{0.003})CO_3$ (Logvinenko *et al.*, 1972).

Rhodochrosite stalagmites from the Oe mine, Hokkaido, Japan (Kojima *et al.*, 1983) show a columnar shape, 2–30 mm in diameter and 1–12 cm in length. The crystals are various colours, from rose pink to pale purple and grey, and consist of solution-growth textures such as concentric banding. The rhodochrosite stalagmites show a considerable increase in $MnCO_3$ content from core to rim.

References

Abrecht, J., 1988. Experimental evaluation of the $MnCO_3 + SiO_2 + CO_2$ equilibrium at 1 kbar. *Amer. Min.*, **73**, 1285–91.

Alikhanov, R. A., 1959. Neutron diffraction investigation of the antiferromagnetism of the carbonates of manganese and iron. *J. Exp. Theoretical Phys.*, **9**, 1204–8.

Allsman, P., 1956. Oxidation and enrichment of manganese deposits of Butte, Montana. *Mining Eng.*, **8**, 1110–2.

Berman, H. and **Gonyer, F. A.**, 1930. Pegmatite minerals from Poland, Maine. *Amer. Min.*, **15**, 375–87.

Birch, W. D., 1986. Zinc–manganese carbonates from Broken Hill, New South Wales. *Min. Mag.*, **50**, 49–52.

Boström, K., 1973. The origin and fate of ferromanganoan active ridge sediments. *Stockholm Contrib. Geol.*, **24**, 149–243.

Böttcher, M. E., **Gehlken, P. L.** and **Usdowski, E.**, 1991. Infrared spectroscopic investigations of the calcite–rhodochrosite and parts of the calcite–magnesite mineral series. *Contrib. Min. Petr.*, **109**, 304–6.

Calvert, S. S. and **Price, N. B.**, 1970. Composition of manganese nodules and manganese carbonates from Loch Fyne, Scotland. *Contrib. Min. Petr.*, **29**, 215–33.

Casadevall, T. and **Ohmoto, H.**, 1977. Sunnyside mine, Eureka mining district, San Juan County, Colorado: geochemistry of gold and base metal ore deposition in a volcanic environment. *Econ. Geol.*, **72**, 1285–320.

Damiani, V., **Morton, T. W.** and **Thomas, R. I.**, 1973. Freshwater ferromanganese nodules from the Big Bay Section of Quint, Northern Lake, Ontario. *Proc. Conf. Great Lakes Res.*, **16**, 397–403.

De Capitani, C. and **Peters, Tj.**, 1981. The solvus in the system $MnCO_3$–$CaCO_3$. *Contrib. Min. Petr.*, **76**, 394–400.

Della Sala, J., **Fuentes, J. C.**, **Herrero, M.**, **Laserre, N.** and **Viand, J.**, 1973. Contribution to the knowledge of the Argentine rhodochrosite. *Gems Gemol.*, **14**, 141–3.

Dickson, J. A. D., 1966. Carbonate identification and genesis as revealed by staining. *J. Sed. Petr.*, **36**, 491–505.

Dzyaloshinsky, I., 1958. A thermodynamic theory of 'weak' ferromagnetism of antiferromagnetics. *J. Phys. Chem. Solids*, **4**, 241–55.

Eckel, E. R., 1961. Minerals of Colorado – a 100 Years Record. *U.S. Geol. Surv. Bull.*, **1114**, 399 pp.

Effenberger, H., **Merreiter, K.** and **Zemann, J.**, 1981. Crystal structure refinements of magnesite, rhodochrosite, siderite, smithsonite, and dolomite, with discussion of some aspects of the stereochemistry of calcite-type carbonates. *Z. Krist.*, **156**, 233–43.

Essene, E. J., 1983. Solid solutions and solvi among metamorphic carbonates with applications to geologic thermobarometry. In *Carbonates: Mineralogy and Chemistry*: Ed. R. J. Reeder, *Reviews in Mineralogy*, **11**, 77–96, Mineralogical Society of America, Washington, DC.

Fan, D., **Liu, T.** and **Ye, T.**, 1992. The process of formation of manganese carbonate deposits hosted in Black Shale Series. *Econ. Geol.*, **87**, 1419–29.

Galloni, E. E., 1950. The crystal structure of ferroan zincian rhodochrosite. *Amer. Min.*, **35**, 562–70.

Garrels, R. M., **Thompson, M. E.** and **Siever, R.**, 1960. Stability of some carbonates at 25 °C and one atmosphere total pressure. *Amer. J. Sci.*, **258**, 402–18.

Gies, H., 1975. Activation possibilities and geochemical correlations of photoluminescing carbonates, particularly calcite. *Min. Deposita*, **10**, 216–27.

Goldsmith, J. R., 1983. Phase relations of rhombohedral carbonates. In *Carbonates: Mineralogy and Chemistry*: Ed. R. J. Reeder, *Reviews in Mineralogy*, **11**, 49–76, Mineralogical Society of America, Washington, DC.

Goldsmith, J. R. and **Graf, D. L.**, 1957. The system CaO–MnO–CO_2: solid solutions and decomposition reactions. *Geochim. Cosmochim. Acta*, **11**, 310–34.

Goldsmith, J. R. and **Graf, D. L.**, 1960. Subsolidus phase relations in the system $CaCO_3$–$MgCO_3$–$MnCO_3$. *J. Geol.*, **68**, 324–35.

Greenwood, H. J., 1967. Wollastonite: stability in H_2O–CO_2 mixture and occurrence in a contact metamorphic aureole near Salmo, British Columbia, Canada. *Amer. Min.*, **52**, 1669–80.

Hewitt, D. F., 1966. Stratified deposits of the oxides and carbonates of manganese. *Econ. Geol.*, **61**, 431–61.

Henry, D. A., 1988. Minerals from the Clunes Goldfield. *Austr. Min.*, **3**, 43–52.

Huebner, J. S., 1969. Stability relations of rhodochrosite in the system manganese–carbon–oxygen. *Amer. Min.*, **54**, 457–81.

Isherwood, B. J. and **James, J. A.**, 1976. Structural dependence of the optical birefringence of crystals with calcite- and aragonite-type structures. *Acta Cryst.*, **A32**, 340–1.

Johnson, K. S., 1982. Solubility of rhodochrosite ($MnCO_3$) in water and seawater. *Geochim. Cosmochim. Acta*, **46**, 1805–9.

Kashima, N. and **Motomura, Y.**, 1977. [Mn-carbonate nodule from the Susaki formation of the Shimanto belt, Shikoku.] *Sci. Rept. Dept. Geol., Kyushu Univ.*, **12**, 273–8 (in Japanese with English summary).

Kim, S. J., 1972. Manganese deposits of Korea. *Acta Min. Petrogr. Szeged.*, **20**, 255–64.

Kojima, S. and **Sugaki, A.**, 1983. [On the rhodochrosite stalagmite from the Oe mine, Hokkaido, Japan.] *J. Japan Assoc. Mineralogists, Petrologists & Economic Geologists*, **78**, 491–6 (in Japanese with English abstract).

Lo, B. W. N., 1973. The polarizability of planar nitrate, carbonate and borate anions in crystals. *J. Phys. Chem. Solids*, **34**, 513–20.

Logvinenko, N. V., **Volkov, I. I.** and **Sokolova, Ye. G.**, 1972. Rhodochrosite in deep sea sediments of the Pacific Ocean. *Dokl. Acad. Sci. USSR, Earth Sci. Sect.*, **203**, 178–81.

Lynn, D. C. and **Bonatti, E.**, 1965. Mobility of manganese in diagenesis of deep-sea sediments. *Marine Geol.*, **3**, 457–74.

Maartense, I., 1969. Magnetism of synthetic and natural $MnCO_3$. *Phys. Rev.*, **188**, 924–9.

Maartense, I., 1971. A note on the critical behavior of $MnCO_3$ and NiF_2. *Solid State Comm.*, **9**, 2071–6.

Manheim, F. T., 1961. A geochemical profile in the Baltic Sea. *Geochim. Cosmochim. Acta*, **25**, 52–61.

Meijer, H. C., Pimmelaar, L. M. W. A., Brouwer, S. R. and **Handel, Van den, Jr.,** 1970. Some magnetic properties of $MnCO_3$ at low field strength. *Physica*, **46**, 279–90.

Morgan, J. J., 1967. Chemical equilibria and kinetic properties of manganese in natural waters. In *Principles and Applications of Water Chemistry*: Eds. S. D. Faust and J. V. Hunter, New York (John Wiley & Sons), pp. 561–622.

Okita, P. M., 1992. Manganese carbonate mineralization in the Molango District, Mexico. *Econ. Geol.*, **87**, 1345–66.

Park, C. F., 1946. The spilite and manganese problems of the Olympic Peninsula, Washington. *Amer. J. Sci.*, **244**, 305–23.

Peacor, D. R., Essene, E. J. and **Gaines, A. M.,** 1987. Petrologic and crystal-chemical implications of cation order–disorder in kutnahorite [$CaMn(CO_3)_2$]. *Amer. Min.*, **72**, 319–28.

Pedersen, T. F. and **Price, N. B.,** 1982. The geochemistry of manganese carbonate in Panama Basin sediments. *Geochim. Cosmochim. Acta*, **46**, 59–68.

Peters, Tj., Schwander, H. and **Trommsdorf, V.,** 1973. Assemblages among tephorite, pyromangite, rhodochrosite, quartz: experimental data and occurrences in the Rhetic Alps. *Contrib. Min. Petr.*, **42**, 325–32.

Peters, Tj., Trommsdorf, V. and **Sommerauer, J.,** 1978. Manganese pyroxenoids and carbonates: critical phase relations in metamorphic assemblages from the Alps. *Contrib. Min. Petr.*, **66**, 383–8.

Prinz, W. C., 1967. Geology and ore deposits of the Philipsburg District, Granite County, Montana. *U.S. Geol. Surv. Bull.*, **1237**, 66 pp.

Rao, K. V. K. and **Murthy, S.,** 1970. Thermal expansion of manganese carbonate. *J. Materials Sci.*, **5**, 82–3.

Robie, R. A., Hemingway, B. S. and **Fisher, J. R.,** 1978. Thermodynamic properties of minerals and related substances at 298.15K and 1 bar (10^5 Pa) pressure and at higher temperatures. *U.S. Geol. Surv. Prof. Paper*, **525D**, 65–72.

Robie, R. A., Haselton, H. T., Jr. and **Hemingway, B. S.,** 1984. Heat capacities and entropies of rhodochrosite ($MnCO_3$) and siderite ($FeCO_3$) between 5 and 600K. *Amer. Min.*, **69**, 349–57.

Rosenberg, P. E., 1963. Synthetic solid solution in the system $MgCO_3$–$FeCO_3$ and $MnCO_3$–$FeCO_3$. *Amer. Min.*, **48**, 1396–400.

Rozhanov, A. G., Sokolov, V. S. and **Volkov, I. I.,** 1973. Forms of iron and manganese in sediments of northwest Pacific Ocean. *Lithol. Min. Res.*, **7**, 423–33.

Seguin, M. K., 1972. Study of the stability of $MnCO_3$ in inert atmosphere and in air. *Amer. Min.*, **57**, 571–23.

Shikazono, N., 1977. Composition of sediments and the environments of formation of vein-type deposits in Japan. *Econ. Geol.*, **12**, 632–41.

Sivaprakash, C., 1980. Mineralogy of manganese deposits of Koduru and Garbyam, Andhra Pradesh, India. *Econ. Geol.*, **75**, 1083–104.

So, C. S., Yun, S. T. and **Koh, Y. K.,** 1993. Mineralogic, fluid inclusion, and stable isotope evidence for the genesis of carbonate-hosted Pb–Zn(–Ag) orebodies of the Taebaek deposit, Republic of Korea. *Econ. Geol.*, **88**, 855–72.

Sundius, N., 1962. Rhodochrosite from Långban. *Arkiv. Min. Geol.*, **3**, 293–5.

Tsusue, A., 1967. Magnesian kutnahorite from Ryujima mines, Japan. *Amer. Min.*, **52**, 1751–61.

Warne, S. St. J., 1962. A quick field or laboratory staining scheme for the differentiation of the major carbonate minerals. *J. Sed. Petr.*, **32**, 29–38.

Wayland, R. G., 1942. Composition, specific gravity and refractive indices of rhodochrosite; rhodochrosite from Butte, Montana. *Amer. Min.*, **27**, 614–28.

Weber, H. J., 1988. Anisotropic bond polarizabilities in birefringent crystals. *Acta Cryst.*, **A44**, 320–6.

Wilson, W. E. and **Dunne, D. J.,** 1978. Famous mineral localities: the Kalahari manganese field. *Min. Record*, **9**, 137–53.

Winchell, A. N. and **Meek, W. B.,** 1947. Birefringence–dispersion ratio as a diagnostic. *Amer. Min.*, **32**, 336–43.

Winter, G. A., Essene, E. J. and **Peacor, D. R.,** 1981. Carbonates and pyroxenoids from the manganese deposit near Bald Knob, North Carolina. *Amer. Min.*, **66**, 278–89.

Wyckoff, R. W. G., 1920. The crystal structure of some carbonates of the calcite group. *Amer. J. Sci.*, 4th Ser., **50**, 317–60.

Yudovich, Ya. E. and **Ketris, M. P.,** 1982. Manganese carbonate ores of the Pay Khoy. *Dokl. Acad. Sci. USSR, Earth Sci. Sect.*, **257**, 139–42.

Žák, L. and **Povondra, P.,** 1980. Vein manganese carbonates from the Chvaletice deposit (East Bohemia). *Casopis pro Min. Geol.*, **25**, 369–76.

Siderite $FeCO_3$

Trigonal (−)

ε	1.635
ω	1.875
δ	0.240
Dispersion:	Strong.
D	3.970
H	4–4½
Cleavage[1]:	$\{10\bar{1}1\}$ perfect.
Twinning[1]:	Occasional lamellar twinning on $\{01\bar{1}2\}$, rare simple twinning on $\{0001\}$.
Colour:	Yellowish brown, brown or dark brown; colourless to yellowish brown in thin section.
Unit cell:	a_{hex} 4.691 Å, c_{hex} 15.379 Å, $Z_{hex}=6$. a_{rh} 5.795 Å, α 47° 43′, $Z_{rh}=2$. Space group $R\bar{3}c$

Slowly soluble in cold dilute HCl, dissolves with effervescence in hot acid.

Siderite occurs in bedded sedimentary iron ores, usually massive, but also in botryoidal and globular forms and in earthy masses. It is also found in hydrothermal sulphide veins and in some metamorphic deposits. The name is derived from the Greek *sideros*, for iron, in reference to its composition. Various names have been given to siderite that contains appreciable amounts of manganese, magnesium, calcium and cobalt. The name chalybite (Greek, of steel) has been used by many in the past.

Structure

Siderite is isostructural with calcite (Wyckoff, 1920). Refinements by Effenberger *et al.* (1981) assign a space group of $R\bar{3}c$ to siderite with *a* 4.6916, *c* 15.3796 Å and $Z=6$. The interatomic distances are: C–O, 1.2869(5) Å (1.2895 Å with a correction for riding motion); Fe–O, 2.1445(5) Å; d_1 (edge of the FeO_6 octahedron inclined to (0001)), 3.0864(6) Å; d_2 (edge of the FeO_6 octahedron parallel to (0001)), 2.9781(9) Å; and d_3 (shortest O–O distance outside the coordination figure), 2.9080(5) Å. The C–O distances of siderite (1.2869 Å) and calcite (1.2815 Å) represent two of the most extreme values obtained for the calcite-type carbonates and illustrate the small variation in the C–O bond length for minerals of this group, including calcite, magnesite, rhodochrosite, siderite and smithsonite.

Rosenberg (1963) examined the solid solution series in the system $MgCO_3$–$MnCO_3$, showing a linear relationship between $d_{21\bar{3}1}$ and composition. From the

[1] Cleavage, twinning and morphology are indexed using the hexagonal morphological cell with quadrupled *a* parameter (see Table 16, p. 103).

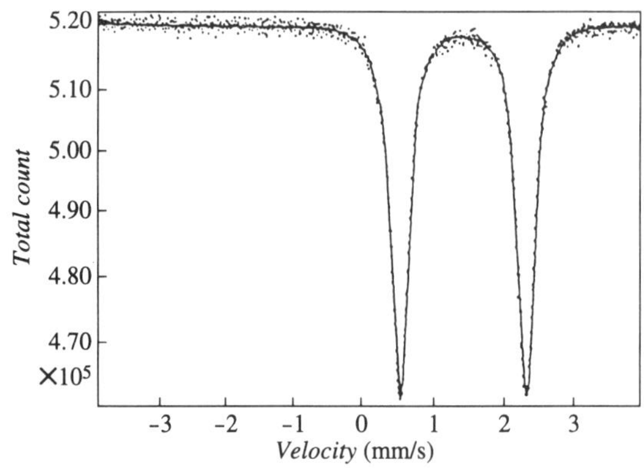

Fig. 55. Mössbauer spectrum of polycrystalline $FeCO_3$ at 295 K (after Nagy *et al.*, 1975).

Cracow–Silesian (Poland) Zn–Pb deposits, Bak and Żabiński (1981) found carbonate minerals with compositions ranging from siderite to smithsonite ($ZnCO_3$). In the series, cell dimensions vary linearly with composition; *c* shows a more pronounced change than *a*, probably because of the deformation of the FeO_6 octahedron as a result of the Jahn–Teller effect (Rosenberg and Foit, 1979).

Mössbauer measurements on $FeCO_3$ have been made by Ôno and Ito (1964), and subsequently by Nagy *et al.* (1975), Amigo and Fortune (1981), Gallagher *et al.* (1981) and others. The spectrum with an isomer shift of 1.38 ± 0.02 mm/s and quadrupole splitting of 1.87 ± 0.10 mm/s is shown in Fig. 55. Between 0 and 500 K, both isomer shift and quadrupole splitting show a linear decrease with temperature above the Néel temperature (about 32–40 K) (Nagy *et al.*, 1975). From the known temperature dependence of quadrupole splitting, Nagy *et al.* (1975) obtained a high saturation value for $FeCO_3$ (~ 2.1 mm/s), which suggests a predominant ionic $FeCO_3$.

Chemistry

Pure siderite, $FeCO_3$, is rare, because the substitution of Fe^{2+} by other metallic ions, especially Mg^{2+} and Mn^{2+}, is common. Magnesian varieties have been called sideroplesite for 5–30 per cent $MgCO_3$, and pistomesite for 30–50 per cent $MgCO_3$. A range of chemical compositions of siderite is given in Table 23.

Phase relationships in the system $MgCO_3$–$FeCO_3$ were studied by Goldsmith *et al.* (1962) and Rosenberg (1963, 1967). A complete series of solid solution has been well established. Representatives of this complete series were found by Buckley and Woolley (1990) in their study of four carbonatite complexes and their data are given in Fig. 47 (p. 140). An examination of the analyses listed in Table 23 shows that MgO is present in siderite from various geological environments. Specimens from siderite–sulphide nodules of low-grade Archaean banded iron formation (Table 23, anal. 23) and from cryolite–siderite dykes at Ivigtut, Greenland (anal. 10), perhaps represent siderite closest to its ideal composition.

Like Mg, manganese is commonly present in siderite as can be seen in Table 23. A collection of published chemical data from 1912 to 1978 by Essene (1983) illustrates the natural occurrence of a complete series from Fe-rich to Mn-rich

Table 23. Siderite Analyses

	1	2	3	4	5	6
FeO	53.69	46.01	52.81	51.78	52.22	54.00
MnO	0.50	1.57	2.41	0.98	2.53	0.01
MgO	5.25	9.24	0.28	4.68	3.58	1.03
CaO	0.13	0.66	0.71	2.91	2.35	5.08
	2.10 (oth.)	1.48 (oth.)	5.46 (oth.)			
			0.76 (Fe_2O_3)			
CO_2	40.25(c)	39.76(c)	35.47	39.72(c)	39.31(c)	38.19(c)
Total	101.92	98.72	98.17	100.07	99.99	98.31
	Numbers of ions on the basis of six (O)					
Fe	1.685	1.417	1.769	1.597	1.627	1.732
Mn	0.016	0.049	0.081	0.031	0.080	—
Mg	0.294	0.507	0.016	0.257	0.199	0.059
Ca	0.005	0.026	0.030	0.115	0.094	0.209
C	2.000	1.999	2.051	2.000	2.000	2.000

(c) Calculated from stoichiometry.

1 Diagenetic siderite from ironstone in northwest Hebei, China (Chen *et al.*, 1982).
2 Sedimentary siderite from Guizhou, China (Liao, 1984).
3 Chemically precipitated siderite in Jurassic greywackes and siltstones, Otway Coast, Victoria, Australia (Gill and Segnit, 1980).
4, 5 Sideritic concretions, Yorkshire, England (Pearson, 1974).
6 Siderite from the Clunies gold-bearing reefs in Ordovician sediments, Clunies Goldfield, Australia (Henry, 1988).

Table 23. *Continued*

	7	8	9	10	11	12
FeO	43.80	37.47	41.32	59.08	52.3	36.4
MnO	0.38	1.27	1.28	2.95	9.8	0.97
MgO	13.73	17.04	14.31	—	0.29	16.80
CaO	0.66	1.14	0.75	—	0.17	—
						2.9 (oth.)
CO_2	42.57(c)	43.23(c)	42.31(c)	37.88	37.4	42.4
Total	101.14	100.15	99.97	99.91	99.96	99.47
	Numbers of ions on the basis of six (O)					
Fe	1.260	1.062	1.196	1.908	1.695	1.061
Mn	0.011	0.037	0.037	0.096	0.321	0.029
Mg	0.704	0.861	0.738	—	0.017	0.873
Ca	0.024	0.041	0.028	—	0.007	—
C	1.999	2.001	1.999	1.997	1.980	2.018

6 Siderite from the Clunies gold-bearing reefs in Ordovician sediments, Clunies Goldfield, Australia (Henry, 1988).
7–9 Siderite from differentiated dolerite, Kalgoorlie gold deposit, Australia (Phillips and Brown, 1987).
10 Siderite from Ivigtut, Greenland, US National Museum R2460 (Jarosewich and Macintyre, 1983).
11, 12 Siderite from pegmatite dykes from Dolni Bory, West Moravia (Nemec, 1977).

Table 23. *Continued*

	13	14	15	16	17	18
FeO	40.56	30.43	44.58	31.3	34.6	56.18
MnO	1.56	21.08	2.00	23.0	20.1	1.00
MgO	12.47	—	10.99	0.6	1.3	2.71
CaO	2.07	3.32	1.06	1.1	1.3	1.19
		7.06 (ZnO)		5.6 (ZnO)	4.7 (ZnO)	
CO_2	41.05(c)	38.11(c)	41.37(c)	38.5	38.6	38.92(c)
Total	97.71	100.00	100.00	100.1	100.6	100.00
	Numbers of ions on the basis of six (O)					
Fe	1.210	0.978	1.320	1.017	1.097	1.768
Mn	0.047	0.686	0.060	0.747	0.645	0.032
Mg	0.663	—	0.580	0.034	0.073	0.152
Ca	0.079	0.136	0.040	0.045	0.053	0.048
		0.200 (Zn)		0.158 (Zn)	0.132 (Zn)	
C	2.000	2.000	2.000	2.009	1.999	2.000

13 Magnesian siderite from carbonatites, Chipman Lake, Ontario (Buckley and Woolley, 1990).
14 Zincian oligonspar from iron–manganese ores of the Atasu deposits, USSR (Shchibrik and Taranushich, 1971).
15 Siderite from sulphide-bearing hydrothermal vein, Stara Voda-Svedlar, Czechoslovakia (Václav, 1982).
16, 17 Siderite coexisting with sphalerite in a hydrothermal vein, Japan (Shikozono, 1977).
18 Siderite at bottom horizon of sideritic ores with claystones, Czestochowa (Zakzewski, 1974).

Table 23. *Continued*

	19	20	21	22	23	24
FeO	53.39	40.91	33.46	50.62	59.70	46.20
MnO	1.41	7.75	7.58	1.28	0.05	0.44
MgO	2.96	9.06	14.07	6.80	0.15	10.13
CaO	3.08	1.42	2.01	0.42	0.04	0.48
CO_2	39.22(c)	40.86(c)	42.88(c)	39.55(c)	36.79(c)	40.01(c)
Total	100.06	100.00	100.00	98.67	96.73	97.26
	Numbers of ions on the basis of six (O)					
Fe	1.667	1.226	0.973	1.568	1.988	1.415
Mn	0.045	0.235	0.223	0.040	0.002	0.014
Mg	0.165	0.484	0.729	0.376	0.009	0.553
Ca	0.123	0.054	0.075	0.017	0.002	0.019
C	2.000	1.999	2.000	2.001	2.001	2.001

19 Siderite from a concretionary clay-ironstone band, Yorkshire, England (Curtis *et al.*, 1975).
20 Siderite from metamorphosed iron formation, Quebec, Canada (Butler, 1969).
21 Siderite from Proterozoic iron formation, Labrador Trough, Canada (Klein, 1978).
22 Medium-grained siderite in finely banded stilpnomelane–siderite–magnetite assemblages from low-grade metamorphic iron formation, Howells River area, Quebec (Klein, 1974).
23 Siderite from siderite–sulphide nodule in low-grade Archaean banded iron formation, Weld Range, Western Australia (Gole, 1980).
24 Siderite from siderite–sulphide chamosite–minnesotaite banded iron formation, Weld Range, Western Australia (Gole, 1980).

end-members (Fig. 52). Phase relationships in the system $FeCO_3$–$MnCO_3$ were studied by Rosenberg (1963), and results obtained at 450 °C support the existence of a complete series.

Bak and Żabiński (1981) found carbonate minerals with compositions ranging from siderite to smithsonite ($ZnCO_3$) in the Cracow–Silesian (Poland) Zn–Pb deposits. Monheimite (Palache *et al.*, 1951) was found by Bak and Niec (1978) with a composition of $(Fe_{0.55}Zn_{0.45})CO_3$ representing a member of the siderite–smithsonite series. From its occurrence in a part of the deposits not affected by weathering, Bak and Niec suggest that it originated by endogenic ore mineralization. Siderites with zinc content up to 5–7 wt. per cent (Table 23, anals. 14, 16 and 17) were reported by Shchibrik and Taranushich (1971) and Shikazono (1977).

Experimental Work

In the system $CaCO_3$–$FeCO_3$, limited ranges of solid solution exist at both ends, and $CaFe(CO_3)_2$ is not a stable phase (Fig. 37, p. 112). At 450 °C, the series can accommodate a maximum of 32 mol. per cent $FeCO_3$, and compositions richer in $FeCO_3$ produce binary and ternary assemblages (see Figs 71 and 83). Compositional data for co-existing Fe-bearing calcite + Ca-bearing siderite and a thermodynamic model for the $(Ca,Fe)CO_3$ join were presented by Davidson *et al.* (1993), who found that the stability of ordered $CaFe(CO_3)_2$ is limited to below ~450 °C by the breakdown to calcite + siderite; increasing pressure may stabilize the disordered phase.

A differential thermal analysis (DTA) of siderite has been performed in many studies. However, dispute still exists because the DTA characteristics are strongly variable. Typical DTA curves for siderite show a single endothermic peak (520–650 °C) followed by a strong exothermic peak (560–600 °C) and a minor exothermic peak (around 830 °C) (Kerr and Kulp, 1947; Rowland and Jonas, 1949; Smykatz-Kloss, 1964). The endothermic peak is due to decomposition of the mineral, while the strong exothermic peak is due to oxidation of the FeO produced. The minor exothermic peak represents another oxidation reaction to produce hematite. The temperatures at which endothermic and exothermic peaks occur and the presence or absence of peaks depend on the experimental conditions such as grain size of the sample, conditions of packing, rate of heating and atmosphere provided in the reaction chamber. A series of DTA curves is shown in Fig. 56. Curves 'A' and 'B' were produced by firmly packing the ground siderite into the DTA sample holder without the cover so that the sample was exposed to a furnace atmosphere of air and nitrogen, respectively. The exothermic peak due to the oxidation of FeO is suppressed in 'B'. Quenched samples show an assemblage of spinel (60 per cent) + hematite (40 per cent) from 'A' and magnetite + spinel from 'B'. Curve 'C' was produced from a similarly packed sample to 'A' but treated in a covered sample holder; it represents the net heat effect resulting from the loss of CO_2 and the oxidation of FeO. Quenched sample 'C' consists of >90 per cent of spinel and hematite. Curve 'D' was produced when the ground siderite was loosely packed in an uncovered sample holder, and it represents a heat balance between the decomposition of $FeCO_3$ and the immediate oxidation of FeO. Quenched sample 'D' has 30 per cent spinel and 70 per cent hematite.

Thermogravimetry, mass spectrographic evolved gas analysis, thermomagnetometry and X-ray diffraction were used by Gallagher and Warne (1981) to examine

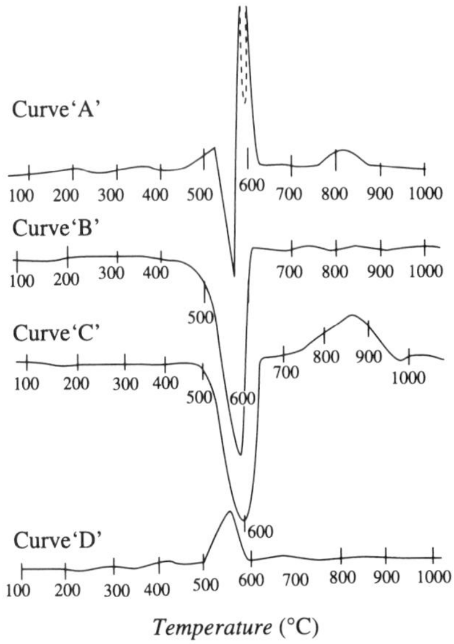

Fig. 56. DTA curves for siderite (after Rowland and Jonas, 1949). Curves A, B, C and D represent different experimental conditions.

the thermal decomposition of siderite. The weight loss was definitely detected at 400 °C, and reached a constant level near 600 °C. In a vacuum, FeO was the dominant solid decomposition product, but at even very low P_{O_2} magnetite is produced. In air or oxygen, the oxidation is so rapid that hematite is the only solid decomposition product. Intermediate spinel-type phases deriving from the presence of Mn and Mg in siderite were identified in the decomposition products from their Curie temperatures, as measured by thermomagnetometry. In the evolved gas analysis, no evolution of oxygen was detected in the decomposition of siderite.

Mössbauer spectra of siderite at various stages of thermal decomposition in a vacuum were measured at room temperature by Gallagher *et al.* (1981). At the early stages of decomposition in the temperature range 385–410 °C, iron, magnetite and carbon dioxide are the major products because the temperatures are well below that for the disproportionation of wüstite (FeO). Between 450 and 600 °C the decomposition of siderite is almost complete, and wüstite and magnetite become predominant as shown in Fig. 57. Magnesium and manganese in siderite are incorporated into the spinel phase. For decomposition in nitrogen, magnetite is the predominant solid decomposition product, whereas in oxygen, hematite is formed.

Stability relationships of siderite in the system Fe–C–O at 500 °C and 2000 bar under conditions of controlled oxygen fugacity in a $CO_2 + CO$ atmosphere were investigated by French (1971). The stability field of siderite in the presence of a gas phase is a narrow wedge bounded by three divariant reaction surfaces, representing the decomposition of siderite to hematite + gas, magnetite + gas, and magnetite + graphite + O_2. An isobaric section at $P_{gas} = 500$ bar is given in Fig. 58. The point siderite + magnetite + graphite + gas at 455 °C (log $f_{O_2} = -25.7$) represents the highest temperature at which, in the presence of a gas, siderite is stable. The stability relationships of siderite at total pressures of 1000 and 2000 bar are virtually

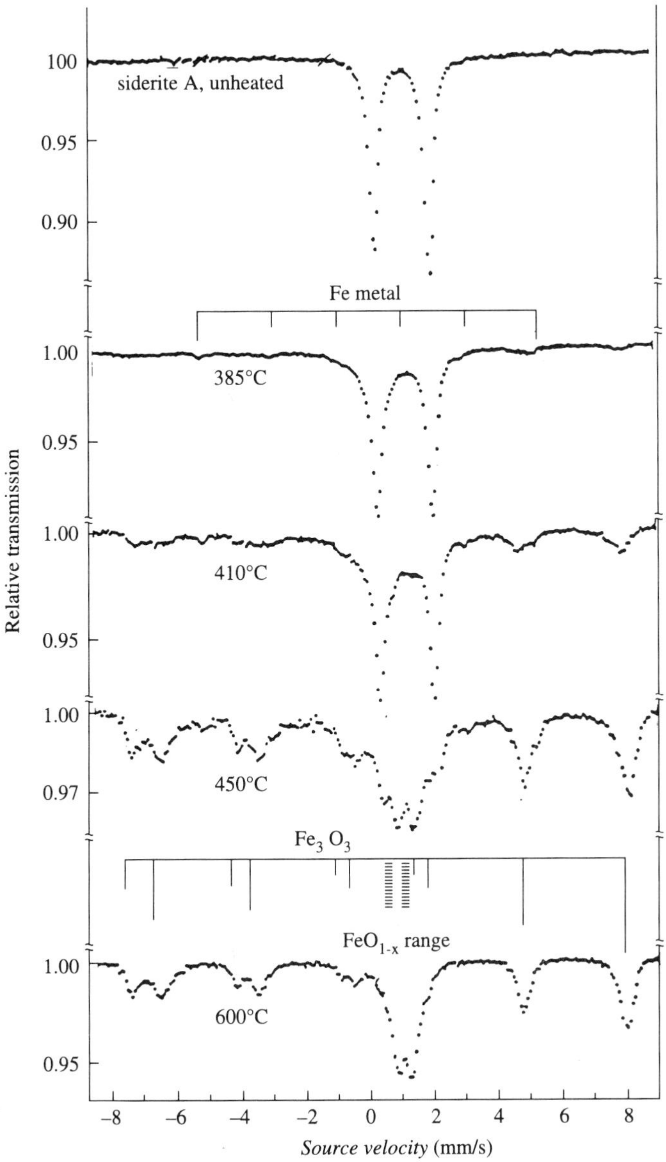

Fig. 57. Mössbauer spectra at room temperature of decomposition products of siderite (after Gallagher *et al.*, 1981).

identical to those at 500 bar, except for the slightly higher temperatures of the invariant points at higher pressures. Application of the experimental data to naturally occurring siderite shows that 465 °C is the upper limit of stability of siderite. The presence of Mg, Mn or Ca replacing Fe will stabilize siderite to higher temperatures and higher oxygen fugacity values.

The heat capacity of siderite with a composition of $(Fe_{0.956}Mn_{0.042}Mg_{0.002})CO_3$ was measured between 5 and 550 K by Robie *et al.* (1984). At 298.15 K, the heat capacity (C_p), entropy ($S^\circ_T - S^\circ_0$), enthalpy ($(H^\circ_T - H^\circ_0)/T$) and Gibbs energy

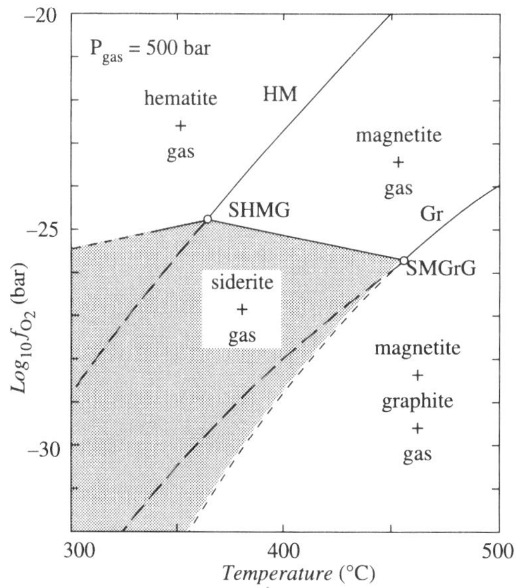

Fig. 58. Isobaric section showing the stability field of siderite + gas as a function of temperature and $\log f_{O_2}$ (after French, 1971). In the figure, HM and Gr indicate hematite and graphite buffer curves; SHMG and SMGrG are univariant equilibrium curves for siderite + hematite + magnetite + gas, and siderite + magnetite + graphite + gas, respectively.

$(-(G^\circ_T - H^\circ_0)/T)$ of siderite (in J/mol.K) are given as 82.44, 95.47, 48.94 and 46.53 respectively. With temperature increasing from 5 to 380 K, the values of all thermodynamic functions increase. The heat capacity between 350 and 550 K can be represented by the equation $C^\circ_p = 257.38 - 0.04620T - 3081.9T^{-0.5} + 1.523 \times 10^6 T^{-2}$. The enthalpy of formation of siderite at 298 K was determined also by Chai and Novrotsky (1994) and used to calculate the univariant curve for the equilibrium: siderite + hematite = magnetite + CO_2.

The solubility of siderite in acid and neutral conditions has been studied by Singer and Stumm (1970) and Reiterer *et al.* (1981); Bruno *et al.* (1992) investigated siderite solubility in acid, neutral and alkaline media. The solubility constants obtained, $\log K$ (experimental) and $\log K^\circ$ (extrapolated to standard state conditions) for the reaction $FeCO_3(s) = Fe^{2+} + CO_3^{2-}$ are, respectively, -9.6 and -10.2, -10.3 and -11.2, and -9.55 and -10.8. There is fairly good agreement between these three studies. Figure 59 shows solubility data for $FeCO_3$ at 25 °C and I (ionic strength) = 1.0 M $NaClO_4$ for two constant partial pressures of CO_2 (Bruno *et al.*, 1992). The characteristics of these data are the strong pH dependence on $[Fe^{2+}]$ in the acidic range, no pH dependence in the neutral pH range, and an increase in $[Fe^{2+}]$ with pH in the alkaline range. Bruno *et al.* (1992) identified two Fe^{2+} carbonate complexes, $FeCO_3(aq)$ and $Fe(CO_3)_2^{2-}$, and established their formation constants. CO_2 was found to be bound to Fe^{2+} as a carbonate ligand in neutral and alkaline environments, and the increased solubility due to the carbonate ligand may explain the high Fe^{2+} concentrations for the siderite equilibrium in anoxic waters.

The commonest type of alteration affecting siderite is to a hydrous ferric oxide, generally goethite, often known loosely as limonite. Such an alteration, as described by the reaction $Fe^{2+} + \frac{1}{4}O_2 + \frac{5}{2}H_2O \rightarrow Fe(OH)_3 + 2H^+$, would result in increased

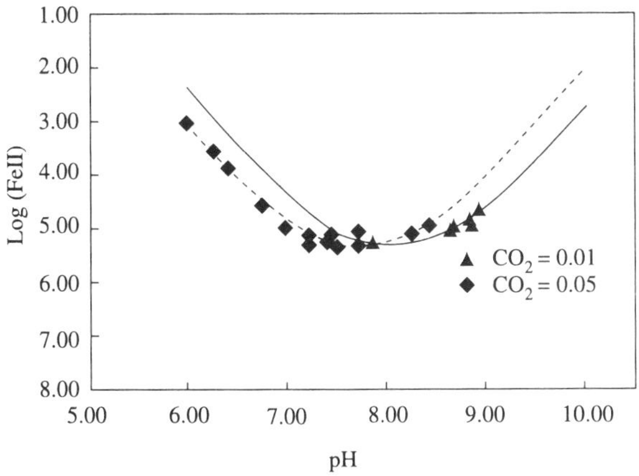

Fig. 59. Solubility curves of siderite $FeCO_3$ (solid) at 25 °C and I = 1.0 M $NaClO_4$ for P_{CO_2} = 0.05 atm (diamond) and P_{CO_2} = 0.01 atm (triangle). Fe II is total Fe II concentration in the solution (after Bruno *et al.*, 1992).

acidity, further accelerating the solution of carbonate. As pointed out by Morris (1983), the complete replacement of siderite by goethite is common both in outcrop banded iron formation and in ore throughout the world. Oxidation and replacement tend to follow cleavage patterns and it is fairly common to find virtually unaltered siderite and goethite on opposite sides of a cleavage plane.

Optical and Physical Properties

Siderite is uniaxial and optically negative with extreme birefringence ($\omega - \varepsilon = 0.240$). Interference colours are typically high-order white and grey, even for orientation near the optic axis. The structural dependence of birefringence of carbonates has been treated by Lo (1973), Isherwood and James (1976) and Weber (1988). The refractive indices decrease with the substitution of Mg, Mn^{2+} and Zn for Fe^{2+} (Kennedy, 1947; Winchell and Meek, 1947). A zincian oligonspar, an intermediate member of the $FeCO_3$–$MnCO_3$ series (Table 23, anal. 14), has ω 1.837 and ε 1.630.

In colour, siderite is sometimes white but more frequently has some shade of yellow or brown. Like magnesite, siderite is not one of the strong luminescent carbonates (Gies, 1975).

Siderite normally occurs as a fine- to coarse-grained anhedral aggregate. It sometimes exists as ooliths, or as nodular or botryoidal forms consisting of radiating coarse fibres. Euhedral crystals are usually rhombohedra $\{10\bar{1}1\}$, on the hexagonal morphological cell, and faces are often curved because of the aggregation of many offset crystals. Both hardness and specific gravity are lowered by the substitution of Mg, Mn^{2+} or Ca for Fe^{2+}. With 50 mol. per cent $MgCO_3$, the specific gravity decreases from 3.96 to 3.48.

The infrared spectra of siderite, like those of calcite, are characterized by three prominent absorption maxima, as shown in Fig. 43 (p. 120). In the siderite–magnesite series, the absorption frequencies shift towards lower wavenumbers with increasing Fe content (Peng *et al.*, 1985).

Amigo and Fortune (1981) have demonstrated that hydrothermal siderites have a more constant peak width in their Mössbauer spectra despite a very large variation in chemical composition.

Distinguishing Features

Within the group of rhombohedral carbonates, siderite has the highest refractive indices and its specific gravity is considerably higher than that of others except smithsonite. Siderite differs from calcite in its relative insolubility in cold dilute HCl, and from magnesite in its fairly common lamellar twinning. If present, the yellowish or brownish colour, commonly a result of alteration, may be distinctive.

Paragenesis

The most common occurrence of siderite is in the banded iron formations and ironstones. The carbonate iron formation, together with the oxide, silicate and sulphide iron formations, form the four major iron-formation facies (James, 1954). Carbonate iron formations generally consist of one or more carbonates (siderite, calcite and members of the dolomite–ankerite series), magnetite, quartz or chert, and some iron-bearing silicates. Textural relationships (Dimroth and Chauvel, 1973; Klein, 1974; Klein and Fink, 1976) and compositions (Floran and Papike, 1975; Klein and Fink, 1976) indicate that the carbonate minerals are formed as sedimentary or diagenetic products under low-temperature conditions, and iron oxides are considered to be phases recrystallized from hydrous iron oxides. The stability of siderite relative to other iron-bearing minerals in aqueous media is shown in Fig. 60 with variable pE and pH (Klein and Bricker, 1977). If the pH of the system $Fe–H_2O–O_2–SiO_2$–dissolved carbon was within the range of present-day sea-water, the pE for a stable assemblage greenalite–siderite–magnetite must be <0 and the activity of total carbon approximately 10^0 (Fig. 60a). With the presence of a stilpnomelane-like phase, the stable field of siderite disappears when the activity of total dissolved carbon is $<10^{-1}$ (Fig. 60(b)). In Fig. 60(c), a more oxidized stilpnomelane-like phase is present which eliminates the stability field of magnetite in the system. During subsequent metamorphism, if the chemical potential of carbon dioxide is high enough to prevent the breakdown of the carbonates, the mineral assemblages of the carbonate facies will not produce any chemical reactions. If reduction of the chemical potential of CO_2 occurs, quartz or chert will react with the carbonates to form clinopyroxene or actinolite.

Banded iron formations are known in every Precambrian continental region and occur in abundance in almost every case. Major formations were found in the Labrador–Quebec–Ontario–Minnesota belt of northeastern North America (Bayley and James, 1973; Goodwin, 1973); in the greenstone belt of Kaapvaal (South Africa) and Rhodesian cratons, the Damara Mobile belt (southwest Africa), the Limpopo metamorphic belt (Botswana) and the cratonic basin (South Africa) (Beukes, 1973); in the Minas Gerais, Serra dos Carajus and Uracum districts of Brazil (Dorr, 1973); in the Yilgarn block and Hamersley Basin of Western Australia (Trendall, 1973); and in the Singhbhum region of eastern India (Banerji, 1977).

Siderite deposits of Ouenza, Algeria, and Jerissa, Tunisia, North Africa, were considered by Pohl *et al.* (1986) to be of metasomatic origin. The early diagenetic

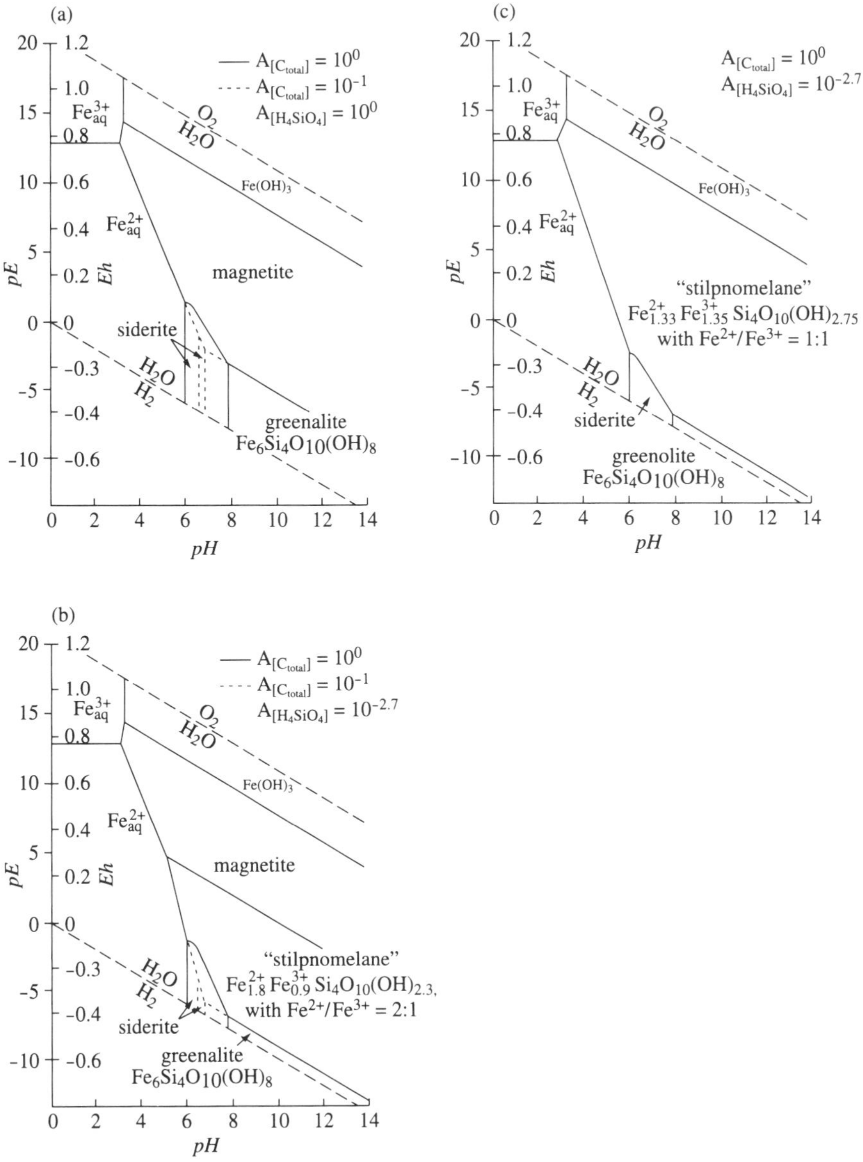

Fig. 60. pE–pH diagrams for the stability fields of siderite, magnetite, greenalite and stilpnomelane in the system Fe–H_2O–O_2–SiO_2–dissolved carbon at 25°C and 1 atm total pressure: (a) without stilpnomelane, (b) with stilpnomelane having a composition $Fe^{2+}/Fe^{3+} = 2/1$ and (c) with stilpnomelane having a composition of $Fe^{2+}/Fe^{3+} = 1/1$ (after Klein and Bricker, 1977).

reaction of hot brine with the host rocks allowed acid transport of Fe and other cations to be channelled to the apical parts of diapirs, and deposited there by metasomatism in fractured roof carbonates.

Carbonates of the siderite–magnesite series have been found in carbonatites from the Lueshe (Zaïre), Newania (India), Kangankunde (Malawi) and Chipman

Lake (Canada) complexes (Buckley and Woolley, 1990). Phillips and Brown (1987) described the occurrence of siderite, ankerite and calcite in differentiated tholeiitic dolerite sills; their presence is related to carbonation and sulphidation processes that accompanied gold mineralization in the Kalgoorlie gold deposit, Australia.

The occurrence of siderite in the Ivigtut cryolite deposit, Greenland, is well known. The pegmatitic dykes of Dolni Bory, Western Moravia, carry Mn-rich siderite and ankerite accompanied by sulphides (Nemec, 1977).

Siderite is one of the principal gangue minerals of the Pb–Zn–Ag hydrothermal vein deposits in the Coeur d'Alene district in northern Idaho, USA (Fryklund, 1964). Siderite with an average composition of $(Fe_{0.66}Mg_{0.29}Mn_{0.03}Ca_{0.02})CO_3$ was described by Václav (1982) from hydrothermal sulphide-bearing veins in the central part of the Spissko-gemerske Rudnohory mountains. Shikazono (1977) analysed siderite from two different types of hydrothermal vein, the Au–Ag type and the Pb–Zn type, and revealed a remarkable decrease in zinc content. Siderite from the former contains an appreciable amount of zinc, ranging from 0.8 to 5.8 per cent ZnO, while the zinc content in the latter is below levels detectable by an electron microprobe analysis. Shikazono concluded that Au–Ag mineralization occurred under relatively high oxidation conditions in comparison with Pb–Zn mineralization, and zinc content increases with increasing f_{O_2} in the oxidized sulphur and carbon species.

Abundant siderite concretions occur in the Upper Cretaceous Gammon shale in the northern Great Plains (Gautier, 1982). They were formed during the early compactional process of the fine-grained marine sediments. Two types of siderite concretion were recognized: (a) those formed at shallow depths in the sediment column with a large volume of carbonate (75–85 per cent) with oxygen isotope ratios similar to those of carbonates precipitated in equilibrium with sea-water and (b) those concretions that formed longer after deposition and at greater depths in the accumulating sediments with lower oxygen isotopic values.

Large amounts of siderite and pyrite were found in freshwater and brackish sediments, respectively, in Skjernå delta swamp, Denmark (Postma, 1982). Berner (1964) has theoretically demonstrated that siderite is only stable relative to pyrite at extremely low dissolved sulphide activities. If sufficient hydrogen sulphide is produced by bacterial sulphate reduction, the precipitation of pyrite and other iron sulphides will not allow ferrous iron concentrations to reach a level whereby siderite becomes stable. Mozley (1989) examined early diagenetic sediments from various depositional environments and established a distinct compositional trend. Siderite from freshwater environments is relatively pure with >90 per cent $FeCO_3$, whereas that from marine environments is always impure with extensive substitution of Fe by Mg. This variations appears to result from differences in the chemistry of marine and meteoric pore waters.

Siderite concretions occur in coal measures in the central Pennine region of England. There, the siderites between two marine horizons were deposited from freshwater low in dissolved sulphate. The siderite concretions have high Mn/Fe ratios and are enriched in ^{13}C (Curtis *et al.*, 1986). The association of siderite with certain phosphate deposits was discussed by Kazakov *et al.* (1957). Siderite forms at pH values (6.2–7.0) substantially lower than those (8.0–8.5) for the formation of phosphates. Both cannot be stable or be formed at the same time. Therefore, Kazakov *et al.* stated that siderite formed in the phosphates should be regarded as a diagenetic product. Postma (1981) described a bog deposit consisting of siderite and vivianite in Denmark. Both minerals contain a solid solution with Mn^{2+} and Ca^{2+}, and Mn^{2+} is preferentially concentrated in the siderite.

References

Amigo, J. M. and **Fortune, J. P.**, 1981. Mineralogical application of the Mössbauer effect to the study of siderites from Batére (France): comparison with siderite of hydrothermal origin. *Neues Jahrb. Min., Mh.*, 237–44.

Bak, B. and **Niec, M.**, 1978. The occurrence of monheimite in the Boleslaw Zn–Pb ore deposit near Olkusz. *Min. Polonica*, **9**, 123–8.

Bak, B. and **Żabiński, W.**, 1981. On the continuity of the solid solution series smithsonite–siderite. *Min. Polonica*, **12**, 75–9.

Banerji, A. K., 1977. On the Precambrian banded iron-formation and the manganese ores of the Singhbhum region, Eastern India. *Econ. Geol.*, **72**, 90–8.

Bayley, R. W. and **James, H. J.**, 1973. Precambrian iron-formation of the United States. *Econ. Geol.*, **68**, 934–59.

Berner, R. A., 1964. Stability fields of iron minerals in anaerobic marine sediments. *J. Geol.*, **72**, 826–34.

Beukes, N. J., 1973. Precambrian iron-formations of Southern Africa. *Econ. Geol.*, **68**, 960–1004.

Bruno, J., **Wersin, P.** and **Stumm, W.**, 1992. On the influence of carbonate in mineral dissolution. II. The solubility of $FeCO_3$(s) at 25 °C and 1 atm. total pressure. *Geochim. Cosmochim. Acta*, **56**, 1149–55.

Buckley, H. A. and **Woolley, A. R.**, 1990. Carbonates of magnesite–siderite series from four carbonatite complexes. *Min. Mag.*, **54**, 413–8.

Butler, P., 1969. Mineral composition and equilibria in the metamorphosed iron-formation of the Ganon region, Quebec, Canada. *J. Petr.*, **10**, 56–101.

Chai, L. and **Navrotsky, A.**, 1994. Enthalpy of formation of siderite, and its application to phase equilibrium calculation *Amer. Min.*, **79**, 921–9.

Chen, Z., **Yu, J.** and **Hon, K.**, 1982. [The genesis of siderite in the Xuanlong area of northwest Hebei.] *Scientia Geol. Sinica*, **4**, 393–402 (in Chinese with English abstract).

Curtis, C. D., **Pearson, M. J.** and **Somogy, V. A.**, 1975. Mineralogy, chemistry, and origin of a concretionary siderite sheet (clay-ironstone band) in the Westphalian of Yorkshire. *Min. Mag.*, **40**, 385–93.

Curtis, C. D., **Coleman, M. L.** and **Love, L. G.**, 1986. Pore water evolution during sediment burial from isotopic and mineral chemistry of calcite, dolomite and siderite concretions. *Geochim. Cosmochim. Acta*, **50**, 2321–34.

Davidson, P. M., **Symmes, G. H.**, **Cohen, B. A.**, **Reeder, R. J.** and **Lindsley, D. H.**, 1993. Synthesis of the new compound $CaFe(CO_3)_2$ and experimental constraints on the $(Ca,Fe)CO_3$ join. *Geochim. Cosmochim. Acta*, **57**, 5105–9.

Dimroth, E. and **Chauvel, J. J.**, 1973. Petrography of the Sokoman iron formation in part of the central Labrador Trough, Quebec, Canada. *Geol. Soc. Amer. Bull.*, **84**, 111–34.

Dorr, J., II, Von, 1973. Iron-formation in South Africa. *Econ. Geol.*, **68**, 1005–22.

Effenberger, H., **Mereiter, K.** and **Zemann, J.**, 1981. Crystal structure refinements of magnesite, calcite, rhodochrosite, siderite, smithsonite, and dolomite, with discussion of some aspects of the stereochemistry of calcite-type carbonates. *Z. Krist.*, **156**, 233–43.

Essene, E. J., 1983. Solid solutions and solvi among metamorphic carbonates with applications to geologic thermobarometry. In *Carbonates: Mineralogy and Chemistry*: Ed. R. J. Reeder, *Reviews in Mineralogy*, **11**, 77–96, Mineralogical Society of America, Washington, DC.

Floran, R. J. and **Papike, J. J.**, 1975. Petrology of the Gunflint iron formation, Ontario–Minnesota: Pt. I. The low grade rocks. *Geol. Soc. Amer. Bull.*, **86**, 1169–90.

French, B. M., 1971. Stability relations of siderite ($FeCO_3$) in the system Fe–C–O. *Amer. J. Sci.*, **271**, 37–78.

Fryklund, V. C., *Jr.*, 1964. Ore deposits of the Coeur d'Alene district, Shoshone County, Idaho. *U.S. Geol. Surv. Prof. Paper*, **445**, 103.

Gallagher, P. K. and **Warne, S. St J.**, 1981. Thermomagnetometry and thermal decomposition of siderite. *Thermochim. Acta*, **43**, 253–67.

Gallagher, P. K., **West, K. W.** and **Warne, S. St. J.**, 1981. Use of the Mössbauer effect to study the thermal decomposition of siderite. *Thermochim. Acta*, **50**, 41–7.

Gautier, D. L., 1982. Siderite concretions: indicators of early diagenesis in the Gammon shale (Cretaceous). *J. Sed. Petr.*, **52**, 859–71.

Gies, H., 1975. Activation possibilities and geochemical correlations of photoluminescing carbonates, particularly calcite. *Min. Deposita*, **10**, 216–27.

Gill, E. D. and **Segnit, E. R.**, 1980. Siderite from the Otway coast of Victoria, S. E. Australia. *Aust. Min.*, **31**, 152–4.

Goldsmith, J. R., **Graf, D. L.**, **Witters, J.** and **Northrop, D. A.**, 1962. Studies in the system $CaCO_3$–$MgCO_3$–$FeCO_3$: 1. Phase relations; 2. A method for major element spectrochemical analysis; 3. Compositions of some ferroan dolomites. *J. Geol.*, **70**, 659–88.

Gole, M. J., 1980. Mineralogy and petrology of very-low-metamorphic grade Archaean banded iron-formations, Weld Range, Western Australia. *Amer. Min.*, **65**, 8–25.

Goodwin, A. M., 1973. Archaean iron formations and tectonic basins of the Canadian shield. *Econ. Geol.*, **68**, 915–33.

Henry, D. A., 1988. Mineralogy from the Clunes Goldfield. *Aust. Min.*, **2**, 43–52.

Isherwood, B. J. and **James, J. A.**, 1976. Structural dependence of the optical birefringence of crystals with calcite- and aragonite-type carbonates. *Acta Cryst.*, **A32**, 340–1.

James, H. L., 1954. Sedimentary facies of iron-formation. *Econ. Geol.*, **49**, 235–93.

Jarosewich, E. and **Macintyre, I. G.**, 1983. Carbonate reference samples for electron microprobe and scanning electron microscope analysis. *J. Sedim. Petr.*, **53**, 677–8.

Kazakov, A. L., **Tikhomorova, M. M.** and **Plotnikova, V. I.**, 1957. The system ferrous oxide–carbon dioxide–water and conclusions in the paragenesis of siderite and phosphorites. *Internat. Geol. Review*, **1(10)**, 1–39.

Kennedy, G. C., 1947. Charts for correlation of optical properties with chemical composition of some rock-forming minerals. *Amer. Min.*, **32**, 561–73.

Kerr, P. F. and **Kulp, J. L.**, 1947. Differential thermal analysis of siderite. *Amer. Min.*, **32**, 678–80.

Klein, C., Jr., 1974. Greenalite, stilpnomelane, minnesotaite, crocidolite, and carbonate in a very low-grade metamorphic Precambrian iron-formation. *Can. Min.*, **12**, 475–98.

Klein, C., Jr., 1978. Regional metamorphism of Proterozoic iron-formation, Labrador Trough, Canada. *Amer. Min.*, **63**, 898–912.

Klein, C., Jr. and **Bricker, O. P.**, 1977. Some aspects of the sedimentary and diagenetic environment of Proterozoic banded iron-formation. *Econ. Geol.*, **72**, 1457–504.

Klein, C., Jr. and **Fink, R. P.**, 1976. Petrology of the Sokoman iron formation in the Howells River area, at the western edge of the Labrador Trough. *Econ. Geol.*, **71**, 453–88.

Liao, S. F., 1982. *A study on the reworked sedimentary siderite deposits in west Guizhou, China.* Geol. Publ. House, Beijing, 116 pp.

Lo, B. W. N., 1973. The polarizabilities of planar nitrate, carbonate, and borate anions in crystals. *J. Phys. Chem. Solids*, **34**, 513–20.

Morris, R. C., 1983. Supergene alteration of banded iron formation, in 'Iron Formation Facts and Problems' Trendall, A. F. and Morris, R. C., Eds., *Developments in Precambrian Geology*, **6**, Amsterdam, New York, Oxford and Tokyo (Elsevier), 513–34.

Mozley, P. S., 1989. Relation between depositional environment and the elemental composition of early diagenetic siderite. *Geology*, **17**, 704–6.

Nemec, D., 1977. Carbonates of the pegmatites from Dolni Bory (West Moravia). *Neues Jahrb. Min., Mh.*, 517–21.

Ôno, K. and **Ito, A.**, 1964. Mössbauer study of magnetic properties in ferrous compounds. *J. Phys. Soc. Japan*, **19**, 899–915.

Palache, C., **Berman, H.** and **Frondel, C.**, 1951. *The System of Mineralogy.* New York (John Wiley & Sons), 1124 pp.

Pearson, M. J., 1974. Sideritic concretions from the Westphalian of Yorkshire: a chemical investigation of the carbonate phase. *Min. Mag.*, **39**, 696–9.

Peng, W., **Liu, G.** and **Ke, L.**, 1985. [Infrared spectra study of magnesite–siderite series.] *Acta Min. Sinica*, **5**, 229–33 (in Chinese with English abstract).

Phillips, G. N. and **Brown, I. J.**, 1987. Host rock and fluid control on carbonate assemblages in the Golden Mile dolerite, Kalgoorlie gold deposit, Australia. *Can. Min.*, **25**, 265–73.

Pohl, W., **Anouri, M.**, **Kolli, O.**, **Scheffer, R.** and **Zachmann, D.**, 1986. A new genetic model for the North African metasomatic siderite deposits. *Min. Deposita*, **21**, 228–33.

Postma, D., 1981. Formation of siderite and vivianite and the pore-water composition of Recent bog sediment in Denmark. *Chem. Geol.*, **31**, 225–44.

Postma, D., 1982. Pyrite and siderite formation in brackish and freshwater swamp sediments. *Amer. J. Sci.*, **282**, 1151–83.

Reiterer, F., **Johannes, W.** and **Gamsjäger, H.**, 1981. Semimicro determination of solubility constants: copper(II) carbonate and iron(II) carbonate. *Mikrochim. Acta*, **1**, 63–72.

Robie, R. A., **Haselton, H. T., Jr.** and **Hemingway, B. S.**, 1984. Heat capacities and entropies of rhodochrosite ($MnCO_3$) and siderite ($FeCO_3$) between 5 and 600K. *Amer. Min.*, **69**, 349–57.

Rosenberg, P. E., 1963. Synthetic solid solutions in the systems $MgCO_3$–$FeCO_3$ and $MnCO_3$–$FeCO_3$. *Amer. Min.*, **48**, 1396–400.

Rosenberg, P. E., 1967. Subsolidus phase relations in the system $CaCO_3$–$MnCO_3$–$FeCO_3$ between 350° and 500 °C. *Amer. Min.*, **52**, 787–96.

Rosenberg, P. E. and **Foit, F. F.**, 1979. The stability of transition metal dolomites in carbonate systems: a discussion. *Geochim. Cosmochim. Acta*, **43**, 951–5.

Rowland, R. A. and **Jonas, E. S.**, 1949. Variation in differential thermal analysis curves of siderite. *Amer. Min.*, **34**, 550–8.

Shchibrik, V. I. and **Taranushich, F. F.**, 1971. First find of zinc oligonspar in the USSR. *Dokl. Acad. Sci. USSR, Earth Sci. Sect.*, **197**, 136–9.

Shikazono, N., 1977. Composition of siderite and the environments of formation of vein-type deposits in Japan. *Econ. Geol.*, **77**, 632–41.

Singer, P. C. and **Stumm, W.**, 1970. The solubility of ferrous iron in carbonate waters. *J. Amer. Water Works Assoc.*, **62**, 198–212.

Smykatz-Kloss, W., 1964. Differential-Thermal-Analyse von einigen Karbonat-Mineralen. *Beit. Min. Petr.*, **9**, 481–502.

Trendall, A. F., 1973. Precambrian iron-formation of Australia. *Econ. Geol.*, **68**, 1023–34.

Václav, J., 1982. [Chemical composition of siderites of hydrothermal veins in the area of Stara Vodasvedlar.] *Geol. Prace*, **78**, 195–202 (in Czech with English abstract).

Weber, H. J., 1988. Anisotropic bond polarizabilities in birefringent crystals. *Acta Cryst.*, **A44**, 320–6.

Winchell, A. N. and **Meek, W. R.**, 1947. Birefringence–dispersion ratio as a diagnostic. *Amer. Min.*, **32**, 336–43.

Wyckoff, R. W. G., 1920. The crystal structure of some carbonates of the calcite group. *Amer. J. Sci., 4th Ser.*, **50**, 317–60.

Zakrzewski, M., 1974. Mineralogy of bottom horizon of sideritic ores in Czestochowa. *Polska. Akad. Nauk, Prace Min.*, **38**, 41–74.

Smithsonite $ZnCO_3$

Trigonal (−)

ε	1.625
ω	1.850
δ	0.225
Dispersion:	Strong.
D	4.43
H	$4–4\frac{1}{2}$
Cleavage[1]:	$\{10\bar{1}1\}$ perfect.
Twinning[1]:	Not observed.
Colour:	Grey, greyish white, also brown, green, blue; colourless in thin section.
Unit cell:	a_{hex} 4.652 Å, c_{hex} 15.052 Å, $Z_{hex}=6$. a_{rh} 5.683 Å, α_{rh} 48°19′, $Z_{rh}=2$. Space group $R\bar{3}c$

Dissolves with effervescence in weak acids.

Smithsonite is normally a secondary mineral found in the oxidized zones of ore deposits containing primary zinc minerals, especially sphalerite. It usually appears as botryoidal or stalactitic masses and often occurs as a secondary encrustation. Distinct crystals of smithsonite are rare. 'Dry bone' is a term given to cellular and porous varieties. The mineral is named in honour of James Smithson (1754–1829), founder of the Smithsonian Institution, Washington, DC, USA.

Structure

Smithsonite is isostructural with calcite. Refinements by Effenberger *et al.* (1981) assign a space group of $R\bar{3}c$ to smithsonite with *a* 4.6526, *c* 15.0257 Å and Z=6. The interatomic distances C–O and Zn–O are 1.2859 and 2.1167 Å, respectively. The length of the O–O edge of the ZnO_6 octahedron inclined to (0001) is 3.0263 Å (d_1) and that parallel to (0001) is 2.9430 Å (d_2). The shortest O–O distance outside the coordination figure is 2.8524 Å, which is comparable with that in magnesite but is unusually short in comparison with other calcite-type carbonates (Zemann, 1983). All calcite-type carbonates, including calcite, magnesite, rhodochrosite, siderite and smithsonite, have elongated octahedra parallel to [0001], and the smallest elongation, as measured by d_1/d_2, is shown by smithsonite. It is known that the elongation is not just a function of the *M*–O distance of the MO_6 octahedron (Effenberger *et al.*, 1981).

In the smithsonite–siderite series, cell dimensions increase linearly with $FeCO_3$ content (Bak and Żabiński, 1981), whereas the increase of $d_{10\bar{1}4}$ with $MnCO_3$

[1] Cleavage, Twinning and morphology are indexed using the hexagonal morphological cell with quadrupled *a* parameter (see Table 16, p. 103).

content was used by Birch (1986) to illustrate the effect of substitution upon the cell dimensions in the smithsonite–rhodochrosite series.

Chemistry

In smithsonite, Fe^{2+} is often present in considerable quantities. Results of chemical analysis of carbonates from the Cracow–Silesian (Poland) lead–zinc deposit are given in Fig. 61, which illustrates a complete series of substitution between zinc and ferrous iron. According to Bak and Żabiński (1981), the contents of Mn and Pb in those carbonates do not exceed 1 per cent in any of 32 analyses. The mineral with a composition nearly intermediate between smithsonite and siderite has been named monheimite (Palache *et al.*, 1951; Bak and Niec, 1978).

Mn^{2+} is generally present in smithsonite at a few per cent, but naturally occurring carbonates with compositions covering the entire smithsonite–rhodochrosite series have been reported by Birch (1986) from the oxidized zone at Broken Hill, New South Wales, Australia. Analytical results are given in Fig. 62 in a plot of $ZnCO_3$ versus $MnCO_3$. Such a plot illustrates a comparatively greater divergence from the 1:1 line for pure Mn–Zn carbonates at the Mn-rich end, where the ease of substitution of Pb and Ca appears to be enhanced. Equilibrium relationships of $ZnCO_3$ with either $FeCO_3$ or $MnCO_3$ have not been determined (Goldsmith, 1983).

Smithsonite with >1 per cent MgO is not common. A high Mg-smithsonite, however, was reported from Broken Hill, New South Wales, Australia, by Chen and Chang (1975). It contains 7.56–7.65 per cent MgO, 0.90–1.24 per cent CaO, 0.12–0.34 per cent MnO and 53.16–53.26 per cent ZnO with no detectable FeO.

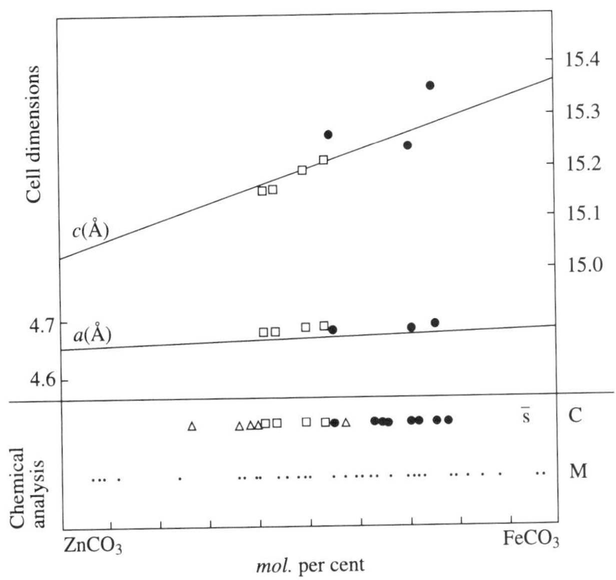

Fig. 61. Cell dimensions and chemical compositions of solid solutions in the smithsonite–siderite series (after Bak and Żabiński, 1981). Triangle, square and S represent published data taken by Bak and Żabiński from various sources; large and small solid circles are Bak and Żabiński's data; C and M denote analyses by chemical and electron probe methods, respectively.

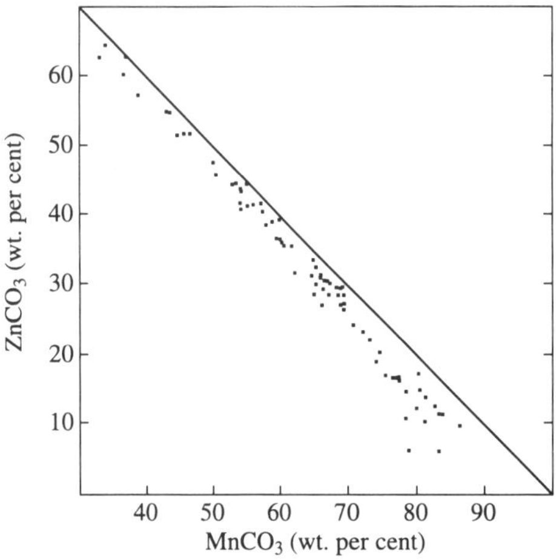

Fig. 62. Chemical compositions of zinc-manganese carbonates from Broken Hill, New South Wales, Australia (after Birch, 1986). The straight line represents the 1:1 line for pure stoichiometric Zn–Mn carbonates.

A formula of $(Ca_{0.02-0.03}Mg_{0.22}Zn_{0.76-0.75})CO_3$ is suggested, and cell dimensions measured are *a* 5.630, *c* 14.994 Å. A sample of smithsonite from Tsumeb, Namibia, analysed by Effenberger *et al.* (1981) to have a composition of 63.3 per cent ZnO, 0.9 per cent FeO, 0.3 per cent MgO and 0.2 per cent MnO, may represent a relatively pure smithsonite.

Garavelli *et al.* (1982) described a Zn-dolomite from the oxidation zone in Tsumeb, Namibia, deposits. It has the composition $Ca_{0.98}Zn_{0.92}Mg_{0.07}Fe_{0.03}(CO_3)_2$, with *a* 4.8183 and *c* 16.0295 Å, and was named minrecordite. Limited experimental data obtained at relatively high temperatures of 600–750 °C show the absence of any intermediate compound between $CaCO_3$ and $ZnCO_3$ (Goldsmith and Northrop, 1965). It is possible, as Goldsmith (1983) pointed out, that Zn-dolomite may have a maximum temperature of stability. The potential formation of a Zn-dolomite is suggested by Rosenberg and Foit (1979) on the basis of octahedral distortion considerations. A series has been established for the decreasing stability of divalent cations in the formation of dolomite-type phases: $Mg \gg Mn > Zn > Fe > Co > Ni > Cu$. Zn occupies a higher position in the series than does Fe^{2+}.

Smithsonite decomposes at relatively low temperatures. Harker and Hutta (1956) presented a univariant P_{CO_2}–T curve for the reaction $ZnCO_3 \rightleftharpoons ZnO + CO_2$. It passes through 310, 410 and 460 °C at 0.69, 2.07 and 3.45 kbar, respectively. Their P_{CO_2}–T curve has been modified and extended to higher pressures by Haselton and Goldsmith (1987), and is shown in Fig. 63. By using a sample of smithsonite from Tsumeb, Namibia, Smykatz-Kloss (1974) recorded an endothermic peak at 499 °C in differential thermal analysis.

Thermodynamic data, including heat capacity, entropy, enthalpy function and Gibbs energy function, have been measured or calculated by Haselton and Goldsmith (1987). They are (in J/mol.K) at 298.15 K, respectively, 80.05, 81.19, 0.000 and 81.19. The heat capacity between 298.15 and 1200.00 K can be represented by the equation:

$$C_p^0 = 148.38 + 0.02835T - 1419T^{0.5} + 4.796 \times 10^5 T^{-2}.$$

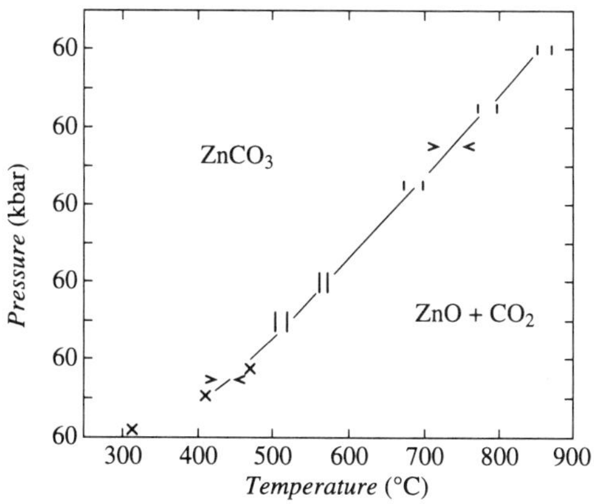

Fig. 63. P_{CO_2}–T **curve for the reaction** $ZnCO_3 = ZnO + CO_2$ **(after Haselton and Goldsmith, 1987). Solid symbols and arrows represent results from experiments using unsealed and sealed capsules, respectively; the crosses are data points taken from Harker and Hutta (1956).**

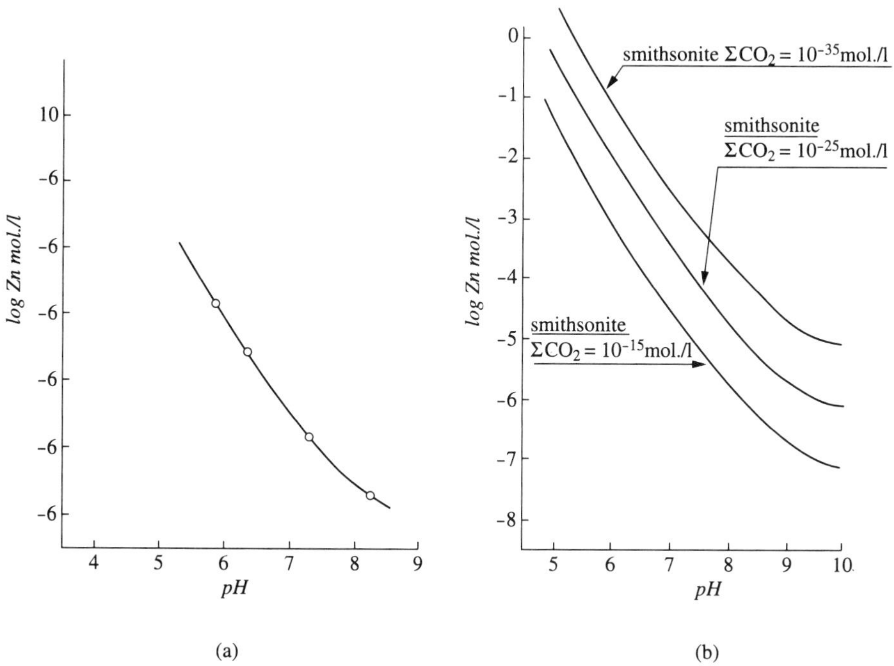

Fig. 64. Solubility curves of smithsonite (a) in solutions containing 0.03 mol./l CO_2 **at 25 °C and 1 atm, and (b) calculated from thermodynamic data (after Takahashi, 1960).**

In aqueous solutions smithsonite is slightly soluble. Its solubility as a function of pH was measured by Takahashi (1960) at 1 atm and 25 °C in solutions saturated with CO_2 ($\Sigma CO_2 = 10^{-1.5}$ mol./l) (Fig. 64). By assuming that the concentrations of undissolved carbonate and various basic complex ions are negligibly small compared with the concentration of Zn^{2+} in the pH range between 5 and 10, and that the total carbon dioxide in solution is approximately equal to the summation of the activities of undissociated carbonic acid, bicarbonate ions and carbonate ions, Takahashi (1960) further calculated solubilities of smithsonite in solutions of saturated CO_2 and lower CO_2 concentrations. As shown in Fig. 64, the calculated solubility curves of $\sum CO_2 = 10^{-1.5}$ mol./l match well with the measured curve. In a recent measurement of solubility of smithsonite in NaCl solution, Afanas'ev *et al.* (1987) have found that the solubility increases with temperature and ionic strength of the solution, but is independent of the hydrostatic pressure.

Optical and Physical properties

Smithonsite is uniaxial and optically negative, with strong birefringence (ω–ε=0.225). Interference colours seen in thin sections or grain mounts are very high-order white, even if the optic axis is near the microscope axis. As for other carbonates, the structural dependence of birefringence has been discussed by Lo (1973), Isherwood and James (1976) and Weber (1988). Although the effect of Fe^{2+} and Mn^{2+} on the refractive indices of smithsonite has not been determined, Birch (1986) suggests that the variation with composition in the smithsonite–rhodochrosite series is linear. The high Mg-smithsonite (7.56–7.65 per cent MgO) from Broken Hill, New South Wales, Australia, has ε 1.601 and ω 1.815, less than the values for pure smithsonite (Chen and Chang, 1975). Refractive indices of minrecordite, $CaZn(CO_3)_2$, are ε 1.550 and ω 1.750, and are reduced to ε 1.542 and ω 1.734 in the presence of Mg for the composition $Ca_{1.07}(Zn_{0.70}Mg_{0.13}Fe_{0.07}Mn_{0.01})(CO_3)_2$ (Garvelli *et al.*, 1982). Smithsonite is one of the strongest luminescent carbonates, but the presence of Fe^{2+} in it much reduces its intensity (Gies, 1975).

Commonly, smithsonite is greyish white to dark grey. It may also be brown, green, pink or blue. Distinct crystals are rare, and faces are generally curved and rough. It normally occurs as botryoidal reniform masses or as stalactites, crystalline encrustations or earthy to fibrous granular masses. Massive stalactitic or botryoidal smithsonite usually displays concentric colour banding. The specific gravity of smithsonite is greater than that of other calcite-type carbonates, so a decrease is expected when Zn is substituted by Fe^{2+}, Mn^{2+}, Ca or Mg. Birch (1986) suggests a linear relationship between specific gravity and composition for the series smithsonite–rhodochrosite. The specific gravity of minrecordite is 3.45, whereas that of Mg-minrecordite is 3.32, both lower than the specific gravity of smithsonite.

The infrared spectra of smithsonite, like those of calcite, are characterized by three prominent absorption maxima as shown in Fig. 43 (p. 120). The frequencies of absorption maxima of smithsonite, rhodochrosite and siderite correlate well with the relative size of the respective divalent cation. A comparison of vibrational wave numbers (cm^{-1}) for the calcite-type carbonates is as follows (White, 1974):

	Calcite	Magnesite	Siderite	Rhodochrosite	Smithsonite
ν_1	876	887	866	867	870
ν_2	1434	1450	1422	1433	1440
ν_3	712	748	737	727	734

Distinguishing Features

Smithsonite is distinguished from other rhombohedral carbonates by its high specific gravity and solution in weak acids with effervescence. The botryoidal form and lack of twinning are also distinctive of smithsonite. X-ray powder diffraction was used by Birch (1986) for the characterization of members of the smithsonite–rhodochrosite series.

Warne (1962) has recommended a series of staining procedures for the differentiation of smithsonite from other rhombohedral carbonates. Using alizarin red S, calcite and ankerite are stained a deep red and a purple colour, respectively, while smithsonite is not stained. Using alizarin red S with a 30 per cent NaOH solution, siderite is stained dark brown to black and smithsonite a purple colour. Using alizarin red S with a 5 per cent NaOH solution, rhodochrosite is stained a purple colour. Magneson with a 30 per cent NaOH solution has no effect on smithsonite, but on magnesite a deep blue colour forms. Smithsonite and magnesite may also be differentiated using tropaeolin 00, which stains smithsonite yellow and has no effect on magnesite.

Paragenesis

Smithsonite is predominantly a secondary mineral in oxidized sulphide veins, and is mainly derived from the alteration of sphalerite in the presence of limestone. It is commonly associated with other secondary minerals such as cerussite, hemimorphite, malachite, azurite, anglesite, hydrozincite and pyromorphite.

The lead–zinc deposits in the Moresnet District, Belgium, contain masses of smithsonite of enormous size, extending from the surface to depths of 50–100 m (Berg *et al.*, 1950). The lead–zinc deposits in Upper Silesia, Poland, like the Moresnet deposits, are of sedimentary origin (Gürich, 1903). The replacement ore, consisting of galena, sphalerite and abundant marcasite, occurs along two horizons; the lower horizon is in a bed of dolomite and the upper is in a bed of smithsonite. Both horizons are of considerable thickness. Throughout the northwestern half of Tunisia (Sainfeld, 1956), iron–lead–zinc mineralization is represented in scattered deposits. The oxidized zones are generally very important in a limestone environment, and are composed essentially of smithsonite and cerussite.

Smithsonite is the predominant supergene zinc mineral throughout the Upper Mississippi Valley zinc–lead mining district (Bastin, 1939; Heyl *et al.*, 1955), whereas in the southern Appalachian district both smithsonite and hemimorphite are abundant (Brown, 1935; Brokaw, 1948; Oder and Hook, 1950; Oder and Ricketts, 1961). In East Tintic, Utah, USA, supergene processes have greatly modified the results of hypogene mineralization. The oxidized minerals occur chiefly above the water table, but the zone of oxidation extends approximately 30 m below the present water table. Two carbonates characteristic of the oxidation zone are cerussite and smithsonite (Bush *et al.*, 1960). Near East Harptree, Mendip Hills, south of Bristol, UK, secondary aluminium minerals, allophane, kaolinite, halloysite, gibbsite and others, are associated with deeply oxidized smithsonite-bearing veins in limestone (Alabaster, 1990).

In the lead–zinc deposits of Goodsprings District, Nevada, USA, hydrozincite is the most abundant supergene mineral and tends to decrease in abundance with depth. Smithsonite and hemimorphite are distributed widely, and their presence increases with depth (Hewett, 1931; Takahashi, 1960). The stability and paragenetic sequence of formation of smithsonite, hydrozincite and hemimorphite in supergene processes have been evaluated by Takahashi (1960) in terms of pH, P_{CO_2} and silica concentration in solution.

References

Afanas'ev, Yu. A., Konovalov, S. K. and **Ryabibin, A. I.**, 1987. Solubility in the $ZnCO_3$–NaCl–H_2O and $PbCO_3$–NaCl–H_2O systems at different hydrostatic pressures. *Russ. J. Inorg. Chem.*, **32**, 270–2.

Alabaster, C., 1990. New occurrences of secondary aluminium minerals from the Bristol district. *J. Russell Soc.*, **3**, 49–59.

Bak, B. and **Niec, M.**, 1978. The occurrence of monheimite in the Boleslaw Zn–Pb ore deposit near Olkusz. *Min. Polonica*, **9**, 123–8.

Bak, B. and **Żabinski, W.**, 1981. On the continuity of the solid solution series smithsonite–siderite. *Min. Polonica*, **12**, 75–80.

Bastin, E. S., 1939. Contributions to a knowledge of the lead and zinc deposits of the Mississippi Valley region. *Geol. Soc. Amer. Spec. Paper*, **24**, 1–156.

Berg, G., Friedensburg, F. and **Sommerlatte, H.**, 1950. Blei und Zink. In *Die Metallischen Rohstoffe*, no. 9, Stuttgart (Ferdinand Enke), 468 pp.

Birch, W. D., 1986. Zinc–manganese carbonates from Broken Hill, New South Wales. *Min. Mag.*, **50**, 49–53.

Brokaw, A. L., 1948. Geology and mineralogy of the East Tennessee zinc district, *Proc. 18th Internal. Geol. Congress, London,* 46–61.

Brown, W. H., 1935. A quantitative study of zoning of ores at the Austinville mine, Wythe County, Virginia. *Econ. Geol.*, **30**, 425–33.

Bush, J. B., Cook, D. R., Lovering, T. S. and **Morris, H. T.**, 1960. The Chief Oxide–Burgin area discoveries, East Tintic District, Utah, a case history. *Econ. Geol.*, **55**, 1507–40.

Chen, T. T. and **Chang, L. L. Y.**, 1975. High Mg-smithsonite from Broken Hill, New South Wales, Australia. *Min. Mag.*, **40**, 307–8.

Effenberger, H., Mereiter, K. and **Zemann, J.**, 1981. Crystal structure refinements of magnesite, calcite, rhodochrosite, siderite, smithsonite and dolomite, with discussion of some aspects of the stereochemistry of calcite-type carbonates. *Z. Krist.*, **156**, 233–43.

Garavelli, C. L., Vurro, F. and **Fioravanti, G.**, 1982. Minrecordite, a new mineral from Tsumeb. *Min. Record*, **13**, 131–6.

Gies, H., 1975. Activation possibilities and geochemical correlations of photoluminescing carbonates, particularly calcite. *Min. Deposita*, **10**, 216–27.

Goldsmith, J. R., 1983. Phase relations of rhombohedral carbonates. In *Carbonates: Mineralogy and Chemistry:* Ed. R. J. Reeder, *Reviews in Mineralogy*, **11**, 49–76, Mineralogical Society of America, Washington, DC.

Goldsmith, J. R. and **Northrop, D.**, 1965. Subsolidus phase relations in the systems $CaCO_3$–$MgCO_3$–$CoCO_3$ and $CaCO_3$–$MgCO_3$–$NiCO_3$. *J. Geol.*, **73**, 817–29.

Gürich, G., 1903. Zur Genesis der oberschlesischen Erzlagerstätten. *Z. prakt. Geol.*, **11**, 202–5.

Harker, R. I. and **Hutta, J. J.**, 1956. The stability of smithsonite. *Econ. Geol.*, **51**, 375–81.

Haselton, H. T. and **Goldsmith, J. R.**, 1987. The decarbonation and heat capacity of $ZnCO_3$. *Geochim. Cosmochim. Acta*, **51**, 261–5.

Hewett, D. F., 1931. Geology and ore deposits of the Goodsprings quadrangle, Nevada. *U.S. Geol. Surv. Prof. Paper*, **162**, 1–172.

Heyl, A. U., Lyons, E. J., Agnew, A. F. and **Behre, C. H., Jr.**, 1955. Zinc–lead–copper resources and general geology of the Upper Mississippi Valley district. *U.S. Geol. Surv. Bull.*, **1015–6**, 227–45.

Isherwood, B. J. and **James, J. A.**, 1976. Structural dependence of the optical birefringence of crystals with calcite- and aragonite-type carbonates. *Acta Cryst.*, **A32**, 340–1.

Lo, B. W. N., 1973. The polarizabilities of planar nitrate, carbonate, and borate anions in crystals. *J. Phys. Chem. Solids*, **37**, 513–20.

Oder, C. R. L. and **Hook, J. W.**, 1950. Zinc deposits of the southeastern States. In *Symposium on Mineral Resources of the Southeastern United States.* Knoxville (Univ. Tennessee Press). pp. 72–87.

Oder, C. R. L. and **Ricketts, J. E.**, 1961. Geology of the Mascot–Jefferson City zinc district, Tennessee. *Rept. Invest. N12, Div. Geol., Nashville, Tenn.*, 29 pp.

Palache, C., Berman, H. and **Frondel, C.**, 1951. *The System of Mineralogy,* New York (John Wiley & Sons), 1124 pp.

Rosenberg, P. E. and **Foit, F. F.**, 1979. The stability of transition metal dolomites in carbonate systems: a discussion. *Geochim. Cosmochin. Acta*, **43**, 951–5.

Sainfeld, P., 1956. The lead–zinc-bearing deposits of Tunisia. *Econ. Geol.*, **51**, 150–77.

Smykatz-Kloss, W., 1974. *Differential Thermal Analysis: Application and Results in Mineralogy*. New York, Heidelberg and Berlin (Springer-Verlag), 185 pp.

Takahashi, T., 1960. Supergene alteration of zinc and lead deposits in limestone. *Econ. Geol.* **55**, 1083–115.

Warne, S. St J., 1962. A quick field or laboratory staining scheme for the differentiation of the major carbonate minerals. *J. Sed. Petr.*, **32**, 29–38.

Weber, H. J., 1988. Anisotropic bond polarizabilities in birefringent crystals. *Acta Cryst.*, **A44**, 320–26.

White, W. B., 1974. The carbonate minerals. In *Infrared Spectra of Minerals:* Ed. V. C. Farmer, *Min. Soc. Monograph*, **4**, 227–84.

Zemann, J., 1983. Ungewöhnlich kleine interpolyedrische Sauerstoff Sauerstoff-Abstände in anorganischen Kristallstrukturen. *Österreichische Akad. Wissenschaften, Math.-Naturwiss. Kl., Anzeiger*, **120**, 81–2.

Dolomite and Dolomite-type Carbonates

Dolomite $CaMg(CO_3)_2$
Ankerite $Ca(Mg,Fe,Mn)(CO_3)_2$

Dolomite $CaMg(CO_3)_2$

Trigonal (−)

ε	1.500
ω	1.679
δ	0.179
Dispersion:	Strong.
D	2.86
H	$3\frac{1}{2}$
Cleavage[1]:	$\{10\bar{1}1\}$ perfect.
Twinning[1]:	Simple twinning common on $\{0001\}$, $\{10\bar{1}0\}$ and $\{11\bar{2}0\}$; rare on $\{10\bar{1}1\}$. Glide twinning on $\{02\bar{2}1\}$.
Colour:	White to yellow or brown; colourless in thin section or grain mount.
Unit cell:	a_{hex} 4.807 Å, c_{hex} 16.00 Å, $Z_{hex}=3$. a_{rh} 6.015 Å, α_{rh} 47°6′, $Z_{rh}=1$. Space group $R\bar{3}$.

Soluble with difficulty in cold dilute HCl.

Dolomite is a common mineral of sedimentary rocks, where it is formed mostly as a diagenetic product in limestone. It is also found in marble and other metamorphic rocks, and in evaporites. Carbonatites may contain dolomite as a primary igneous mineral.

The dolomite considered here has a composition close to $CaMg(CO_3)_2$. It is common for its Mg to be substituted by Fe and/or Mn to produce ankerite $Ca(Mg,Fe,Mn)(CO_3)_2$. The name dolomite is after a French engineer, Déodat de Dolomieu (1750–1801), who described the occurrence and characteristics of this mineral.

Structure

The crystal structure of dolomite was first examined by Wyckoff and Merwin (1924). They established the unit cell dimensions and showed that the symmetry of the dolomite structure was lower than that of calcite, and that four atomic parameters were needed to evaluate the structure. The atomic parameters were not, however, investigated until X-ray powder data were obtained (Bradley *et al.*, 1953; Howie and Broadhurst, 1958). Structural refinements were made by Steinfink and Sans (1959), Althoff (1977), Beran and Zemann (1977), Effenberger *et al.* (1981) and Reeder and Wenk (1983). Lippmann (1973) and Reeder (1983) have given comprehensive reviews of the structure.

In general, the crystal structure of dolomite may be described by retaining the

[1] Cleavage, twinning and morphology are indexed using the hexagonal morphological cell with quadrupled *a* parameter (see Table 16, p. 103).

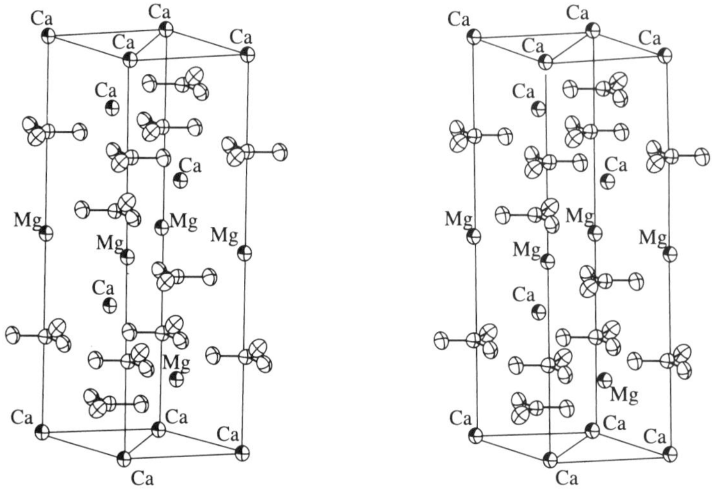

Fig. 65. Stereoscopic projection of the hexagonal unit cell for dolomite (after Reeder, 1983). The CO_3 groups are in a six-layer sequence along the *z* direction (vertical).

calcite structure but substituting every other calcium layer with a magnesium layer. This arrangement eliminates the two-fold rotation axis which intersects the carbon atoms in the calcite structure. The carbonate layers remain unchanged (Fig. 65) (Reeder, 1983) in their six-layer sequence along the *z* direction, and the unit cell configuration is comparable with that of calcite (Fig. 28, p. 102). Each oxygen of the carbonate group coordinates with one calcium and one magnesium. The cations coordinate with six oxygens, each from a different carbonate group. Reeder (1983) presented the parameters of the calcite- and dolomite-type structure as follows:

	Calcite	Dolomite
Space group:	$R\bar{3}c$ (No. 167)	$R\bar{3}$ (No. 148)
Unit cell content:	$6CaCO_3$	$3MgCa(CO_3)_2$
Unit cell parameter:		
a (Å)	4.9896	4.8069
c (Å)	17.0610	16.0034
Atom position:		
Ca	in 6(b): 0, 0, 0	in 3(a): 0, 0, 0
Mg	—	in 3(b): 0, 0, $\frac{1}{2}$
C	in 6(a): 0, 0, $\frac{1}{2}$	in 6(c): 0, 0, z_c
O	in 18(e): $x, 0, \frac{1}{2}$	in 18(f): x_o, y_o, z_o

In the CO_3 group, the C–O distance, the C–O–C angles and the O–O distances are consistent with those in calcite, magnesite and other rhombohedral carbonates. Althoff (1977) gave respective values of 1.2835(15) Å, 119.95(1)° and 2.222(2) Å, and values from Beran and Zemann (1977) are 1.2826(5) Å, 119.983(6)° and 2.228(2) Å. Corresponding values given by Reeder and Wenk (1983) are 1.2826(6) Å, 119.97(1)° and 2.2220(9) Å. The CO_3 group in dolomite is non-planar. The minimum distance of a carbon atom from the plane of oxygen atoms is 0.018 Å (Effenberger *et al.*, 1981). All the CO_3 groups within a given layer are rotated around the three-fold

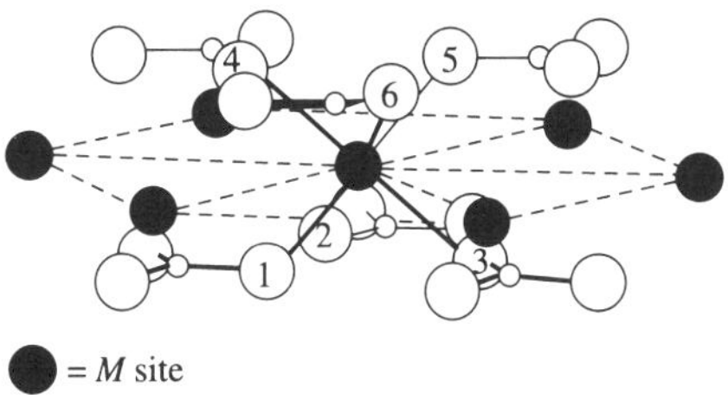

Fig. 66. A sketch of the octahedral site in the Ca–Mg carbonates (after Oh *et al.*, 1973).

rotation axis relative to their position in calcite so as to accommodate the difference in bond lengths between Ca–O and Mg–O bonds, with the angle of rotation being determined as 6.5 (Reeder and Wenk, 1983) or 6.35° (Beran and Zemann, 1977). CO_3 groups in adjacent layers are no longer related by a *c* glide, as in calcite.

The Ca–O bond in dolomite is 2.3802 (Reeder and Wenk, 1983) or 2.381(1) Å (Beran and Zemann, 1977), larger than that in calcite (2.3598 Å; Effenberger *et al.*, 1981); the Mg–O bond is 2.0839(5) (Reeder and Wenk, 1983) or 2.087(1) Å (Beran and Zemann, 1977), smaller than that in magnesite (2.105(1) Å; Oh *et al.*, 1973). The interatomic angles, O_1–M–O_2, O_1–M–O_6, O_1–M–O_5 about Ca and Mg in dolomite are 87.66(5), 92.33(5) and 180.00(0)°, and 88.38(6), 91.61(6) and 180.00(0)°, respectively (Althoff, 1977), giving an octahedron which is more ideal than those in calcite and magnesite. A sketch of the octahedral arrangement is shown in Fig. 66 (Oh *et al.*, 1973). Additional crystallographic parameters are given for dolomite in Table 29, p. 220, where comparison is made with ankerite.

The crystallographic properties of dolomite at high temperature were determined by Reeder and Markgraf (1986). Both *a* and *c* dimensions increase in magnitude with increasing temperature. As shown in Fig. 67, the variation curves for dolomite are quite similar to those for magnesite. For calcite, however, *c* is appreciably greater, and *a* contracts with increasing temperature. The mean thermal expansion coefficients for *a* and *c* of dolomite, magnesite and calcite are as follows:

		a (Å/°C)	*c* (Å/°C)
Dolomite	24–500 °C		22.9×10^{-6} (calculated)
	24–700 °C	6.2×10^{-6}	25.8×10^{-6}
Magnesite	24–500 °C	6.75×10^{-6}	22.9×10^{-6}
Calcite	25–700 °C		32.0×10^{-6} (calculated)
	24–900 °C	-2.8×10^{-6}	32.3×10^{-6}

A comparison of *c* for these three carbonates in the corresponding temperature ranges further emphasizes the similarity between dolomite and magnesite and the distinction between dolomite and calcite.

The variations of the Ca–O and Mg–O bond lengths with temperature and their comparison with those in calcite and magnesite are shown in Fig. 68. Volume expansion of the CaO_6 octahedron in dolomite is slightly greater than in calcite, while that of the MgO_6 octahedron is similar to that in magnesite. A small increase in distortion was found in the octahedron MgO_6 but not in CaO_6. In the temperature range between 24 and 600 °C, the mean thermal expansion coefficient

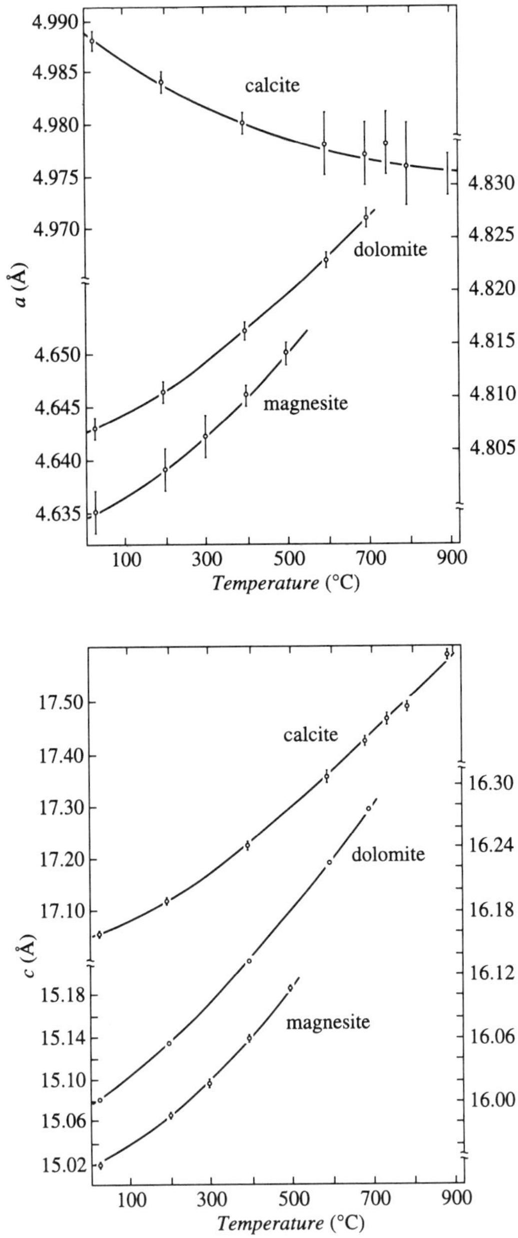

Fig. 67. Variations of unit-cell dimensions with temperature for dolomite, calcite and magnesite (after Reeder and Markgraf, 1986). In both figures, the scale for calcite is given on the left and the scale for dolomite is on the right. Error bars represent two standard deviations of the cell dimensions.

for the Ca–O bond in dolomite is 17.6×10^{-6}/°C, higher than that for calcite (12.2×10^{-6}/°C). The mean thermal expansion coefficients for the Mg–O bond are comparable in dolomite and magnesite (16.5×10^{-6}/°C).

For the CO_3 group, the C–O and O–O bond lengths in the temperature range between 24 and 600 °C increase from 1.2900(5) to 1.2930(8) Å and from 2.234(1)

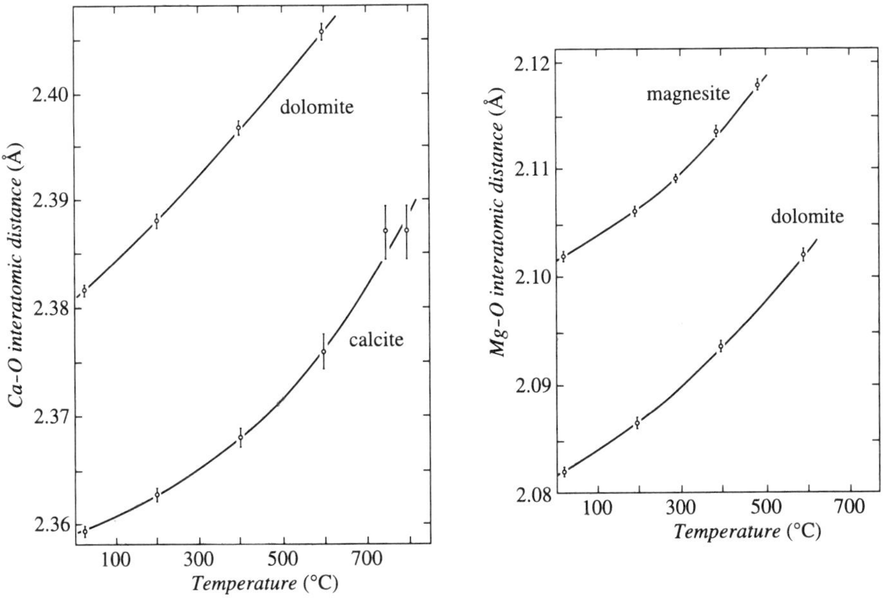

Fig. 68. Variations of the Ca–O and Mg–O bond lengths with temperature in dolomite and their comparison with those in (a) calcite and (b) magnesite (after Reeder and Markgraf, 1986).

Table 24. Unit Cell Dimensions of Dolomite at Various Pressures from 0.0 to 46.9 kbar (from Ross and Reeder, 1992)

Pressure (kbar)	*a* (Å)	*c* (Å)	*V* (Å)
0.0	4.8064(5)	16.006(2)	320.22(8)
4.7	4.8028(7)	15.692(2)	318.87(9)
15.0	4.7910(7)	15.856(2)	315.20(9)
23.4	4.7843(9)	15.785(2)	312.90(11)
29.0	4.7777(5)	15.730(1)	310.97(6)
37.0	4.7703(7)	15.653(2)	308.47(10)
42.0	4.7672(6)	15.611(2)	307.24(8)
46.9	4.7636(5)	15.582(3)	306.21(8)

Standard deviations in parentheses.

to 2.39(1) Å respectively after correction for libration. No change in the aplanarity of the CO_3 group with temperature was observed.

Ross and Reeder (1992) examined dolomite under high pressure and showed that the major changes are the compression of the CaO_6 and MgO_6 octahedra. The linear compressibilities of Ca–O and Mg–O bonds in dolomite are 3.77×10^{-4} and 3.52×10^{-4}/kbar, respectively. In the pressure range between 0.0 and 46.9 kbar, the CO_3 group is a rigid unit and remains essentially unchanged. Unit cell dimensions of dolomite at various pressures are listed in Table 24. The *z* axis is approximately three times as compressible as the *x* axis. No phase change was detected in the pressure range. Measurements of cell dimensions at higher pressures were made by Fiquet *et al.* (1994). Between 55 and 109 kbar, *a* decreases from

4.780(5) to 4.753(8) Å, *c* from 15.66(1) to 15.29(2) Å, and *V* from 309.8(6) to 299.0(8) $Å^3$. In comparison with the results of Ross and Reeder (1992) (Table 24) for cell dimension changes in the pressure range between 0 and 46.9 kbar, *a*, *c* and *V* all show a smaller decrease with increasing pressure (see also Fig. 82, p. 221).

Disordering in dolomite occurs at high temperatures (Goldsmith, 1959, 1980) inducing gradual changes in the crystal structure. Cell dimensions, selected interatomic distances and angles (°), and site occupancy fractions for untreated and treated dolomites were measured by Reeder and Wenk (1983), and a selection is presented below:

	Ordered	Disordered
a(Å)	4.8038	4.799 (1200°C)
c(Å)	16.006	15.986 (1200°C)
M_A–O(Å)	2.3802	2.350 (1100°C)
M_B–O(Å)	2.0839	2.114 (1100°C)
A site Ca	0.9989	0.8773 (1100°C)
A site Mg	0.0011	0.1227 (1100°C)
B site Ca	0.0025	0.1241 (1100°C)
B site Mg	0.9975	0.8759 (1100°C)

Graf and Goldsmith (1956) introduced the term 'protodolomite' for a rhombohedral carbonate which deviates from the ideal dolomite composition and has significant disordering of its cations. At equilibrium, protodolomite transforms to dolomite. Gaines (1977) redefined this term by adding 'metastable' and 'high degree of cation order' to the description of protodolomite. This term has received wide recognition but it is not without objection. Deelman (1977) argued that 'protodolomite' conflicts with the earlier definition of magnesian calcite by Chave (1952), and there is no experimental proof of the transformation of protodolomite to dolomite *sensu stricto* at atmospheric temperatures. Gidman (1977) stresses that a protodolomite should be a Ca–Mg carbonate having the major dolomite X-ray diffraction lines but lacking the ordered reflections. Reeder (1983) states that if the ordered reflections are present in protodolomite, no matter how weak, it is dolomite. If there is no ordered reflection, the protodolomite is not dolomite.

Evidence for superstructures in dolomite has been provided by electron diffraction and microscopy (Reeder and Wenk, 1983; Van Tendeloo *et al.*, 1985; Wenk and Zhang, 1985). The superstructure reflections have been termed *c*- and *d*-type, and both reflections in calcian dolomites arise from small local domains in the host crystals (Wenk and Zhang, 1985). The *c*-type reflections occur midway between $(10\bar{1}4)$-, $(11\bar{2}0)$- and $(01\bar{1}2)$-type reflections (indexed for the hexagonal structural cell), whereas the *d*-type reflections are present as weak satellites on both sides of $(000l)$ reflections with $l=3n$ in the c^* direction.

Chemistry

Normally, the composition of dolomite is fairly close to its formula $CaMg(CO_3)_2$, but many dolomites do contain some Fe in place of Mg. The Fe substitution produces a complete series to ankerite, $Ca(Fe,Mg)(CO_3)_2$, which is considered separately (p. 219). Ferroan dolomite is the term used to describe dolomite having an Fe content up to an Fe/(Mg+Fe) ratio of 0.2. Selected analyses of dolomite and ferroan dolomite are listed in Table 25. Those from Oberdorf, Austria (anal. 12), from the Platteville Formation, Wisconsin, USA (anal. 5) and from the Monterey Formation, California, USA (anal. 7) may represent the purest dolomites.

Mn may also replace Mg in dolomite, and the replacement may extend to $CaMn(CO_3)_2$, kutnahorite, a Mn-dolomite described by Frondel and Bauer (1955). The replacement is commonly made in the presence of Fe, as shown in the analyses of manganoan dolomite in Table 25 (anal. 11).

Zincian dolomites have been described from Warynski mine, Upper Silesia, Poland (Zasieńska and Żabiński, 1972; Żabiński, 1980; Rosenberg and Champness, 1989) and Broken Hill, New South Wales, Australia (Birch, 1983), and compositions are plotted in the system $CaCO_3$–$ZnCO_3$–$(Mg,Fe,Mn)CO_3$ (Fig. 69) (Rosenberg and Champness, 1989). The chemical compositions of Zn-dolomite from Tsumeb, Namibia (Garavelli *et al.*, 1982) are listed in Table 26. The zincian analogue of dolomite (Table 26 and Fig. 69) is named minrecordite.

Experimental Work

Subsolidus phase relationships in the system $CaCO_3$–$MgCO_3$–$MnCO_3$ are shown in Fig. 70 (Goldsmith and Graf, 1960). The $CaMg(CO_3)_2$–$CaMn(CO_3)_2$ join is binary only at temperatures $>700\,^\circ C$. The greater solid solubility in this system than in the system with $FeCO_3$ is shown by the size and shape of the three-phase region (see also Mucci, 1991).

While the general form of the phase relationships is similar from both studies, the three-phase region for coexisting calcite + siderite + ankerite solid solution established by Goldsmith *et al.* (1962) is significantly wider than that obtained by

Table 25. Dolomite Analyses

	1	2	3	4	5	6
CaO	29.59	30.46	33.83	30.55	32.69	33.32
MgO	21.39	21.76	18.95	21.57	19.67	19.38
MnO	0.14	—	—	—	—	—
FeO	0.63	0.08	—	0.23	0.15	—
CO_2	47.05(c)	49.02(c)	47.23(c)	47.66(c)	47.21(c)	47.30(c)
Total	98.80	101.32	100.01	100.01	99.72	100.00
	Numbers of ions on the basis of six (O)					
Ca	0.897	1.002	1.124	1.006	1.087	1.105
Mg	0.993	0.906	0.876	0.988	0.909	0.895
Mn	0.004	—	—	—	—	—
Fe	0.016	0.002	—	0.006	0.004	—
C	2.000	2.000	2.000	2.000	2.000	2.000

(c) Calculated from stoichiometry.

1 Dolomite from the carbonate-cemented sandstone cap-rock in the Salton Sea geothermal system, California, USA (McDowell and Paces, 1985).

2 Reddish dolomite embedded in gypsum from Lake Arthur, New Mexico, USA (Reeder and Sheppard, 1984); *a* 4.8069, *c* 16.0034 Å.

3–5 Dolomite from Eocene rocks of the Floridan Aquifer and from Ordovician Platteville Formation, Wisconsin, USA (Reeder and Sheppard, 1984). Recalculated from mol. per cent.

6 Dolomite from the Miocene Monterey Formation, California, USA (Burns and Baker, 1987). Recalculated from mol. per cent. Fe, Mn and Sr are present in trace amounts.

Table 25. *Continued*

	7	8	9	10	11	12
CaO	30.66	30.83	30.62	31.36	28.81	30.56
MgO	21.65	21.20	19.73	15.95	13.40	22.04
MnO	—	—	—	—	6.37	—
FeO	—	0.39	2.54	6.62	5.98	—
CO_2	47.69(c)	47.57(c)	47.12(c)	46.07(c)	44.80(c)	46.93
Total	100.00	99.99	100.01	100.00	99.36	99.53
	Numbers of ions on the basis of six (O)					
Ca	1.009	1.017	1.020	1.068	1.009	1.014
Mg	0.991	0.973	0.914	0.756	0.653	1.018
Mn	—	—	—	—	0.176	—
Fe	—	0.010	0.066	0.176	0.163	—
C	2.000	2.000	2.000	2.000	1.999	1.984

7 Dolomite from the Miocene Monterey Formation, California, USA (Burns and Baker, 1987). Recalculated from mol. per cent. Fe, Mn and Sr are present in trace amounts.

8–10 Ferroan dolomite from the Miocene Monterey Formation, California, USA (Burns and Baker, 1987). Recalculated from mol. per cent. Mn and Sr are present in trace amounts.

11 Mn-bearing ferroan dolomite from crystalline stalactite-like pipes in a volcanotectonic basin, Uzbekistan (Talantsev, 1983).

12 Dolomite from Oberdorf, Austria; USNM 10057 (Jarosewich and Macintyre, 1983).

Table 25. *Continued*

	13	14	15	16	17	18
CaO	31.03	30.3	28.65	30.65	30.30	31.27
MgO	19.39	19.4	17.63	21.70	21.20	21.12
MnO	0.46	0.41	0.28	0.08	—	0.00
FeO	0.63	2.5	6.94	0.41	0.46	0.22
	0.64 (SrO)	0.27 (oth.)			0.07 (SrO)	0.12 (SiO_2)
CO_2	46.46(c)	46.74(c)	46.14(c)	48.04(c)	47.00	47.22
Total	98.61	99.62	99.64	100.88	99.03	99.97
	Numbers of ions on the basis of six (O)					
Ca	1.048	1.017	0.974	1.001	1.010	1.036
Mg	0.911	0.906	0.834	0.986	0.983	0.973
Mn	0.012	0.011	0.007	0.002	—	—
Fe	0.017	0.066	0.184	0.010	0.012	0.006
	0.012 (Sr)				0.001 (Sr)	—
C	2.000	2.000	2.000	2.000	1.997	1.993

13 Dolomite from carbonatite dyke, Ontario, Canada (Hogarth *et al.*, 1988).

14 Dolomite vein cutting through clinopyroxene in Bellsbank Kimberlite, South Africa (Hervig and Smith, 1981).

15 Ferroan dolomite from ultrabasic lamprophyres, McKellar Harbour, Northwest Ontario, Canada (Platt and Mitchell, 1982).

16 Dolomite from a metamorphic carbonate complex at Eugui, Spain (Reeder and Wenk, 1983); *a* 4.8038, *c* 16.006 Å.

17 Dolomite microcrystals in Grenville marble, north of Ottawa River, Grenville Province, Ontario (Kretz, 1988).

18 White dolomite, Ross Township, Haley, Ontario, Canada (Harker and Tuttle, 1955) (total includes H_2O^- 0.02).

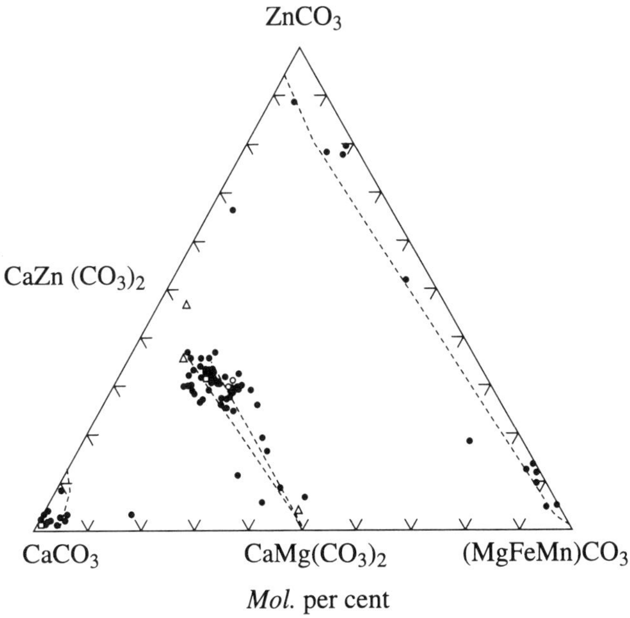

Fig. 69. Comparisons of zincian carbonates in the system $CaCO_3$–(Mg,Fe,Mn)CO_3–$ZnCO_3$ (after Rosenberg and Champness, 1989). ●, Waryński mine, single analysis; ○, Waryński mine, four coincident analyses; △, Tsumeb, Namibia (Garavelli *et al.*, 1982); – – –, one-phase areas in the system $CaCO_3$–$MgCO_3$–$FeCO_3$ at 450 °C (Rosenberg, 1967).

Table 26. Zincian Dolomite Analyses

	1	2	3	4
CaO	25.79	26.66	22.17	20.92
MgO	16.61	20.06	2.16	0.83
MnO	0.03	0.16	0.07	0.09
FeO	0.02	1.40	0.57	1.06
	7.55 (ZnO)	1.86 (ZnO)	25.15 (ZnO)	26.93 (ZnO)
CO_2	42.49(c)	44.78(c)	50.12(c)	49.83(c)
Total	92.49	94.92	100.24	96.66
	Numbers of ions on the basis of six (O)			
Ca	0.953	0.934	0.997	0.942
Mg	0.853	0.978	0.113	0.044
Mn	0.001	0.004	0.003	0.003
Fe	—	0.038	0.022	0.041
	0.192 (Zn)	0.045 (Zn)	0.885 (Zn)	0.968 (Zn)
C	2.000	2.000	2.000	2.000

(c) Calculated from stoichiometry.

1, 2 Zincian dolomite from Tsumeb, Namibia (Garavelli *et al.*, 1982).
3, 4 Minrecordite from Tsumeb, Namibia (Garavelli *et al.*, 1982).

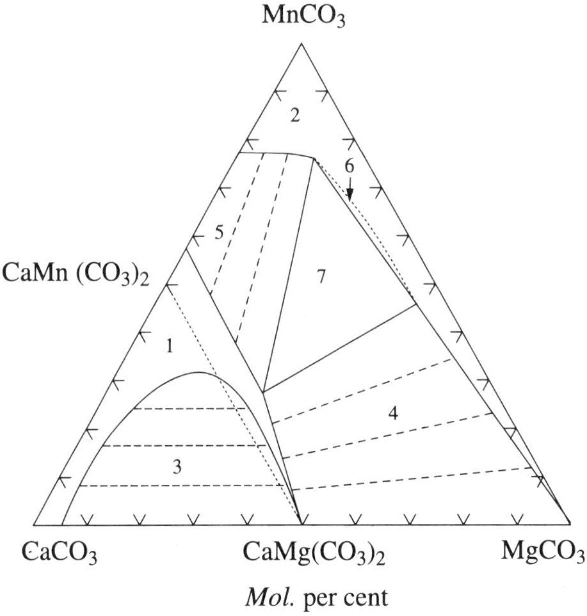

(a)

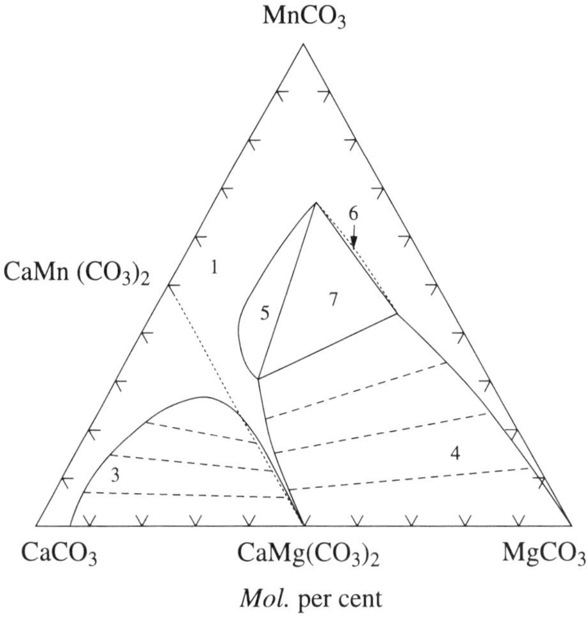

(b)

Fig. 70. Subsolidus phase relationships in the system $CaCO_3$–$MgCO_3$–$MnCO_3$ at (a) 500 °C and 10 kbar and (b) 600 °C and 10 kbar (after Goldsmith and Graf, 1960). The join $CaMg(CO_3)_2$–$CaMn(CO_3)_2$ is shown by a dotted line. Phase assemblages are represented by (1) calcite–dolomite$_{ss}$ in (a) and (Ca,Mg,Mn)CO_3 solid solution in (b), (2) magnesite–rhodochrosite$_{ss}$, (3) calcite$_{ss}$ + dolomite$_{ss}$, (4) dolomite$_{ss}$ + magnesite–rhodochrosite$_{ss}$, (5) dolomite$_{ss}$ + magnesite–rhodochrosite$_{ss}$, (6) two magnesite–rhododochrosite$_{ss}$ and (7) dolomite$_{ss}$ + two magnesite–rhodochrosite$_{ss}$.

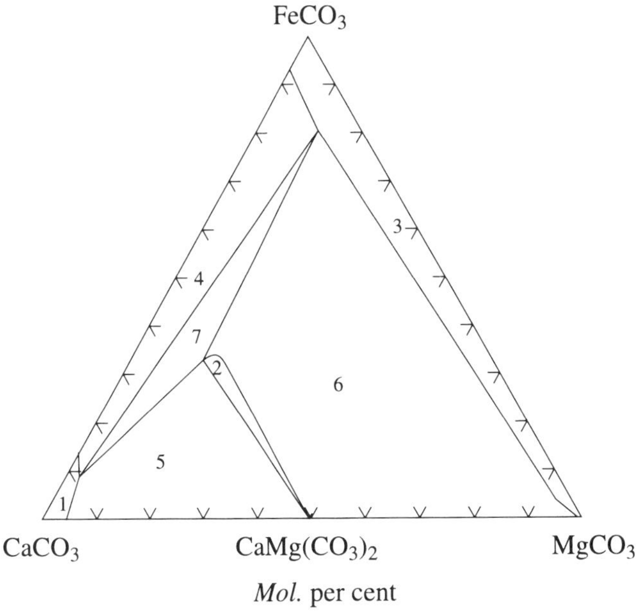

Fig. 71. Subsolidus phase relationships in the system $CaCO_3$–$MgCO_3$–$FeCO_3$ at 450 °C (after Rosenberg, 1967). Phase assemblages are represented by (1) calcite and calcite$_{ss}$, (2) dolomite and dolomite$_{ss}$, (3) magnesite–siderite$_{ss}$, (4) calcite$_{ss}$ + magnesite–siderite$_{ss}$, (5) calcite$_{ss}$ + dolomite$_{ss}$, (6) dolomite$_{ss}$ + magnesite–siderite$_{ss}$ and (7) calcite$_{ss}$ + dolomite$_{ss}$ + magnesite–siderite$_{ss}$.

Rosenberg (1967) (Fig. 71). Both studies found that $CaFe(CO_3)_2$ was not stable, and that the dolomite–ferroan dolomite–ankerite solid solution lies on the Ca-rich side of the $CaMg(CO_3)_2$–$CaFe(CO_3)_2$ join. Goldsmith *et al.* (1962) also conducted experimental runs at $\geqslant$700 °C in the system. Fig. 72 shows the disappearance of the three-phase region.

The thermal stability of dolomite was determined by Goldsmith and Heard (1961). The observed limits of ordering in dolomite as a function of composition and temperature in the system $CaCO_3$–$MgCO_3$ are outlined in Fig. 73. No experimental evidence was obtained to indicate the presence of a two-phase region consisting of both ordered and disordered $CaMg(CO_3)_2$. and the phase diagram is constructed on the basis of a non-first-order transition. The top of the calcite–dolomite solvus is at 1075 °C and a composition of $(Ca_{0.57}Mg_{0.43})CO_3$. The stability of dolomite in the system $CaCO_3$–$MgCO_3$ under high pressures was studied by Goldsmith and Newton (1969), Irving and Wyllie (1975) and Byrnes and Wyllie (1981). At 30 kbar (Irving and Wyllie, 1975; correction made by Byrnes and Wyllie, 1981, to 27 kbar), the melting loop extends from 1610 ($CaCO_3$) to 1585 °C ($MgCO_3$) through a liquid minimum at 1290 °C and 42 mol. per cent $MgCO_3$, as shown in Fig. 74. The dolomite–magnesite solvus intersects the melting loop to produce a peritectic reaction at 1385 °C.

The pressure–temperature projection for reactions in the system $CaCO_3$–$MgCO_3$ (Fig. 75(a)) was constructed by Irving and Wyllie (1975) on the basis of experimental data obtained and deduced from the *P*–*T*–*X* relationships in the systems by Goldsmith and Heard (1961) and Goldsmith and Newton (1969). The dashed line C_c-C_d drawn through the point of 1075 °C and 8.5 kbar, the crest of the calcite–dolomite solvus (Goldsmith and Heard, 1961), was assumed to be parallel

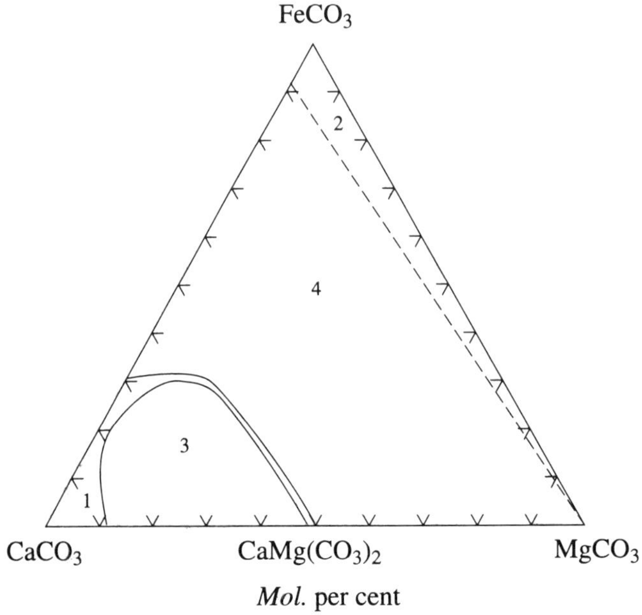

Fig. 72. Phase relationships in the system $CaCO_3$–$MgCO_3$–$FeCO_3$ at 700 °C and 15 kbar illustrating the disappearance of the three-phase region (after Goldsmith *et al.*, 1962). Phase assemblages are presented by (1) calcite–dolomite$_{ss}$, (2) magnesite–siderite$_{ss}$, (3) calcite$_{ss}$ + dolomite$_{ss}$ and (4) calcite–dolomite$_{ss}$ + magnesite–siderite$_{ss}$.

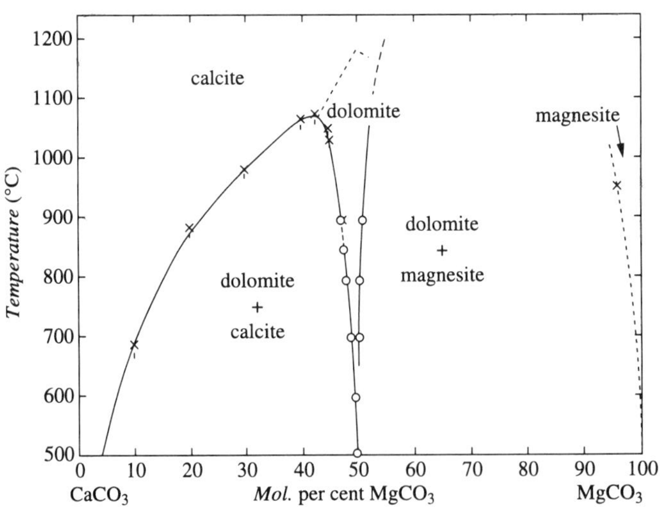

Fig. 73. The system $CaCO_3$–$MgCO_3$ (after Goldsmith and Heard, 1961). The observed limits of ordering in dolomite as a function of composition and temperature are outlined by a dashed line in the central portion of the diagram.

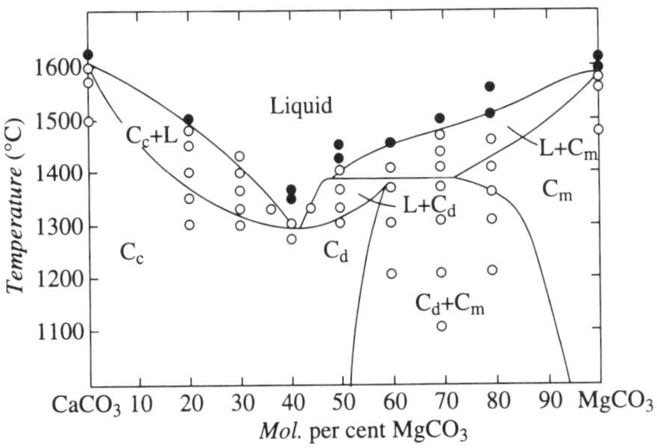

Fig. 74. Phase relationships in the system $CaCO_3$–$MgCO_3$ at 27 kbar (after Irving and Wyllie, 1975). C_c, C_d and C_m indicate calcite solid solution, carbonate solid solution near dolomite composition, and magnesite solid solution, respectively.

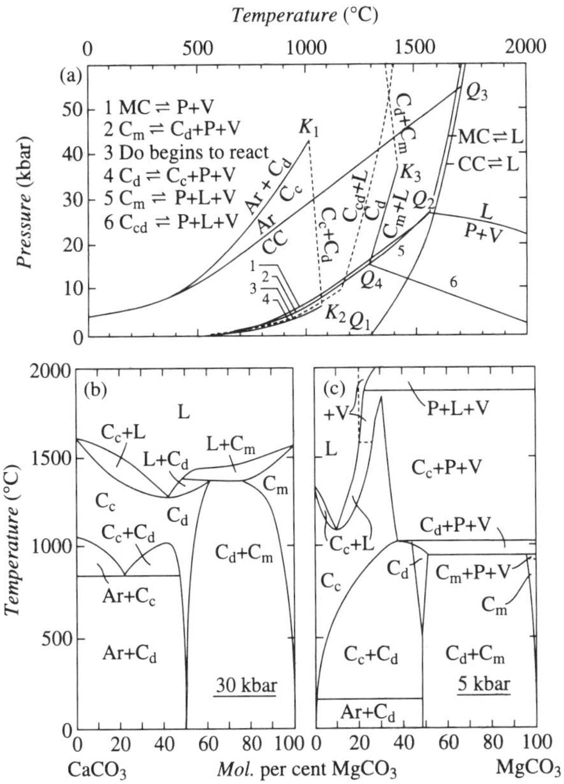

Fig. 75. (a) Pressure–temperature projection for reactions in the system $CaCO_3$–$MgCO_3$ constructed by Irving and Wyllie on the basis of experimental data and published results (after Irving and Wyllie, 1975). (b) Isobaric reactions at 30 kbar and (c) isobaric reactions at 5 kbar. Symbols used, in addition to C_c, C_d and C_m (see Fig. 74), are V (vapour), L (liquid), P (periclase), Ar (aragonite), Do (dolomite), CC (calcite), MC (magnesite) and C_{cd} (calcite–dolomite solid solution).

to a line joining the 9 and 20 kbar solvus temperatures for the composition of 70 mol. per cent $CaCO_3$ and 30 mol. per cent $MgCO_3$ (Goldsmith and Newton, 1969). This line gives the temperature of the C_c–C_d solvus crest at various pressures including those of Figs. 75(b) and 75(c). The formation of magnesian calcite from aragonite and dolomite ($Ar + C_d \rightleftharpoons C_c$) was studied experimentally up to 22 kbar by Goldsmith and Newton (1969), and the univariant curve in Fig. 75(a) was extrapolated to higher pressures when it meets the critical curve $C_c = C_d$ and terminates at a critical end-point, K_1. The incongruent melting reaction for the dolomite solid solution $C_c \rightleftharpoons C_m$ + liquid was assumed by Irving and Wyllie (1975) to have a slope similar to that for calcite. It extends to lower pressure and terminates at an invariant point, Q_4. To higher pressure, it terminates at a critical end-point, K_3, when the C_d–C_m solvus becomes separated from the solidus. The temperature at K_3 (1410 °C) was estimated for the metastable solvus crest $C_d = C_m$ in Fig. 75(b). The line $C_{cd} = L$, drawn through the 30 kbar point to the melting regions reported at 10 (Goldsmith and Heard, 1961) and 12 kbar (Goldsmith and Newton, 1969), is approximately parallel to the melting curves of calcite and magnesite at high pressures. At 5 kbar (Fig. 75(c)), dissociation of carbonates occurs in the system $CaCO_3$–$MgCO_3$. The isothermal curve truncating the solvus C_c–C_d represents the dissociation of calcic dolomite, $C_d \rightleftharpoons C_c + P + V$, which must be at a temperature higher than that for pure dolomite. It extends to the critical curve C_c–C_d and terminates at a critical end-point, K_2. The isothermal curve truncating the solvus C_d–C_m represents the reaction $C_m \rightleftharpoons C_d + P + V$, which must occur at a temperature higher than the dissociation temperature for pure magnesite but lower than that for pure dolomite. The magnesite dissociation curve extends upwards to meet the curve for the incongruent melting of dolomite at Q_4. Also extending from this invariant point are the incongruent melting curves for magnesite solid solution ($C_m \rightleftharpoons P + L + V$) and for carbonate solid solution ($C_{cd} \rightleftharpoons P + L + V$).

A series of isobaric sections was prepared by Irving and Wyllie (1975) showing successive changes in the phase relationships as pressure increases from 6 to 25 kbar (Fig. 76). Byrnes and Wyllie (1981) established the subsolidus and melting relations in the system $CaCO_3$–$MgCO_3$ at 10 kbar (Fig. 77), and the general geometry is consistent with that deduced for 12 kbar by Irving and Wyllie (1975). In comparison with phase relationships at 27 kbar, the carbonate solidus is closer to the crest of the calcite–dolomite solvus at the lower pressure. If the trend persists to still lower pressures, co-precipitation of calcite and dolomite solid solutions from the carbonate liquid might become a possibility. Navrotsky and Loucks (1977) calculated subsolidus phase relationships in the system $CaCO_3$–$MgCO_3$ by formulating the free energies of mixing in terms of parameters related to the stability of dolomite, the free energies of transformation of $CaCO_3$ and $MgCO_3$ to the dolomite structure, and an energy term describing Bragg–Williams-type disordering in dolomite. The calculated phase relationships reproduce the essential topologies of experimentally determined phase diagrams, but do not exactly match in the location of the boundaries.

Dolomite shows two differential thermal analysis endotherms in CO_2 (above 300 mm Hg) at 775 and 870 °C, representing $CaMg(CO_3)_3 \rightarrow CaCO_3 + MgO + CO_2$ and $CaCO_3 \rightarrow CaO + CO_2$, respectively (Otsuka, 1986). With decrease of CO_2 pressure, the 775 °C endotherm is shifted to a higher temperature and the 870 °C endotherm moves in the opposite direction, until at pressures <260 mm Hg the two endotherms merge and the decomposition of dolomite proceeds in one step:

$$CaMg(CO_3)_2 \rightarrow CaO + MgO + 2CO_2.$$

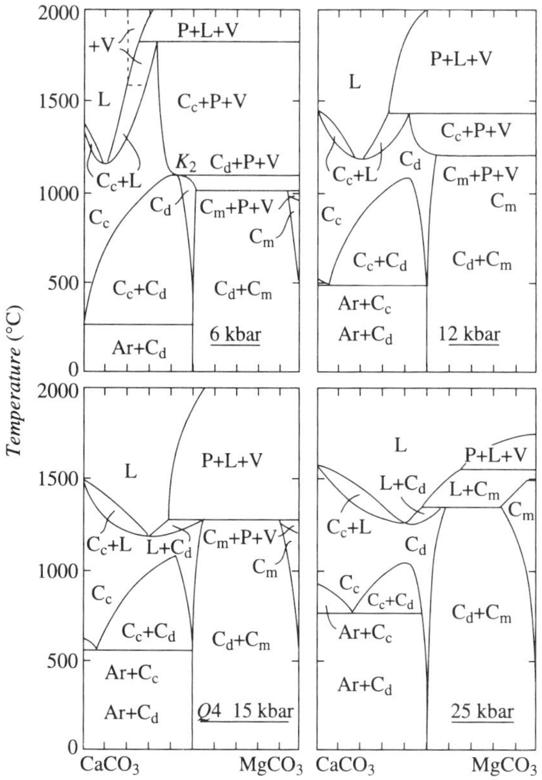

Fig. 76. Isobaric sections showing successive changes in phase relationships in the system $CaCO_3$–$MgCO_3$ (after Irving and Wyllie, 1975); symbols as Figs. 74 and 75.

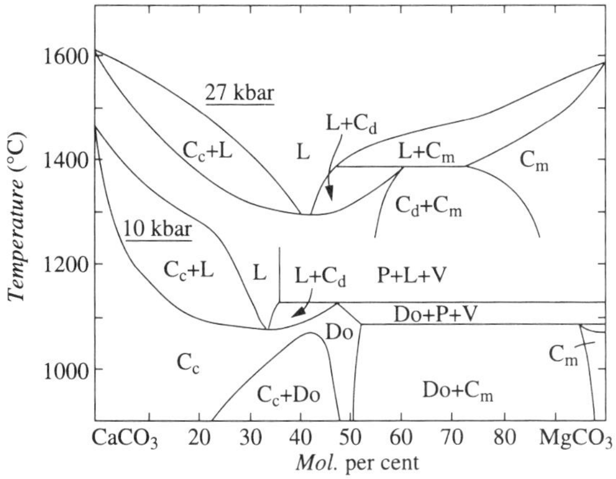

Fig. 77. Subsolidus and melting relationships in the system $CaCO_3$–$MgCO_3$ at 10 kbar and a comparison with results of Irving and Wyllie (1975) for 27 kbar (after Byrnes and Wyllie, 1981).

Experimental studies on dolomite–calcite equilibrium at 220–240 °C at saturation vapour pressure have shown that dolomites precipitated in equilibrium with calcite in this temperature range may be metastable with respect to ideal dolomite (Morrow *et al.*, 1994).

The hypothesis that Ostwald's Step Rule in dolomitization (in which a reaction proceeds through a series of intermediate stages, each thermodynamically more stable than that preceding) is determined by nucleation kinetics has been verified by Nordeng and Sibley (1994), who found that significant nucleation occurs early in the reaction and that the length of the induction period is due to the slow initial growth of metastable phases.

The thermodynamic properties of dolomite have been determined by Stout and Robie (1963). Values for the heat capacity, entropy and free energy at 298.15 K are 37.65 cal/kmol., 37.09 cal/kmol. and 6210 cal/mol., respectively. Navrotsky and Capobianco (1987) measured the enthalpy of formation of dolomite as expressed by $0.5CaCO_3+0.5MgCO_3=Ca_{0.5}Mg_{0.5}CO_3$. For ordered dolomite from Eugui, Navarra, Spain, the value is -5.74 kJ/mol., which compares well with the value of -6.07 kJ/mol. of Stout and Robie (1963). For the same dolomite treated at 1250 °C and showing no ordering reflections, the value is 1.23 ± 0.32 kJ/mol. Thus, the enthalpy of disordering is 6.97 ± 0.22 kJ/mol.

There is no evidence for precipitation of dolomite directly from sea-water of normal salinity, even though sea-water is about 20 times supersaturated with respect to it (Bathurst, 1976). It is always the more soluble calcite, with metastable aragonite and high-Mg-calcite which is precipitated despite dolomite being thermodynamically the most stable carbonate in sea-water. It seems clear that the reasons for this paradox are kinetic (Hsü, 1967; Land, 1985; Sibley *et al.*, 1987). The major kinetic factors that were identified to affect dolomite precipitation below 100 °C can be summarized as follows. (a) The reaction rate is very temperature sensitive (Land, 1967) and the reaction pathway may be complex, with the formation of intermediate precursor phases (see also Sibley et al., 1994). The dissolution rate of the precursor phase determines the rate of dolomite formation (Katz and Matthews, 1977; Bullen and Sibley, 1983; Mucci and Morse, 1983). (b) The pronounced difference in the hydration energy between Mg^{2+} and Ca^{2+} reduces the chances for magnesium ions to be dehydrated for incorporation into carbonate formation (Lippmann, 1973). The proportion of Mg ions (or the Mg/Ca ratio) must be increased so as to have sufficient energy to shed their strong hydration sheath (Morrow, 1982). An increase in the CO_3/Ca ratio will produce similar results. In most natural solutions the activity of CO_3^{2-} is low and this inhibits the precipitation of Mg-rich carbonates (Lippmann, 1973). (c) The segregation of Ca^{2+} and Mg^{2+} into their respective layers of an ordered dolomite is not facilitated during the rapid crystallization of supersaturated solutions (Lippmann, 1973; Morrow, 1982). (d) Certain organic compounds (Gaines, 1980) and dissolved SO_4^{2-} (Baker and Kastner, 1981) inhibit dolomite precipitation. Dolomite has been synthesized at 60°C by the reaction of $CaCO_3$ with saturated $CaCl_2$–$MgCl_2$ solutions for periods of up to seven years (Usdowski, 1994). The phase boundaries of calcite–dolomite and magnesite–dolomite are at Ca^{2+}/Mg^{2+} ratios of 0.445 and 0.107, respectively.

Morrow and Ricketts (1986) studied the influence of sulphate ion concentration on the formation of the more readily precipitated analogues of dolomite. $PbMg(CO_3)_2$ was found to precipitate with equal ease from both sulphate-bearing and sulphate-free alkaline solutions at 25–80 °C. The variations of sulphate concentration have no discernible influence. This is unexpected in view of the experiments of Baker and Kastner (1981) on dolomite precipitation at high temperatures. The precipitation of $BaMg(CO_3)_2$, norsethite, only in sulphate-free solutions is consistent with the results of Baker and Kastner (1981).

The kinetic study by Sibley *et al.* (1987) at 200 °C demonstrates that in the dolomitization of calcite there are two distinct stages: an induction stage, with no detectable formation of dolomite, and a nucleation–growth stage, during which dolomite forms at the expense of calcite. As the Mg/Ca ratio of the solution increases, the length of the induction stage decreases. A two-stage process has also been observed by Busenberg and Plummer (1982) in their measurements of weight loss in the dissolution of dolomite at various temperatures and CO_2 pressures. In the first stage, the $CaCO_3$ component dissolves faster and the surface composition of the dolomite becomes enriched with $MgCO_3$. In the second stage, both components of the solid are released stoichiometrically. It was found that the dissolution rate is apparently more dependent on crystallographic order than on compositional variations. Reviews of dolomite formation have been given by Zenger and Durham (1980), Wells (1986), Machel and Mountjoy (1986) and Morse and Mackenzie (1990).

The major uses of dolomite include construction aggregates, fluxstone, agricultural stone, fillers and refractory dolomite bricks.

Optical and Physical Properties

Dolomite is uniaxial and optically negative. Like calcite, rare occurrences of biaxial dolomite have been recorded, and the cause of such occurrences is attributed to the presence of twin lamellae (Rath and Pohl, 1971). The size of the optic axial angle depends on the proportion of lamellae in the two orientations and the twin angle.

The maximum birefringence of dolomite, ω–ε 0.179, is very high, and it increases further to 0.202 as Fe^{2+} replaces Mg. Interference colours seen in thin section or grain mount are typically creamy high-order white, even for orientations near the optic axis. Analyses of the relationship between birefringence and structure by Lo (1973), Isherwood and James (1976) and Weber (1988) are as applicable to dolomite as they are to calcite.

Refractive indices increase with the substitution of Fe^{2+} for Mg, as shown in Fig. 42 (p. 118); the birefringence is also increased. A universal-stage technique for the determination of the $MgCa(CO_3)_2/FeCa(CO_3)_2$ ratio in iron-bearing dolomite has been described (Howell and Dawson, 1958) using the angle of rotation required to change the transmitted ray from ε to ε', where $\varepsilon' = 1.54$, the refractive index of the mounting medium. The substitution of Mn for Mg affects the refractive indices to a much lesser extent, ω being increased more than ε, leading to an increase in the birefringence. The substitution of Zn for Mg or of Pb for Ca also increases the refractive indices, as shown in Table 27 (Hurlbut, 1957).

Dolomite possesses fewer twin forms than calcite, and only one type of deformation twinning is known. It occurs on $\{\bar{2}021\}$ between 300 and ~600 °C, with a shear direction parallel to the intersection of the twin plane with the $\{\bar{1}2\bar{1}0\}$ plane (Higgs and Handin, 1959). Unusual stress–temperature behaviour was observed in dolomite single crystals, with a yield by *c*-slip occurring at a stress which is twice as large at 500 °C than it is at room temperature (Barber *et al.*, 1981). Long faults occur in dolomite crystals on the *c* and *f* planes between 300 and 500 °C (Barber *et al.*, 1978). The former always terminate at dislocations, while the latter do so much less frequently.

Data on Ca/Mg variation between non-equivalent growth sectors in sedimentary dolomite crystals show that sector zoning involves variations in $CaCO_3$ content

Table 27. Specific Gravity, Hardness and Refractive Indices of Zincian and Plumbian Dolomite from Tsumeb (from Hurlbut, 1957)

	1	2	3	4	5
ZnO content (wt. per cent)	4.48	3.23	8.74	4.96	8.31
PbO content (wt. per cent)	0.58	3.75	6.80	0.70	1.06
Specific gravity					
Measured	2.90	2.93	3.08	2.93	2.95
Calculated	2.91	2.97	3.08	2.93	3.01
Hardness	4	$3\frac{1}{2}$	$3\frac{1}{2}$	$3\frac{1}{2}$	4
Refractive indices					
ω	1.690	1.698	1.703	1.699	1.700
ε	1.520	1.519	1.520	1.519	1.521

of $\leqslant 3$ mol. per cent between coeval $10\bar{1}4$ and $10\bar{1}1$ growth sectors (Fouke and Reeder, 1992). These sectors were found to have different cathodoluminescence, suggesting differences in Mn and/or Fe contents between them. The variations in $CaCO_3$ and $MgCO_3$ between the growth sectors are associated with differences in surface structure and possibly growth mechanism.

Dolomite when pure is white or colourless, but it often has a yellow or brown colour due to the presence of Fe^{2+} or a rose-pink colour from the Mn content. Allen and Fahey (1952) reported a dolomite that has the colour of rhodochrosite with only 1.45 per cent MnO. The presence of cobalt may give a vivid pink colouration. Dolomite of organic origin may be light brown due to its hydrocarbon content (Spotts and Silverman, 1966).

Dolomite is not one of the strongly luminescing carbonates (Gies, 1975). Przibram (1956), however, reported a violet dolomite from the Belgian Congo which showed a pale-blue fluorescence in ultraviolet light. The thermoluminescence glow curve of dolomite was studied by Lewis (1956), who reported characteristic peaks resulting from groupings of energy levels in the crystals. These energy levels can be populated by metastable electrons by exposure of the mineral to cobalt-60 γ-radiation. The X-ray-induced thermoluminescence of dolomite from Eugui, Navarra, was investigated by Calderon *et al.* (1982). The emission spectra of thermoluminescent peaks contain bands at 250, 280, 310, 380 and 430 nm, and were interpreted in terms of an electron (CO_3^{3-})–hole (X^{3+}) (X=impurity) recombination. The cathodoluminescence of dolomite is controlled chemically by the Fe and Mn contents, as shown in Fig. 78 (Fairchild, 1983). The Fe-quench line shown is at 2.4 wt. per cent $FeCO_3$. At lower Fe concentrations, luminescence is activated when the $MnCO_3$ concentration is greater than about 0.03 wt. per cent. Cathodoluminescence growth zones were observed in dolomite associated with sphalerite mineralization in the Mascot-Jefferson City district, Tennessee, USA (Ebers and Kopp, 1979). With luminescence from very bright to very dark red, a total of six major growth zones were recognized, which could be used for microstratigraphic correlation.

Infrared and Raman spectra of dolomite are shown in Fig. 79 (White, 1974). They are comparable with those of calcite. The internal mode ν_1 is not discernible in the infrared spectra and ν_2 does not appear in the Raman spectra.

The specific gravity of pure dolomite is 2.86, but it is increased by the substitution of Fe^{2+} and/or Mn^{2+} for Mg. In all cases the increase appears to be proportional to the specific gravities and the amounts of end-members substituted. Specific

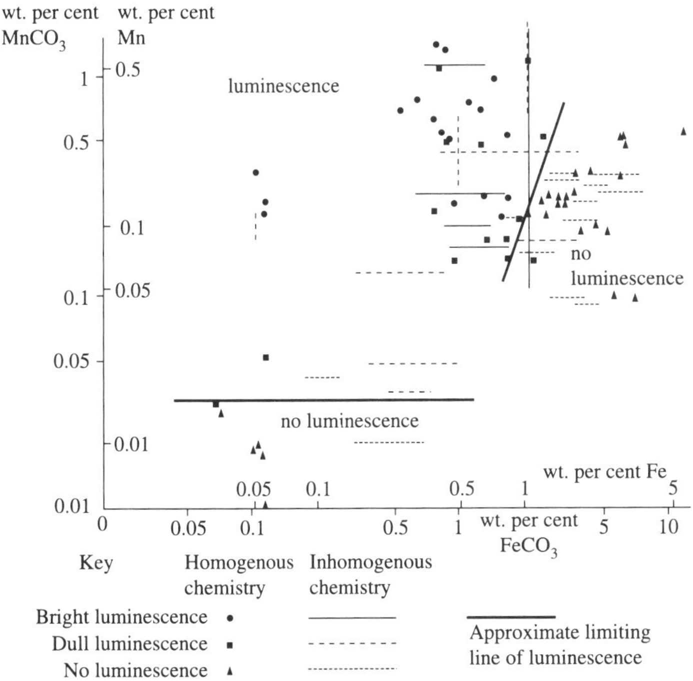

Fig. 78. $FeCO_3$ and $MnCO_3$ contents and luminescence in dolomites from the Bonhaven Formation (after Fairchild, 1983). Samples with apparently homogeneous luminescence, but chemically inhomogeneous on the scale of a few microns, are represented by lines instead of points.

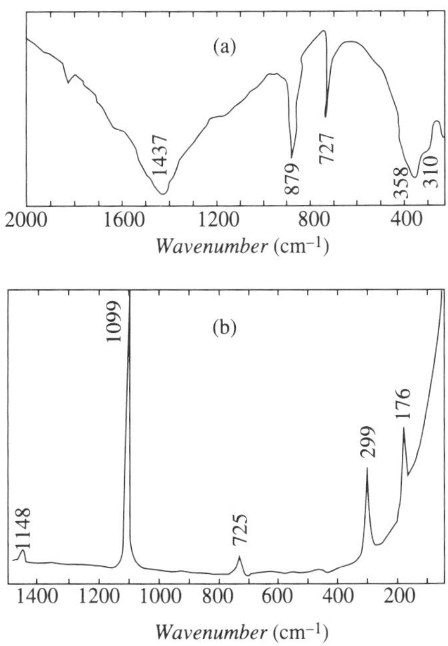

Fig. 79. (a) Infrared spectrum and (b) Raman spectrum of dolomite measured on powders (after White, 1974).

gravities and hardness of zincian and plumbian dolomites from Tsumeb, Namibia, are given in Table 27.

Dolomite appears most commonly as a fine to coarse cleavage aggregate; euhedral crystals, however, are rather common as simple rhombohedra often with curved faces. Some crystal faces are so curved that the name saddle-shaped dolomite is given to them. Saddle dolomite occurs in limestones and dolomites as cavity fillings and as replacements, and is associated with base-metal sulphides, fluorite, sulphates and hydrocarbons. Radke and Mathis (1980) suggest the following conditions for its development: temperatures between 60 and 150 °C, sulphate-reduction conditions, and pyroelectric effects under fluctuating temperature conditions which result in selective ion adsorption at different sites on the crystal surface. A transmission electron microscopic study by Barber *et al.* (1985) revealed the presence of abundant thin (< 100 nm), coherent calcic laths, with predominant $\{10\bar{1}4\}$ (hexagonal structural cell) twins within the dolomite which are believed to contribute to the overall distortion because of differences in lattice spacings between calcite and dolomite. Searl (1989) used scanning electron microscopy with back-scattered imaging and analysis (BSEM) to study saddle dolomite, and presented evidence that lattice distortion is caused by the incorporation of extra lattice planes at propagating crystal edges, relative to face centres. These extra planes are visible in BSEM as wedges which are always Mg-enriched with respect to the bulk crystal. It is suggested that the wedges develop during rapid, transport-controlled crystal growth.

A crystal-growth model has been proposed in which, beginning with the central rhombohedron, growth occurs by rhombohedral faces that advance outwards, and by nucleation and growth of new rhombohedra at the six lateral vertices (Kretz, 1992). Continued growth and repeated nucleation would then produce chains of rhombohedra emanating from the central rhombohedron. It is suggested that slight and progressive misorientation of nuclei along the length of a chain gives rise to numerous blocky sub-crystals and to sweeping extinction.

The mean thermal expansion coefficient for dolomite is $4.1 \times 10^{-6}/°C$, which agrees well with that for magnesite ($3.4 \times 10^{-6}/°C$) but not with that for calcite ($19.8 \times 10^{-6}/°C$) (Reeder and Markgraf, 1986). The mean axial compressibilities are 1.48×10^{-4}/kbar for a and 4.01×10^{-4}/kbar for c (Figuet *et al.*, 1994).

Distinguishing Features

Dolomite may be distinguished as a rhombohedral carbonate by its extremely high birefringence and perfect rhombohedral $\{10\bar{1}1\}$ cleavage. It is commonly found together with calcite in limestone, dolomite rock and marble, and the following features may be used to distinguish dolomite from calcite: (a) dolomite is more commonly euhedral and calcite is more commonly twinned, (b) dolomite has higher refractive indices and higher specific gravity, (c) dolomite, unlike calcite, is only slowly soluble in cold dilute acids, and does not effervesce, and (d) glide twinning occurs on $\{02\bar{2}1\}$ (hexagonal morphological cell) rather than on $\{01\bar{1}2\}$ as in calcite; thus in a cleavage rhomb of dolomite the twin lamellae lie parallel to the short diagonal as well as parallel to the long diagonal, as in calcite.

To distinguish between dolomite and calcite in carbonate rocks, differential staining methods are commonly used. Adams *et al.* (1984) have summarized the characteristics of staining procedures as described in Table 19 (p. 123). The staining method is generally based upon the fact that organic dyes stain calcite in acid solution and dolomite in alkaline solution (Friedman, 1959). The alkaline solution is prepared by adding the organic dye in methanol and caustic soda solution. The stain develops after about 5 minutes' immersion dependent upon the dye used. There are at least 20 organic dyes which give excellent results. Examples are titan yellow for a deep orange-red stain, alizarin red S for a purple stain, trypan blue for a blue stain, and neutral red for an orange stain. Inorganic staining solutions have also been widely used. In the Lemberg method (Rodgers, 1940), the carbonate sample is immersed in a 2.5 per cent ferric chloride solution for a few seconds, thoroughly washed, and then immersed in an ammonium sulphide solution for a few seconds. Calcite will stain black, whereas dolomite remains unaffected. A copper nitrate and ammonium hydroxide treatment gives a blue-green stain with calcite, but not with dolomite.

Dolomite and calcite can also be distinguished, and their ratio when present together can be semi-quantitatively determined, by X-ray powder diffraction using the principal reflections for calcite at 3.03 Å and for dolomite at 2.88 Å (Royse *et al.*, 1971). Calibration curves given by Tennant and Berger (1957), Weber and Smith (1961) and Royse *et al.* (1971) agree well with each other. However, the application of these calibration curves to carbonate rocks is not without some difficulties. Lumsden (1979) pointed out that systematic errors in the X-ray powder diffraction method can be caused by the variation in crystallite size, lattice deformation and interference by quartz. Because all three effects may occur together, correction can become complex. Gensmer and Weiss (1980), however, established a set of procedures for X-ray determination. For rocks with carbonate grains generally $>100\ \mu m$, the procedure yields consistent results in comparison with a point-counting method. For grains $<100\ \mu m$ differences in results occur, and Gensmer and Weiss concluded that the errors lie in the point-counting not in the X-ray method. Runnells (1970) cautioned that the presence of solid solutions in calcite and dolomite could affect the intensities of the strongest reflections. According to Lumsden (1979), each mol. per cent of excess Ca in the dolomite solid solution causes an approximately 2 per cent overestimate of the amount of dolomite.

Paragenesis

Dolomite typically occurs as the major constituent of extensive bedded sedimentary formations in association with calcite. Most dolomite is considered to be sedimentary in origin and formed from calcite and aragonite by the action of magnesium-rich solutions from marine or hydrothermal sources. A few sedimentary rocks are known to contain dolomite precipitated directly from solutions, such dolomite being termed primary. Highly saline, isolated marine waters or saline lakes are normally the environments for primary dolomite precipitation.

Dolomite is being precipitated with high intertidal–supratidal sediments of the Trucial Coast sabkhas, Arabian Gulf (Illing *et al.*, 1965; McKenzie *et al.*, 1980), where evaporation-induced capillary action is the predominant factor in changing pore-water chemistry to promote the precipitation. The sabkha dolomite is Ca-rich (52.0–54.6 mol. per cent $CaCO_3$) and not well ordered (McKenzie, 1981). A suite of magnesium and calcium carbonates is being formed in the lagoons and ephemeral lakes of Coorong, South Australia. The suite includes magnesite, hydromagnesite, aragonite, high- and low-magnesian calcite and dolomite (Borch, 1976; Borch and

Lock, 1979). The variation in pH of the water is remarkable, ranging from 8.2 to 10.3. The Mg^{2+}/Ca^{2+} ratio in the various waters can be as low as 2.5 and as high as 20.0, and salinities range from 10 to 274‰. The dolomite forms mostly in the more distal lakes, which annually pass through a desiccation period. In the continental saline lakes of Victoria, Australia, poorly ordered dolomite appears to be a direct precipitate from waters with high alkalinity and high Mg/Ca ratio (Deckker and Last, 1988). Dolomites precipitating in hypersaline marine lagoons are recorded from Baffin Bay, Texas, USA (Behrens and Land, 1972) and Kuwait (Gunatilaka *et al.*, 1984). Both have the required salinity and typical marine Mg/Ca ratios.

The Mg/Ca ratios of the most abundant solutions and the field boundary between calcite and dolomite suggest that the most frequent temperatures of 'burial' dolomitization are 80–90°C (Usdowski, 1994). Mineral assemblages and compositions of solutions from sabhka-type environments reflect specific reaction paths: however, as reaction rates are slow, sufficient time is necessary.

Dolomite crusts were found on the tidal flats of Andros Island, Bahamas (Shinn *et al.*, 1965), Sugarloaf Key, Florida, USA (Shinn, 1968) and Bonaire, Netherlands Antilles (Deffeyes *et al.*, 1965). The typical crust is dark grey to black, brittle and has a pitted surface with variable dolomite content. The Florida Key crust has been interpreted as being precipitated from only slightly modified Florida Bay water pumped through the sediments by tidal action (Carballo *et al.*, 1987).

By the integration of stratigraphic, petrographic, geochemical and hydrological parameters, several models for the formation of secondary dolomite have been proposed and critically discussed (Zenger, 1972; Machel and Mountjoy, 1986; Wells, 1986; Hardie, 1987). The 'seepage–reflux' model relies upon the generation of Mg-rich fluids through evaporation and sulphate precipitation from lagoon water or tidal flat pore-waters, and the descent of these fluids into underlying carbonate sediments, where dolomitization occurs. Dolomites formed by this process are always in association with evaporites. Examples are dolomite of the Lower Cretaceous Edwards Formation in Texas, USA (Fisher and Rodda, 1969), the Zechstein dolomite of northwest Europe (Clark, 1980), and dolomite in the Northern Michigan Reef Belt, USA (Sears and Lucia, 1980).

The concept of 'mixing-zone' dolomitization originated from the study by Hanshaw *et al.* (1971) on the groundwaters in Tertiary limestones of the Florida Aquifer. They showed that mixtures of meteoric waters and sea-water could be potential dolomitization fluids. This concept was developed into various 'mixing-zone' models, including the 'Dorag' model by Badiozamani (1973) and the 'meteoric' dolomitization model by Land (1973a). The 'Dorag' model postulates that the mixing of meteoric waters and sea-water can result in a solution that is undersaturated with respect to calcite but supersaturated with respect to dolomite. In the meteoric dolomitization model the mixing of sea-water with meteoric water results in solutions of lowered ionic strength and the required Mg^{2+}/Ca^{2+} ratios. These solutions resulting from mixing migrate both laterally and vertically in the carbonate sediments and promote dolomitization on a regional scale. Dolomites formed by these processes are generally not accompanied by evaporites. Examples are dolomite of the Hope Gate Formation of Jamaica (Land, 1973a,b), the non-supratidal dolomite of the Ste Genevieve limestone of the Illinois Basin, USA (Choquette and Steinen, 1980), and the non-ferroan dolomite in the Gully oolite, Glamorgan, South Wales, UK (Searl, 1988).

The 'burial–compaction' model of dolomitization was first introduced by Illing (1959) with the idea that permeable reef limestones could be dolomitized by connate water derived from the compaction of adjacent or enclosing argillaceous sediments. Dolomitization by this process should proceed more easily at depth, where

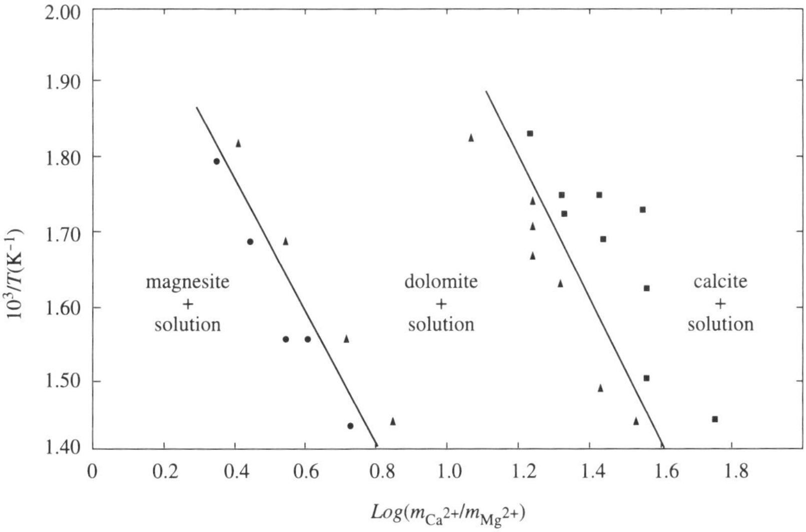

Fig. 80. Reactions involving calcite, dolomite and magnesite in solutions as a function of $m_{Ca^{2+}}/m_{Mg^{2+}}$ and temperature (after Holland, 1968). In the figure, ■, ● and ▲ indicate experimental runs in which dolomite was replaced by calcite, dolomite was replaced by magnesite, and calcite or magnesite was replaced by dolomite, respectively.

temperature is higher and the proportions of hydrated Mg^{2+} ions are smaller. Examples are dolomite of the Upper Devonian Mietts Build-up, Jasper National Park, Alberta, Canada (Mattes and Mountjoy, 1980), dolomite in the Pendleside limestone, Bowland Basin, northern England (Gawthorpe, 1987), and the Cambrian Bonneterre dolomite in southeast Missouri, USA (Gregg, 1985).

Dolomitization by normal sea-water without modification was proposed by Land (1985) as long as there is a driving mechanism to pump the fluids through the carbonate sediments. Dolomite was formed in the Eocene coralline algal-rotaline foraminiferal grainstone of Enewetak Atoll, Pacific Ocean, by ocean water circulating freely deep within the atoll (Saller, 1984). The driving mechanisms are likely to have been oceanic tides and thermal convection. Mullins *et al.* (1985) have applied the 'Kohout convection' (Kohout, 1967) of sea-water to the dolomitization in the periplatform oozes off the Little Bahama Bank.

Dolomite also commonly occurs in hydrothermal veins in association with fluorite, baryte and quartz. It is the most important carbonate gangue mineral next to calcite. The dolomitization of limestones around many high-temperature hydrothermal orebodies can be readily understood in terms of the experimental data shown in Fig. 80 (Holland, 1968). The age relationships to orebodies in various Cordilleran mines show that nearly all of the hydrothermal dolomite formed earlier than the ores (Lovering, 1969).

The authigenic minerals formed from volcanic emanations at Soussaki, West Attica Peninsula, Greece, have been described in detail (Kyriakopoulos *et al.*, 1990). Reactions between H_2S-, CO_2- and H_2O-bearing hydrothermal fluids and pre-existing serpentine produced magnesite, dolomite and sulphates on top of partly altered basement rocks.

In igneous rocks, dolomite usually occurs in carbonatites with calcite, ankerite and siderite. The typical sequence of crystallization of carbonate minerals with decreasing temperature in a number of carbonatites is calcite, dolomite, ankerite and siderite (Heinrich, 1966). The principal minerals of the Blackburn carbonatite dykes, Ontario, Canada, are dolomite with subordinate calcite (Hogarth *et al.*, 1988). The carbonates are fine-grained and spherulitic; some spheroids consist entirely of dolomite, and others have a dolomite core with calcite in optical continuity. Both dolomite and calcite have an FeO content <1 per cent (Table 25, anal. 13). The carbonatite dykes of the Maimecha-Kouti region, northern Siberia, consist of 60–80 per cent euhedral calcite phenocrysts set in a groundmass of apatite, calcite and dolomite (Zhabin and Cherepivskaya, 1965). The Jacupiranga carbonatite plug, Brazil (Gaspar and Wyllie, 1987), has five stages of intrusion. The first, third and fourth are calcitic, the second contains both calcite and dolomite, and the fifth is dolomitic. The application of calcite–dolomite geothermometry to carbonatites has been discussed by Gittins (1979). He questions the value of this geothermometry in the estimation of carbonatite temperature, because in the presence of an alkali–H_2O–CO_2–halogen fluid the Mg diffusion from calcite into co-existing dolomite, which continues during subsolvus cooling, is far more extensive in carbonatite than in metamorphic marble.

Regional or contact metamorphism may cause the recrystallization of dolomite to form marble, and according to Hutcheon and Moore (1973) and Puhan (1976) the metamorphic temperature trend of marble formation can be inferred from the compositions of the calcite–dolomite assemblages. Powell *et al.* (1984) undertook an experimental study to calibrate the iron correction to the calcite–dolomite geothermometer, and presented a relationship of the mol. fraction of iron in dolomite and the mol. fraction of magnesium in calcite for equilibrium between calcite and dolomite in the temperature range 300–550 °C (Fig. 81). This has been used extensively to evaluate conditions of metamorphism because it is independent of fluid composition and any pressure correction is small. Garde (1977), however, in his study of marble of the Marmorilik Formation, west Greenland, cautioned concerning the use of calcite–dolomite geothermometry.

For siliceous dolomites, a progressive series of reactions occurs with rising

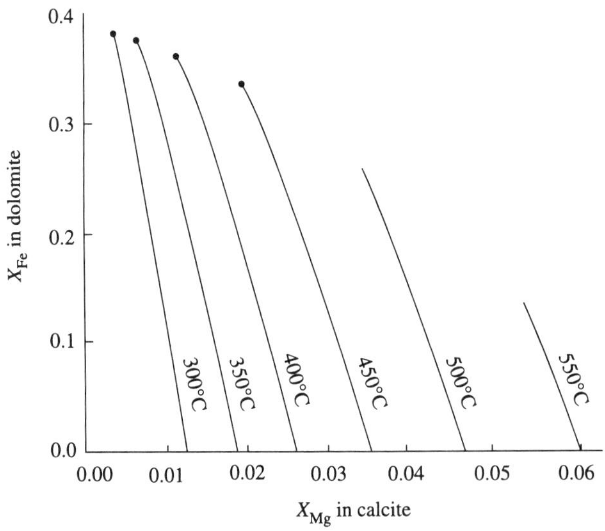

Fig. 81. A plot of the mol. fraction of iron in dolomite versus the mol. fraction of magnesium in calcite for temperatures of 300–550 °C for equilibrium between calcite and dolomite (after Powell *et al.*, 1984). Solid circles represent compositions for the three-phase assemblage calcite–siderite–dolomite.

temperature, from which talc, tremolite, forsterite, diopside, periclase, wollastonite and monticellite are formed. At higher temperatures, adjacent to igneous intrusives, äkermannite, tilleyite, spurrite, rankinite, merwinite and larnite occur as reaction products. The reaction:

$$3CaMg(CO_3)_2 + 4SiO_2 + H_2O = \underset{\text{talc}}{Mg_3Si_4O_{10}(OH_2)} + \underset{\text{calcite}}{3CaCO_3} + 3CO_2$$

was considered to be the initial step (Tilley, 1948), which produces talc and calcite with the release of CO_2 (Gordon and Greenwood, 1970; Puhan and Metz, 1987). At higher temperatures, talc reacts with calcite to produce tremolite according to the reaction:

$$\underset{\text{talc}}{5Mg_3Si_4O_{10}(OH)_2} + 6CaCO_3 + 4SiO_2 = \underset{\text{tremolite}}{3Ca_2Mg_5Si_8O_{22}(OH)_2} + 6CO_2 + 2H_2O.$$

This reaction yields both CO_2 and H_2O (Slaughter *et al.*, 1975; Puhan and Metz, 1987). In turn, tremolite reacts with dolomite to produce forsterite (Metz, 1976):

$$\underset{\text{dolomite}}{11CaMg(CO_3)_2} + \underset{\text{tremolite}}{Ca_2Mg_5Si_8O_{22}(OH)_2} = \underset{\text{forsterite}}{8Mg_2SiO_4} + 13CaCO_3 + 9CO_2 + H_2O.$$

Forsterite may also form from the reaction between pyroxene and dolomite (Käse and Metz, 1980):

$$\underset{\text{pyroxene}}{CaMg(SiO_3)_2} + \underset{\text{dolomite}}{3CaMg(CO_3)_2} = \underset{\text{forsterite}}{2Mg_2SiO_4} + 4CaCO_3 + 2CO_2.$$

The major phases formed in the progressive metamorphism of siliceous dolomite may be discussed with reference to the five-component system $CaO–MgO–SiO_2–CO_2–H_2O$. Details are summarized in Skippen (1971), Winkler (1979) and Slaughter *et al.* (1975). In the presence of aluminium and potassium, the metamorphism of dolomite leads to the formation of phlogopite (Puhan, 1978):

$$\underset{\text{dolomite}}{3CaMg(CO_3)_2} + KAlSi_3O_8 + H_2O = \underset{\text{phlogopite}}{KMg_3AlSi_3O_{10}(OH)_2} + 3CaCO_3 + 3CO_2.$$

Calculations of mass transfer during the burial of siliceous dolomites assuming various metamorphic gradients were formulated by Flowers and Helgeson (1983).

Under high pressures (23–77 kbar), calcite is unstable in the presence of enstatite, and dolomite and magnesite are the stable carbonates (Kushiro *et al.*, 1975). It is suggested that in the upper mantle, where enstatite is present, dolomite is stable to a depth of approximately 150 km, assuming that the partial pressure of CO_2 is close to the total pressure (see also Biellmann *et al.*, 1993).

References

Adams, A. E., **MacKenzie, W. W.** and **Guilford, C.**, 1984. *Atlas of Sedimentary Rocks under the Microscope*. London (Longman) and New York (John Wiley & Sons), 104 pp.

Allen, V. T. and **Fahey, J. J.**, 1952. New occurrences of minerals at Iron Mountain, Missouri. *Amer. Min.*, **37**, 736–42.

Althoff, P. L., 1977. Structural refinement of dolomite and a magnesian calcite and implications for dolomite formation in the marine environment. *Amer. Min.*, **62**, 772–83.

Badiozamani, K., 1973. The Dorag dolomitization model—application to the Middle Ordovician of Wisconsin. *J. Sed. Petr.*, **43**, 965–84.

Baker, P. A. and **Kastner, M.**, 1981. Constraints on the formation of sedimentary dolomite. *Science*, **213**, 214–16.

Barber, D. J., **Heard, H. C.**, **Paterson, M. S.** and **Wenk, H. R.**, 1978. Stacking faults in dolomite. *Nature*, **269**, 789–90.

Barber, D. J., Heard, H. C. and **Wenk, H. R.**, 1981. Deformation of dolomite single crystals from 20° to 800°C. *Phys. Chem. Min.*, **7**, 271–86.

Barber, D. J., Reeder, R. J. and **Smith, D. J.**, 1985. A TEM microstructural study of dolomite with curved faces (saddle dolomite). *Contrib. Min. Petr.*, **91**, 82–92.

Bathurst, R. G. C., 1976. *Carbonate Sediments and their Diagenesis* (Developments in Sedimentology, **12**), Amsterdam, Oxford and New York (Elsevier), 658 pp.

Behrens, E. W. and **Land, L. S.**, 1972. Subtidal Holocene dolomite, Baffin Bay, Texas, *J. Sed. Petr.*, **42**, 155–61.

Beran, A. and **Zemann, J.**, 1977. Refinement and comparison of the crystal structures of a dolomite and of a Fe-rich ankerite. *Tschermaks Min. Petr. Mitt.*, **24**, 279–86.

Biellmann, C., Gillet, P., Guyot, F., Peyronneau, J. and **Reynard, B.**, 1993. Experimental evidence for carbonate stability in the Earth's lower mantle. *Earth Planet. Sci. Let.*, **118**, 31–41.

Birch, W. D., 1983. Zincian dolomite from Broken Hill, New South Wales. *J. Geol. Soc. Aust.*, **30**, 85–7.

Borch, C. C. Von Der, 1976. Stratigraphy and formation of Holocene dolomitic carbonate deposits of the Coorong area, South Australia. *J. Sed. Petr.*, **46**, 952–66.

Borch, C. C. Von Der and **Lock, D.**, 1979. Geological significance of Coorong dolomites. *Sedimentology*, **26**, 813–24.

Bradley, W. F., Burst, J. F. and **Graf, D. L.**, 1953. Crystal chemistry and differential thermal effects of dolomite. *Amer. Min.*, **38**, 207–17.

Bullen, S. B. and **Sibley, D. F.**, 1983. Dolomite selectivity, an experimental approach. *Amer. Assoc. Petrol. Geol. Bull.*, **67**, 433–4.

Burns, S. J. and **Baker, P. A.**, 1987. A geochemical study of dolomite in the Monterey formation, California. *J. Sed. Petr.*, **57**, 128–39.

Busenberg, E. and **Plummer, L. N.**, 1982. The kinetics of dissolution of dolomite in CO_2–H_2O systems at 1.5 to 65°C and 1 atm P_{CO_2}. *Amer. J. Sci.*, **282**, 45–78.

Byrnes, A. P. and **Wyllie, P. J.**, 1981. Subsolidus and melting relations for the join $CaCO_3$–$MgCO_3$ at 10 kbars. *Geochim. Cosmochim. Acta*, **45**, 321–8.

Calderon, K. T., Torres, R. and **Coy-Yll, R.**, 1982. Luminescenia de la dolomita irradiada at 20K. *Bol. Soc. Española Min.*, **5**, 99–105.

Carballo, J. D., Land, L. S. and **Miser, D. E.**, 1987. Holocene dolomitization of supratidal sediments by active tidal pumping, Sugarloaf Key, Florida. *J. Sed. Petr.*, **57**, 153–65.

Chave, K. E., 1952. A solid solution between calcite and dolomite. *J. Geol.*, **60**, 190–2.

Choquette, P. W. and **Steinen, R. P.**, 1980. Mississippian nonsupratidal dolomite, Ste. Genevieve limestone, Illinois Basin: evidence for mixed-water dolomitization. In *Concepts and Models of Dolomitization*: Eds. D. H. Zenger, J. B. Durham and R. C. Ethington, Special Publ., *Soc. Econ. Paleont. Min.*, **28**, 163–96.

Clark, D. N., 1980. The diagenesis of Zechstein carbonate sediments. *Contrib. Sediment.*, **9**, 167–203.

Deckker, P. De and **Last, W. M.**, 1988. Modern dolomite deposition in continental, saline lakes, western Victoria, Australia. *Geology*, **16**, 29–32.

Deelman, J. C., 1977. Protodolomite redefined. *J. Sed. Petr.*, **48**, 1004–7.

Deffeyes, K. S., Lucia, F. J. and **Weyl, P. K.**, 1965. Dolomitization of Recent and Plio-Pleistocene sediments by marine evaporite water on Bonaire, Netherlands Antilles. In *Dolomitization and Limestone Diagenesis, A Symposium:* Eds. L. C. Pray and R. C. Murray, *Special Publ., Soc. Econ. Paleo. Min.*, **13**, 71–88.

Ebers, M. L. and **Kopp, O. C.**, 1979. Cathodoluminescent microstratigraphy in gangue dolomite, the Mascot-Jefferson City district, Tennessee. *Econ. Geol.*, **74**, 908–18.

Effenberger, H., Mereiter, K. and **Zemann, J.**, 1981. Crystal structure refinements of magnesite, calcite, rhodochrosite, siderite, smithsonite and dolomite, with discussion of some aspects of the stereochemistry of calcite-type carbonates. *Z. Kirst.*, **156**, 233–43.

Fairchild, I. J., 1983. Chemical controls of cathodoluminescence of natural dolomites and calcites: new data and review. *Sedimentology*, **30**, 579–83.

Figuet, G., Guyot, F. and **Itié, J.**, 1994. High-pressure X-ray diffraction study of carbonates; $MgCO_3$, $CaMg(CO_3)_2$, and $CaCO_3$. *Amer. Min.*, **79**, 15–23.

Fisher, W. L. and **Rodda, P. U.**, 1969. Edwards Formation (Lower Cretaceous), Texas: dolomitization in a carbonate platform system. *Bull. Amer. Assoc. Petrol. Geol.*, **53**, 55–72.

Flowers, G. C. and **Helgeson, H. C.**, 1983. Equilibrium and mass transfer during progressive metamorphism of siliceous dolomites. *Amer. J. Sci.*, **283**, 230–86.

Foyke, B. W. and **Reeder, R. J.**, 1992. Surface structural controls on dolomite composition: evidence from sector zoning. *Geochim. Cosmochim. Acta*, **56**, 4015–24.

Friedman, G. M., 1959. Identification of carbonate minerals by staining methods. *J. Sed. Petr.*, **29**, 87–97.

Frondel, C. and **Bauer, H. H.**, 1955. Kutnahorite: a manganese dolomite, $CaMn(CO_3)_2$. *Amer. Min.*, **40**, 748–50.

Gaines, A. M., 1977. Protodolomite redefined. *J. Sed. Petr.*, **47**, 543–6.

Gaines, A. M., 1980. Dolomitization kinetics: recent experimental studies. *Soc. Econ. Paleont. Min.*, Special Publ., **28**, 81–6.

Garavelli, C. G., Vurro, F. and **Fioravanti, G. C.,** 1982. Minrecordite, a new mineral from Tsumeb. *Min. Rec.*, **13**, 131–6.

Garde, A. A., 1977. Calcite–dolomite thermometry: negative evidence from the Marmorilik Formation, west Greenland. *Contrib. Min. Petr.*, **62**, 265–70.

Gaspar, J. L. and **Wyllie, P. J.,** 1987. The phlogopites from the Jacupiranga carbonatite intrusions. *Min. Petr.*, **36**, 121–34.

Gawthorpe, R. L., 1987. Burial dolomitization and porosity development in a mixed carbonate–clastic sequence: an example from the Bowland Basin, northern England. *Sedimentology*, **34**, 533–58.

Gensmer, R. P. and **Weiss, M. P.,** 1980. Accuracy of calcite/dolomite ratios by X-ray diffraction and comparison with results from staining techniques. *J. Sed. Petr.*, **50**, 626–9.

Gidman, J., 1977. Protodolomite redefined. *J. Sed. Petr.*, **47**, 1007–8.

Gies, H., 1975. Activation possibilities and geochemical correlations of photoluminescing carbonates, particularly calcite. *Min. Deposita*, **10**, 216–27.

Gittins, J., 1979. Problems inherent in the application of calcite–dolomite geothermometry to carbonatites. *Contrib. Min. Petr.*, **69**, 1–4.

Goldsmith, J. R., 1959. Some aspects of the geochemistry of carbonates. In *Researches in Geochemistry*: Ed. P. H. Abelson, New York (John Wiley), pp. 336–58.

Goldsmith, J. R., 1980. Thermal stability of dolomite at high temperatures and pressures. *J. Geophys. Res.*, **85**, 6949–54.

Goldsmith, J. R. and **Graf, D. L.,** 1960. Subsolidus relations in the system $CaCO_3$–$MgCO_3$–$MnCO_3$. *J. Geol.*, **68**, 324–35.

Goldsmith, J. R. and **Heard, H. C.,** 1961. Subsolidus phase relations in the system $CaCO_3$–$MgCO_3$. *J. Geol.*, **69**, 45–74.

Goldsmith, J. R. and **Newton, R. C.,** 1969. *P–T–X* relations in the system $CaCO_3$–$MgCO_3$ at high temperatures and pressures. *Amer. J. Sci.*, **267A**, 160–90.

Goldsmith, J. R., Witters, J. and **Northrop, D. A.,** 1962. Studies in the system $CaCO_3$–$MgCO_3$–$FeCO_3$: 1. Phase relations; 2. A method for major element spectrochemical analysis; 3. Compositions of some ferroan dolomite. *J. Geol.*, **70**, 659–88.

Gordon, T. M. and **Greenwood, H. J.,** 1970. The reaction: dolomite + quartz + water = talc + calcite + carbon dioxide. *Amer. J. Sci.*, **268**, 225–42.

Graf, D. L. and **Goldsmith, J. R.,** 1956. Some hydrothermal syntheses of dolomite and protodolomite. *J. Geol.*, **64**, 173–86.

Gregg, J. M., 1985. Regional epigenetic dolomitization in the Bonneterre dolomite (Cambrian), southeastern Missouri. *Geology*, **13**, 503–6.

Gunatilaka, A., Saleh, A., Al-Temeemi, A. and **Nasser, A.,** 1984. Occurrence of subtidal dolomite in a hypersaline lagoon, Kuwait. *Nature*, **311**, 450–2.

Hanshaw, B. B., Back. W. and **Dieke, R. G.,** 1971. A geochemical hypothesis for dolomitization by groundwater. *Econ. Geol.*, **66**, 710–24.

Hardie, L. A., 1987. Perspectives on dolomitization: a critical view of some current views. *J. Sed. Petr.*, **57**, 166–83.

Harker, R. I. and **Tuttle, O. F.,** 1955. Studies in the system CaO–MgO–CO_2. Part I, the thermal dissociation of calcite, dolomite and magnesite. *Amer. J. Sci.*, **253**, 209–24.

Heinrich, E. W., 1966. *The Geology of Carbonatites*, Skokie, Illinois (Rand McNally), 555 pp.

Hervig, R. L. and **Smith, J. V.,** 1981. Dolomite–apatite inclusion in chrome–diopside crystal, Bellshank kimberlite, South Africa. *Amer. Min.*, **66**, 346–9.

Higgs, D. V. and **Handin, J. W.,** 1959. Experimental deformation of dolomite single crystals. *Bull. Geol. Soc. Amer.*, **70**, 245–78.

Hogarth, D. D., Rushforth, P. and **McCorkell, R. H.,** 1988. The Blackburn carbonatites, near Ottawa, Ontario: dykes with fluidized emplacement. *Can. Min.*, **26**, 377–90.

Holland, H. D., 1968. Gangue minerals in hydrothermal deposits. In *Geochemistry of Hydrothermal Ore Deposits*: Ed. H. L. Barnes, Holt, New York (Reinhart and Winston), pp. 382–436.

Howell, J. E. and **Dawson, K. R.,** 1958. Technique for optical determination of iron-bearing dolomites. *Can. Min.*, **6**, 292–4.

Howie, R. A. and **Broadhurst, F. M.,** 1958. X-ray data for dolomite and ankerite. *Amer. Min.*, **43**, 1210–4.

Hsü, K. J., 1967. Chemistry of dolomite formation. In *Carbonate Rocks, Origin, Occurrence and Classification*: Eds. H. J. Bissel and R. W. Fairbridge, Amsterdam (Elsevier), pp. 169–91.

Hurlbut, C. S., Jr., 1957. Zincian and plumbian dolomite from Tsumeb, South-west Africa. *Amer. Min.*, **42**, 798–803.

Hutcheon, I. and **Moore, J. M.,** 1973. The tremolite isograd near Marble Lake, Ontario. *Can. J. Earth Sci.*, **10**, 936–47.

Illing, L. V., 1959. Deposition and diagenesis of some Upper Paleozoic carbonate sediments in western

Canada. *Proc. 5th World Petroleum Congress, New York,* Sect. **1**, 23.

Illing, L. V., Wells, A. J. and **Tayler, J. C. M.,** 1965. Penecontemporaneous dolomite in the Persian Gulf. In *Dolomitization and Limestone Diagenesis:* Eds. R. C. Pray and R. C. Murray, Special Publ., *Soc. Econ. Paleont. Min.*, **13**, 89–111.

Irving, A. J. and **Wyllie, P. J.,** 1975. Subsolidus and melting relationships for calcite, magnesite and the join $CaCO_3$–$MgCO_3$ to 36 kbars. *Geochim. Cosmochim. Acta*, **39**, 35–53.

Isherwood, B. J. and **James, J. A.,** 1976. Structural dependence of the optical birefringence of crystals with calcite- and aragonite-type structure. *Acta Cryst.*, **A32**, 340–2.

Jarosewich, E. and **Macintyre, I. G.,** 1983. Carbonate reference samples for electron microprobe and scanning electron microprobe analyses. *J. Sed. Petr.*, **53**, 677–8.

Käse, H. R. and **Metz, P.,** 1980. Experimental investigation of metamorphism of siliceous dolomite: IV. Equilibrium data for the reaction,: 1 diopside + 3 dolomite = 2 forsterite + 4 calcite + $2CO_2$. *Contrib. Min. Petr.*, **73**, 151–9.

Katz, A. and **Matthews, A.,** 1977. The dolomitization of $CaCO_3$: an experimental study at 252°–295°C. *Geochim. Cosmochim. Acta*, **41**, 297–308.

Kohout, F. A., 1967. Groundwater flow and the geothermal regime of Florida plateau. *Trans. Gulf Coast Assoc. Geol. Soc.*, **17**, 339–54.

Kretz, R., 1988. SEM study of dolomite microcrystals in Grenville marble. *Amer. Min.*, **73**, 619–31.

Kretz, R., 1992. Carousel model for the crystallization of saddle dolomite. *J. Sed. Petr.*, **62**, 190–5.

Kushiro, I., Satake, H. and **Akimoto, S.,** 1975. Carbonate–silicate reactions at high pressures and possible occurrence of dolomite and magnesite in the upper mantle. *Earth Planet. Sci. Lett.*, **28**, 116–20.

Kyriakopoulos, K. G., Kanaris-Sotiriou, R. and **Stamatakis, M. G.,** 1990. The authigenic minerals formed from volcanic emanations at Soussaki, West Attica Peninsula, Greece. *Can. Min.*, **28**, 363–8.

Land, L. S., 1967. Diagenesis of skeletal carbonates. *J. Sed. Petr.*, **37**, 914–30.

Land, L. S., 1973a. Holocene meteoric dolomitization of Pleistocene limestone, North Jamaica. *Sedimentology*, **20**, 411–24.

Land, L. S., 1973b, Contemporaneous dolomitization of Middle Pleistocene reefs by meteoric water, North Jamaica. *Bull. Marine Sci.*, **23**, 64–92.

Land, L. S., 1985. The origin of massive dolomite. *J. Geol. Educ.*, **33**, 112–25.

Lewis, D. R., 1956. The thermoluminescence of dolomite and calcite. *J. Phys. Chem.*, **60**, 698–701.

Lippmann, F., 1973. *Sedimentary Carbonate Minerals.* New York (Springer-Verlag) 228 pp.

Lo, B. W. M., 1973. The polarizability of planar nitrate, carbonate and borate anions in crystals. *J. Phys. Chem. Solids*, **34**, 513–20.

Lovering, T. S., 1969. The origin of hydrothermal and low temperature dolomite. *Econ. Geol.*, **64**, 743–54.

Lumsden, D. N., 1979. Discrepancy between thin section and X-ray estimates of dolomite in limestone. *J. Sed. Petr.*, **49**, 429–35.

Machel, H. G. and **Mountjoy, E. W.,** 1986. Chemistry and environments of dolomitization – a reappraisal. *Earth Sci. Rev.*, **23**, 175–222.

Mattes, B. W. and **Mountjoy, E. W.,** 1980. Burial dolomitization of the Upper Devonian Miette Build-up, Jasper National Park, Alberta. in *Concepts and Models of Dolomitization:* Eds. D. H. Zenger, J. B. Durham and R. L. Ethington, Special Publ., *Soc. Econ. Paleont. Min.*, **28**, 259–97.

McDowell, S. D. and **Paces, J. B.,** 1985. Carbonate alteration minerals in the Salton Sea geothermal system. California, U.S.A. *Min. Mag.*, **49**, 469–79.

McKenzie, J. A., 1981. Holocene dolomitization of calcium carbonate sediments from coastal sabkhas of Abu Dhabi, U.A.E.: a stable isotope study. *J. Geol.*, **89**, 185–98.

McKenzie, J. A., Hsü, K. J. and **Schneider, J. F.,** 1980. Movement of subsurface water under the sabkha, Abu Dhabi, U.A.E., and its relation to evaporative dolomite genesis. In *Concepts and Models of Dolomitization:* Eds. D. H. Zenger, J. B. Durham and R. L. Ethington, Special Publ., *Soc. Econ. Paleont. Min.*, **28**, 11–30.

Metz, P., 1976. Experimental investigation of the metamorphism of siliceous dolomites. III. Equilibrium data for the reaction: 1 tremolite + 11 dolomite = 8 forsterite + 13 calcite + 9 $CO_2 + H_2O$. *Contrib. Min. Petr.*, **58**, 137–48.

Morrow, D. W., 1982. Diagenesis of dolomite. Part I: The chemistry of dolomitization and dolomite precipitation. *Geosci. Can.*, **9**, 95–107.

Morrow, D. W. and **Ricketts, B. D.,** 1986. Chemical controls on the precipitation of mineral analogues of dolomite. *Geology*, **14**, 408–10.

Morrow, D. W., Gorham, B. L. and **Wong, J. N. W.,** 1994. Dolomite–calcite equilibrium at 220 to 240°C at saturation vapour pressure: experimental data. *Geochem. Cosmochim. Acta*, **58**, 169–77.

Morse, J. W. and **McKenzie, F. T.,** 1990. *Geochemistry of Sedimentary Carbonates* (Developments in Sedimentology, **48**), Amsterdam, Oxford and New York (Elsevier), 706 pp.

Mucci, A. and **Morse, J. W.,** 1983. The incorporation of Mg^{+2} and Sr^{+2} into calcite overgrowth: influences of growth rate and solution composition. *Geochim. Cosmochim. Acta*, **47**, 217–33.

Mullins, H. T., Wise, S. W., Garcluski, A. F., Hinchey, E. J., Master, P. M. and **Giegel, D. I.,** 1985. Shallow subsurface diagenesis of Pleistocene periplatform ooze, northern Bahamas. *Sedimentology,* **32,** 473–94.

Navrotsky, A. and **Capobianco, C.,** 1987. Enthalpies of formation of dolomite and of magnesian calcite. *Amer. Min.,* **72,** 782–7.

Navrotsky, A. and **Loucks, D.,** 1977. Calculation of subsolidus phase relations in carbonates and pyroxenes. *Phys. Chem. Min.,* **1,** 109–27.

Nordeng, S. H. and **Sibley, D. F.,** 1994. Dolomite stoichiometry and Ostwald's Step Rule. *Geochim. Cosmochim. Acta,* **58,** 191–6.

Oh, K. D., Moriakawa, H., Iwai, S. and **Aoki, H.,** 1973. The crystal structure of magnesite. *Amer. Min.,* **58,** 1029–33.

Otsuka, R., 1986. Recent studies on the decomposition of the dolomite group by thermal analysis. *Thermochim. Acta,* **100,** 69–80.

Platt, R. G. and **Mitchell, R. H.,** 1982. The Marathon dikes: ultrabasic lamprophyres from the vicinity of McKellar Harbour, N. W. Ontario. *Amer. Min.,* **67,** 907–16.

Powell, R., Condliffe, D. M. and **Condliffe, E.,** 1984. Calcite–dolomite geothermometry in the system $CaCO_3$–$MgCO_3$–$FeCO_3$: an experimental study. *J. Metamorph. Geol.,* **2,** 33–41.

Przibram, K., 1956. *Irradiation Colors and Luminescence.* London (Pergamon Press), 332 pp.

Puhan, D., 1976. Metamorphic temperature determined by means of the dolomite–calcite solvus geothermometer—examples from the central Damara orogeny (South West Africa). *Contrib. Min. Petr.,* **58,** 23–8.

Puhan, D., 1978. Experimental study of the reaction: dolomite + K-feldspar + H_2O = phlogopite + calcite + CO_2 at the total gas pressures of 4000 and 6000 bars. *Neues Jahrb. Min., Mh.,* 110–27.

Puhan, D. and **Metz, P.,** 1987. Experimental equilibrium data for the reactions: 3 dolomite + 4 quartz + H_2O = talc + 3 calcite + $3CO_2$ and 5 talc + 6 calcite + 4 quartz = 3 tremolite + $6CO_2$ + $2H_2O$ at a total gas pressure of 5000 bars. *Neues Jahrb. Min., Mh.,* 515–20.

Radke, B. M. and **Mathis, R. L.,** 1980. On the formation and occurrence of saddle dolomite. *J. Sed. Petr.,* **50,** 1149–60.

Rath, R. and **Pohl, D.,** 1971. Zwei achsig erscheinende Calcite und Dolomite. *Contrib. Min. Petr.,* **32,** 74–8.

Reeder, R. J., 1983. Crystal chemistry of the rhombohedral carbonates. In *Carbonates: Mineralogy and Chemistry;* Ed. R. J. Reeder, *Reviews in Mineralogy,* **11,** 1–47, *Mineralogical Society of America,* Washington, DC.

Reeder, R. J. and **Markgraf, S. A.,** 1986. High-temperature crystal chemistry of dolomite. *Amer. Min.,* **71,** 795–804.

Reeder, R. J. and **Sheppard, C. E.,** 1984. Variation of lattice parameters in some sedimentary dolomites. *Amer. Min.,* **69,** 520–7.

Reeder, R. J. and **Wenk, H. R.,** 1979 Microstructures in low temperature dolomites. *Geophys. Res. Letters,* **6,** 77–80.

Reeder, R. J. and **Wenk, H. R.,** 1983. Structure refinements of some thermally disordered dolomites. *Amer. Min.,* **68,** 769–76.

Rodgers, J., 1940. Distinction between calcite and dolomite on polished surfaces. *Amer. J. Sci.,* **238,** 788–98.

Rosenberg, P. E., 1963. Subsolidus phase relations in the system $CaCO_3$–$FeCO_3$. *Amer. J. Sci.,* **261,** 683–90.

Rosenberg, P. E., 1967. Subsolidus relations in the system $CaCO_3$–$MgCO_3$–$FeCO_3$ between 350° and 550°C. *Amer. Min.,* **52,** 787–97.

Rosenberg, P. E. and **Champness, P. F.,** 1989. Zincian dolomites and associated carbonates from the Warynski mine, Poland: an AEM investigation. *Amer. Min.,* **74,** 461–5.

Ross, N. L. and **Reeder, R. J.,** 1992. High-pressure structural study of dolomite and ankerite. *Amer. Min.,* **77,** 412–21.

Royse, C. E., Jr., Wadell, J. S. and **Petersen, L. E.,** 1971. X-ray determination of calcite–dolomite: an evaluation. *J. Sed. Petr.,* **41,** 483–8.

Runnells, D. A., 1970. Errors in X-ray analysis of carbonates due to solid solution variation in composition of component minerals. *J. Sed. Petr.,* **40,** 1156–66.

Saller, A. H., 1984. Petrologic and geochemical constraints on the origin of subsurface dolomite, Enewetak Atoll: an example of dolomitization by normal seawater. *Geology,* **12,** 217–20.

Searl, A., 1988. Mixing-zone dolomites in the Gully oolite, Lower Carboniferous, South Wales. *J. Geol. Soc.,* **145,** 891–9.

Searl, A., 1989. Saddle dolomite: a new view of its nature and origin. *Min. Mag.,* **53,** 547–55.

Sears, S. O. and **Lucia, F. J.,** Dolomitization of northern Michigan Niagara reefs by brine refluxion and fresh water/seawater mixing. In *Concepts and Models of Dolomitization:* Eds. D. H. Zenger, J. B. Durham and R. L. Ethington, Special Publ., *Soc. Econ. Paleont. Min.,* **28,** 215–35.

Shinn, E. A., 1968. Selective dolomitization of recent sedimentary structures. *J. Sed. Petr.*, **38**, 612–6.

Shinn, E. A., Ginsberg, R. N. and **Lloyd, R. M.,** 1965. Recent supratidal dolomite from Andros Island, Bahamas. In *Dolomitization and Limestone Diagenesis, A Symposium:* Eds. L. C. Pray and R. D. Murray, Special Publ., *Soc. Econ. Paleont. Min.*, **13**, 112–23.

Sibley, D. F., Dedoes, R. E. and **Bartlett, T. R.,** 1987. Kinetics of dolomitization. *Geology*, **15**, 1112–4.

Sibley, D. F., Nordeng, S. H. and **Borkowski, M. L.,** 1994. Dolomitization kinetics in hydrothermal bombs and natural settings. *J. Sedim. Res.*, **A64**, 630–7.

Skippen, G. B., 1971. Experimental data for reactions in siliceous marble. *J. Geol.*, **79**, 457–81.

Slaughter, I., Kerrick, D. M. and **Wall, V. T.,** 1975. Experimental and thermodynamic study of equilibria in the system $CaO–MgO–SiO_2–H_2O–CO_2$. *Amer. J. Sci.*, **275**, 143–62.

Spotts, J. H. and **Silverman, S. R.,** 1966. Organic dolomite from Point Fermin, California. *Amer. Min.*, **51**, 1144–55.

Steinfink, H. and **Sans, F. J.,** 1959. Refinement of the crystal structure of dolomite. *Amer. Min.*, **44**, 679–82.

Stout, J. W. and **Robie, R. A.,** 1963. Heat capacities from 11 to 300K, entropy and heat of formation of dolomite. *J. Phys. Chem.*, **67**, 2248–52.

Talantsev, A. S., 1983. Geothermobarometry based on parageneses of dolomite–ankerite and calcite with isomorphously admixed rhodochrosite. *Dokl. Akad. Sci. USSR, Earth Sci. Sect.*, **260**, 171–5.

Tennant, C. B. and **Berger, R. W.,** 1957. X-ray determination of dolomite–calcite ratio of a carbonate rock. *Amer. Min.*, **42**, 23–9.

Tilley, C. E., 1948. Earlier stages in the metamorphism of siliceous dolomites. *Min. Mag.*, **28**, 272–4.

Usdowski, E., 1994. Synthesis of dolomite and geochemical applications. *Spec. Publ. Intern. Assoc. Sedimentology*, **21**, 345–60.

Van Tendeloo, G., Wenk, H. R. and **Gronsky, R.,** 1985. Modulated structures in calcian dolomites: a study by electron microscopy. *Phys. Chem. Min.*, **12**, 333–41.

Weber, H. J., 1988. Anisotropic bond polarizability in birefingent crystals. *Acta Cryst.*, **A44**, 320–6.

Weber, J. N. and **Smith, F. G.,** 1961. Rapid determination of calcite–dolomite ratios in sedimentary rocks. *J. Sed. Petr.*, **31**, 130–1.

Wells, A. J., 1986. The dolomite enigma. In *Geology in the Real World, the Kingsley Dunham Volume:* Eds. R. W. Nesbitt and I. Nicol, London (Inst. Mining Metall.). pp. 465–73.

Wenk, H. R. and **Zhang, F.,** 1985. Coherent transformations in calcian dolomites. *Geology*, **13**, 457–60.

White, W. B., 1974. The carbonate minerals. In *The Infrared Spectra of Minerals:* Ed. V. C. Farmer, Min. Soc. Monograph **4**, 227–84.

Winkler, H. G. F., 1979. *Petrogenesis of Metamorphic Rocks*. 5th edn, New York (Springer-Verlag), 347 pp.

Wyckoff, W. G. and **Merwin, H. S.,** 1924. The crystal structure of dolomite. *Amer. J. Sci.*, **8**, 447–61.

Żabiński, W., 1980. Zincian dolomite: the present state of knowledge. *Min. Polonica*, **11**, 19–32.

Zasieńska, S. and **Żabiński, W.,** 1972. Electron microprobe investigation of an unusual zincian dolomite from the Wyrynski mine (Upper Silesia). *Min. Polonica*, **3**, 67–70.

Zenger, D. H., 1972. Significance of supratidal dolomitization in the geologic record. *Geol. Soc. Amer. Bull.*, **83**, 1–12.

Zenger, D. H. and **Durham, J. B.,** 1980. Concepts and models of dolomitization, an introduction. In *Concepts and Models of Dolomitization:* Eds. D. H. Zenger, J. B. Durham and R. L. Ethington, Special Publ., *Soc. Econ. Paleont. Min.*, **28**, 1–9.

Zhabin, A. G. and **Cherepivskaya, G. Y.,** 1965. Carbonatite dikes as related to ultrabasic–alkalic extrusive igneous activity. *Dokl. Akad. Nauk USSR, Earth Sci. Sect.*, **160**, 135–8

Ankerite $Ca(Mg,Fe^{2+},Mn)(CO_3)_2$

Trigonal (−)

ε	1.510–1.548
ω	1.690–1.750
δ	0.182–0.202
Dispersion:	Strong.
D	2.93–3.10
H	$3\frac{1}{2}$–4
Cleavage[1]:	$\{10\bar{1}1\}$ perfect.
Twinning[1]:	Simple twinning common on {0001}, $\{10\bar{1}0\}$ and $\{11\bar{2}0\}$.
Colour:	White, yellow or brown. Colourless in thin section or grain mount.
Unit cell:	a_{hex} 4.811–4.831 Å, c_{hex} 16.042–16.166 Å, $Z_{hex}=3$. a_{rh} 6.032–6.070 Å, α_{rh} 47°00′, $Z_{rh}=1$. Space group $R\bar{3}$

Soluble with difficulty in cold dilute HCl.

Ankerite is considered an Fe-bearing dolomite with an $Fe^{2+}/(Mg+Fe^{2+})$ ratio $\geqslant 0.2$. The term ferroan dolomite is used for minerals of composition between ankerite and dolomite. Like dolomite, ankerite characteristically occurs in sedimentary rocks. It is also formed in metamorphosed carbonates and in igneous rocks. Kutnahorite is a Mn-dolomite ($CaMn(CO_3)_2$).

The mineral was named after E. J. Anker (1772–1843), an Austrian mineralogist.

Structure

Ankerite is isostructural with dolomite, and structural refinements were performed by Beran and Zemann (1977) and in more detail by Reeder and Dollase (1989). The ankerite used for these refinements have the compositions and cell dimensions as listed in Table 28.

The interatomic distances and bond angles for ankerite are listed in Table 29, in which data for a dolomite are also given for comparison. For ankerite, the *M*(A site)–O bond is shorter and the *M*(B site)–O bond is longer than in dolomite. With increasing Fe^{2+} content, the *M*(A site)–O distance decreases and the *M*(B site)–O distance increases. The C–O bond length is nearly constant in these ankerites, but the distance of the carbon atom from the plane of the oxygen atoms decreases with increasing Fe content. For an ankerite from the same locality and of the same composition as that studied by Beran and Zemann (1977), Jarosch (1985) measured the aplanarity of the carbonate group to be 0.013(2) Å. Measurement of the octahedral distortion using quadratic elongation values (Robinson *et al.*, 1971) indicates that the distortion is very small and the octahedra are essentially quite regular (Reeder and Dollase, 1989). Refined occupancies show

[1]Cleavage, twinning and morphology are indexed using the hexagonal morphological cell with quadrupled *a* parameter (see Table 16, p. 103).

Table 28. Compositions and Cell Dimensions of the Ankerites Used for Structural Refinements

	Beran and Zemann (1977)	Reeder and Dollase (1989)		
Sample number		BM1937-294	AMNH6376	AMNH8059
Sample locality	Erzberg, Styria, Austria	Minas, Gerais, Brazil	Goldbrath, Styria, Austria	Erzberg, Styria, Austria
Composition				
$CaCO_3$	1.05	0.997	1.007	0.997
$MgCO_3$	0.27	0.769	0.451	0.273
$FeCO_3$	0.63	0.221	0.502	0.676
$MnCO_3$	0.05	0.013	0.040	0.054
Cell dimension				
a (Å)	4.830(1)	4.8116(2)	4.8240(2)	4.8312(2)
c (Å)	16.167(3)	16.0421(3)	16.1217(3)	16.1663(3)

Table 29. Interatomic Distances (Å) and Angles (Degrees) for Four Ankerites and One Dolomite (from Reeder and Dollase, 1989)

	Dolomite	BM1931-294	AMNH6376	AMNH8059	BZ Ank
A site					
M–O	2.3816(5)	2.3781(5)	2.3734(5)	2.3716(4)	2.371(1)
O_1–O_2	3.298(1)	3.2925(9)	3.283(1)	3.2788(6)	3.277(2)
O_1–O_6	3.437(1)	3.4324(9)	3.429(1)	3.4275(6)	3.428(2)
O_1–*M*–O_2	87.63(2)	87.62(2)	87.50(2)	87.46(1)	87.42(5)
O_1–*M*–O_6	92.37(2)	92.38(2)	92.50(2)	92.54(1)	92.58(5)
B site					
M–O	2.0821(5)	2.0939(5)	2.1158(5)	2.1266(4)	2.126(1)
O_1–O_2	2.902(1)	2.9164(9)	2.946(1)	2.9598(6)	2.959(2)
O_1–O_6	2.987(1)	3.0054(9)	3.037(1)	3.0542(6)	3.053(2)
O_1–*M*–O_2	88.34(2)	88.28(2)	88.25(2)	88.20(1)	88.21(5)
O_1–*M*–O_6	91.66(2)	91.72(2)	91.75(2)	91.80(1)	91.79(6)
CO_2					
C–O	1.2858(5)	1.2843(5)	1.2836(6)	1.2843(4)	1.284(1)
Apl.[a]	0.017(1)	0.020(1)	0.018(1)	0.0163(8)	0.011(3)

[a]Aplanarity of carbon atom in the CO_3 group.

O_1 and O_2 are oxygens in the same layer; O_1 and O_6 are oxygens in different layers. Descriptions of BM, AMNH and BZ ankerites are given in Table 28.

that all ankerites studied in the structure refinement (Reeder and Dollase, 1989) are ordered. Ca occupies the A site (Table 29) from 94 per cent (BM1931-294) to 99 per cent (AMNH6376), and the B site is from 27 per cent (AMNH8059) to 77 per cent (BM1931-294) occupied by Mg. The difference at both sites is made up by (Fe + Mn). Beran and Zemann (1977) also found that their Fe-rich ankerite was ordered. Microstructures in all samples were essentially homogeneous, except for some localized defects (Reeder and Dollase, 1989).

A structural analysis of an ankerite with approximately 70 mol. per cent $CaFe(CO_3)_2$ under high pressure (Ross and Reeder, 1992) showed that the major change is the compression of both CaO_6 and $(Mg,Fe)O_6$ octahedra with increasing pressure. Variations in Ca–O and (Mg,Fe)–O bond lengths with pressure are shown in Fig. 82, along with those of dolomite. The compressibilities of A(Ca)–O and B(Mg,Fe)–O bonds are 3.93×10^{-2} and 3.51×10^{-2}/kbar, respectively, and are

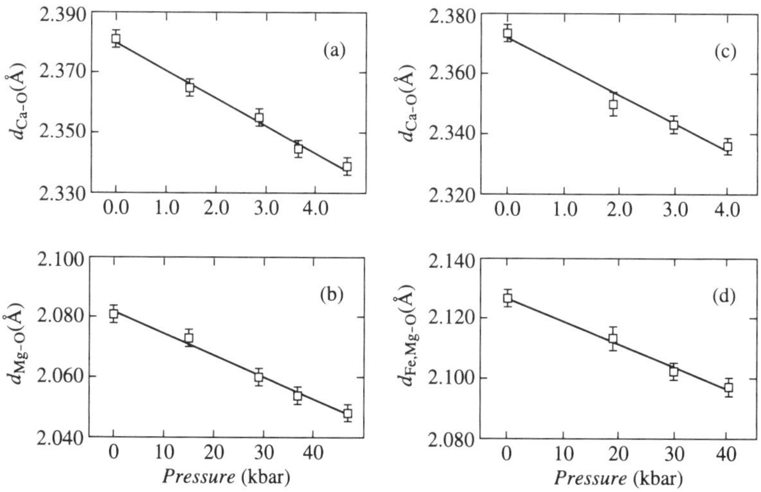

Fig. 82. Variations of bond lengths with pressure: (a) Ca–O bond in dolomite, (b) Mg–O bond in dolomite, (c) Ca–O bond in ankerite and (d) (Mg,Fe)–O bond in ankerite (after Ross and Reeder, 1992).

Table 30. Unit Cell Dimensions of Ankerite at Pressures from 0 to 40 kbar (from Ross and Reeder, 1992)

Pressure (kbar)	*a* (Å)	*c* (Å)	*V* (Å^3)
0.0	4.8360	16.186	327.82
9.7	4.8265	16.085	324.51
19.0	4.8163	15.992	321.26
25.6	4.8109	15.924	319.18
29.7	4.8070	15.881	317.81
34.0	4.8038	15.847	316.71
40.0	4.7990	15.792	314.97

comparable with those for dolomite. The C–O bond length varies only slightly with pressure, decreasing from 1.290(2) Å at 0 kbar to 1.285(2) Å at 40 kbar. The unit cell dimensions of ankerite at various pressures are given in Table 30.

The ^{57}Fe Mössbauer spectra of four ankerites containing 17, 29, 54 and 66 mol. per cent $CaFe(CO_3)_2$ show only one moderately split quadruple doublet with an isomer shift of 1.24(1) mm/s (Reeder and Dollase, 1989). This is in good agreement with the results reported by DeGrave and Vochten (1985) and is consistent with an FeO_6 octahedron that is only slightly distorted.

As for dolomite, electron diffraction from calcian ankerite formed at low temperatures reveals weak *c*- and *d*-type reflections, indicating that superstructures occur in ankerite (Reksten, 1990). A model for the *c*-domain superstructure was proposed by Reksten which has space group *P*2/*c*.

Kutnahorite, like ankerite, is isostructural with dolomite. Refinements were made by Peacor *et al.* (1987) for two different kinds of sample. The kutnahorite from Bald Knob, North Carolina, USA, is disordered in Ca–Mn distribution, with a composition of $(Ca_{0.49}Mn_{0.50}Mg_{0.01})CO_3$, and the Sterling Hill, New Jersey, USA, sample is substantially ordered, with a composition of

$(Ca_{0.84}Mn_{0.16})(Ca_{0.27}Mn_{0.73})(CO_3)_2$. Peacor *et al.* (1987) stated that the long-range order in kutnahorite could only be measured by single-crystal techniques. For the Bald Knob and Sterling Hill kutnahorites, crystallographic data are *a* 4.8732(8), 4.894(1) Å and *c* 16.349(1), 16.50(2) Å, respectively. Selected cation–oxygen distances are: Ca–O 2.277(2) and 2.351(2) Å; Mn–O, 2.267(2) and 2.217(2) Å; and C–O 1.285(1) and 1.296(1) Å, respectively. For the A octahedron, the O–O′, O–O″, O–Ca–O and O–Ca–O″ values are 3.152(3) and 3.244(3) Å, 3.288(3) and 3.404(3) Å, 87.59(9) and 87.76(8)°, and 94.41(9) and 92.74(8)°, respectively. For the B octahedron, the values in corresponding order are 3.146(3) and 3.047(3) Å, 3.264(3) and 3.196(3) Å, 87.90(9) and 87.76(7)°, and 92.10(9) and 92.24(8)°. The O–O and O–C–O measurements for the CO_3 group are 2.225(2) and 2.244(2) Å, and 120.00(1) and 119.98(2)°. The displacement of the carbon atom from the plane of three coordinating oxygen atoms is nearly zero (0.002 Å) in the Bald Knob kutnahorite; and in the Sterling Hill kutnahorite it is 0.016 Å. Peacor *et al.* (1987) suggested that the displacement provides a measure of the degree of cation order in dolomite-type carbonates.

Chemistry

The chemical composition of ankerite varies from $Ca(Mg_{0.8}Fe_{0.2})(CO_3)_2$ at the Mg-rich end to approximately $Ca(Mg_{0.3}Fe_{0.7})(CO_3)_2$ at the Fe-rich end of the $Ca(Mg,Fe)(CO_3)_2$ series, as shown in Table 31. Fe-dolomite of the composition $CaFe(CO_3)_2$ has been neither found in nature nor synthesized experimentally. Rosenberg and Foit (1979) related the instability of $CaFe(CO_3)_2$ to a Jahn-Teller effect of Fe^{2+}, which causes excessive octahedral distortion. However, recent structural refinement and Mössbauer studies (Reeder and Dollase, 1989) have demonstrated that both octahedral positions, the A and B sites, are nearly regular in ankerite as well as in the entire solid solution range to dolomite. Ankerites containing Ca in excess of that required by its structure to completely fill the A site were reported by Goldsmith *et al.* (1962) and Murata *et al.* (1972).

Phase relationships in the system $CaCO_3$–$MgCO_3$–$FeCO_3$ established by Goldsmith *et al.* (1962) and Rosenberg (1967) have been discussed in previous chapters. From analyses of natural carbonates and established equilibrium relations, Anovitz and Essene (1987) illustrate carbonate mineral assemblages in the system $CaCO_3$–$MgCO_3$–$FeCO_3$ in a series of phase diagrams, each representing a grade of metamorphism including subgreenschist at $T = 250\,°C$ (Fig. 83(a)), biotite grade at $T = 400\,°C$ (Fig. 83(b)), amphibolite grade at $T = 550\,°C$ (Fig. 83(c)), and granulite grade at $T = 700\,°C$ (Fig. 83(d)). Field boundaries for the subgreenschist and granulite-grade conditions were constructed graphically, while those for biotite and amphibolite-grade conditions were calculated from a solution model. The crossing tie-lines in the diagrams, as suggested by Anovitz and Essene (1987), may be due to temperature variations, analysis of disequilibrium pairs and analytical inaccuracies.

A new set of equations was derived by McSwiggen (1993a,b) to describe the equilibrium ordering of three cations mixing on two sites, and applied to ternary carbonates. The model generates valid compositional relationships for the equilibrium relationships within the $CaCO_3$–$MgCO_3$–$FeCO_3$ system, and gives reasonable free energies of mixing; however, more work needs to be carried out on the significance of Mg/Fe ordering in this system.

Davidson (1994) uses a ternary expansion of the generalized point approximation

Table 31. Ankerite Analyses

	1	2	3	4	5	6	7
CaO	29.23	32.45	29.43	28.93	28.10	27.24	27.50
MgO	12.85	13.77	14.52	9.11	9.68	5.37	7.56
FeO	12.06	7.70	8.37	17.51	17.68	23.93	20.29
MnO	0.77	0.11	2.15	0.67	0.50	0.96	1.31
	0.62 (oth.)						0.21 (oth.)
CO_2	44.70	45.28(c)	45.41(c)	43.78(c)	44.03	42.49(c)	43.13(c)
Total	100.23	99.31	99.88	100.00	99.99	99.99	100.00
	Numbers of ions on the basis of six (O)						
Ca	1.024	1.125	1.017	1.027	1.004	1.006	1.001
Mg	0.626	0.664	0.698	0.454	0.480	0.276	0.383
Fe	0.330	0.208	0.059	0.490	0.493	0.690	0.576
Mn	0.022	0.003	0.059	0.019	0.014	0.028	0.038
C	1.995	2.000	2.000	2.000	2.004	2.000	2.000

(c) Calculated from stoichiometry.

1 Ankerite, Oak Victoria Colliery, Oldham, Lancashire, England (Broadhurst and Howie, 1958); ε 1.515, ω 1.710, D 2.97.
2 Ankerite from borehole, Salton Sea geothermal field, California, USA (McDowell and Paces, 1985).
3 Ankerite from Tintagel, Cornwall, England (Dubrawski and Warne, 1985).
4 Ankerite from Blue Rock Tunnel, Pennsylvania, USA (Dubrawski and Warne, 1985).
5, 6 Ankerite from differentiated dolerite, Kalgoorlie, Australia (Phillips and Brown, 1987).
7 Ankerite from banded iron formation, Western Australia (Miyano, 1982).

Table 31. *Continued*

	8	9	10	11	12	13
CaO	33.15	26.80	29.3	28.02	26.97	29.18
MgO	12.39	6.33	9.2	9.02	5.31	10.94
FeO	6.07	22.77	14.9	17.89	23.43	7.41
MnO	9.08	1.25	2.6	1.41	1.85	7.67
CO_2	39.31(c)	42.85(c)	44.0	43.66(c)	42.44(c)	44.14
Total	100.00	100.00	100.0	100.00	100.00	99.34
	Numbers of ions on the basis of six (O)					
Ca	1.064	0.986	1.047	1.007	0.997	1.033
Mg	0.553	0.324	0.457	0.451	0.273	0.539
Fe	0.152	0.654	0.415	0.502	0.676	0.207
Mn	0.237	0.036	0.073	0.040	0.054	0.212
C	2.000	2.000	2.003	2.000	2.000	2.004

8, 9 Ankerite from low-grade metamorphic iron formation, Howells River, Quebec (Klein, 1974).
10 Ankerite from Iron Hill carbonatite complex, Colorado, USA (Nash, 1972).
11, 12 Ankerite from Styria, Austria (Reeder and Dollase, 1989); see Table 28.
13 Ankerite from crystalline stalactite-like pipes in a volcanic tectonic basin, Uzbekistan (Talantsev, 1983).

to model the order–disorder, phase equilibria and solution enthalpy relationships for (Ca,Mg,Fe)CO_3 solutions. Calculations show that the stability of ordered CaFe(CO_3) is limited to about <450 °C by breakdown to calcite + siderite, this temperature increasing with Mg content. A metastable CaFe(CO_3) may, however, exist up to 767 °C, this temperature also rising with increasing substitution of Mg for Fe.

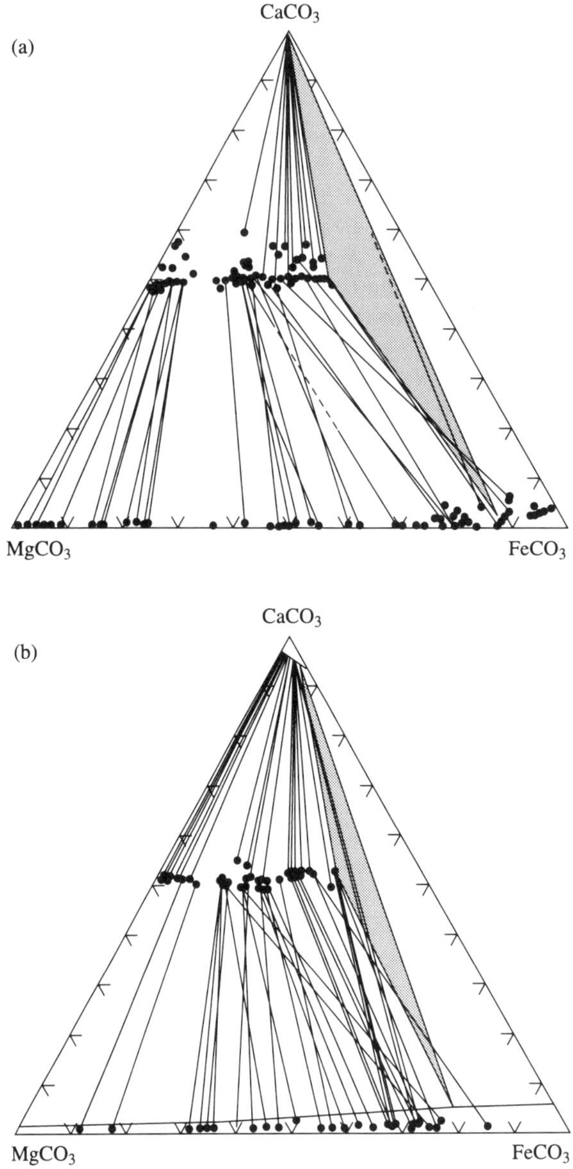

Fig. 83. Analysed carbonates in the system $CaCO_3$–$MgCO_3$–$FeCO_3$: (a) for subgreenschist conditions with $T \sim 250\,^\circ C$, (b) for biotite-grade conditions with $T \sim 400\,^\circ C$, (c) for amphibolite-grade conditions with $T \sim 550\,^\circ C$ and (d) for granulite-grade conditions with $T \sim 700\,^\circ C$ (after Anovitz and Essene, 1987). Two- and three-phase field boundaries in (b) and (c) were calculated from the solution model, whereas those in (a) and (d) were fitted graphically to correspond with experimental results and by analogy with calculated 400 and 550 °C results. The three-phase field is shaded.

The presence of Mn in ankerite is illustrated in the analyses listed in Table 31. The Mn-dolomite, kutnahorite, has compositions varying over a wide range, from a composition close to $CaMn(CO_3)_2$ (Frondel and Bauer, 1955; Peacor *et al.*, 1987; Mucci, 1991) to those with high FeO (Frondel and Bauer, 1955; Generalov and

(c)

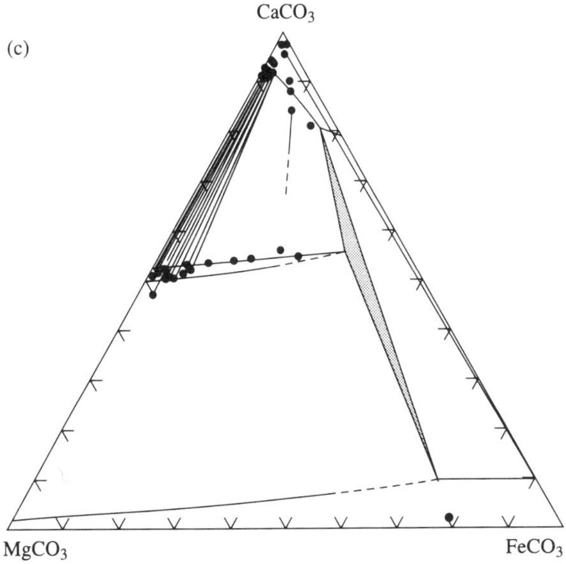

(d)

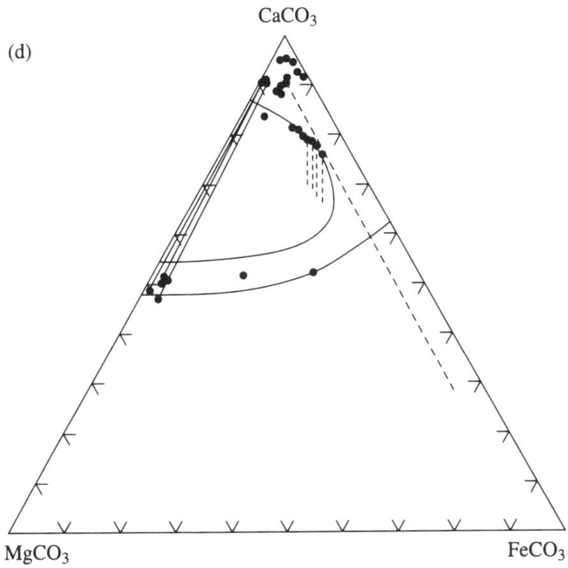

Fig. 83. (*contd*)

Moraler, 1989) and high MgO (Frondel and Bauer, 1955; Tsusue, 1967; Wenk and Maurizio, 1978; Iwafuchi *et al.*, 1983, Table 32). Miyahisa *et al.* (1975) reported kutnahorite from Hoei, Japan, with a range of compositions expressed as $Ca_{1.0-1.2}Mn_{0.5-0.8}Mg_{0.1-0.4}Fe_{0.4}(CO_3)_2$. Essene (1983) made an extensive compilation of compositions within the system $CaMg(CO_3)_2$–$CaFe(CO_3)_2$–$CaMn(CO_3)_2$ as shown in Fig. 84. Analysed compositions demonstrate a region of solid solution existing among dolomite, kutnahorite and ankerite. Carbonate minerals with compositions of > 70 per cent $CaFe(CO_3)_2$ are very rare. The absence

Table 32. Kutnahorite Analyses

	1	2	3	4	5
CaO	20.90	24.66	28.91	25.1	20.33
MgO	0.79	5.18	8.39	4.3	1.47
MnO	35.70	23.76	14.07	25.2	34.81
FeO	0.34	4.27	4.99	3.4	1.25
ZnO	—	—	—	—	—
CO_2	42.27	42.62	43.62(c)	42.0(c)	39.91(c)
Total	100.00	100.49	99.98	100.0	97.77
	Numbers of ions on the basis of six (O)				
Ca	0.828	0.910	1.040	0.936	0.799
Mg	0.043	0.266	0.420	0.223	0.080
Mn	1.118	0.693	0.400	0.742	1.082
Fe	0.010	0.123	0.194	0.099	0.038
Zn	—	—	—	—	—
C	2.000	2.004	2.000	2.000	2.000

(c) Calculated from stoichiometry.

1 Kutnahorite from Sterling Hill, Franklin, New Jersey, USA (Mucci, 1991).
2 Kutnahorite from Kutna Hora, Czechoslovakia (Frondel and Bauer, 1955); D 3.06.
3 Kutnahorite in quartzite from Piz Cam, Bergell Alps, Switzerland (Wenk and Maurizio, 1978).
4 Kutnahorite from the Chvaletice pyrite and Mn deposit, East Bohemia, Czech Republic (Žák and Povondra, 1981).
5 Kutnahorite from metamorphosed Mn-deposit at Bald Knob, North Carolina, USA (Winter *et al.*, 1981).

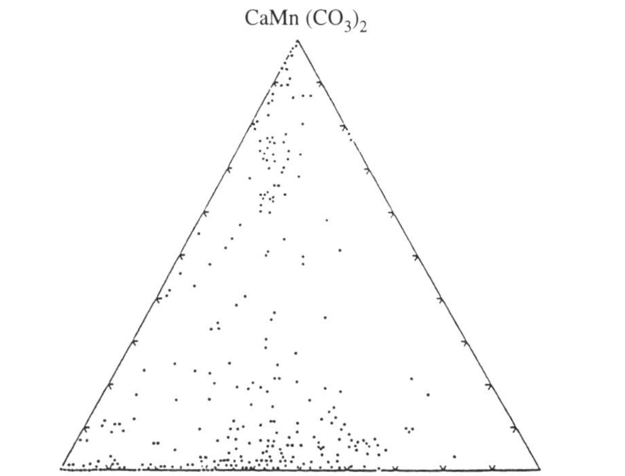

Fig. 84. Carbonate compositions in the system $CaMg(CO_3)_2$–$CaFe(CO_3)_2$–$CaMn(CO_3)_2$ (after Essene, 1983).

of carbonates with compositions along the join $CaMn(CO_3)_2$–$CaFe(CO_3)_2$ and in the Mg-poor region of the system correlates with experimental results (Rosenberg, 1968), which show the existence of an immiscibility gap from $CaFe(CO_3)_2$ to about 97 mol. per cent $CaMn(CO_3)_2$. Using the variation of ω and $d_{10\bar{1}4}$ with composition in the system $CaMg(CO_3)_2$–$CaFe(CO_3)_2$–$CaMn(CO_3)_2$, Smetannikova (1979) constructed a network which can be used for the determination of compositions.

Differential thermal analysis (DTA) of ankerite was fully described by Otsuka (1986) who summarized previous studies. Three endotherms were produced when an ankerite of composition $Ca_{1.025}(Mg_{0.329}Fe_{0.588}Mn_{0.060})(CO_3)_2$ was treated in flowing CO_2. The first endotherm is attributed to the decomposition of $FeCO_3$ and $MgCO_3$; which produces $CaCO_3$, MgO and $FeO{\cdot}Fe_2O_3$ by the oxidation of FeO. $FeO{\cdot}Fe_2O_3$ reacts further with MgO to form $MgO{\cdot}Fe_2O_3$. The CO evolved during the first endotherm results from the oxidation of FeO by CO_2 to form $FeO{\cdot}Fe_2O_3$. The second endotherm is caused by the reaction of $MgO{\cdot}Fe_2O_3$ with $CaCO_3$ to form $2CaO{\cdot}Fe_2O_3$ and MgO, and the CO evolved is ascribed to the oxidation of the remaining Fe^{2+} by CO_2 released during the second reaction. The third endotherm results from the decomposition of $CaCO_3$. The two-step evolution of CO was also observed in an N_2 atmosphere. Similar results were obtained from a sample of composition $Ca_{1.015}(Mg_{0.515}Fe_{0.225}Mn_{0.245}(CO_3)_2$. However, with an atmosphere of mixed CO and CO_2, only two endotherms were obtained during its thermal decomposition: the first is due to the formation of $CaCO_3$, (Mg,Mn)O and Fe_2O_3, and the second is ascribed to the reaction of $CaCO_3$ with (Mg,Mn)O and Fe_2O_3 to produce (Ca,Mn)O, (Mg,Mn)O and $Ca_2MgFe_2O_6$. When the CO_2/CO ratio is 1, the two-step decomposition of ankerite produces $CaCO_3$ and (Mg,Mn,Fe)O in the first reaction, and the equilibrium distribution of Fe and Mn occurs between (Mg,Mn,Fe)O and (Ca,Mn,Fe)O in the second step.

DTA was carried out by Shibuya and Harada (1976) on kutnahorite and by Iwafuche *et al.* (1983) on magnesian kutnahorite. Under low partial pressures of oxygen, Mn remains bivalent, and kutnahorite decomposes in two steps like dolomite. The solubility products of kutnahorite expressed in the logarithmic form at 5, 25 and 40 °C are -21.88 ± 0.06, -21.8 ± 0.07 and -21.60 ± 0.04 (Mucci, 1991).

The decomposition enthalpies of dolomite–ankerite carbonates were measured by differential scanning calorimetry in flowing nitrogen (Warne and Dubrawski, 1987) and in flowing CO_2 (Dubrawski and Warne, 1985). The results obtained demonstrate a linear decrease in enthalpy with increasing iron substitution in the series.

Optical and Physical Properties

With increasing replacement of Mg by Fe^{2+}, the optical and physical properties of the ankerite series grade continuously from those of dolomite and ferroan dolomite. Indices of refraction and birefringence vary in a linear manner as shown in Fig. 42 (p. 118). An ankerite of composition $Ca_{1.024}(Mg_{0.626}Fe_{0.330}Mn_{0.022})(CO_3)_2$ has ω 1.710 and ε 1.515 (Howie and Broadhurst, 1958). Howell and Dawson (1958) describe a universal stage technique which allows the determination of the Mg/Fe ratio of an ankerite from the measurement of the angle of rotation necessary to change ε to ε', where $\varepsilon' = 1.54$, which is the refractive index of the mounting medium.

The colour of ankerite is variable. Commonly white or brownish white, it may also be yellowish brown or brown when the oxidation of iron has taken place; varieties containing Mn tend to be pinkish. The specific gravity of ankerite ranges from 3.10 for the most Fe-rich member to 2.86 for the Fe-free dolomite.

Kutnahorite is translucent with a pale-pink colour. Its optical and physical properties are similar to those of ankerite. Its specific gravity ranges from 2.98 to 3.18, and values of ω and ε are 1.717–1.731 and 1.524–1.529, respectively.

An infrared spectroscopic evaluation of ankerite by Farmer and Warne (1978) demonstrated that the iron substitution for Mg in the series $Ca(Mg,Fe)(CO_3)_2$ causes shifts in the frequencies of three bands, 880, 730 and 150–160 cm^{-1}. Of these, the 150–160 cm^{-1} band has the largest shift, which can be measured with precision and related to Mg/(Mg+Fe+Mn). Using Fourier transform infrared spectroscopy, Dubrawski *et al.* (1989) established a linear relationship between the frequency of the 1340 cm^{-1} band and the mol. fractions Fe/(Fe+Mn+Ca+Mg) and (Fe+Mn)/(Fe+Mn+Ca+Mg).

Distinguishing Features

Ankerite may be distinguished as a rhombohedral carbonate by its extremely high birefringence and its perfect $\{10\bar{1}1\}$ cleavage, as in the case of dolomite. It is denser than calcite and has higher refractive indices.

The identification of carbonate minerals by staining methods has been described by Friedman (1959). For ankerite, Warne (1962) detailed a procedure with alizarin red S first in acid, and then in alkaline solution, which can effectively differentiate ankerite from other carbonates.

Paragenesis

Ankerite occurs in sedimentary rocks as a result of diagenetic processes and hydrothermal alteration. Organic geochemical influences are important in some instances. In the subsurface sandstone, southwest Texas, USA, Boles (1978) reported ankerite cements occurring at depths from 2560 (125 °C) to at least 4650 m (210 °C), whereas calcite cements were observed at depths <2315 m. Some shallow ankerites have appreciable excess Ca, but those at depths >7300 m have compositions close to $Ca(Mg_{0.5}Fe_{0.5})(CO_3)_2$. Ankerite is believed to form from calcite hydrothermally by the reaction:

$$4CaCO_3 + Fe^{2+} + Mg^{2+} = 2Ca(Mg_{0.5}Fe_{0.5})(CO_3)_2 + 2Ca^{2+}.$$

Fe and Mg for this reaction were apparently released by the breakdown of smectite to illite in mixed-layer clays. A carbonate vein in the northern Upper Harz, Germany, consists of ankerite at the margins with sharp boundaries to calcite in the central portion (Gundlach, 1970). The ankerite has an Fe/Mg ratio of 3.2–4.0, and the calcite has an Fe/Mg ratio of 0.4–0.9. In the core of the vein the calcite approximates the composition of ankerite. Gundlach (1970) suggested a minor rejuvenation of the mineralizing fluid.

The magnesite–siderite–ankerite mineralization in the northern greywacke zone of the Eastern Alps has a $^{87}Sr/^{86}Sr$ ratio in magnesite similar to that of the host carbonate rocks, whereas siderite and ankerite show significantly higher ratios. Frimmel (1988) attributed the former to a syngenetic or early diagenetic Mg cementation, and the latter to an epigenetic hydrothermal origin. Vein ankerite from the Styrian Erzberg, Austria, has an $FeCO_3$ content up to 33 mol. per cent (Beran, 1977). This is near the upper limit observed for an ankerite phase and requires a minimum formation temperature of 450 °C at 2–3 kbar.

In the Salton Sea geothermal field, California, USA, ankerite occurs in the carbonate-rich sandstones. Detrital dolomite is rimmed by calcian ankerite in optical continuity at shallow depth, by aggregates of small euhedral ankerite rhombs at intermediate depth, and by anhedral aggregates of calcite at greater depths (McDowell and Paces, 1985). Calcic dolomite–ankerite was found in the pores of uncompacted mud of argillaceous Kimmeridge clay at Kimmeridge Bay, Dorset, England (Irwin, 1981).

Ferroan dolomite and ankerite are commonly found in ancient lacustrine deposits but are apparently less common in present-day lakes. Desborough (1978) described their occurrences in the Green River Formation sediments of the western USA. Cole and Picard (1978) made a comparative mineralogical study of nearshore and offshore lacustrine lithofacies, including ankerite, dolomite, calcite, K- and Na-feldspars, quartz and analcite. The origin of authigenic ankerite from the Ninian Field, UK North Sea, is probably synchronous with hydrocarbon emplacement and related to clay mineral diagenesis (Kantorowicz, 1985). This ankerite occurs throughout the reservoir sandstone, but is particularly abundant below the oil–water contact.

Calcite, dolomite, ankerite and siderite were found in the Iron Hill carbonatites, Colorado, USA (Nash, 1972). A typical sequence of crystallization is calcite–dolomite–ankerite–siderite with decreasing temperature. Ankerite was also discovered in other carbonatites and lamprophyre dykes (Platt and Mitchell, 1982; Hogarth *et al.*, 1985). Among the four carbonate minerals in the differentiated tholeiitic Golden Mile dolerite sill, Kalgoorlie, Australia (Phillips and Brown, 1987), ankerite is the most widespread, occurring throughout the dolerite and in the zones of chlorite, carbonate and pyrite alteration. In Dolni Bory, western Moravia (Némec, 1977), the pegmatite dykes carry Mn-rich siderite, ferroan magnesite and ankerite accompanied by sulphides. They were formed by a late hydrothermal fluid which was not derived from the original pegmatitic magma. Micro-inclusions of ankerite (with Mg/Fe/Ca ratios of 28/18/54) have been recorded in the coatings of coated diamonds from Zaïre (Walmsley and Lang, 1992).

Members of the dolomite–ankerite series and siderite are abundant in low-grade metamorphic iron-rich rocks (Klein, 1974). In the Howells River area, at the western edge of the Labrador Trough, the finely banded silicate–carbonate members of the Sokoman Formation consist of ankerite, siderite and dolomite. The ankerites have a maximum of 22.77 wt. per cent FeO and show a range in MnO from 1.25 to 4.63 wt. per cent. Ankerite occurs in association with stilpnomelane, iron oxide and quartz in the banded iron formation of the Dales Gorge Member, Hamersley Group, Western Australia (Miyano, 1982). In the sideritic iron ore of Canigou, French Pyrenees (Guitard, 1970) a distinct pattern of carbonate association with depth was observed, with the sequences siderite–ankerite–dolomite, ankerite–dolomite and siderite–dolomite exemplifying a metasomatic zonation. In the Archaean Helen siderite deposits, Wawa, Ontario, Canada, the underlying felsic lavas and pyroclastics were altered to form a sericite–chlorite–pyrophyllite zone capped by ankerite replacement (Morton and Nebel, 1984). The phase relationships illustrated correlate with the biotite grade of Anovitz and Essene (1987) (Fig. 83(b)). Devaraju and Murthy (1982) reported the occurrence of ferroan carbonate from the Chiknayakanhalii schist belt, Karnataka, India, with a composition of $Ca_{0.93}(Fe,Mn)_{0.67}Mg_{0.38}(CO_3)_2$.

References

Anovitz, L. M. and **Essene, E. J.**, 1987. Phase equilibria in the system $CaCO_3$–$MgCO_3$–$FeCO_3$. *J. Petr.*, **28**, 389–414.

Beran, A., 1977. Die Kluftankerite des Steirischen Erzberges und ihre Mogliche Verwendung als Geothermometer. *Min. Deposita*, **12**, 90–5.

Beran, A. and **Zemann, J.**, 1977. Refinement and comparison of the crystal structures of a dolomite and of an Fe-rich ankerite. *Tschermaks Min. Petr. Mitt.*, **24**, 279–86.

Boles, J. R., 1978. Active ankerite cementation in the subsurface Eocene of southwest Texas. *Contrib. Min. Petr.*, **68**, 13–22.

Broadhurst, F. M. and **Howie, R. A.**, 1958. On the occurrence of ankerite in a band of non-marine lamellibranchs from the Lancashire Coal Measures. *Geol. Mag.*, **95**, 397–402.

Cole, R. D. and **Picard, M. D.**, 1978. Comparative mineralogy of nearshore and offshore lacustrine lithofacies, Parachute Creek Member of the Green River Formation, Piceance Creek basin, Colorado, and eastern Uinta basin, Utah. *Bull. Geol. Soc. Amer.*, **89**, 1441–54.

Davidson, P. M., 1994. Ternary iron, magnesium, calcium carbonates: a thermodynamic model for dolomite as an ordered derivative of calcite structure solid solutions. *Amer. Min.*, **79**, 332–9.

DeGrave, E. and **Vochten, R.**, 1985. An ^{57}Fe Mössbauer study of ankerite. *Phys. Chem. Min.*, **12**, 108–13.

Desborough, G. A., 1978. A biogenic–chemical stratified lake model for the origin of the oil shale of the Green River Formation: an alternative to the playa-lake model. *Bull. Geol. Soc. Amer.*, **89**, 961–71.

Devaraju, T. C. and **Murthy, K. S. A.**, 1982. Ferroan ankerite from the Chiknayakanhalli schist belt, Karnataka. *J. Geol. Soc. India*, **23**, 381–6.

Dubrawski, J. V. and **Warne, S. St. J.**, 1985. Differential scanning calorimetry of minerals of the dolomite–ferroan dolomite–ankerite series in flowing carbon dioxide. *Min. Mag.*, **52**, 627–35.

Dubrawski, J. V., **Channon, A. L.** and **Warne, S. St. J.**, 1989. The effects of substituion in the dolomite–ferroan dolomite–ankerite series as illustrated by FTIR. *Neues Jahrb. Min., Mh.*, 337–44.

Essene, E. J., 1983. Solid solutions and solvi among metamorphic carbonates with applications to geological thermobarometry. In *Carbonates; Mineralogy and Chemistry:* Ed. R. J. Reeder, *Reviews in Mineralogy*, **11**, 77–96, Mineralogical Society of America, Washington, DC.

Farmer, V. C. and **Warne, S. St J.**, 1978. Infrared spectroscopic evaluation of iron contents and excess calcium in minerals of the dolomite–ankerite series. *Amer. Min.*, **63**, 779–81.

Friedman, G. N., 1959. Identification of carbonate minerals by staining methods. *J. Sed. Petr.*, **29**, 87–97.

Frimmel, H., 1988. Strontium isotopic evidence for the origin of siderite, ankerite and magnesite mineralization in the Eastern Alps. *Min. Deposita*, **23**, 268–75.

Frondel, C. and **Bauer, H. H.**, 1955. Kutnahorite: a manganese dolomite, $CaMn(CO_3)_2$. *Amer. Min.*, **40**, 748–50.

Generalov, M. E. and **Moraler, S. V.**, 1989. [Kutnahorite from a near-surface gold-silver deposit.] *Zap. Vses. Min. Obshch.*, **118**, 42–7 (in Russian with English abstract).

Goldsmith, J. R., **Graf, D. L.**, **Wittens, J.** and **Northrup, D. A.**, 1962. Studies in the system $CaCO_3$–$MgCO_3$–$FeCO_3$: 1. Phase relations; 2. A method for major element spectrochemical analysis; 3. Composition of some ferroan dolomites. *J. Geol.*, **70**, 659–88.

Guitard, G., 1970. Sur des relations entre la sidérite, l'ankérite et la dolomie dans les gisements de fer métasomatiques du Canigou (Pyrénéesorientales). *C. R. Acad. Sci., Paris*, **271D**, 1233–6.

Gundlach, H., 1970. Einbisher unbekannter Karbonatzgang im nördlichen Oberhorz. *Neues Jahrb. Min., Abh.*, **113**, 215–8.

Higgs, D. V. and **Handin, J. W.**, 1959. Experimental deformations of dolomite single crystals. *Geol. Soc. Amer. Bull.*, **70**, 245–78.

Hogarth, D. D., **Hartree, R.** and **Loop, J.**, 1985. Rare-element minerals in four carbonatites near Gatineau, Quebec. *Amer. Min.*, **70**, 1135–42.

Howell, J. E. and **Dawson, K. R.**, 1958. Technique for optical determination of iron-bearing dolomite. *Can. Min.*, **6**, 292–4.

Howie, R. A. and **Broadhurst, F. M.**, 1958. X-ray data for dolomite and ankerite. *Amer. Min.*, **43**, 1210–4.

Irwin, H., 1981. On calcic dolomite–ankerite from the Kimmeridge clay. *Min. Mag.*, **44**, 105–7.

Iwafuche, K., **Watanabe, C.** and **Otsuka, R.**, 1983. Thermal decomposition of magnesian kutnahorite. *Thermochim. Acta*, **60**, 361–81.

Jarosch, D., 1985. Confirmation of the aplanarity of the carbonate group in ankerite from X-ray four-circle diffractometer data. *Östereichische Akad. Wissenschaften, Math. Naturwiss. Kl., Anzeiger*, **122**, 61–2.

Kantorowicz, J. D., 1985. The origin of authigenic ankerite from the Ninian Field, UK North Sea. *Nature*, **315**, 214–6.

Klein, C., Jr., 1974. Greenalite, stilpnomelane, minnesotaite, crocidolite and carbonates in a very low-grade metamorphic Precambrian iron formation. *Can. Min.*, **12**, 475–98.

McDowell, S. D. and **Paces, J. R.**, 1985. Carbonate alteration mineralogy in the Salton Sea geothermal system, California, U.S.A. *Min. Mag.*, **49**, 469–79.

McSwiggen, P. L., 1993a. Alternative solution model for the ternary carbonate system $CaCO_3$–$MgCO_3$–$FeCO_3$. I: A ternary Bragg–Williams ordering model. *Phys. Chem. Min.*, **20**, 33–41.

McSwiggen, P. L., 1993b. Alternative solution model for the ternary carbonate system

$CaCO_3$–$MgCO_3$–$FeCO_3$. II: Calibration of a combined ordering and mixing model. *Phys. Chem. Min.*, **20**, 42–55.

Miyahisa, M., Harada, S., Ishibashi, K., Shibnma, S. and **Motomura, M.**, 1975. Kutnahorite from Hoei tin mine, with special concern to its chemical composition and mineral paragenesis. *Mining Geol. (Japan)*, **25**, 347–57.

Miyano, T., 1982. Stilpnomelane, iron-rich mica, K-feldspar and hornblende in banded iron-formation assemblages of the Dales Gorge Member, Hamersley Group, Western Australia. *Can. Min.*, **20**, 189–202.

Morton, R. L. and **Nebel, M. L.**, 1984. Hydrothermal alteration of felsic volcanic rocks at the Helen siderite deposit, Wawa, Ontario. *Econ. Geol.*, **79**, 1319–33.

Mucci, A., 1991. The solubility and free energy of formation of natural kutnahorite. *Can. Min.*, **29**, 113–21.

Murata, K. J., Friedman, I. and **Cremer, M.**, 1972. Geochemistry of diagenetic dolomites in Miocene marine formations of California and Oregon. *U. S. G. S. Prof. Paper*, **724C**, C1–C12.

Nash, W. P., 1972. Mineralogy and petrology of the Iron Hill carbonatite complex, Colorado. *Bull. Geol. Soc. Amer.*, **83**, 1361–82.

Némec, D., 1977. Carbonates of the pegmatites from Dolni Bory (western Moravia). *Neues Jahrb. Min., Mh.*, 517–21.

Otsuka, R., 1986. Recent studies on the decomposition of the dolomite group by thermal analysis. *Thermochim. Acta*, **100**, 69–80.

Peacor, D. R., Essene, E. J. and **Gaines, A. M.**, 1987. Petrologic and crystal–chemical implication of cation order–disorder in kutnahorite [$CaMn(CO_3)_2$]. *Amer. Min.*, **72**, 319–28.

Phillips, G. N. and **Brown, I. J.**, 1987. Host rock and fluid control on carbonate assemblages in the Golden Mile dolerite, Kalgoorlie gold deposit, Australia. *Can. Min.*, **25**, 265–73.

Platt, R. G. and **Mitchell, R. H.**, 1982. The Marathon dikes: ultrabasic lamprophyres from the vicinity of McKeller Harbour, N. W. Ontario. *Amer. Min.*, **67**, 907–16.

Reeder, R. J. and **Dollase, W. A.**, 1989. Structural variation in the dolomite–ankerite solid solution series: an X-ray, Mössbauer, and TEM study. *Amer. Min.*, **74**, 1159–69.

Reksten, K., 1990. Superstructure in calcian ankerites. *Phys. Chem. Min.*, **17**, 266–70.

Robinson, K., Gibbs, G. V. and **Ribbe, P. H.**, 1971. Quadratic elongation: a quantitative measure of distortion in coordination polyhedra. *Science*, **172**, 567–70.

Rosenberg, P. E., 1967. Subsolidus relations in the system $CaCO_3$–$MgCO_3$–$FeCO_3$ between 350° and 550°C. *Amer. Min.*, **52**, 787–97.

Rosenberg, P. E., 1968. Subsolidus relations on the dolomite join, $CaMg(CO_3)_2$–$CaFe(CO_3)_2$–$CaMn(CO_3)_2$. *Amer. Min.*, **53**, 881–9.

Rosenberg, P. E. and **Foit, F. F.**, 1979. The stability of transition metal dolomites in carbonate systems: a discussion. *Geochim. Cosmochim. Acta*, **43**, 951–5.

Ross, N. L. and **Reeder, R. J.**, 1992. High-pressure structural study of dolomite and ankerite. *Amer. Min.*, **77**, 412–21.

Shibuya, G. and **Harada, S.**, 1976. Mineralogical studies on kutnahorite and other carbonate minerals from the Hoei mine, Oita Prefecture. *J. Min. Soc. Japan*, **13**, 1–25.

Smetannikova, O. G., 1979. [Determination of the composition of carbonates of the dolomite–ferrodolomite series.] *Zap. Vses. Min. Obshch.*, **108**, 244–8 (in Russian with English abstract).

Talantsev, A. S., 1983. Geothermobarometry based on parageneses of dolomite–ankerite and calcite with isomorphously admixed rhodochrosite. *Dokl. Acad. Sci. USSR, Earth Sci. Sect.*, **260**, 171–5.

Tsusue, A., 1967. Magnesian kutnahorite from Ryujima mine, Japan. *Amer. Min.*, **52**, 1751–61.

Walmsley, J. C. and **Lang, A. R.**, 1992. On sub-micrometre inclusions in diamond coat: crystallography and composition of ankerites and related rhombohedral carbonates. *Min. Mag.*, **56**, 533–43.

Warne, S. St J., 1962. A quick field or laboratory staining scheme for the differentiation of the major carbonate minerals. *J. Sed. Petr.*, **32**, 29–38.

Warne, S. St J. and **Dubrawski, J. V.**, 1987. Differential scanning calorimetry of the dolomite–ankerite mineral series in flowing nitrogen. *Thermochim. Acta*, **121**, 39–49.

Wenk, H. R. and **Maurizio, R.**, 1978. Kutnahorite, a rare Mn mineral from Piz Cam (Bergell Alps). *Schweiz. Min. Petr. Mitt.*, **58**, 97–100.

Winter, G. A., Essene, E. J. and **Peacor, D. R.**, 1981. Carbonates and pyroxenoids from the manganese deposit near Bald Knob, North Carolina. *Amer. Min.*, **66**, 278–89.

Žák, L. and **Povondra, P.**, 1981. Kutnahorite from the Chvaletice pyrite and manganese deposit. *Tschermaks Min. Petr. Mitt.*, **28**, 55–63.

Aragonite and Aragonite-type Carbonates

Aragonite $CaCO_3$
Strontianite $SrCO_3$
Witherite $BaCO_3$
Cerussite $PbCO_3$

Aragonite $CaCO_3$

Orthorhombic (−)

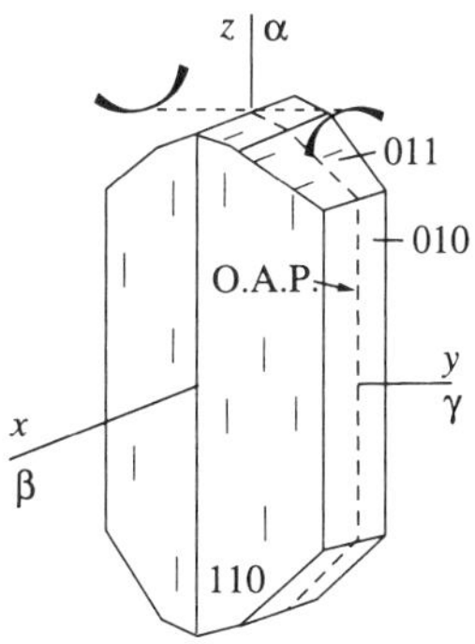

α	1.531
β	1.681
γ	1.685
δ	0.154
$2V_\alpha$	18–18°30′
Orientation:	
$\alpha=z$, $\beta=x$, $\gamma=y$, O.A.P. (100)	
Dispersion:	$r<v$, weak.
D	2.95
H	$3\frac{1}{2}$–4
Cleavage:	{010} imperfect, {110} poor and not usually seen.
Twinning:	Common cyclic, twin plane {110}, producing pseudohexagonal columns that show striations parallel to *z*.
Colour:	Typically colourless or white; colourless in thin section or grain mount.
Unit cell:	*a* 4.960, *b* 7.964, *c* 5.738 Å Z=4. Space group *Pmcn*

Easily soluble with effervescence in cold dilute HCl. When immersed in hot $Co(NO_3)_2$ solution, aragonite chips or grains become lilac-coloured.

Aragonite is one of the three naturally occurring polymorphic forms of $CaCO_3$; the other two are calcite and vaterite. It is found in marine sediments, and the shells of some marine organisms are made of aragonite. Aragonite occurs in amygdales in basalt and andesite and in the oxidized zones of some ore deposits. It is a characteristic mineral of high-pressure, low-temperature metamorphism.

The name is from the locality in Aragon Province, Spain.

Structure

The structure of aragonite was determined by Bragg (1924) and Wyckoff (1925), and refinements were made by Dickens and Bowen (1971), De Villiers (1971) and Dal Negro and Ungaretti (1971).

The structure can be described as layers of calcium atoms arranged with a hexagonal ABAB... sequence parallel to (001) along the *z* axis and separated by CO_3 layers, which are also arranged hexagonally in the *ab* plane, as shown in Fig. 85 (De Villiers, 1971). This is in contrast with the cubic ABCABC... arrangement of the calcium atoms in calcite. The carbon atoms do not stack on top of each other along the *z* axis and are displaced relative to each other in the *y* direction. A projection of layer sequence onto (100) was made by Speer (1983) to illustrate the stacking of both Ca and CO_3 layers in the aragonite structure (Fig. 86). There

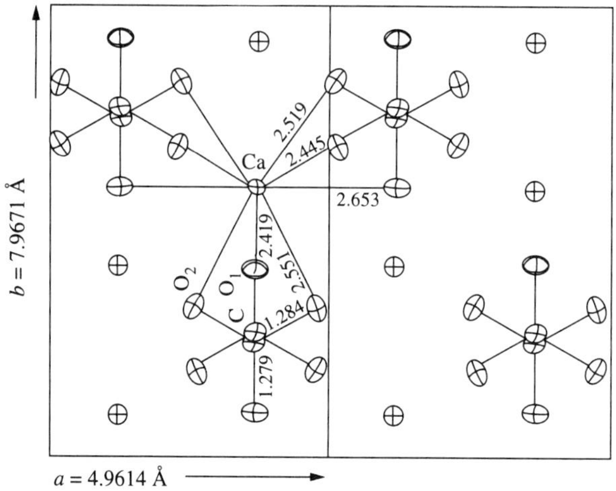

Fig. 85. The aragonite structure viewed along the *z* axis (after De Villiers, 1971).

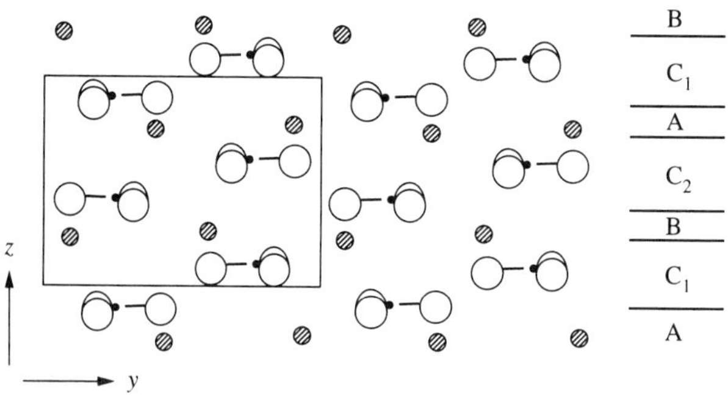

Fig. 86. A projection of the aragonite structure onto (100) showing its layer sequence (after Speer, 1983).

are two distinct CO_3 layers at different heights along the *z* axis, which gives a stacking order of $AC_1BC_2AC_1BC_2$.

The Ca atom lies on the mirror plane at $x=\frac{1}{4}$ and is coordinated by nine oxygens from three CO_3 edges, O(1,2), O(1^{I},2^{I}) and O(2^{IV},2^{V}) and three apices, O(1^{II},2^{II},2^{III}), as shown in Fig. 87 (Speer, 1983). Ca–O bond distances with oxygens of the CO_3 edges range from 2.519 to 2.653 Å, whereas oxygens of the apices range from 2.419 to 2.445 Å (De Villiers, 1971).

The CO_3 group is non-planar, with the carbon atom being 0.026 (Dickens and Bowen, 1971) or 0.020 Å (De Villiers, 1971) out of the oxygen plane of the CO_3 group. This oxygen plane is tilted about 2.5° from the (001) plane (De Villiers, 1971). The average C–O distances and O–C–O angles of 1.282 Å and 119.86°, respectively, correlate well with the average distance and angle of 1.28 ± 0.02 Å and $120 \pm 2°$ of 34 different CO_3 group minerals (Zemann, 1981). Crystal data and atomic parameters of the three refinements are tabulated in Table 33.

The crystal structure of aragonite has also been refined by neutron diffraction (Jarosch and Heger, 1985). Jarosch and Heger confirmed earlier measurements

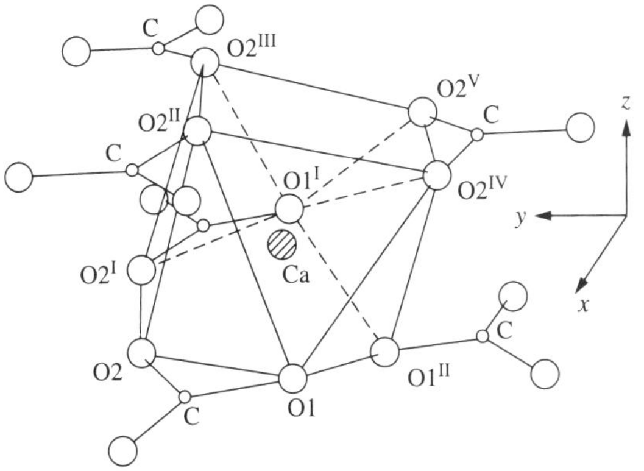

Fig. 87. The CaO_3 group of the aragonite-type structure (after Speer, 1983).

Table 33. Crystallographic Data for Aragonite

	1	2	3
Crystal data			
a (Å)	4.9598	4.9614	4.9616
b (Å)	7.9641	7.9671	7.9705
c (Å)	5.7379	5.7404	5.7394
Interatomic distance (Å)			
Ca–O(1^{II})	2.414	2.419	2.411
Ca–O(2^{II},2^{III})	2.445	2.445	2.448
Ca–O(2,2^{I})	2.520	2.519	2.526
Ca–O(2^{IV},2^{V})	2.544	2.551	2.542
Ca–O(1,1^{I})	2.655	2.653	2.656
C–O(1)	1.288	1.279	1.280
C–O(2)	1.283	1.284	1.287
Bond angle (°)			
O(1)–C–O(2)	120.1	120.2	120.7
O(2)–C–O(2)	119.6	119.6	118.5

1 Dickens and Bowen (1971).
2 De Villiers (1971)
3 Dal Negro and Ungaretti (1971).

and refined the aplanarity of the carbonate group to 0.0280(5) Å. The interatomic distances and bond angles obtained from neutron diffraction are, in agreement with those listed in Table 33, 2.4157, 2.4502, 2.5197, 2.5454 and 2.6538 Å for Ca–O, 1.2847 and 1.2772 Å for C–O, and 120.28 and 119.30° for O–C–O angles.

Aragonite is the high-pressure form of $CaCO_3$. Because of the common occurrence of both polymorphs of $CaCO_3$, the equilibrium relationships between aragonite and calcite have received perhaps the most extensive attention of all mineral pairs. A thorough review of earlier investigations on the calcite–aragonite equilibrium has been made by Boettcher and Wyllie (1968), and a comprehensive evaluation of the relationships based on studies to 1983 has been given by Carlson (1983). The relationships are considered to be established from three contemporary investigations (Boettcher and Wyllie, 1968; Goldsmith and Newton, 1969; Johannes and Puhan, 1971) in the temperature and pressure ranges 400–700 °C and 5–20 kbar. Results are shown in Fig. 88. Both Boettcher and Wyllie (1968) and

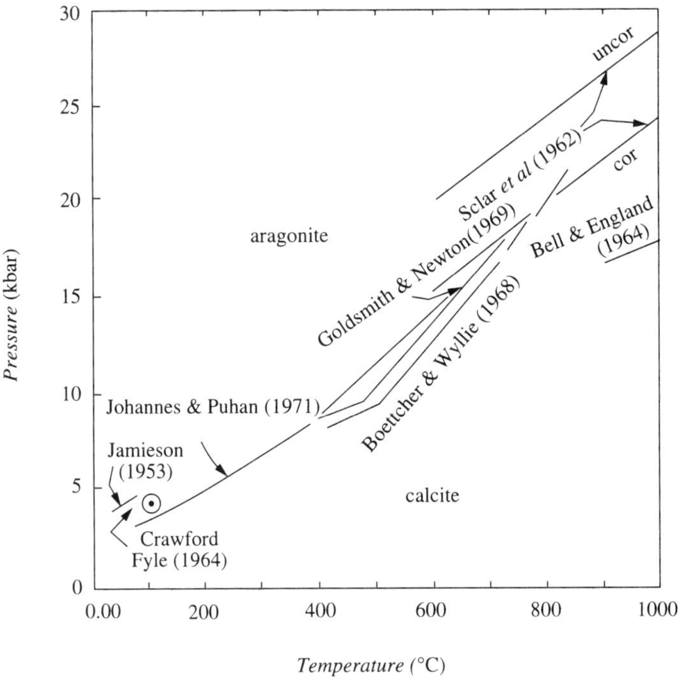

Fig. 88. Aragonite and calcite equilibrium curves (after Johannes and Puhan, 1971).

Goldsmith and Newton (1969) showed the existence of a break in the calcite–aragonite equilibrium curve at around 500 °C and 10 kbar. Johannes and Puhan (1971) obtained a continuous curve between 300 and 500 °C, but above and below this range the curve is more or less a straight line. Carlson (1983) proposed that the change in slope of the aragonite–calcite equilibrium curve is due to anion rotational disordering, the same transition that Chang (1965) observed for the binary Ca–Sr carbonates. From the measurement of phonon spectra, Salje and Viswanathan (1976) calculated phase relationships between aragonite and calcite and obtained a break in their equilibrium curve which correlates with the experimentally determined break. They explained that the break represents the region where the energy attributed to phonon processes reaches the saturation point and where the effect of the anion disorder predominates.

Chemistry

Most aragonites are relatively pure and have compositions close to $CaCO_3$. Strontium and lead are the two common substitutions for Ca, and varieties rich in these elements are known as mossottite and tarnowitzite, respectively. Eight chemical analyses of aragonites selected from schists, sediments and vein fillings are listed in Table 34. Analyses 2 and 5 may represent the maxima of strontium and lead, respectively, recorded in naturally occurring aragonite. An authigenic ferriferous aragonite was obtained from the bottom sediments of the Adriatic Sea (anal. 1) (Favretto, 1966) with *a* 4.933, *b* 7.950, *c* 5.731 Å.

Table 34. Aragonite Analyses

	1	2	3	4	5	6	7	8
CaO	47.22	51.72	52.90	56.10	51.97	49.34	55.26	54.40
SrO	—	4.63	0.98	—	—	0.51	0.76	1.18
BaO	—	—	0.03	0.98	—	—	—	—
MgO	—	0.09	0.19	—	—	0.08	—	—
FeO	2.07	—	0.48	—	—	0.02	—	—
Insol.	1.41	—	3.50	—	5.29 (PbO)	0.04 (NiO)	1.30	1.62
CO_2	49.29	43.13	41.50	44.31(c)	42.24	50.00	42.70	42.70
Total	99.99	99.57	99.58	101.39	99.50	99.99	100.02	99.90
	Numbers of ions on the basis of six (O)							
Ca	1.624	1.889	1.986	1.987	1.937	1.670	1.981	1.992
Sr	—	0.092	0.020	—	—	0.010	0.015	0.023
Ba	—	—	—	0.013	—	—	—	—
Mg	—	0.005	0.010	—	—	0.004	—	—
Fe	0.056	—	0.014	—	—	0.001	—	—
					0.050 (Pb)			
C	2.160	2.007	1.985	2.000	2.006	2.157	1.992	1.992

(c) Calculated from stoichiometry.

1 Authigenic ferriferous aragonite, bottom sediments, Adriatic Sea (Favretto, 1966).
2 Aragonite deposited by hot spring, Cuanza Sul District, Angola (De Figueiredo Gomes, 1972).
3 Aragonite vein fillings in marine Mn nodules, Blake Plateau, off the SE coast of the USA (McFarlin, 1967).
4 Bladed aragonite in association with calcite from Knickerbocker quarry, Chester, Pennsylvania, USA (Crawford and Hoersch, 1972).
5 Plumbian aragonite, Tsumeb Pb–Cu–Zn mine, Namibia (O'Daniel, 1930); α 1.5397, β 1.6950, γ 1.7026, D 3.015.
6 Aragonite from blueschists, Washington, WA, USA (Evans and Misch, 1976).
7, 8 Metamorphic aragonite from glaucophane schists, California, USA (Coleman and Lee, 1962).

Experimental Work

Phase relationships in the system $CaCO_3$–$SrCO_3$ were studied by Froese and Winkler (1966), Chang (1971) and Carlson (1980, 1983). At 25 kbar, aragonite and strontianite form a complete series of solid solution with cell dimensions varying linearly with composition. Carlson has made a comprehensive evaluation of phase relationships in the system and the effect of $SrCO_3$ upon the calcite–aragonite transition. He presented a series of isothermal sections, three of which at 600, 500 and 400 °C are shown in Fig. 89. Isoplethal contours on the Sr-rich limit of the miscibility gap are shown in Fig. 90. As pointed out by Carlson, the amount of $SrCO_3$ has a negligible effect on the pressure of the calcite–aragonite transition in the temperature range of blueschist metamorphism. A projection of the order–disorder transition boundary onto the pressure–temperature plane is also shown in Fig. 90. The intersections with the isoplethal contours mark the change from increasing slopes in the Sr-rich solid solutions to constant slopes in the Sr-poor solid solutions.

Phase relationships between $CaCO_3$ of the aragonite form and $PbCO_3$ are shown in Fig. 91 (Chang and Brice, 1972). The complete series of solid solution dissociates below 600 °C, and the solid solubility of $PbCO_3$ in aragonite is 5 mol. per cent at 450 °C. Lead-bearing aragonite was precipitated from artificial sea-water under normal conditions (Nechiporenko and Gul'ko, 1984). The resulting zoned crystals are made up, from centre to rim, of cerussite, tarnowitzite and pure aragonite.

Strontium and magnesium are commonly found in skeletal aragonites. In

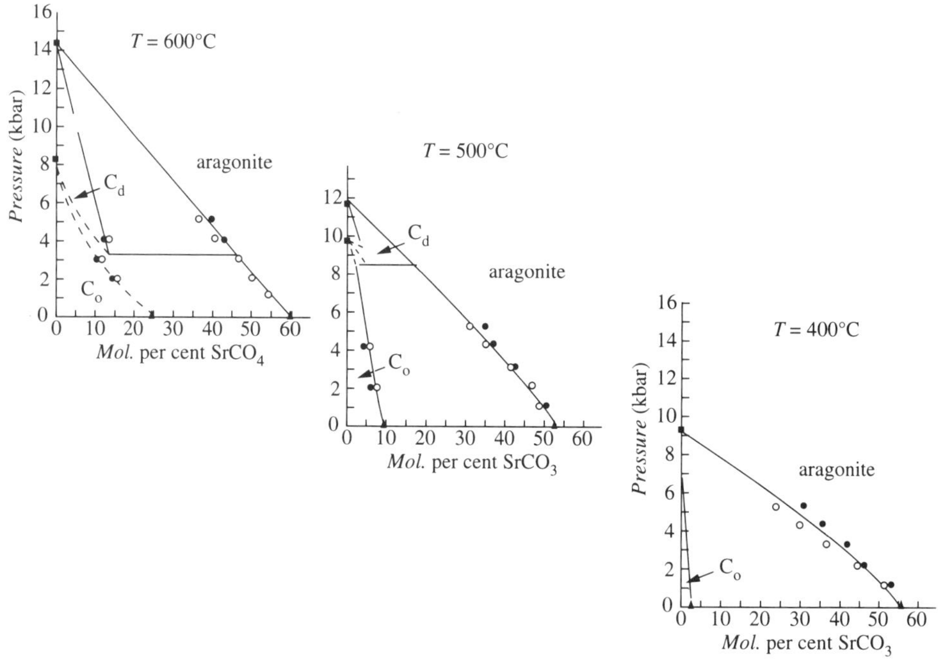

Fig. 89. Isothermal sections in the system $CaCO_3$–$SrCO_3$ at 600, 500 and 400 °C (after Carlson, 1980). Symbols used are: C_o, ordered calcite; C_d, disordered calcite (after Johannes and Puhan, 1971; Goldsmith and Newton, 1969; Chang, 1965; and Carlson, 1980).

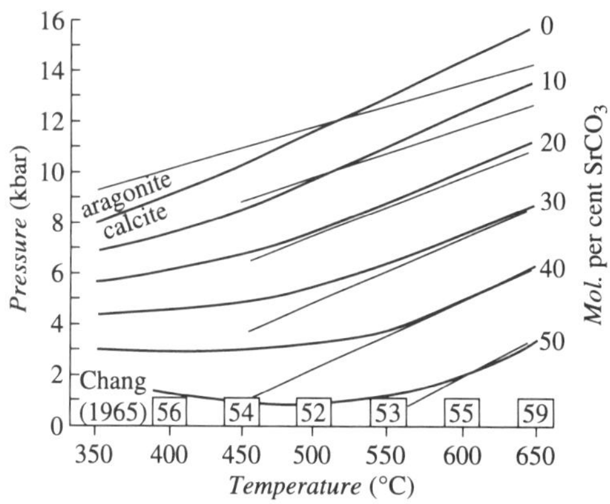

Fig. 90. Isothermal contours of Sr-rich limits of the miscibility gap in the system $CaCO_3$–$SrCO_3$ (after Carlson, 1980). Heavy lines (after Carlson, 1980); light lines (after Froese and Winkler, 1966); composition in rectangles from Chang (1965).

average weight percentage, their concentrations in various marine carbonates are shown in Fig. 92 (Morse and Mackenzie, 1990). By using an average sea-water composition and the range of empirically determined distribution coefficients in inorganic aragonite, Morse and Mackenzie calculated the ranges of concentration of both Sr and Mg in aragonite, which are also shown in Fig. 92. Both correlate well with the analytical data, except in the case of mollusca, which

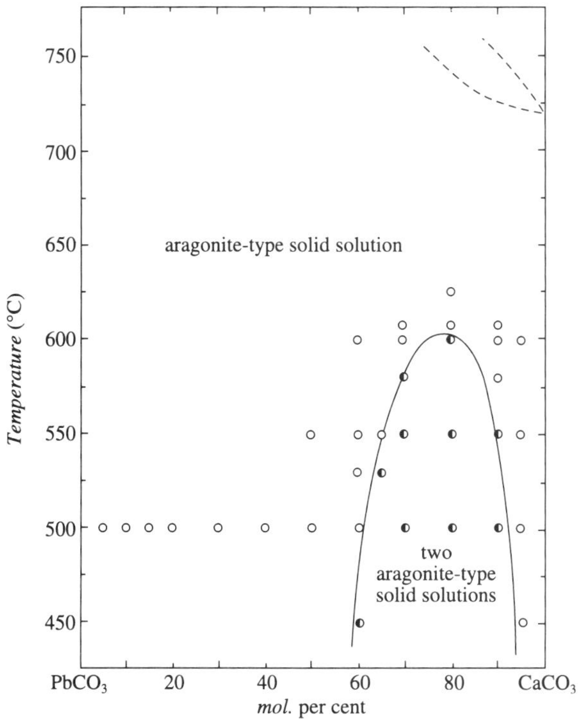

Fig. 91. Subsolidus phase relationships in the system $CaCO_3$–$PbCO_3$ at 15 kbar (after Chang and Brice, 1972).

strongly discriminate against Sr. Smith *et al.* (1979) demonstrated experimentally that the relationship between Sr/Ca ratio and water temperature exists in three genera of aragonitic corals providing a thermometry known as sclerothermometry. The ratio increases with decreasing temperature of sea-water, with a distribution coefficient of $K = 1.30 - 0.0094T$ (T in Kelvin). The coefficient for inorganic aragonite is $K = 1.24 - 0.0045T$ (T in Kelvin).

Direct solubility measurements for aragonite by Plummer and Busenberg (1982) in CO_2–H_2O solutions between 0 and 90 °C indicate the following values of the equilibrium constant K for the reaction $CaCO_{3(s)} = Ca^{2+} + CO_3^{2-}$: $\log K_a = -171.9773 - 0.07793T + 2903.293/T + 71.595 \log T$, where T is in Kelvin. At 25 °C, the range of measured P_{CO_2} is between 0.9528 and 0.9554 atm, that of total calcium is between 10.51 and 10.63 mmol./kg of the solution, that of calculated pH is between 6.078 and 6.074, and that of $\log K_a$ is between -8.319 and -8.343. The solubility of aragonite in sea-water has been measured in a number of studies, but results are not in good agreement (MacIntyre, 1965; Ingle *et al.*, 1973; Berner, 1976; Morse *et al.*, 1980). It has been suggested by Morse *et al.* (1980) that the time of equilibration is a major variable since in their experiments they found that the aragonite solubility decreased with increased equilibration time. The apparent solubility constants of aragonite at 25 °C and 35‰ salinity are $6.65 \pm 0.12 \times 10^{-7}$ (measured) and $8.40 \pm 0.86 \times 10^{-7}$ mol.^{2}/kg^2 (calculated). The difference between measured and calculated solubilities may be caused by the formation of surface layers of different composition and lower solubility than the bulk solid.

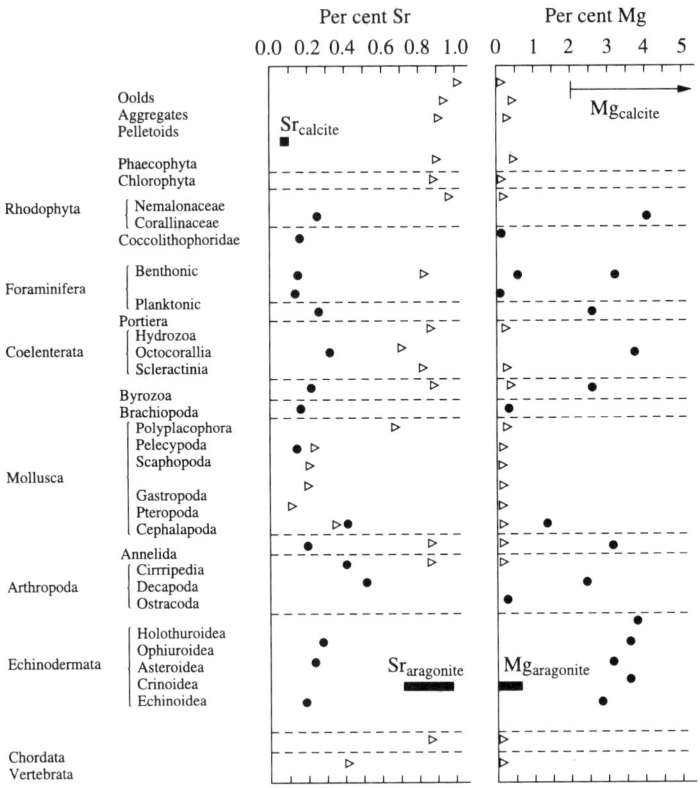

Fig. 92. Average Sr and Mg (in wt. per cent) and range of values of Sr and Mg in various marine aragonites(△) and calcites (●) (after Morse and Mackenzie, 1990).

The influence of pH on the rate of dissolution of aragonite follows the same kinetic mechanism and has similar dissolution rates to calcite, as shown in Fig. 93 (Chou *et al.*, 1989). The rate constants for the three steps involved in the dissolution reaction are:

(a) $MCO_3 + H^+ \rightleftharpoons M^{2+} + HCO_3^-$, $K_1 = 1.2 \times 10^{-4}$ mol./cm²s,

(b) $MCO_3 + H_2CO_3^* \rightleftharpoons M^{2+} + 2HCO_3^-$, $K_2 = 4.0 \times 10^{-8}$ mol./cm²s,

and

(c) $MCO_3 \rightleftharpoons M^{2+} + CO_3^{2-}$, $K_3 = 1.0 \times 10^{-10}$ and $K_{-3(\text{backward})} = 1.7 \times 10^{-2}$ mol./cm²s.

Acker *et al.* (1987) investigated aragonite dissolution in sea-water under low-temperature and high-pressure conditions which simulate the deep-ocean environment. Relationships were estimated for dissolution rate and the depth for the North Pacific water column at 50°N, 145°W. The degree of undersaturation of aragonite was found to be greater at 1000, 1500 and 2000 m depths than at 3000 m. The dissolution rate at 3000 m depth exceeds the rates measured at shallower depths. Ogino *et al.* (1987) examined the formation and transformation of $CaCO_3$ in highly supersaturated solutions of Ca^{2+} and CO_3^{2-}. Amorphous calcium carbonate rapidly precipitates first, and within several minutes transforms

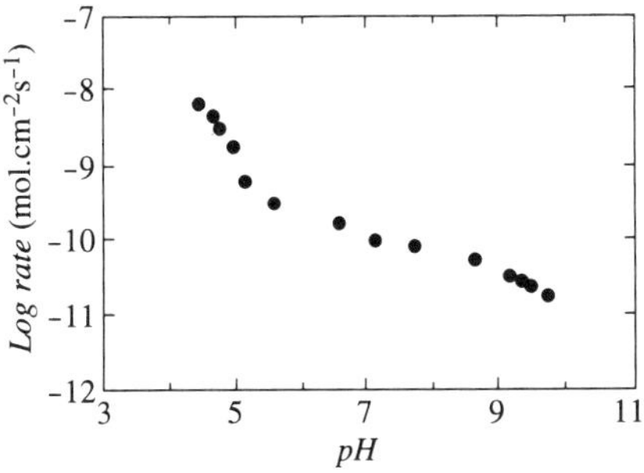

Fig. 93. Dissolution of aragonite (after Chou *et al.*, 1989).

to a mixture of crystalline forms of $CaCO_3$. At temperatures between 60 and 80 °C, the transformed polymorphs are aragonite and calcite; at temperatures between 14 and 30 °C, calcite and vaterite are the products. At intermediate temperatures all three polymorphs are formed. Both aragonite and vaterite are metastable and gradually transform to the stable form, calcite; 1000–1300 minutes are required for the transformation of aragonite to calcite in the temperature range 60–80 °C. The polymorphic transformation of vaterite and aragonite to calcite takes place through dissolution of the metastable phases and growth of the stable phase, calcite. The rate-determining step of the transformation is the growth of calcite.

Aragonite is a metastable phase relative to calcite under Earth surface conditions. The kinetics of the aragonite–calcite transition in solid state were examined by Kunzler and Goodell (1970) using high-temperature X-ray diffraction. Results indicate that chemical impurities appear to have less effect on the transition than the physical structuring of domains of the aragonite crystallites. The rate of transition correlates with the age of aragonite, with the older crystals transforming at a slower rate at a given temperature. Kunzler and Goodell proposed that the low-temperature annealing process coalesces the domains within aragonite crystallites. All reaction curves (X-ray diffraction peak height versus log t (time)) show the sigmoidal character that may be represented by the Avrami equation (Avrami, 1939, 1940). Results of an Avrami analysis indicate that both the shape of the reaction curve and the length of time for a given fraction of a sample to transform at a given temperature depend upon the variety of aragonite. The variables include origin (biochemical and inorganic) geological age, grain size and impurity.

Brar and Schloessin (1979) made a thorough study of the effect of pressure, temperature and grain size on the kinetics of the calcite–aragonite transition. They used both single crystals and polycrystalline samples, and selected conditions of experimentation well within the stability field of aragonite from 14 to 25 kbar and from 300 to 600 °C. Values for the rate constant (k) and the exponent (n) characterizing the transition were computed from the transformed mass fractions ($x(t)$) data using Cahn's (1956) nucleation and growth model for solid state transformation. At 17 kbar, k (s^{-1}) increases from 2.00×10^{-5} at 300 °C to 1.39×10^{-3} at 600 °C, and the apparent activation energy for the transition is 16 kcal/mol. The exponent (n) has values mostly close to unity, indicating that the transition is growth-controlled, as expected from Cahn's models. For single crystals the relationship between $x(t)$ and P is linear for a period of 1 hour, and for a given time x increases exponentially with temperature (T). For polycrystalline samples

$x(t)$ decreases as the square root of grain size ($D^{1/2}$). This last observation was at variance with the model, which predicts a decrease proportional to grain size (D), not to its square root. Brar and Schloessin have suggested that this discrepancy may be related to deviatoric stresses and stress concentrations by the already transformed fractions in the polycrystalline samples.

In aqueous solutions the fraction of aragonite transformed to calcite above 50 °C is proportional to the time squared, indicating that the rate of reaction is dependent on the nucleation and growth rates of calcite, not the dissociation rate of aragonite (Bischoff, 1969). The rate constant decreases with temperature, corresponding with an activation energy of 57.4 kcal/mol. Mg^{2+}, SO_4^{2-} and amino acids can inhibit the transition of aragonite to calcite (Bischoff and Fyfe, 1968; Jackson and Bischoff, 1971). The inhibiting function of Mg^{2+} is interpreted as resulting from difficulties in dehydration (Mucci and Morse, 1984) or crystal poisoning by the absorption of Mg^{2+} at reactive sites (Reddy and Wang, 1980). No major influence of Mg^{2+} on aragonite precipitation kinetics was observed (Berner, 1975). The morphology and trace element composition of calcite that is produced from aragonite is largely controlled by the porosity of the aragonite and the volume of solution involved in the transition (Martin *et al.*, 1986). In the study of calcite produced from the transformation of aragonite in a marine gastropod, the well-preserved microstructure and modal compositions of aragonite in the resulting calcite led Wardlaw *et al.* (1978) to believe that only a thin film (< 100 Å) of water may be necessary for the transformation in a closed system.

A theoretical approach to the kinetics of the calcite⇌aragonite transformation has been presented by Madon and Gillet (1984). Assuming that the transformation is achieved by a dislocation glide mechanism obeying first-order kinetics, they computed the transformation domains of the two polymorphs and discussed the influence on kinetics by the deviatoric stress, the grain size and the (P, T, t) paths. These domains are different from the stability fields defined with respect to Clapeyron's equation. Madon and Gillet (1984) have shown that there could exist a domain where both directions of transformation are conceivable, and it could also be possible to transform calcite or aragonite in its own stability field.

The thermodynamic properties of aragonite have been established. Values of heat capacity and of derived thermodynamic properties at 298.15 K are as follows (Staveley and Linford, 1969): $C_p = 19.675$ cal/kmol.; $S_T^o - S_o^o = 21.03$ cal/kmol.; $(H_T^o - H_o^o)/T = 11.47$ cal/kmol. and $-(G_T^o - G_o^o)/T = 9.56$ cal/kmol. All properties listed were measured in the temperature range 10–300 K, and all their values increased with increasing temperature.

Optical and Physical Properties

Aragonite is biaxial and optically negative, with $2V_\alpha$ 18–18°30′. Basal sections show centred acute bisectrix figures with numerous isochromes. Isogyres are relatively thin at the metatopes but broaden substantially towards the periphery of the figures. Aragonite has an extreme birefringence which, as for calcite and calcite-type carbonates, is structurally dependent. Lo (1973) found that the polarizability of the polyatomic anions, such as the CO_3 groups in crystals, is dependent on the nature of the constituent atoms as well as their arrangement. Isherwood and James (1976) suggested that the birefringence and oxygen bond length are influenced by the electronegativity of the cation, and birefringence decreases with an increase of the C–O bond length. Weber (1988) developed a model to relate birefringence to anisotropic bond polarizabilities.

The refractive indices of aragonite vary very little. The substitution of Sr for Ca lowers the refractive index and optic axial angle and slightly lowers the birefringence, while the substitution of Pb brings about an increase in the values of all these properties. With increasing temperature, refractive indices and 2V of aragonite at a wavelength of 587.6 μm decrease as follows (Speer, 1983):

	α	β	γ	2V
20 °C	1.5306	1.6807	1.6852	18.3°
320 °C	1.5263	1.6731	1.6768	16.8°

The aragonite from the glaucophane schist of the Franciscan Formation has approximately 1 wt. per cent SrO (Table 34, anal. 8), which has no effect on its refractive indices and 2V value (Coleman and Lee, 1962).

The infrared spectra of aragonite and aragonite-type carbonate minerals are shown in Fig. 94 (White, 1974) based upon measurements of Huang and Kerr (1960) and Adler and Kerr (1963). In comparison with the infrared spectra of calcite, the out-of-plane bending mode (ν_2) changes to a lower frequency at 851 cm^{-1}, while the stretching mode (ν_3) shifts to a greater value of 1465 cm^{-1}. The high frequency band (ν_4) remains essentially unchanged except for the splitting. The vibration frequencies corresponding with the three major absorption bands are

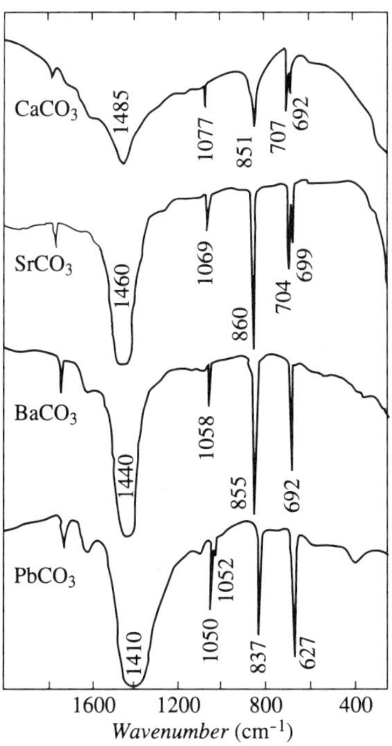

Fig. 94. Infrared spectra of aragonite and aragonite-type carbonate minerals (after White, 1974).

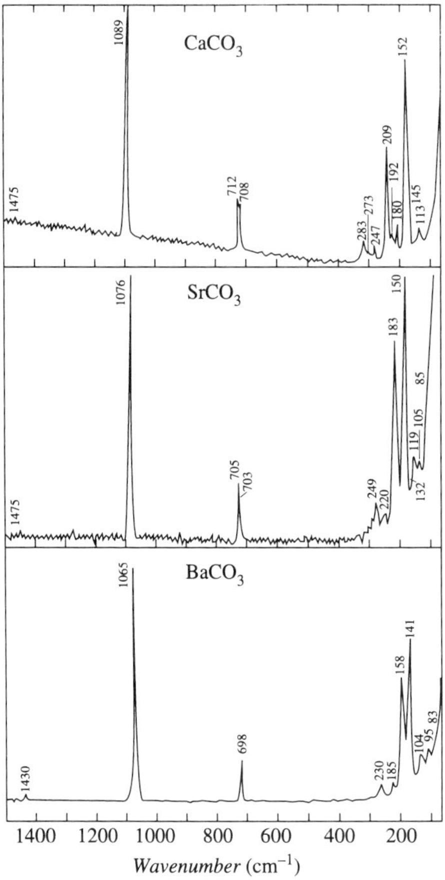

Fig. 95. Raman spectra of aragonite and aragonite-type carbonate minerals (after White, 1974).

primarily related to the vibration energies of the CO_3^{2-} ion, which has a more or less identical environment in the two polymorphic forms of $CaCO_3$.

The Raman spectra of aragonite-type minerals are given in Fig. 95 (White, 1974). The symmetric stretch (ν_1) is extremely intense and the other modes are rather weak; ν_2 is too weak to measure in aragonite and becomes progressively more intense as the cation mass increases.

Usually aragonite is colourless, but sometimes it is pale yellow and has various tints. Luminescence of considerable intensity is exhibited by aragonite (Gies, 1975). With a *REE* activation, aragonite emits three small sharp bands between 470 and 510 nm. Phosphorescing aragonites all have higher concentrations of Zn. Two Pb-rich aragonites (tarnowitzite) were found to exhibit a pale golden-yellow luminescence. Aragonite from Girgenti, Sicily, which gives a rose-red fluorescence, encloses sulphur containing appreciable amounts of selenium (Dériběré, 1942). The presence of various organic compounds has been postulated for the occurrence of fluorescence in aragonite. Thermoluminescence data show the presence of five thermoluminescence peaks at 35, 47, 63, 170 and 230 K (Calderon *et al.*, 1981).

Aragonite occurs in a great variety of forms. It is commonly found in radiating or columnar aggregates, in divergent bundles of fibres, as a needle-like form, in stalactites and in crusts. Crystal morphology is affected not only by structure and composition (Schneer, 1968, 1970), but also by the conditions in which the crystal is grown (Franke *et al.*, 1981). Morphological variations of aragonite formed in an organic environment have been reviewed by Watabe (1981). Simple crystals are much less common than pseudohexagonal columns formed by cyclic twinning on {110}, which is extremely common. Basal sections of cyclic twins show six-fold radiating sectors. Polysynthetic twinning on {110} which produces parallel twin lamellae is also common. Both cyclic and polysynthetic twinnings yield striations parallel to the z axis.

Because the compositional range of aragonite is small, the variation in both hardness and specific gravity is likewise limited. Speer (1983) calculated densities of aragonite and other orthorhombic carbonates and presented them as a function of cation atomic mass. Densities of aragonite–strontianite solid solutions fall along the $CaCO_3$–$SrCO_3$ join.

Distinguishing Features

The features that distinguish aragonite from calcite are its biaxial character, lack of rhombohedral cleavage and slightly higher refractive indices. With a density of 2.95 g/cm^3 in comparison with 2.71 for calcite and 2.86 for dolomite, aragonite sinks in bromoform, while calcite and dolomite float. In metamorphic rocks aragonite, calcite and dolomite commonly display lamellar twinning. Brown *et al.* (1962) suggest that these three carbonate minerals can be readily distinguished microscopically by using the flat stage. Twinning is widespread in calcite and the twin plane is $\{01\bar{1}2\}$, while in dolomite many grains may show no twinning, and in twinned grains lamellae tend to be few. Aragonite is almost universally twinned on {110}.

By using Feigl's solution, aragonite is stained to black whereas calcite and dolomite remain unaffected (Friedman, 1959). An alternative method known as the Meigen's reaction is to boil the carbonates in cobalt nitrate solution for about 20 minutes and to separate from the solution by filtering. Aragonite assumes a lilac-purple colour, whereas calcite and dolomite are not stained.

Aragonite may be mistaken for zeolites in cavities and vesicles in volcanic rocks. Zeolites have low birefringence and do not effervesce in acid. By using a spiking technique, Gunatilaka and Till (1971) proposed an X-ray diffraction method for the quantitative determination of aragonite and magnesian calcite in carbonate rocks.

Paragenesis

Aragonite is common in recent marine carbonate sediments as a product of direct precipitation from sea-water and as the detritus of the hard parts of some organisms. In modern oceans aragonite is formed as needles or ooliths in calcareous muds, such as those occurring in the Great Bahama Bank and Florida Bay, USA. Its origin has been much discussed, with the dispute centring on inorganic precipitation versus algal formation (Bathurst, 1976). Cloud (1962) argued for a direct precipitation of the aragonite needles in the Bahamian lime mud because of the low standing crop of calcareous algae in the area. Studies in south Florida by Stockman *et al.* (1967) and in the Bight of Abaco, Bahamas, by Neumann and

Land (1975) have shown that more than enough aragonite is produced by the disintegration of calcareous codiaccan algae to account for lime mud in the lagoons. Loreau (1982) concludes that muds of the lime of the Great Bahama Bank are precipitated inorganically, whereas those of Florida Bay are largely algal in origin. His evidence is morphological and chemical. Morphologically, the lime muds of the Bahama platform consist of 90 per cent needles, whereas those of Florida Bay have much less. Only 25–40 per cent of crystals in codiaccan algae are needles, the rest are equant grains. Chemically, the Sr/Mg ratio of algae is <2, and that of inorganic aragonite is >4. The great Bahama Bank aragonite mud has the higher Sr/Mg ratio, whereas the Florida Bay mud has the ratio corresponding to the low Sr/Mg. A survey of the fabrics of Phanerozoic oolites made by Sandberg (1983) shows that there is a secular variation in their composition, with aragonite ooids occurring in the late Precambrian to Cambrian, mid-Carboniferous to Triassic, and Tertiary to Recent. This appears to be a direct consequence of the subtle changes in composition of the ocean water. Low levels of partial pressure of CO_2 and a high Mg/Ca ratio promote the precipitation of aragonite and high-Mg-calcite, whereas under reversed conditions low-Mg-calcite is precipitated (Mackenzie and Piggotte, 1981). Aragonite is also found associated with gypsum and halite in many evaporites, such as the Trucial Coast, Arabian Gulf (Butler, 1969), Bonaire Island, Netherlands Antilles (Deffeyes *et al.*, 1965), and Ojodeliebre Lagoon, Baja California, Mexico (Phileger, 1969). Abundant precipitation of aragonite is suggested by the well-preserved dolomite fabrics in the Lower Proterozoic Rockport Formation, northwest Canada (Grotzinger and Reed, 1983). These fabrics consist of silica and dolomite pseudomorphs after botryoidal aragonite associated with tepees, botryoidal sea-floor cement fans and cement crusts in cryptalgal tufas.

Acicular and bladed aragonite and rhombic magnesian calcite occur in gravelly sediments from temperate waters off the Scottish Hebrides (Adams and Schofield, 1983). The carbonates are associated with the wreck of a ship that sank in 1857, and have similar morphology to those found in modern tropical and subtropical environments. It is suggested that sulphate-reducing bacteria both corroded the iron of the ship and increased the alkalinity of the water sufficiently to permit the precipitation of carbonates.

Many organisms with calcareous skeletons build their shells of aragonite. Rucker and Carver (1969) examined the mineralogy of bryozoa and found that in the order of Anasca the encrusting species are calcitic but the free-living species have aragonite in the outer basal lamina. Calcareous algae contain both high-Mg-calcite and aragonite. Observations made by Bathurst (1976) on the Codiacean halimeda in the Bahamas show that in the fresh plant the aragonite is a compact mass of needles. Of the five suborders of foraminiferida, four are calcareous. Only one of them, Rotalina, has the test wall built with both calcite and aragonite. All others are predominantly calcitic (Loeblish and Tappan, 1964a,b). Molluscan shells are built of organized aggregates of microcrystals dispersed in layers. These layers are either all aragonite or interlayered aragonite and calcite (Cloud, 1962). The two carbonate minerals are never mixed, which is a basic characteristic of the molluscan shell (Stenzel, 1963). As it is unstable, the aragonite of fossil shells is gradually altered to calcite, and aragonitic shells older than Mesozoic have not been identified.

The instability of aragonite in marine sediments is illustrated by the absence of aragonite in sediments 2–4 cm below the water–sediment interface at the Rio Grande Rise (Berner and Honjo, 1981). From 0 to 2 cm, the pelagic flux of aragonite accounts for as much as 21 per cent of the total carbonate in surficial sediments.

Bruni and Wenk (1985) analysed the fossils of the patch reefs of the San Cassiano Formation, northern Italy, and found aragonite still preserved in thin shells. Transmission electron microscopical examination reveals that recrystallization occurs primarily along crystal defects. Bruni and Wenk suggest that the kinetics of replacement are controlled mainly by the microstructure of aragonite under slow diagenesis. In pearls, the $CaCO_3$ is normally in the form of aragonite (Alexander, 1940).

Aragonite is deposited as sinter in hot springs and as travertine dripstone in caves. Aragonite vein fillings in marine manganese nodules have been reported from the Blake Plateau off the SE coast of the USA by McFarlin (1967). The morphology of the aragonite, its complete enclosure in Mn nodules and the depth of formation suggest that the aragonite is of inorganic origin. Oxygen isotope analyses suggest that the temperature of formation is 4–6 °C.

The existence of metamorphic aragonite was first reported by McKee (1962) and Coleman and Lee (1962) from glaucophane schist within the Franciscan Formation of northern California, USA. The metamorphic aragonite occurs in veins with pumpellyite, lawsonite and clinozoisite, whereas in the carbonate lenses aragonite is associated with calcite, amphibole, garnet and stilpnomelane. Such aragonite-bearing rocks appear over large areas along the Pacific coast of North America (Ernst, 1965; Blake *et al.*, 1967; Platt, 1975). Aragonite is also widely distributed in extensively deformed rocks in northwest Washington state, USA. It was originally assigned by Vance (1968) to the prehnite–pumpellyite facies, indicating that the aragonite was not formed at the high pressures of the experimentally determined calcite–aragonite transition. The discovery of lawsonite + quartz in these rocks, however, implies pressures in excess of 3 kbar, and the aragonite may have crystallized within the field of its stability (Glassley *et al.*, 1976). In northernmost New Caledonia, aragonite occurs in lawsonite–albite schist (Brothers, 1970). Metamorphic aragonite was also reported from Turkey (Cogulu, 1967) and Italy (Hoffman, 1970).

Aragonite may occur in amygdales in basalt or andesite with zeolites, and in veins in serpentine with dolomite, magnesite or brucite. Bonatti *et al.* (1980) reported aragonite in veins and cavities of serpentinized peridotites recovered from the ocean floor in the Atlantic and Indian Oceans. The aragonite was shown to have crystallized from sea-water, on the basis of oxygen isotope data, and the temperature of formation was found to be 1–5 °C. These aragonites have high Sr and low Mg contents relative to shallow-water marine aragonite.

References

Acker, J. G., Byrne, R. H., Ben-Yaakov, S., Feely, R. A. and **Betzer, P. R.,** 1987. The effect of pressure on aragonite dissolution rates in seawater. *Geochim. Cosmochim. Acta*, **51**, 2171–5.

Adams, A. E. and **Schofield, K.,** 1983. Recent submarine aragonite, magnesian calcite, and hematite cements in a gravel from Islay, Scotland. *J. Sed. Petr.*, **53**, 417–21.

Adler, H. H. and **Kerr, P. F.,** 1963. Infrared absorption frequency trends for anhydrous normal carbonates. *Amer. Min.*, **47**, 124–37.

Alexander, A. E., 1940. An X-ray study of aragonite in natural and cultured pearls. *Amer. J. Sci.*, **238**, 366–71.

Avrami, M., 1939. Kinetics of phase change I: General theory. *J. Chem. Phys.*, **7**, 1103–12.

Avrami, M., 1940. Kinetics of phase change II: Transformation–time relations for random distribution of nuclei. *J. Chem. Phys.*, **8**, 212–24.

Bathurst, R. G. C., 1976. *Carbonate Sediments and Their Diagenesis* (*Developments in Sedimentology*, **12**), Amsterdam, Oxford and New York (Elsevier), 658 pp.

Bell, P. M. and **England, J. L.,** 1964. High-pressure differential thermal analysis of a fast reaction with $CaCO_3$. *Carnegie Inst. Washington Year Book,* **63**, 176–8.

Berner, R. A., 1975. The role of magnesium in the crystal growth of calcite and aragonite from seawater. *Geochim. Cosmochim. Acta,* **39**, 498–504.

Berner, R. A., 1976. The solubility of calcite and aragonite in seawater at atmospheric pressure and 34.5‰ salinity. *Amer. J. Sci.,* **276**, 713–30.

Berner, R. A. and **Honjo, S.,** 1981. Pelagic sedimentation of aragonite: its geochemical significance. *Science,* **211**, 940–2.

Bischoff, J. L., 1969. Temperature controls on aragonite–calcite transformation in aqueous solution. *Amer. Min.,* **54**, 149–55.

Bischoff, J. L. and **Fyfe, W. S.,** 1968. Catalysis, inhibition, and the calcite–aragonite problem. I. The aragonite–calcite transformation. *Amer. J. Sci.,* **266**, 65–79.

Blake, M. C., Irwin, W. P. and **Coleman, P.,** 1967. Upside-down metamorphic zonation, blueschist facies, along a regional thrust in California and Oregon. *U. S. Geol. Surv. Prof. Paper,* **575**, C1–C9.

Boettcher, A. L. and **Wyllie, P. J.,** 1968. The calcite–aragonite transition in the system $CaO–CO_2–H_2O$. *J. Geol.,* **76**, 314–40.

Bonatti, E., Lawrence, J. R., Hamlyn, P. R. and **Berger, D.,** 1980. Aragonite from deep sea ultramafic rocks. *Geochim. Cosmochim. Acta,* **44**, 1207–14.

Bragg, W. L., 1924. The structure of aragonite. *Proc. Royal. Soc. London,* **A105**, 16–39.

Brar, N. S. and **Schloessin, H. H.,** 1979. Effects of pressure, temperature, and grain size on the kinetics of the calcite–aragonite transformation. *Can. J. Earth Sci.,* **16**, 1402–18.

Brothers, R. N., 1970. Lawsonite–albite schists from northernmost New Caldedonia. *Contrib. Min. Petr.,* **25**, 185–202.

Brown, W. H., Fyfe, W. S. and **Turner, F. J.,** 1962. Aragonite in California glaucophane schists, and the kinetics of the aragonite–calcite transformation, *J. Petr.,* **3**, 566–85.

Bruni, S. F. and **Wenk, H.-R.,** 1985. Replacement of aragonite by calcite in sediments from the San Cassiano Formation (Italy). *J. Sed. Petr.,* **55**, 159–70.

Butler, G. P., 1969. Modern evaporite deposition and geochemistry of coexisting brines, the Sabkha, Trucial Coast, Arabian Gulf. *J. Sed. Petr.,* **39**, 70–81.

Cahn, J. W., 1956. The kinetics of grain boundary nucleation reactions. *Acta Metal.,* **4**, 449–59.

Calderon, T., DeLaHera, M. F., Aquilar, M. and **Loy-Yll, R.,** 1981. Luminiscencia en el aragonito de minglanilla (Cuenca). *Soc. Española Min.* (Vol. Extraordinario), **2**, 53–65.

Carlson, W. D., 1980. The calcite–aragonite equilibrium: effect of Sr substitution and anion orientational disorder. *Amer. Min.,* **65**, 1252–62.

Carlson, W. D., 1983. The polymorphism of $CaCO_3$ and the aragonite–calcite transformation. In *Carbonates: Mineralogy and Chemistry:* Ed. R. J. Reeder, *Reviews in Mineralogy,* **11**, 191–225, Mineralogical Society of America, Washington, DC.

Chang, L. L. Y., 1965. Subsolidus phase relations in the systems $BaCO_3$–$SrCO_3$, $SrCO_3$–$CaCO_3$, and $BaCO_3$–$CaCO_3$. *J. Geol.,* **73**, 346–68.

Chang, L. L. Y., 1971. Subsolidus phase relations in the aragonite-type carbonates: I. The system $CaCO_3$–$SrCO_3$–$BaCO_3$. *Amer. Min.,* **56**, 1660–73.

Chang, L. L. Y. and **Brice, W. R.,** 1972. Subsolidus phase relations in the aragonite-type carbonates: II. The systems $CaCO_3$–$SrCO_3$–$PbCO_3$ and $CaCO_3$–$BaCO_3$–$PbCO_3$. *Amer. Min.,* **57**, 155–68.

Chou, L., Garrels, R. M. and **Wollast, R.,** 1989. Comparative study of the kinetics and mechanisms of dissolution of carbonate minerals. *Chem. Geol.,* **78**, 269–82.

Cloud, P. E., 1962. Environment of calcium carbonate deposition west of Andros Island, Bahamas. *U. S. Geol. Surv. Prof. Paper,* **350**, 138 pp.

Cogulu, E., 1967. Étude petrographique de la region de Mihaliccik (Turquie). *Schweiz. Min. Petr. Mitt.,* **47**, 683–825.

Coleman, R. G. and **Lee, D. E.,** 1962. Metamorphic aragonite in the glaucophane schists of Cazadero, California. *Amer. J. Sci.,* **260**, 577–95.

Crawford, W. A. and **Fyfe, W. S.,** 1964. Calcite–aragonite equilibrium at 100°C. *Science,* **144**, 1569–70.

Crawford, W. A. and **Hoersch, A. L.,** 1972. Calcite–aragonite equilibrium from 50°C to 150°C. *Amer. Min.,* **57**, 995–8.

Dal Negro, A. and **Ungaretti, L.,** 1971. Refinement of the crystal structure of aragonite. *Amer. Min.,* **56**, 768–72.

De Figueiredo Gomes, C., 1972. Strontian aragonite deposited by hot springs in the Cuanza Sul District (Angola). *Proc. 24th Int. Geol. Congr., Montreal,* **14**, 210–4.

De Villiers, J. P. R., 1971. Crystal structures of aragonite, strontianite, and witherite. *Amer. Min.,* **56**, 758–67.

Deffeyes, K. S., Lucia, F. J. and **Weyl, P. K.,** 1965. Dolomitization of recent and pliopleistocene sediments by marine evaporite waters on Bonaire, Netherlands Antilles. In *Domomitization and Limestone Diagenesis. Soc. Explor. Paleon. Min. Spec. Publ.,* **13**, 71–88.

Déribéré, M., 1942. Les minéraux luminescents: l'aragonite Girgenti. *Ann. Soc. géol. Belgique*, **65**, 272–4.

Dickens, B. and **Bowen, J. S.**, 1971. Refinement of the crystal structure of the aragonite phase of $CaCO_3$. *J. Res. Natl Bur. Stands., Phys. Chem.*, **75A**, 27–32.

Ernst, W. G., 1965. Mineral parageneses in Franciscan metamorphic rocks, Panoche Pass, California. *Geol. Soc. Amer. Bull.*, **54**, 152–222.

Evans, B. W. and **Misch, P.**, 1976. A quartz–aragonite–talc schist from the lower Skagit Valley, Washington. *Amer. Min.*, **61**, 1005–8.

Favretto, L., 1966. Authigenic ferriferous aragonite from bottom sediments of the Adriatic Sea. *Min. Mag.*, **35**, 781–3.

Franke, W., **Lenk, K.**, **Ittyachen, M. A.** and **Mohanan Pillai, K.**, 1981. The morphology of cerussite $PbCO_3$ grown in silica gel and on hydrothermal conditions. *J. Cryst. Growth*, **51**, 309–13.

Friedman, G. M., 1959. Identification of carbonate minerals by staining methods. *J. Sed. Petr.*, **29**, 87–97.

Froese, E. and **Winkler, H. G. F.**, 1966. The system $CaCO_3$–$SrCO_3$ at high pressures and 500° to 700°C. *Can. Min.*, **8**, 551–66.

Gies, H., 1975. Activation possibilities and geochemical correlations of photoluminescing carbonates, particularly calcite. *Min. Deposita*, **10**, 216–27.

Glassley, W. E., **Whetten, J. T.**, **Cowan, D.** and **Vance, J. A.**, 1976. Significance of coexisting lawsonite, prehnite, and aragonite in the San Juan Island, Washington. *Geology*, **4**, 301–2.

Goldsmith, J. R. and **Newton, R. C.**, 1969. *P–T–X* relations in the system $CaCO_3$–$MgCO_3$ at high temperatures and pressures. *Amer. J. Sci.*, **267A**, 160–90.

Grotzinger, J. P. and **Reed, J. F.**, 1983. Evidence for primary aragonite precipitation, Lower Proterozoic (1.9 Ga) Rockport dolomite, Wopmay orogen, northwest Canada. *Geology*, **11**, 710–3.

Gunatilaka, H. A. and **Till, R.**, 1971. A precise and accurate method for the quantitative determination of carbonate minerals by X-ray diffraction using a spiking technique. *Min. Mag.*, **38**, 481–7.

Hoffman, C., 1970. Die Glaukophangesteine, ihre stofflichen Aquivalente und Umwandlungsprodukte in Nordcalabrien (Suditalien). *Contrib. Min. Petr.*, **27**, 283–320.

Huang, C. K. and **Kerr, P. K.**, 1960. Infrared study of the carbonate minerals. *Amer. Min.*, **45**, 311–24.

Ingle, S. E., **Culberson, C. H.**, **Hawley, J. E.** and **Pytkowicz, R. M.**, 1973. The solubility of calcite in seawater at atmospheric pressure and 35‰ salinity. *Marine Chem.*, **1**, 295–307.

Isherwood, B. J. and **James, J. A.**, 1976. Structural dependence of the optical birefringence of crystals with calcite- and aragonite-type structures. *Acta Cryst.*, **A32**, 340–1.

Jackson, T. A. and **Bischoff, J. L.**, 1971. The influence of amino acids on the kinetics of the recrystallization of aragonite to calcite. *J. Geol.*, **79**, 493–7.

Jamieson, J. C., 1953. Phase equilibrium in the system calcite–aragonite. *J. Chem. Phys.*, **21**, 1385–90.

Jarosch, D. and **Heger, G.**, 1985. Neutron diffraction refinement of the crystal structure of aragonite. *Tschermaks Min. Petr. Mitt.*, **35**, 127–31.

Johannes, W. and **Puhan, D.**, 1971. The calcite–aragonite transition, reinvestigated. *Contrib. Min. Petr.*, **31**, 28–38.

Kunzler, R. H. and **Goodell, H. G.**, 1970. The aragonite–calcite transformation: a problem in the kinetics of a solid–solid reaction. *Amer. J. Sci.*, **269**, 360–91.

Lo, B. W. N., 1973. The polarizability of planar nitrate, carbonate and borate anions in crystals. *J. Phys. Chem. Solids*, **34**, 513–20.

Loeblish, A. R. Jr. and **Tappan, H.**, 1964a. Foraminiferida, morphology and biology. In *Treatise on Invertebrate Paleontology, C. Protista 2(1):* Ed. R. C. Moore, Boulder, Colorado (Geol. Soc. Amer.), pp. 58–134.

Loeblish, A. R. Jr. and **Tappan, H.**, 1964b. Foraminiferida, classification. In *Treatise on Invertebrate Paleontology, C. Protista 2(1):* Ed. R. C. Moore, Boulder, Colorado (Geol. Soc. Amer.), pp. 140–63.

Loreau, J. P., 1982. Sediments aragonitiques et leur genése. *Mem. Museum d'Histoire Naturelle*, Serie **C47**, 312 pp.

McFarlin, P. F., 1967. Aragonite vein fillings in marine manganese nodules. *J. Sed. Petr.*, **37**, 68–72.

MacIntyre, W. G., 1965. The temperature variation of the solubility product of calcium carbonate in sea water. *Fish. Res. Board (Can.), Rept. Ser.*, **200**, 153 pp.

McKee, B., 1962. Aragonite in the Franciscan rocks of the Pacheco Pass area, California. *Amer. Min.*, **47**, 379–87.

Mackenzie, F. T. and **Piggotte, J. D.**, 1981. Tectonic controls of Phanerozoic sedimentary rock cycling. *J. Geol. Soc.*, **138**, 183–96.

Madon, M. and **Gillet, Ph.**, 1984. A theoretical approach to the kinetics of calcite⇌aragonite transition: application to laboratory experiments. *Earth Planet. Sci. Lett.*, **67**, 400–14.

Martin, G. D., **Wilkinson, B. H.** and **Lohmann, K. C.**, 1986. The role of skeletal porosity in aragonite neomorphism—Strombus and Montastrea from the Pleistocene Key Largo Limestone, Florida. *J. Sed. Petr.*, **56**, 194–203.

Morse, J. W. and **Mackenzie, F. T.**, 1990. *Geochemistry of Sedimentary Carbonates* (*Developments in Sedimentology*, **48**), Amsterdam, Oxford, New York and Tokyo (Elsevier), 707 pp.

Morse, J. W., **Mucci, A.** and **Millero, F. J.**, 1980. The solubility of calcite and aragonite in seawater of 35‰ salinity at 25°C and atmospheric pressure. *Geochim. Cosmochim. Acta*, **44**, 85–94.

Mucci, A. and **Morse, J. W.**, 1984. The solubility of calcite in seawater solutions at various magnesium concentrations. $I_t = 0.697$ m and at 25°C and one atmosphere total pressure. *Geochim. Cosmochim. Acta*, **48**, 815–22.

Nechiporenko, G. O. and **Gul'ko, N. I.**, 1984. [Synthetic lead-bearing aragonite.] *Min. Zhurnal*, **6**, 68–74 (in Russian with English abstract).

Neumann, A. C. and **Land, L. S.**, 1975. Lime mud deposition and calcareous algae in the Bight of Abaco, Bahamas: a budget. *J. Sed. Petr.*, **45**, 763–86.

O'Daniel, H., 1930. Ein neues Tarnowitzitvorkommen von Tsumeb-Otavi und die Frage der wechselseitigen isomorphen Vertretung von Ca und Pb. *Z. Krist.*, **74**, 333–62.

Ogino, T., **Suzuki, T.** and **Sawada, K.**, 1987. The formation and transformation mechanism of calcium carbonate in water. *Geochim. Cosmochim. Acta*, **51**, 2757–67.

Phileger, F. B., 1969. A modern evaporite deposit in Mexico. *Bull. Amer. Assoc. Petrol. Geol.*, **53**, 824–9.

Platt, J. P., 1975. Metamorphic and deformational processes in the Franciscan complex, California: some insights from the Catalina schist terrane. *Geol. Soc. Amer. Bull.*, **86**, 1337–47.

Plummer, L. N. and **Busenberg, E.**, 1982. The solubilities of calcite, aragonite and vaterite in CO_2–H_2O solutions between 0 and 90°C, and an evaluation of the aqueous model for the system $CaCO_3$–CO_2–H_2O. *Geochim. Cosmochim. Acta*, **46**, 1011–40.

Reddy, M. M. and **Wang, K. K.**, 1980. Crystallization of calcium carbonate in the presence of metal ions. I. Inhibition by magnesium ion at pH 8.8 and 25°C. *J. Cryst. Growth*, **50**, 470–80.

Rucker, J. B. and **Carver, R. E.**, 1969. A survey of the carbonate mineralogy of Cheilostome Bryozoa. *J. Paleont.*, **43**, 791–9.

Salje, E. and **Viswanathan, K.**, 1976. The phase diagram calcite–aragonite as derived from the crystallographic properties. *Contrib. Min. Petr.*, **55**, 55–67.

Sandberg, P. A., 1983. An oscillating trend in phanerozoic non-skeletal carbonate mineralogy. *Nature*, **305**, 19–22.

Schneer, C. J., 1968. Crystal form and crystal structure. *Helvetica Phys. Acta*, **41**, 1151–5.

Schneer, C. J., 1970. Morphological basis for the reticular hypothesis. *Amer. Min.*, **55**, 1466–88.

Sclar, C. B., **Carrison, L. C.** and **Schwartz, C. M.**, 1962. The calcite–aragonite transition and the calibration of a belt-type apparatus between 15 and 35 kb, 700° to 1400°C. *Amer. Soc. Mech. Eng. Paper*, **62-WA-248**, 7 pp.

Smith, S. V., **Buddemeier, R. W.**, **Redalje, R. C.** and **Houck, J. E.**, 1979. Strontium–calcium thermometry in coral skeletons. *Science*, **204**, 404–7.

Speer, J. A., 1983. Crystal chemistry and phase relations of orthorhombic carbonates. In *Carbonates: Mineralogy and Chemistry:* Ed. R. J. Reeder, *Reviews in Mineralogy*, **11**, 145–190, Mineralogical Society of America, Washington, DC.

Staveley, L. A. K. and **Linford, R. G.**, 1969. The heat capacity and entropy of calcite and aragonite, and their interpretation. *J. Chem. Thermodyn.*, **1**, 1–11.

Stenzel, H. B., 1963. Aragonite and calcite as constituents of adult oyster shells. *Science*, **142**, 232–3.

Stockman, K. W., **Ginsburg, R. N.** and **Shinn, E. A.**, 1967. The production of lime mud by algae in south Florida. *J. Sed. Petr.*, **37**, 633–48.

Vance, J. A., 1968. Metamorphic aragonite in the prehnite–pumpellyite facies, northwest Washington. *Amer. J. Sci.*, **266**, 299–315.

Wardlaw, N., **Oldershaw, A.** and **Stout, M.**, 1978. Transformation of aragonite to calcite in a marine gastropod. *Can. J. Earth Sci.*, **15**, 1861–6.

Watabe, N., 1981. Crystal growth of calcium carbonate in the invertebrates. *Progr. Cryst. Growth Charact.*, **4**, 99–147.

Weber, H. J., 1988. Anisotropic bond polarizabilities in birefringent crystals. *Acta Cryst.*, **A44**, 320–6.

White, W. B., 1974. The carbonate minerals. In *The Infrared Spectra of Minerals:* Ed. V. C. Farmer, Min. Soc. Monograph **4**, 227–84.

Wyckoff, R. W. G., 1925. Orthorhombic space group criteria and their application to aragonite. *Amer. J. Sci., 5th Ser.*, **9**, 145–75.

Zemann, J., 1981. Zur Stereochimie der Karbonate. *Fortschr. Min.*, **59**, 95–119.

Strontianite $SrCO_3$

Orthorhombic (−)

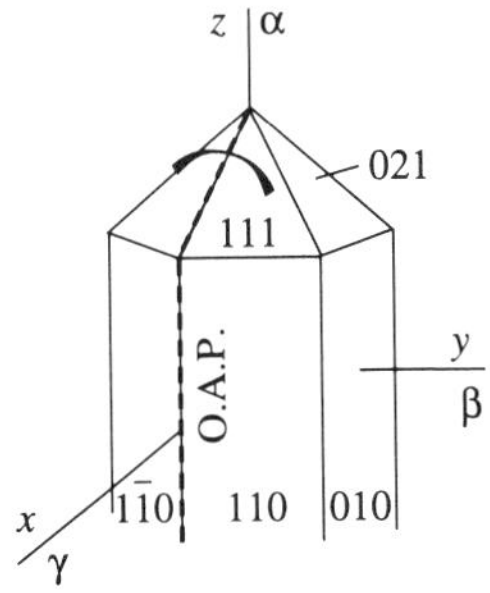

α	1.517
β	1.663
γ	1.667
δ	0.150
$2V_\alpha$	7°
Orientation	
$\alpha = z$, $\beta = y$, $\gamma = x$, O.A.P. (010)	
Dispersion:	$r < v$, very weak.
D	3.72
H	$3\frac{1}{2}$
Cleavage:	{110} distinct, {021} and {010} poor.
Twinning:	Simple, polysynthetic or cyclic, twin plane {110}.
Colour:	Typically colourless or white; colourless in thin section or grain mount.
Unit cell:	*a* 5.106, *b* 8.421, *c* 6.031 Å Z = 4. Space group *Pmcn*

Soluble in dilute HCl. When moistened with HCl, strontianite gives an intense red colour to a flame.

Strontianite is a low-temperature hydrothermal mineral found typically in veins, cavities and concretions in limestone and calcareous clays. It is usually associated with calcite, celestite, baryte or gypsum. It may also occur as a gangue mineral in hydrothermal metalliferous deposits.

The name is from the locality at Strontian, Argyll, Scotland, where the mineral was first found. The element strontium was discovered in this mineral.

Structure

Strontianite is isostructural with aragonite and a member of the group of aragonite-type carbonates (Bragg, 1924). Structural refinement was made by De Villiers (1971) using a crystal of composition $(Sr_{0.8882}Ca_{0.0734}Ba_{0.0371}Mg_{0.0003}Pb_{0.0010})CO_3$. The presence of Ca and Ba may have caused a layer disorder which introduced diffuseness in the high-angle region of the Weissenberg photographs. This effect produces a considerable error in measurement of the back-reflection Weissenberg data.

The crystal structure of strontianite can be described, like that of aragonite, as layers of strontium atoms arranged with a hexagonal ABAB.. sequence along the *z* axis and separated by CO_3 groups which are also arranged hexagonally, as shown in Fig. 85 (p. 236). The *M*–O bond lengths of the polyhedron and the C–O bond lengths of the triangle in strontianite may be expressed by replacing the distance values in Fig. 85 clockwise with the following values: 2.634(5), 2.561(6), 2.725(4), 2.666(5) and 2.552(9) Å, and 1.293(6) and 1.269(10) Å, respectively. The carbon atoms do not project on top of each other along the *z* axis and are displaced relative to each other in the *y* direction.

Like aragonite, there are two distinct CO_3 layers in the structure at different heights along the *z* axis. Speer (1983) illustrates the stacking order as $AC_1BC_2AC_1BC_2$. The CO_3 group appears to be non-planar, with the carbon atom being 0.007 Å out of the plane with the oxygen atoms. The average C–O distance of 1.285 Å and the C–O–C angle of 119.86° correlate well with Zemann's data for a group of 34 carbonate minerals (Zemann, 1981). Pannhorst and Löhn (1970) gave an average value of 1.295 Å for the C–O distance and of 119.75° for the C–O–C angle. From a neutron diffraction analysis, Jarosch and Heger (1988) refined the aplanarity of the carbonate group to 0.0218 Å. A comparison of crystallographic data for strontianite is given below:

	1	**2**	**3**	**4**
a (Å)	5.090(2)	5.093(2)	5.1059(7)	5.1039
b (Å)	8.358(2)	8.376(3)	8.4207(3)	8.4022
c (Å)	5.997(4)	6.020(2)	6.0313(11)	6.0220

1 De Villiers (1971).
2 Pannhorst and Löhn (1970).
3 Speer and Hensley-Dunn (1976).
4 Jarosch and Heger (1988).

$SrCO_3$ is polymorphic. Boeke (1912) first observed an orthorhombic to rhombohedral phase transition, and the transition temperature was later established by Lander (1949) and Chang (1965) under 1 atm CO_2 pressure to be 912 °C. Lander argued, by analogy with nitrate structures, that the rhombohedral $SrCO_3$, as well as the high-temperature form of $CaCO_3$, are disordered, the most likely situation being one in which the number of equivalent orientations obtained by the rotation of a CO_3 group in its plane changes from three to six. Strømme (1975) made a comparison of configurational entropies calculated on the basis of structural models of the disordered phase with those observed in the corresponding phase transition. A large discrepancy between the phases was found. This led Strømme to suggest that the rhombohedral $SrCO_3$ is probably more complex in structure than a disordered phase. Cell dimensions of this high-temperature $SrCO_3$ are *a* 5.092, *c* 9.53 Å at 920 °C (Lander, 1949) and *a* 5.106, *c* 9.516 Å at 965 °C (Moreau, 1964). By further heating of the rhombohedral form of $SrCO_3$ to higher temperatures, Baker (1962) detected another endothermic transition at 1416 °C under 20 atm CO_2 pressure. By analogy with the rhombohedral–cubic transition in $BaCO_3$, Baker suggests that this endothermic transition is to the cubic modification of $SrCO_3$. At 800–1200 °C and 10–40 kbar, Rapoport and Pistorius (1967) examined the orthorhombic– rhombohedral phase transition in $SrCO_3$ and established a transition temperature $T\ (°C) = 930 + 9.10P + 0.0081P^2 - 0.00135P^3$ with a standard deviation of 1.7 °C. Using the latent heat of transition of Lander (1949) (4.70 kcal/mol.) and the initial slope of 9.0 ± 0.6 °C/kbar, Rapoport and Pistorius (1967) found the volume change corresponding with the transition at 1 atm to be $\Delta V = 1.49 \pm 0.10$ cm^3/mol., which correlates well with Moreau's value of $\Delta V = 1.53 \pm 0.38$ cm^3/mol.

$\Delta V = 1.49 \pm 0.10$ cm^3/mol., which correlates well with Moreau's value of $\Delta V = 1.53 \pm 0.38$ cm^3/mol.

Chemistry

Calcium is generally present in strontianite as a substitute for strontium, with a maximum occupation close to 10 wt per cent. (Table 35, anals. 1 and 7). Strontianite from its type locality (anal. 5) has 4.05 wt per cent. CaO. Speer and Hensley-Dunn (1976) analysed strontianites from 12 localities and gave a range of compositions from $(Sr_{0.968}Ca_{0.029})CO_3$ to $(Sr_{0.753}Ca_{0.244})CO_3$, with Ba and Pb present at the p.p.m. level. Cell dimensions vary linearly with composition, and the best measurement of $SrCO_3$ content (in mol.) is given by d_{132} (in Å) according to the equation $SrCO_3(\pm 0.95) = -2116.24 + 1162.84(d_{132})$. The range of strontianite compositions found by Speer and Hensley-Dunn (1976) falls within the compositional limits established experimentally at 96 °C by Holland *et al.* (1963). Phase relationships in the system $CaCO_3$–$SrCO_3$ in the stability field of calcite are shown in Fig. 38, p. 113 (Chang, 1965), illustrating an extensive range of solid solution of strontianite above 400 °C. Isothermal sections at various pressures in the system $CaCO_3$–$SrCO_3$ are shown in Fig. 89, p. 240 (Carlson, 1980).

Both barium and lead may also substitute for strontium in strontianite to a much lesser extent, as shown in Table 35, although a complete series of solid solution was found in each of the systems $SrCO_3$–$BaCO_3$ (Chang, 1965) and

Table 35. Strontianite Analyses

	1	2	3	4	5	6	7
SrO	63.73	66.91	56.31	61.10	65.01	56.41	52.23
CaO	4.20	2.59	11.24	7.02	4.05	6.27	8.68
BaO	0.26	0.28	0.24	0.29	—	1.47	7.04
PbO	0.02	—	—	0.01	—	—	—
					0.06 (SiO_2)	0.48 (MgO) 0.04 (MnO) 0.27 (FeO) 0.37 (oth.)	0.04 (MnO)
CO_2	30.44	30.49	32.77(c)	31.53	30.84(c)	34.79	31.02(c)
Total	98.65	100.27	100.56	99.95	99.96	100.10	99.01
	Numbers of ions on the basis of six (O)						
Sr	1.752	1.863	1.459	1.646	1.806	1.598	1.430
Ca	0.213	0.133	0.538	0.349	0.208	0.327	0.439
Ba	0.005	0.005	0.004	0.005	—	0.028	0.128
Pb	—	—	—	—	—	—	—
						0.034 (Mg) 0.002 (Mn) 0.011(Fe)	
C	2.015	1.999	2.000	2.000	1.993	1.977	2.000

(c) Calculated from stoichiometry.

1 Sedimentary strontianite from Münster district, Westphalia, Germany (Harder, 1964).
2–4 Strontianite and calciostrontianite from carbonate rocks, Pennsylvania, USA (Speer, 1976).
5 Strontianite from Strontian, Argyll, Scotland (Deer *et al.*, 1962).
6 Strontianite from carbonatite, Chipman Lake, Ontario, Canada (Platt and Woolley, 1990).
7 Ca–Ba–Sr carbonate from metamorphosed Mn deposits, Balaghat district, India (Dasgupta *et al.*, 1990).

$SrCO_3$–$PbCO_3$ (Chang and Brice, 1972). A miscibility gap may exist in the systems at temperatures <400 °C, especially between $SrCO_3$ and $BaCO_3$ because of their larger difference in cationic size. Both Chang and Brice (1972) and Speer and Hensley-Dunn (1983) have stated that the lack of more extensive substitution of Ba or Pb in strontianite is not the result of an unavailibility of cations in geological environments. Strontianites are found in association with baryte, witherite and galena. $SrCO_3$ forms a binary compound with $MgCO_3$ (Froese, 1967), $MgSr(CO_3)_2$, which has the norsethite structure with *a* 4.905, *c* 16.44 Å.

The solubility of $SrCO_3$ in water free of atmospheric CO_2 has been determined to be 5.340 ± 0.014 mg/l at 5 °C, 6.940 ± 0.018 mg/l at 15 °C, and 8.950 ± 0.029 mg/l

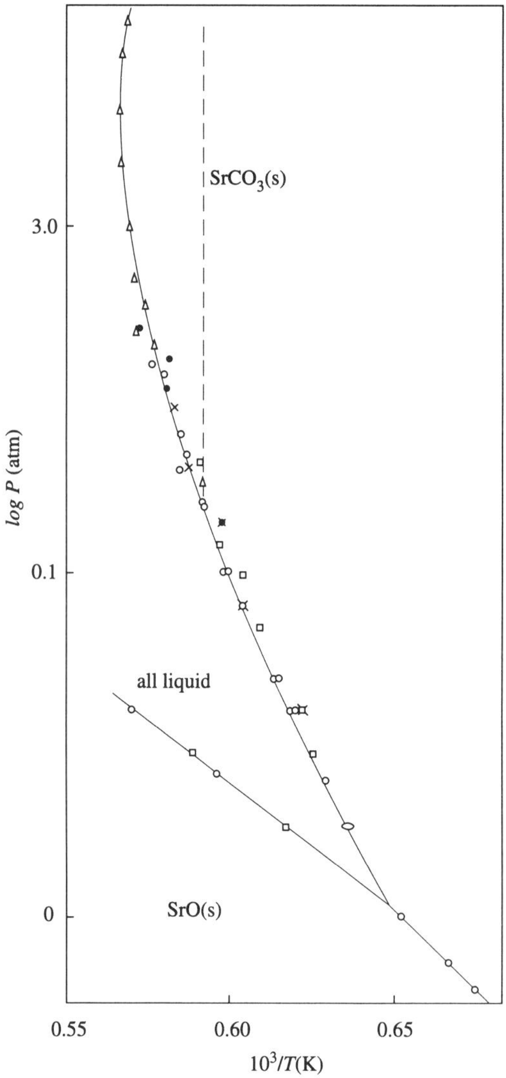

Fig. 96. Melting relations of $SrCO_3$ (after Baker, 1963). Results were obtained by DTA (●), conductance change (×), TGA (○), capillary rise (△) and visual assessment (□).

at 25 °C (Miles and Burton, 1972). For water in equilibrium with the atmosphere at 25 °C, the solubility is 42.7 ± 0.049 mg/l. The solubility of strontianite in CO_2–H_2O solutions was investigated by Busenberg *et al.* (1984). In the temperature range between 2 and 91 °C, the equilibrium constant for the reaction $SrCO_{3(cr)} = Sr^{2+}_{(aq)} + CO^{2-}_{3(aq)}$ is $\log K = 155.0305 - (7239.594/T) - (56.58638 \times \log T)$, where T is in Kelvin. At 25 °C, the log K of strontianite, the Gibbs energy, enthalpy and entropy of the reaction are −9.271 ± 0.020, 52.919 ± 0.080 kJ/mol., −1.67 ± 1.30 kJ/mol. and −183.1 ± 4.0 J/mol. K, respectively.

In an application of the stoichiometric dissolution solubility method to synthetic solids in the strontianite–aragonite solid solution series, Plummer and Busenberg (1987) observed that (a) the substitution of small amounts of Sr in aragonite and Ca in strontianite initially increases the stability of the solid. Aragonite with 0.58 and 3.1 mol. per cent $SrCO_3$ and strontianite with 12.5 and 17.2 mol. per cent $CaCO_3$ are the most stable at 25 and 76 °C, respectively, (b) the substitution of Ca in strontianite and Sr in aragonite can be approximated by Raoult's law for the solvent and Henry's law for the solute, (c) a miscibility gap exists between 0.0058 and 0.875 at 25 °C and 0.03 and 0.828 at 76 °C in terms of mol. fraction of $SrCO_3$, and (d) the equilibrium occurrence of strontianite containing <0.1 mol. per cent $CaCO_3$ would require a Sr/Ca ratio in natural water in excess of 16. Consequently, Ca-free strontianite has not been found in nature.

The dissociation and melting of $SrCO_3$ at high temperatures are shown in Fig. 96. The triple point is defined by 1270 °C and 1.2 atm P_{CO_2}. Dashed lines in the diagram are Baker's proposed rhombohedral–cubic transition in $SrCO_3$ (Baker, 1963). In differential thermal analysis (DTA), $SrCO_3$ shows a single sharp endothermal peak at 930 °C corresponding with the orthorhombic–rhombohedral transition (Morgan and Milodowski, 1982). With $CaCO_3$ in solid solution, two distinct endothermal peaks were obtained, and the transition temperature decreases with increasing $CaCO_3$ content, as shown in Fig. 97. DTA of $SrCO_3$–$BaCO_3$ solid solutions was carried out by Judd and Popp (1973). From $SrCO_3$ to $BaCO_3$ at intervals of 10 mol. per cent, the transition temperatures are 930, 872, 808, 788, 780, 776, 779, 786, 797 and 810 °C. Judd and Popp (1973) recommend that this

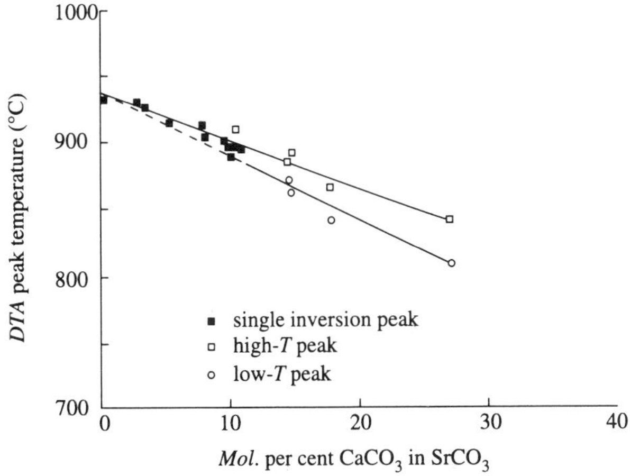

Fig. 97. Plot of DTA peak temperatures as a function of $CaCO_3$ content in (Sr,Ca)CO_3 solid solution (after Morgan and Milodowski, 1982).

series of carbonates may be used as standards for DTA in the temperature range 780–930 °C.

Well-formed single crystals of strontianite have been prepared by Blount (1974) at 80 °C from dilute aqueous solutions containing strontium chloride and sodium bicarbonate. As an industrial mineral, strontianite is used in pyrotechnics and in the manufacture of cathode-ray-screen glass, ferritic magnetic materials and electrolytic zinc.

Optical and Physical Properties

Strontianite is biaxial and optically negative, and has a $2V_\alpha$ of 7° when it is pure and 10° for calciostrontianite. Maximum birefringence, seen on an {010} section, is extreme, and is related to structure, as discussed for aragonite. High-order white interference colours are found for most orientations in both grain mount and thin section. Basal sections show centred acute bisectrix figures with numerous isochromes. Refractive indices increase with increasing Ca content, and variations are approximately linear with composition from strontianite to aragonite (Speer and Hensley-Dunn, 1976).

Strontianite is usually white, but it often has a very pale greyish, yellowish or greenish colour. Luminescence of considerable intensity is exhibited by strontianite (Gies, 1975). The main maximum of the dominating ultraviolet band in strontianite is from 380 to 420 nm. The luminescence colour is whitish with a pale rose or greenish tint. Strontianite has the strongest phosphorescence among carbonates.

Infrared and Raman spectra of strontianite are shown in Figs. 94 and 95 (White, 1974). The similarity to aragonite is clearly illustrated. As cation mass increases from Ca^{2+} to Ba^{2+}, the band width of ν_3 becomes narrower and the symmetric stretch, ν_1, becomes increasingly evident.

Strontianite commonly occurs as granular masses or aggregates of columnar,

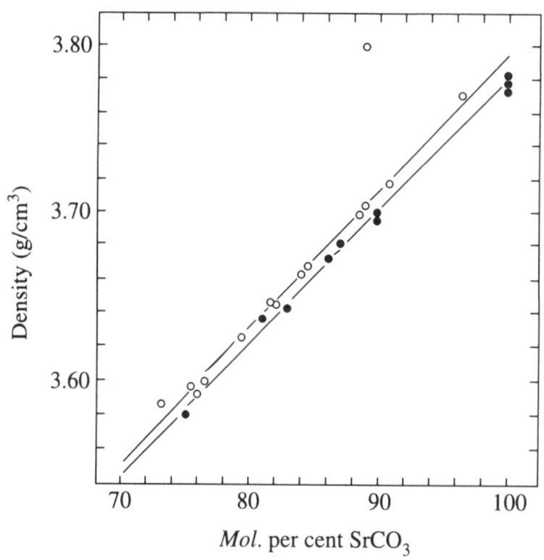

Fig. 98. Variation of calculated density with composition for synthetic strontianite (●) and natural strontianite (○) (after Speer and Hensley-Dunn, 1976).

acicular or fibrous crystals. The acicular and fibrous crystals are elongated along the *z* axis and the columnar crystals are often twinned in sixlings like aragonite.

Densities of natural strontianite with various amounts of calcium were calculated by Speer and Hensley-Dunn (1976). They are shown in Fig. 98 in comparison with calculated densities of synthetic solid solutions in the system $CaCO_3$–$SrCO_3$. The higher values in the natural strontianites are due to the presence of Ba and Pb. The measured densities (*D*) of strontianite and calciostrontianite (composition in wt. per cent) are listed below (Speer and Hensley-Dunn, 1976):

	1	2	3	4	5	6	7
SrO	64.75	63.78	61.52	60.19	60.17	58.33	56.20
CaO	3.36	4.20	6.34	7.13	7.26	8.12	9.85
BaO	0.32	0.26	0.30	0.35	0.31	0.37	0.32
PbO	0.03	0.02	0.04	0.01	0.05	0.03	0.04
D	3.73	3.71	3.62	3.66	3.63	3.59	3.59

Distinguishing Features

Strontianite is similar to the other orthorhombic carbonates, which are distinguished from the rhombohedral carbonates by biaxial character, lack of rhombohedral cleavage, extreme birefringence, parallel extinction and vigorous reaction to weak acids. Strontianite has lower refractive indices than witherite and a higher density than aragonite. Its good prismatic {110} cleavages in two directions which intersect at about 64° are distinctive (Nesse, 1986). The intense red coloration given to a flame is indicative of strontium.

Warne (1962) proposed that etching strontianite with dilute HCl and treating it with alizarin red S produces a purple stain, which separates strontianite, ankerite and cerussite from other carbonates. Treating this group of carbonates again with alizarin red S in NaOH solution differentiates strontianite from the other two: dark purple in ankerite, red-brown in cerussite and no colour in strontianite.

Paragenesis

Strontianite most commonly occurs in veins, vugs and concretions in limestones and calcareous clays. In the Lower Palaeozoic carbonate rocks of the Valley and Ridge Province of Virginia, USA, strontianite with CaO content up to 10 per cent occurs as white to buff acicular crystals and in association with calcite, celestine, fluorite, dolomite, quartz, illite and several other minerals in minor amounts. Textural relationships indicate that it was formed from the alteration of celestine (Mitchell and Pharr, 1961; Speer, 1977; Penick and Haynes, 1983). Similar occurrences of strontianite were found in Pennsylvania (Speer, 1976) and Illinois, USA (Lillie, 1988). Strontianite with a composition of $(Sr_{0.8}Ca_{0.2})CO_3$ was found to encrust blue celestine from the well-known Yate, Avon, England, celestine deposits (Faithfull, 1988). At Durlston Head, Dorset, England, strontianite from the alteration of celestine was in part converted back into celestine (Salter and

West, 1965). The calcite–strontianite veins of Münster, Germany, were suggested to have been formed by the influx of strontium from brine derived from leaching of evaporite beds to the north of Münster (Harder, 1964). Saunders *et al.* (1988) suggest that the formation of strontianite and celestine in the salt dome cap-rocks at Tatum Dome, Mississippi, USA, is the result of replacement of calcite by the action of Sr-enriched fluids introduced from the oilfield. Strontianite as white spherical aggregates 0.5–2.0 mm across was found in the interior of brachiopod shells by Todd (1989). An unusual calcite–strontianite–celestine–magnesian calcite association was found in septarian siderite concretions from late Permian shales and sandstones at Liddell, Hunter Valley, New South Wales, Australia (England, 1988). The minerals are well crystallized, and they appear to be precipitated from externally derived connate waters.

Strontianite also occurs in hydrothermal deposits. It is one of the primary minerals with the widespread baryte and witherite of the northern Pennine orefield, England (Dunham and Wilson, 1985). Other carbonates present are calcite, dolomite, ankerite, siderite and barytocalcite; cerussite, malachite and azurite occur in association as secondary minerals. Both strontianite and calciostrontianite were found in the stibnite deposit at Dúbrava, Slovakia (Chovan and Michálek, 1981). In the paragenetic sequence, both represent late-stage low-temperature products. Strontianite occurs with celestine in native sulphur deposits in the Tarnobrzeg area of southeast Poland (Kowalski *et al.*, 1980). Strontium was supplied from Tertiary gypsum beds and initially crystallized as celestine; strontianite was then formed as a metasomatic alteration product of celestine.

Some carbonatites contain strontianite as a primary mineral constituent of the carbonate assemblage. In the Tajno intrusion, northeast Poland (Dziedzic and Ryka, 1983), three stages of development of the carbonatites are recognized from mineralogical and textural evidence. The first stage crystallized at 340 °C with veins containing burbankite, calcite, albite, pyrite and ilmenite. Strontianite formed in the main stage at temperatures ranging from 320 to 180 °C, along with fluorite, calcite, pyrrhotite and chlorite. The last stage crystallized at temperatures <160 °C. Armbrustmacher (1979) distinguished two types of carbonatite in the Wet Mountains, Colorado, USA: replacement and magmatic. Strontianite occurs in the magmatic carbonatite. Strontianite was also found in the Kangankunde carbonate complex, Malawi, and occurs in potential economic grade and tonnage (Garson and Morgan, 1978). It is a late phase of hydrothermal activity along fracture zones.

The occurrence of strontianite in metamorphic rock was reported by Dasgupta *et al.* (1990) from metamorphosed carbonate-facies rocks of the Precambrian Sausar group; both CaO and BaO (Table 35, anal. 7) co-exist with manganoan calcite, calcite and barytocalcite in metamorphosed manganese deposits. The temperature and pressure under which the deposits formed were 660 °C and 6 kbar.

References

Armbrustmacher, T. J., 1979. Replacement and primary magmatic carbonatites from the Wet Mountains area, Fremont and Custer Counties, Colorado. *Econ. Geol.*, **74**, 888–901.

Baker, E. H., 1962. A high-temperature form of strontium carbonate. *J. Chem. Soc.*, 2525–6.

Baker, E. H., 1963. The strontium oxide–carbon dioxide system in the pressure range 0.3–400 atmospheres. *J. Chem. Soc., Dalton Trans. Inorg. Chem.*, 339–46.

Blount, C. W., 1974. Synthesis of barite, celestite, anglesite, witherite and strontianite from aqueous solutions. *Amer. Min.*, **59**, 1209–19.

Boeke, H. E., 1912. Die Schmelzerscheinun gen und die unkehrbare Umwandlung des Calcium Carbonates. *Neues Jahrb. Min. Geol.*, **1**, 91–121.

Bragg, W. L., 1924. The structure of aragonite. *Proc. R. Soc. London*, **A105**, 16–37.

Busenberg, E., Plummer, L. N. and **Parker, V. B.**, 1984. The solubility of strontianite ($SrCO_3$) in CO_2–H_2O solutions between 2° and 91°C, the association constants of $SrHCO_3^+$(aq) and $SrCO_3^0$(aq) between 5° and 80°C, and an evaluation of the thermodynamic properties of Sr^{2+}(aq) and $SrCO_3$(cr) at 25°C and 1 atm total pressure. *Geochim. Cosmochim. Acta*, **48**, 2021–35.

Carlson, W. D., 1980. The calcite–aragonite equilibrium: effect of Sr substitution and anion orientational disorder. *Amer. Min.*, **65**, 1252–62.

Chang, L. L. Y., 1965. Subsolidus phase relations in the systems $BaCO_3$–$SrCO_3$, $SrCO_3$–$CaCO_3$, and $BaCO_3$–$CaCO_3$. *J. Geol.*, **73**, 346–68.

Chang, L. L. Y., and **Brice, W. R.**, 1972. Subsolidus phase relations in the aragonite-type carbonates. II. The systems $CaCO_3$–$SrCO_3$–$PbCO_3$ and $CaCO_3$–$BaCO_3$–$PbCO_3$. *Amer. Min.*, **57**, 155–68.

Chovan, M. and **Michálek, J.**, 1981. Strontianite from the Dúbrava antimonite deposit (Nizke Tatry Mts., Middle Slovakia). *Min. Slovaca*, **13**, 269–72.

Dasgupta, S., Chakraborti, S., Fukuoka, M., Sengupta, P. and **Roy, S.**, 1990. Ca–Ba–Sr carbonates from metamorphosed manganese deposits of the Sunsar group, India and their petrological significance. *Min. Mag.*, **54**, 511–3.

Deer, W. A., Howie, R. A. and **Zussman, J.**, 1962. *Rock-Forming Minerals*, Vol. **5**, *Non-Silicates*. London (Longman) and New York (John Wiley & Sons), 371 pp.

De Villiers, J. P. R., 1971. Crystal structures of aragonite, strontianite, and witherite. *Amer. Min.*, **56**, 758–67.

Dunham, K. C. and **Wilson, A.A.**, 1985. *Geology of the Northern Pennine Orefield*. Vol. **2**, Economic Memoir, British Geological Survey, 247 pp.

Dziedzic, A. and **Ryka, W.**, 1983. Carbonatites in the Tajno intrusion (NE Poland). *Arch. Min.*, **38**, 4–34.

England, B. M., 1988. A calcite–strontianite–celestite–magnesian calcite association in septarian siderite concretions from late Permian sediments at Liddell, Hunter Valley, New South Wales. *Austral. Min.*, **3**, 19–23.

Faithfull, J., 1988. Strontianite from near Yate (Avon). *J. Russell Soc.*, **2**, 1–3.

Froese, E., 1967. A note on strontium magnesium carbonate. *Can. Min.*, **9**, 65–70.

Garson, M. S. and **Morgan, D. J.**, 1978. Secondary strontianite at Kangankunde carbonatite complex, Malawi. *Trans. Inst. Min. Metall.*, **87B**, 70–3.

Gies, H., 1975. Activation possibilities and geochemical correlations of photoluminescing carbonates, particularly calcite. *Min. Deposita*, **10**, 216–27.

Harder, H., 1964. Geochemische Untersuchungen zur Genese der Strontianitlagerstätten des Münsterlandes. *Contrib. Min. Petr.*, **10**, 198–215.

Holland, H. D., Borcsik, M., Munoz, J. and **Oxburgh, U. M.**, 1963. The coprecipitation of Sr^{2+} with aragonite and of Ca^{2+} with strontianite between 90° and 100°C. *Geochim. Cosmochim. Acta*, **27**, 957–77.

Jarosch, D. and **Heger, G.**, 1988. Neutron diffraction investigation of strontianite $SrCO_3$. *Bull. Min.*, **111**, 139–42.

Judd, M. D. and **Popp, M. I.**, 1973. Transition temperatures for (Sr,Ba)CO_3: a possible DTA temperature standard. *Thermochim. Acta*, **7**, 247–8.

Kowalski, W., Osmôlski, T. and **Pilichowska, E.**, 1980. Strontianite from sulfur deposit of the Machow mine (SE Poland). *Arch. Min.*, **36**, 29–46.

Lander, J. J., 1949. Polymorphism and anion rotational disorder in the alkaline earth carbonates. *J. Chem. Phys.*, **17**, 892–901.

Lillie, R. C., 1988. Minerals of the Harris Creek fluorspar district, Hardin County, Illinois. *Rocks Miner.*, **63**, 210–26.

Miles, D. L. and **Burton, J. D.**, 1972. The solubility of strontium carbonate in water. *J. Chem. Soc., Dalton Trans.*, 1691–2.

Mitchell, R. S. and **Pharr, R. F.**, 1961. Celestite and calciostrontianite from Wise County, Virginia. *Amer. Min.*, **46**, 189–95.

Moreau, R., 1964. Étude diffractométrique du carbonate de strontium et du carbonate de baryum entre 20 et 1000°C. *Bull. Soc. Franç. Min. Crist.*, **87**, 547–52.

Morgan, D. J. and **Milodowski, A. E.**, 1982. Thermal investigations of members of the strontianite–calciostrontianite series. *Proc., 7th Int. Conf. Thermal Analysis:* Ed. B. Miller, **1**, 642–9, John Wiley & Sons, New York.

Nesse, W. D., 1986. *Introduction to Optical Mineralogy*. Oxford and New York (Oxford University Press), 323 pp.

Pannhorst, W. and **Löhn, J.**, 1970. Zur Kristallstrukstur von Strontianit, $SrCO_3$. *Z. Krist.*, **131**, 455–9.

Penick, D. A. and **Haynes, L. D.**, 1983. Strontium minerals from Wise County, Virginia—an update. *Virginia Miner.*, **29**, 13–8.

Platt, R. G. and **Woolley, A. R.**, 1990. The carbonatites and fenites of Chapman Lake, Ontario. *Can. Min.*, **28**, 241–50.

Plummer, L. N. and **Busenberg, E.**, 1987. Thermodynamics of aragonite–strontianite solid solutions: results from stoichiometric solubility at 25° and 76°C. *Geochim. Cosmochim. Acta*, **51**, 1393–411.

Rapoport, E. and **Pistorius, C. W. F. T.**, 1967. Orthorhombic–disordered rhombohedral transition in $SrCO_3$ and $BaCO_3$ to 40 kbars. *J. Geophys. Res.*, **72**, 6353–7.

Salter, D. L. and **West, I. M.**, 1965. Calciostrontianite in the basal Purbeck beds of Durlston Head, Dorset. *Min. Mag.*, **35**, 146–50.

Saunders, J. A., **Prikryl, J. D.** and **Posey, H. H.**, 1988. Mineralogic and isotopic constraints on the origin of strontium-rich cap rock, Tatum Dome, Mississippi, USA. *Chem. Geol.*, **74**, 137–52.

Speer, J. A., 1976. Pennsylvania minerals. II. *Min. Rec.*, **7**, 69–71.

Speer, J. A., 1977. The orthorhombic carbonate minerals of Virginia. *Rocks Miner.*, **52**, 267–74.

Speer, J. A. and **Hensley-Dunn, M. L.**, 1976. Strontianite composition and physical properties. *Amer. Min.*, **61**, 1001–4.

Speer, J. A., and **Hensley-Dunn, M. L.**, 1983. Crystal chemistry and phase relations of orthorhombic carbonates. In *Carbonates: Mineralogy and Chemistry:* Ed. R. J. Reeder, *Rev. Mineral.*, **11**, 149–90. Mineralogical Society of America, Washington, DC.

Strømme, K. O., 1975. On the crystal structures of the high-temperature forms of strontium and barium carbonates and structurally related compounds. *Acta Chem. Scand.*, **A29**, 105–10.

Todd, J. G., 1989. Minerals of Trearne quarry, Beith, Ayrshire, Scotland, *UK J. Mines Miner.*, **6**, 18–20.

Warne, S. St J., 1962. A quick field or laboratory staining scheme for the differentiation of the major carbonate minerals. *J. Sed. Petr.*, **32**, 29–38.

White, W. R., 1974. The carbonate minerals. In *The Infrared Spectra of Minerals:* Ed. V. C. Farmer, Min. Soc. Monograph, **4**, 227–84.

Zemann, J., 1981. Zur Stereochimie der Karbonate. *Fortschr. Min.*, **59**, 95–116.

Witherite $BaCO_3$

Orthorhombic (−)

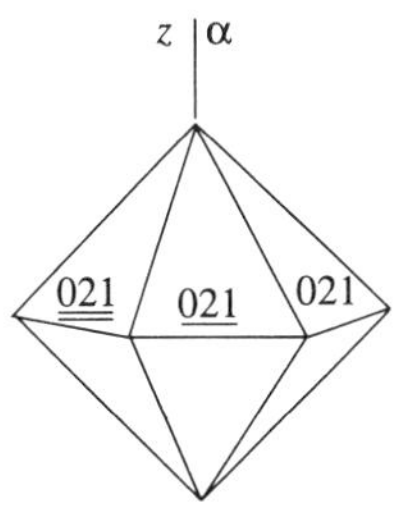

α	1.529
β	1.676
γ	1.677
δ	0.148
$2V_\alpha$	16°
Orientation:	
$\alpha=z$, $\beta=y$, $\gamma=x$, O.A.P. (010)	
Dispersion:	$r>v$, very weak.
D	4.30
H	$3\frac{1}{2}$
Cleavage:	{101} distinct, {110} and {012} poor.
Twinning:	Always present, repeated twins on {110} yielding pseudohexagonal crystal forms.
Colour:	Typically colourless or white; colourless in thin section or grain mount.
Unit cell:	*a* 5.312, *b* 8.896, *c* 6.428 Å Z=4. Space group *Pmcn*

Soluble in dilute HCl, with effervescence. Solution in HCl gives white precipitate on the addition of H_2SO_4.

Witherite is usually found in low-temperature hydrothermal veins and cavities in limestone or other calcareous sedimentary rocks. It is commonly associated with baryte and galena. The mineral is named after Dr W. W. Withering (1741–1799) who originally described it from Alston Moor, Cumbria, UK.

Structure

The crystal structure of witherite was determined by Colby and LaCoste (1935) to be essentially the same structure as for aragonite (Bragg, 1924). Structural refinement was performed by De Villiers (1971) using a crystal which had a composition of $(Ba_{0.9862}Ca_{0.0024}Sr_{0.0106}Pb_{0.0006}Mg_{0.0002})CO_3$ and was free from any visible defects.

The crystal structure of witherite can be described as layers of barium atoms arranged with a hexagonal ABAB.. sequence along the *z* axis and separated by CO_3 groups which are also arranged hexagonally, as shown in Fig. 85 (p. 236). The *M*–O bond lengths of the polyhedron and the C–O bond lengths of the triangle in witherite may be expressed by replacing the distance values in Fig. 85 clockwise with the following values: 2.816(6), 2.742(7), 2.868(5), 2.837(6) and 2.737(11) Å, and 1.289(7) and 1.282(12) Å, respectively. A comparison of crystallographic data with those for aragonite and strontianite is shown below (De Villiers, 1971; Dickens and Bowen, 1971; Speer and Hensley-Dunn, 1976):

(a) The carbon atoms do not project on top of each other along the *z* axis. Their relative displacements in the *y* direction are: aragonite, 0.194 Å; strontianite, 0.169 Å; and witherite, 0.125 Å.

(b) The CO_3 groups are tilted in the aragonite-type structure. The tilts, measured as the angle between the planes through the oxygen atoms and the *ab* plane, are: aragonite, 2°22′; strontianite, 2°40′; and witherite, 2°19′.

(c) The CO_3 groups are non-planar. The distances of the carbon atom from the plane of the oxygen atoms are: aragonite, 0.020 Å; strontianite, 0.007 Å; and witherite, 0.006 Å.

(d) The average C–O distances and C–O–C angles are: aragonite, 1.282 Å, 119.87°; strontianite, 1.285 Å, 119.86°; and witherite, 1.287 Å, 119.86°.

(e) Cell dimensions of the aragonite-type carbonates are for *a*, *b* and *c*, respectively: aragonite, 4.9598(5), 7.9641(9), 5.7379(6); strontianite, 5.1059(7), 8.4207(3), 6.0313(1); and witherite, 5.3126(5), 8.8958(5), 6.4284(5) Å.

$BaCO_3$ is polymorphic, and the nature of its polymorphism is identical to that of $SrCO_3$ (Boeke, 1912; Lander, 1949; Chang, 1965). The temperatures of the orthorhombic–rhombohedral transition and the rhombohedral–cubic transition are, respectively, 802 and 975 °C under 1 atm CO_2 pressure (Lander, 1949; Chang, 1965). Cell dimensions of the rhombohedral $BaCO_3$ determined by Lander at 830 °C and by Chang at 850 °C are *a* 5.205, *c* 10.55 Å. The cell dimension of the cubic form is 6.9543 Å at 1075 °C (Chang, 1965). Rapoport and Pistorius (1967) examined the phase transition in $BaCO_3$ to 40 kbar and established a transition line of $T\,(°C) = 799 + 7.15P - 0.052P^2$ with a standard deviation of 0.7 °C. Using the latent heat of transition of 4.49 kcal/mol. and the initial slope of 7.15 ± 0.5 °C/kbar, Rapoport and Pistorius found the volume change corresponding with the transition at room temperature to be $\Delta V = 1.25 \pm 0.09$ cm^3/mol. compared with the value of $\Delta V = 1.22 \pm 0.35$ cm^3/mol. from Moreau (1964).

There are five other well-characterized $BaCO_3$-bearing double carbonate minerals which do not have the dolomite-type structure. These are alstonite ($CaBa(CO_3)_2$) (Palache *et al.*, 1951), paralstonite ($CaBa(CO_3)_2$) (Roberts, 1978, 1979), barytocalcite ($CaBa(CO_3)_2$) (Palache *et al.*, 1951), benstonite ($Ca_7Ba_6(CO_3)_{13}$) (Lippmann, 1962) and norsethite ($MgBa(CO_3)_2$) (Mrose *et al.*, 1961).

Alstonite has been known to be structurally related to aragonite with *a* 5.00, *b* 8.78, *c* 6.12 Å (Gossner and Mussgnug, 1930). An X-ray investigation by Sartori (1975) shows that the symmetry is triclinic with an orthorhombic pseudocell which has *a*′ 5.02, *b*′ 8.70, *c*′ 6.12 Å and Z = 2. To account for the number of weak superstructure reflections observed, Sartori proposed a true cell of *a* 30.14, *b* 17.40, *c* 6.12 Å, $\alpha = \beta = \gamma = 90°$, Z = 2, and space group $C\bar{1}$ or *C*1. Scheetz and White (1977) concluded from infrared and Raman spectra measurements that alstonite has an aragonite-type structure, but that the two minerals are not isomorphous. In alstonite there are two crystallographically distinct $(CO_3)^{2-}$ groups. The crystal structure of paralstonite (Roberts 1978, 1979) was determined by Effenberger (1980). It is hexagonal with *a* 8.692, *c* 6.148 Å, Z = 3, and space group *P*321. In the structure, Ba is coordinated by ten and Ca by eight oxygen atoms. They are arranged in an ABAB... stacking sequence parallel to (0001). In addition, there are three crystallographically distinct $(CO_3)^{2-}$ groups. Several structural models were proposed by Effenberger to account for the enlarged cell dimension *a* of alstonite with the suggestion that alstonite represents a superstructure based upon paralstonite.

The structure of barytocalcite is monoclinic with *a* 8.092, *b* 5.2344, *c* 6.540 Å, *β* 106.05°, Z = 2 and space group $P2_1/m$ (Dickens and Bowen, 1971). As in the case of calcite- and aragonite-type carbonates, the structure of barytocalcite can also be described in terms of stacking sequences of cation layers. It has an ABAB.. sequence which repeats every three layers. Calcite has the same sequence, but repeats every six layers. There are two crystallographically distinct $(CO_3)^{2-}$ groups in barytocalcite which are like the CO_3 groups in the aragonite-type structure in orientation and differ from the orientation of CO_3 in the calcite-type structure by a rotation of about 30° in the layer. The Ca atom in barytocalcite is coordinated by seven, and the Ba atom by 11 oxygens. At high temperatures, the Ca and Ba atoms become randomly distributed among the cation positions and form a disordered calcite-type arrangement (Chang, 1965).

Norsethite has a unit cell (space group *R*32) similar to that of dolomite (Lippmann, 1968a). Magnesium has a distorted octahedral coordination of oxygens, and barium is surrounded by 12 oxygen atoms at the corners of a trapezohedrally distorted ditrigonal prism. The CO_3 groups in norsethite are related by two-fold rotation axes, whereas in dolomite they are related by a centre of symmetry. The hexagonal cell dimensions are *a* 5.017, *c* 16.77 Å and Z=3. In benstonite (Lippmann, 1962), the arrangement of cations including the carbon is closely related to that in the calcite-type structure (Effenberger, 1979). The coordinations of Ca and Mg atoms (see p. 266) are eight and six, respectively, whereas Ba atoms occupy two positions, 50 per cent in nine coordination and 50 per cent in ten coordination. The superstructure of benstonite thus appears to involve ordering within the cation layers. Crystallographic data given by Effenberger are *a* 18.280, *c* 8.6532 Å, Z=3 and space group *R*3.

Chemistry

Normally witherite is fairly pure $BaCO_3$. Only strontium substitutes barium to some significant extent, as shown in the selected analyses listed in Table 36. Calcium and magnesium may be present in witherite, but only in very minor amounts.

Table 36. Witherite analyses

	1	2	3	4
BaO	77.15	76.55	75.00	72.10
CaO	0.08	0.11	0.24	0.02
SrO	0.68	1.72	3.16	6.06
MgO	—	—	—	—
CO_2	22.50(c)	22.78(c)	23.10(c)	23.28(c)
Total	100.41	101.16	101.50	101.46
	Numbers of ions on the basis of six (O)			
Ba	1.968	1.928	1.864	1.777
Ca	0.006	0.008	0.016	0.002
Sr	0.026	0.064	0.120	0.221
Mg	—	—	—	—
C	2.000	2.000	2.000	2.000

(c) Calculated from stoichiometry.

1, 2 Witherite from Anglezark, Lancashire, and Settlingstones mine, Northumberland, England (Baldasari and Speer, 1979).
3 Witherite from Arknys, Caucasus, USSR (Baldasari and Speer, 1979); *a* 5.297, *b* 8.877, *c* 6.408 Å, D 4.26.
4 Witherite from Cassiar district, British Columbia (Baldasari and Speer, 1979); *a* 5.293, *b* 8.861, *c* 6.404 Å, D 4.22.

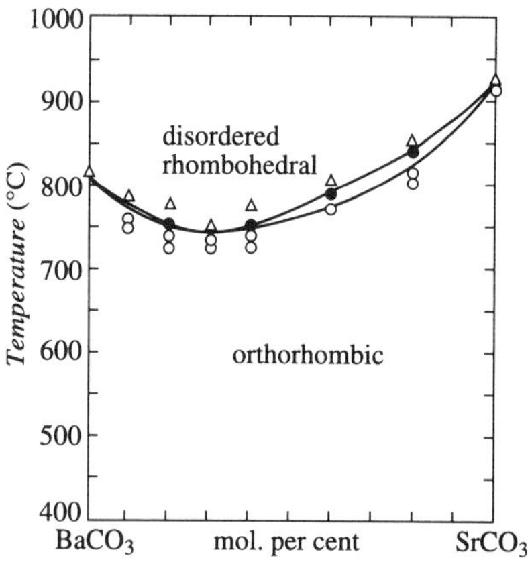

Fig. 99. Subsolidus phase relationships in the system $BaCO_3$–$SrCO_3$ (after Chang, 1965).

Subsolidus phase relationships in the system $BaCO_3$–$SrCO_3$ indicate the existence of a complete solid solution series at high temperatures in both rhombohedral and orthorhombic forms (Fig. 99) (Chang, 1965). Except for the c-dimension of the rhombohedral form, all cell dimensions vary linearly with composition in the system. The c_{rhom}–composition curve has a positive deviation. The restricted range of Sr in witherite and Ba in strontianite in natural occurrences has led both Chang (1971) and Speer and Hensley-Dunn (1976) to suggest that a miscibility gap may exist in the system $BaCO_3$–$SrCO_3$ below 400 °C.

Subsolidus phase relationships in the system $BaCO_3$–$CaCO_3$ are shown in Fig. 39, p. 114 (Chang, 1965). There is no detectable amount of $BaCO_3$ in $CaCO_3$ up to 600 °C, and at 400 °C the amount of $CaCO_3$ in $BaCO_3$ is restricted to <3.5 mol. per cent. $BaCO_3$ is more effective in lowering the temperature of anion disordering in $CaCO_3$ than is $SrCO_3$, and there is a complete series of solid solution in the disordered rhombohedral form between $BaCO_3$ and $CaCO_3$ above 850 °C. In the system, barytocalcite, $CaBa(CO_3)_2$, is a stable phase and is disordered above 520 °C. Neither alstonite nor paralstonite were found up to 40 kbar (Chang, 1965, 1971). Benstonite was not found in the system $BaCO_3$–$CaCO_3$ (Chang, 1965). The chemical composition of benstonite from Hot Spring, Arkansas, USA, its type locality, is expressed in the formula $(Ca_{6.118}Mg_{0.788}Ba_{5.273}Sr_{0.727}(CO_3)_{13}$. Analytical procedures yielded no insoluble residue, and lead and iron were not detected (Lippmann, 1962). Lippmann suggested two possible formulae. One is a calcium–barium carbonate, $Ca_7Ba_6(CO_3)_{13}$, and the other, $MgCa_6Ba_6(CO_3)_{13}$, indicates that magnesium is an essential constituent and not a substitute for calcium. A strontium variety of benstonite, $Sr_{1.71}(Ba_{0.936}Ca_{0.064})_{3.93}Ca_7(CO_3)_{13}$ with a 18.23, c 8.67 Å, was found in the Murun alkali massif, Siberia (Ontoev *et al.*, 1986). It was suggested that the Murun and Arkansas benstonites are part of an isomorphous series $Ba_6Ca_7(CO_3)_{13}$–$Sr_6Ca_7(CO_3)_{13}$. In the system $CaCO_3$–$SrCO_3$–$BaCO_3$, the aragonite-type solid solution forms a complete series from $BaCO_3$ through $SrCO_3$ to $CaCO_3$ at 15 kbar and 550 °C (Chang, 1971).

Although lead does not appear to substitute for barium in witherite, a complete series of solid solution in the system $BaCO_3$–$PbCO_3$ has been established experimentally at relatively high temperatures (Boström *et al.*, 1969; Chang and Brice, 1972). $BaCO_3$ forms double carbonates with both $MnCO_3$ and $MgCO_3$ (Chang, 1964; Lippmann, 1967, 1968b). $MnBa(CO_3)_2$ has the dolomite-type

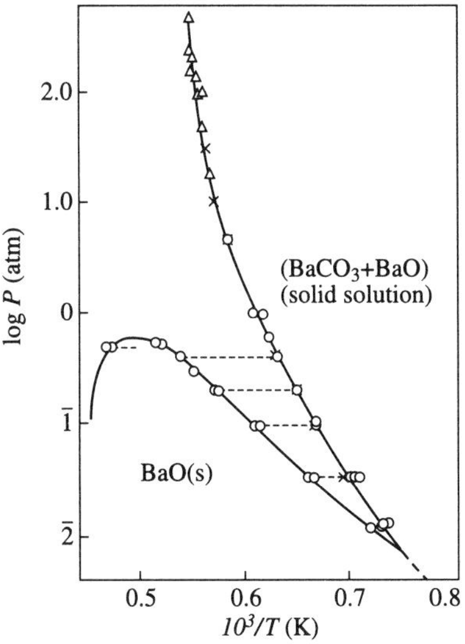

Fig. 100. Melting relationships for $BaCO_3$ (after Baker, 1964). Results were obtained by visual assessment (○), thermogravimetric analysis (×) and capillary (△) methods.

structure, whereas $MgBa(CO_3)_2$ is the synthetic equivalent of norsethite (Mrose *et al.*, 1961). Boström and Joensuu (1974) described a carbonate mineral from unaltered hausmannite ores in Långban, Sweden, which has a composition of $Ca_{0.4}Ba_{0.4}(Pb,Mg)_{0.2}CO_3$ and a disordered calcite-type structure with *a* 5.12, *c* 18.12 Å.

The dissociation and melting of $BaCO_3$ at high temperatures are shown in Fig. 100 (Baker, 1964). The triple point is defined by 1060 °C and 0.0066 atm P_{CO_2}. A differential thermal analysis of witherite and $(Ba,Sr)CO_3$ solid solutions has been performed by Judd and Popp (1973), as described in the section on strontianite, p. 257.

The solubility of $BaCO_3$ in CO_2+H_2O solution between 0 and 90 °C was investigated by Busenberg and Plummer (1986). For the reaction $BaCO_{3(s)}=Ba^{2+}_{(aq)}+CO^{2-}_{3(aq)}$ at 1 atm total pressure, the temperature dependence of the equilibrium constant is given by the equation $\log K = 607.642 + 0.121098T - (20011.25/T) - (236.4948 \log T)$, where T is in Kelvin. At 25 °C, $\log K$ of $BaCO_{3(s)}$, the Gibbs energy, the enthalpy and the entropy of the reaction are: 8.662, 48.87 kJ/mol., 2.94 kJ/mol. and −154.0 J/mol. K, respectively. The influence of pH on the rate of dissolution of witherite has been shown by Chou *et al.* (1989) to be comparable with those of calcite and aragonite. The rate constants for the three steps involved in the dissolution reaction are:

(1) $MCO_3 + H^+ \rightleftharpoons M^{2+} + HCO_3^-$, $k_1 = 7.6 \times 10^{-5}$ mol./cm²s,

(2) $MCO_3 + H_2CO_3^* \rightleftharpoons M^{2+} + 2HCO_3^-$, $k_2 = 3.0 \times 10^{-8}$ mol./cm²s and

(3) $MCO_3 \rightleftharpoons M^{2+} + CO_3^{2-}$, $k_3 = 7.1 \times 10^{-11}$ mol./cm²s.

The rate constant of the backward precipitation of witherite in reaction (3), k_{-3}, is 2.0×10^{-2} mol./cm²s.

Silicate–witherite oxygen isotope fractions ranged from −2‰ at 1000°C to −1‰ at 1400°C (Rosenbaum *et al.*, 1994).

Well-formed single crystals of witherite have been prepared by Blount (1974) at 80 °C from dilute aqueous solutions containing barium chloride and sodium bicarbonate. $BaCO_3$ is used in the manufacture of clay tiles and ceramic products. Other uses are for the production of special glasses, photographic papers, and barium compounds including barium titanate and ferrite.

Optical and Physical Properties

Witherite is biaxial and optically negative with a $2V_\alpha$ of 16°. Maximum birefringence, seen on {010} sections and cleavage fragments, is extreme. Interference colours in thin section and grain mount are usually high-order white and grey for most orientations, except on basal {001} sections, where the birefringence is weak. Witherite is an aragonite-type carbonate, and its extreme birefringence, like that of aragonite and strontianite, follows from its structure. The indices of refraction for witherite do not vary significantly because there is little compositional variation. The substitution of $SrCO_3$ for $BaCO_3$ in witherite should lower the refractive indices, and in general the relationship between composition and refractive indices can be approximated by a straight line between the end-members (Speer, 1983).

Witherite is white or grey to pale yellow, brown or green. Like aragonite and strontianite, it also has a strong luminescence (Gies, 1975). The spectrum consists of a broad band in the blue and long-wave ultraviolet (300–500 nm), and a weak band in the red (600–700 nm). Infrared and Raman spectra of witherite are shown in Figs. 94 and 95, respectively.

Crystals of witherite are always twinned on {110} forming stubby pseudo-hexagonal bipyramids which usually have rough horizontally striated faces. Cell dimensions and density vary regularly with composition in the range up to 11 mol. per cent $SrCO_3$ (Baldasari and Speer, 1979). Using the d_{220} spacing of witherite, the Sr content can be determined by the equation: equivalent $SrCO_3$ mol. per cent $(\pm 1.1) = -1102.58(d_{220}, \text{Å}) + 2515.74$. The calculated densities compared favourably with measured values as listed below:

Sr content				
SrO wt. per cent	0.68	1.72	3.26	6.06
SrO mol. per cent	1.29	3.20	5.99	11.04
Density				
Calculated	4.29	4.28	4.27	4.24
Measured	4.31	4.28	4.26	4.22

The optical and physical properties of $BaCO_3$-containing double carbonates are quite similar in many aspects. Barytocalcite is white, pale grey or pale cream with a resinous to greasy lustre. Alstonite occurs as colourless to white pyramidal crystals with horizontal striae on the pyramidal faces (Young, 1985). Both barytocalcite and alstonite are biaxial and optically negative, with 2V of 15 and 6°, respectively. Refractive indices of barytocalcite are α 1.525, β 1.685, γ 1.686, and those of alstonite are α 1.526, β 1.671, γ 1.672. Specific gravities and hardness are 3.65–3.71 and 4 for barytocalcite and 3.67–3.71 and 4–$4\frac{1}{2}$ for alstonite, respectively. Paralstonite is colourless to greyish white, with a specific gravity of

3.75 and a hardness of 4–4½. It is uniaxial and optically negative, and has ω 1.672, ε 1.527.

Norsethite occurs as clear to milky-white plates and has a vitreous lustre and good rhombohedral cleavage. Its hardness is 3½ and specific gravity is 3.837. Optically, norsethite is uniaxial negative with ω 1.694 and ε 1.519 (Mrose *et al.*, 1961). Benstonite is snow-white to ivory in colour. It has a specific gravity of 3.596 and a hardness between 3 and 4. Benstonite is uniaxial and optically negative with ω 1.690 and ε 1.523. When irradiated with ultraviolet or X-ray radiation, it fluoresces red.

Infrared and Raman spectra of alstonite, barytocalcite, norsethite and benstonite were measured by Scheetz and White (1977). The effects of ordering in this group of double carbonates appear mainly in the infrared spectra.

Distinguishing Features

Witherite shows the extreme birefringence characteristic of carbonates but is distinguished from the calcite-type carbonates by the absence of rhombohedral cleavage. In comparison with strontianite, it has higher refractive indices and a single distinct cleavage, whereas compared with aragonite it has a higher density, with fragments sinking in di-iodomethane. The yellow-green flame test of Ba is diagnostic. By using rhodizonic acid or sodium potassium rhodizonate (Warne, 1962), witherite is stained to an orange-red to red colour.

Paragenesis

Witherite occurs mainly in low-temperature hydrothermal vein deposits, and its formation is usually noted as a replacement of baryte. In the fluorite deposits in Illinois and Kentucky, USA (Fisher, 1977), witherite in the form of pseudohexagonal twinned mosaic crystals lines vugs in limestone. As described by Grawe and Nackowski (1949), the witherite appears to be an alteration product of baryte, although it is not always in direct association with baryte. The presence of witherite in the largest baryte deposit in the former USSR at Turkmen was examined by Mamedov (1963). A gradual replacement of baryte by witherite was observed. In the North Pennine orefield, witherite occurs after baryte, and the baryte–witherite assemblage exhibits a well-defined zonal pattern with fluorite. The overlap of these two zones is limited (Sawkins, 1966). Fluid inclusion measurements indicate that fluorite was deposited at temperatures of between 200 and 100 °C, and baryte from 130 down to <50 °C.

Witherite in sedimentary rocks was found in the Siberian platform (Stepanov and Ilupin, 1973). The calcite–dolomite veins in the dolomitized limestones and unmetamorphosed oil shales contain witherite, pyrite, marcasite and baryte. The veins were formed under low Eh and pH conditions, and the source material for the formation of witherite was suggested to be baryte.

Witherite is abundant in black shale and black chert in the Qinling region of China (Wang and Li, 1991), where it is in association with baryte and barytocalcite. There are carbonaceous mudstone interlayers in the barium ore beds, and the organic carbon content may reach as high as 14 wt. per cent. Radiolaria were found in the black cherts. Sulphur and carbon isotope compositions of baryte and witherite show high $\delta^{34}S$ and low $\delta^{13}C$ values. The abundance of witherite in the Qinling region deposits is a striking anomaly. Stratiform Ba deposits elsewhere are invariably mainl baryte with only traces of witherite (Maynard and Okita, 1991).

The relationships among sulphates and carbonates of barium and calcium were examined by Lydon *et al.* (1985) using a plot of Ba/Ca versus SO_4/CO_3. They concluded that severe depletion of sea-water in SO_4^{2-} was necessary to form witherite rather than baryte.

In hydrothermal veins with sulphide ores, witherite may react with sulphuric acid that was produced by the oxidation of ores to form baryte (Fitch, 1931).

Both alstonite and barytocalcite were first identified in limestone at Alston Moor, England (Palache *et al.*, 1951). Young (1985) described their occurrences in the Northern Pennine orefield, where they are abundant in the limestone wall-rock and are the products of reactions between the limestone and late-stage mineralizing fluids. Associated minerals are calcite, baryte and witherite. At Cave-in-Rock, Illinois, USA, both alstonite (Rossman and Squires, 1974) and paralstonite (Roberts, 1978, 1979) were found as coatings on calcite, witherite or baryte. Norsethite was discovered by Mrose *et al.* (1961) in the dolomite black shale which underlies the main trona bed, Green River, Wyoming, USA. It is associated with shortite, labuntsovite, searlesite, loughlinite, pyrite and quartz. Benstonite occurs in the baryte deposit at Hot Spring, Arkansas, USA, associated with milky quartz, baryte and calcite (Lippmann, 1962).

References

Baker, E. H., 1964. The barium oxide–carbon dioxide system in the pressure range 0.01–450 atmospheres. *J. Chem. Soc., Dalton Trans. Inorg. Chem.*, 699–704.

Baldasari, A. and **Speer, J. A.,** 1979. Witherite composition, physical properties and genesis. *Amer. Min.*, **64**, 742–7.

Blount, C. W., 1974. Synthesis of barite, celestite, anglesite, witherite and strontianite from aqueous solutions. *Amer. Min.*, **59**, 1209–19.

Boeke, H. S., 1912. Die Schmetzerscheinungen und die unkehrbare Umwandlung des Calcium Carbonates. *Neues Jahrb. Min. Geol.*, **1**, 91–121.

Boström, K. and **Joensuu, I.,** 1974. Distribution of Ba and Pb in primary carbonates in Långban, Sweden. *Geol. Fören. Förh.*, **96**, 375–9.

Boström, K., Hanor, J., Blankenburg, J. and **Glaccum, R.,** 1969. Some subsolidus phase relations in the system $BaCO_3$–$CaCO_3$–$PbCO_3$ at 500°C. *Arkiv. Min. Geol.*, **5**, 47–53.

Bragg, W. L., 1924. The structure of aragonite. *Proc. R. Soc. London*, **A105**, 16–37.

Busenberg, E. and **Plummer, L. N.,** 1986. The solubility of $BaCO_3$(cr) (witherite) in CO_2–H_2O solutions between 0° and 90°C, evaluation of the association constants of $BaHCO_3^+$(aq) and $BaCO_3^\circ$(aq) between 5° and 80°C, and a preliminary evaluation of the thermodynamic properties of Ba^{2+}(aq). *Geochim. Cosmochim. Acta*, **50**, 2225–33.

Chang, L. L. Y., 1964. Synthesis of $MBa(CO_3)_2$ compounds. *Amer. Min.*, **49**, 1142–3.

Chang, L. L. Y., 1965. Subsolidus phase relations in the systems $BaCO_3$–$SrCO_3$, $SrCO_3$–$CaCO_3$, and $BaCO_3$–$CaCO_3$. *J. Geol.*, **73**, 346–68.

Chang, L. L. Y., 1971. Subsolidus phase relations in the aragonite-type carbonates. I. The system $CaCO_3$–$SrCO_3$–$BaCO_3$. *Amer. Min.*, **56**, 1660–73.

Chang, L. L. Y. and **Brice, W. R.,** 1972. Subsolidus phase relations in aragonite-type carbonates. II. The systems $CaCO_3$–$SrCO_3$–$PbCO_3$ and $CaCO_3$–$BaCO_3$–$PbCO_3$. *Amer. Min.*, **57**, 155–68.

Chou, L., Garrels, R. M. and **Wollast, R.,** 1989. Comparative study of the kinetics and mechanisms of dissolution of carbonate minerals. *Chem. Geol.*, **78**, 269–82.

Colby, M. Y. and **LaCoste, L. J. B.,** 1935. The crystal structure of witherite. *Z. Krist.*, **90**, 1–7.

De Villiers, J. P. R., 1971. Crystal structures of aragonite, strontianite, and witherite. *Amer. Min.*, **56**, 758–67.

Dickens, B. and **Bowen, J. S.,** 1971. The crystal structure of $BaCa(CO_3)_2$ (barytocalcite). *J. Res. Nat. Bur. Stand.*, **75A**, 197–203.

Effenberger, H., 1979. Kristallstruktur und chemische Formel des Benstonits, $Ba_6Ca_6Mg(CO_3)_{13}$. *Neues Jahrb. Min., Abh.*, **136**, 326–37.

Effenberger, H., 1980. Die Kristallstruktur des Minerals Paralstonite, $BaCa(CO_3)_2$. *Neues Jahrb. Min., Monat.*, 353–63.

Fisher, I. S., 1977. Distribution of Mississippian geodes and geode minerals in Kentucky. *Econ. Geol.*, **72**, 864–9.

Fitch, A. A., 1931. Barite and witherite from near El Portal, Mariposa Co., California. *Amer. Min.*, **16**, 461–8.

Gies, H., 1975. Activation possibilities and geochemical correlations of photoluminescing carbonates, particularly calcite. *Min. Deposita*, **10**, 216–27.

Gossner, B. and **Mussgnug, F.**, 1930. Über Alstonit und Milarit, Ein Beitrag zur Kenntniskomplexgebauter Kristalle. *Centrl. Min., Abt. A*, 220–38.

Grawe, O. R. and **Nackowski, M. P.**, 1949. Strontianite and witherite associated with southern Illinois fluorite. *Science*, **110**, 331.

Judd, M. D. and **Popp, M. I.**, 1973. Transition temperatures for (Sr,Ba)CO_3: a possible DTA temperature standard. *Thermochim. Acta*, **7**, 247–8.

Lander, J. J., 1949. Polymorphism and anion rotational disorder in the alkaline earth carbonates. *J. Chem. Phys.*, **17**, 892–901.

Lippmann, F., 1962. Benstonite, $Ca_7Ba_6(CO_3)_{13}$, a new mineral from the barite deposit in Hot Spring County, Arkansas. *Amer. Min.*, **47**, 585–98.

Lippmann, F., 1967. Die Syntheses des Norsethite bei etwa 20°C und 1 at. Ein Modell zur Dolomitisierung. *Neues Jahrb. Min., Mh.*, 23–9.

Lippmann, F., 1968a. Die Kristallstrujtur des Norsethite, $BaMg(CO_3)_2$, mit einem Strukturverschlag fur $PbMg(CO_3)_2$. *Tschermaks Min. Petr. Mitt.*, **12**, 299–318.

Lippmann, F., 1968b. Synthesis of $BaMg(CO_3)_2$ (norsethite) at 20°C and the formation of dolomite in sediments. In *Recent Developments in Carbonate Sedimentology in Central Europe:* Eds. G. Muller and G. M. Friedman, New York (Springer-Verlag), pp. 33–7.

Lydon, J. W., **Goodfellow, W. D.** and **Jonasson, I. R.**, 1985. A general genetic model for stratiform barite deposits of the Selwyn basin, Yukon Territory and District of Mackenzie. *Can. Geol. Surv. Paper*, **85–1A**, 651–60.

Mamedov, K. M., 1963. Barite–witherite mineralization in mountain regions of the Turkmen, S.S.R. *Ivz. Akad. Nauk Tukm. SSSR, Ser. Fiz.-Tekn. Khim. Geol. Geol. Nauk*, **1**, 78–82.

Maynard, J. B. and **Okita, P. M.**, 1991. Bedded barite deposits in the United States, Canada, Germany, and China: two major types based on tectonic setting. *Econ. Geol.*, **86**, 364–76.

Moreau, R., 1964. Étude diffractométrique du carbonate de strontium et du carbonate de baryum entre 20 et 1000°C. *Bull. Soc. Franç. Min. Crist.*, **87**, 547–52.

Mrose, M. E., **Chao, E. C. T.**, **Fahey, J. J.** and **Milton, C.**, 1961. Norsethite, $BaMg(CO_3)_2$, a new mineral from the Green River formation, Wyoming. *Amer. Min.*, **46**, 420–9.

Ontoev, D. O., **Dmitrieva, M. T.** and **Ontoeva, T. D.**, 1986. Strontian variety of benstonite. *Zap. Vses. Min. Obshch.*, **115**, 496–501.

Palache, C., **Berman, H.** and **Frondel, C.**, 1951. *Dana's System of Mineralogy, vol. II.* 7th edn, New York (John Wiley & Sons), 1124 pp.

Rapoport, E. and **Pistorius, C. W. F. T.**, 1967. Orthorhombic–disordered rhombohedral transition in $SrCO_3$ and $BaCO_3$ to 40 kbars. *J. Geophys. Res.*, **72**, 6353–7.

Roberts, A. C., 1978. Mineralogical study of an unnamed barium calcium carbonate from the Cave-in-Rock district. Illinois. In *Current Research*, Part C, *Geol. Surv. Canada Paper*, **78–1C**, 49–52.

Roberts, A. C., 1979. Paralstonite: a new mineral from the Minerva No. 1 mine, Cave-in-Rock, Illinois. In *Current Research*, Part C, *Geol. Surv. Canada Paper*, **79–1C**, 99–100.

Rosenbaum, J. M., **Kyser, T. K.** and **Walker, D.**, 1994. High Temperature oxygen isotope fractionation in the enstatite-olivine $BaCO_3$ system. *Geochim. Cosmochim Acta*, **58**, 2653–60.

Rossman, G. R. and **Squires, R. L.**, 1974. The occurrence of alstonite at Cave-in-Rock, Illinois. *Min. Rec.*, **5**, 266–9.

Sartori, F., 1975. New data on alstonite. *Lithos*, **8**, 199–207.

Sawkins, F. J., 1966. Ore genesis in the north Pennine orefield, in the light of fluid inclusion studies. *Econ. Geol.*, **61**, 385–401.

Scheetz, B. E. and **White, W. B.**, 1977. Vibrational spectra of the alkaline earth double carbonates. *Amer. Min.*, **64**, 36–50.

Speer, J. A., 1983. Crystal chemistry and phase relations of orthorhombic carbonates. In *Carbonates: Mineralogy and Chemistry:* Ed. R. J. Reeder, *Reviews in Mineralogy*, **11**, 149–190, Mineralogical Society of America, Washington, DC.

Speer, J. A. and **Hensley-Dunn, M. L.**, 1976. Strontianite composition and physical properties. *Amer. Min.*, **61**, 1001–4.

Stepanov, V. L. and **Ilupin, I. P.**, 1973. Find of witherite in sedimentary rocks of the southeastern part of the Siberian platform. *Tr. Min. Mz., Akad. Nauk SSSR*, **22**, 207–10.

Wang, Z. and **Li, G.**, 1991. Barite and witherite deposits in Lower Cambrian shales of south China: stratigraphic distribution and geochemical characterization. *Econ. Geol.*, **86**, 354–63.

Warne, S. St J., 1962. A quick field or laboratory staining scheme for the differentiation of the major carbonate minerals. *J. Sed. Petr.*, **32**, 29–38.

Young, B., 1985. The distribution of barytocalcite and alstonite in the northern Pennine orefield. *Proc. Yorkshire Geol. Soc.*, **95**, 199–206.

Cerussite $PbCO_3$

Orthorhombic (−)

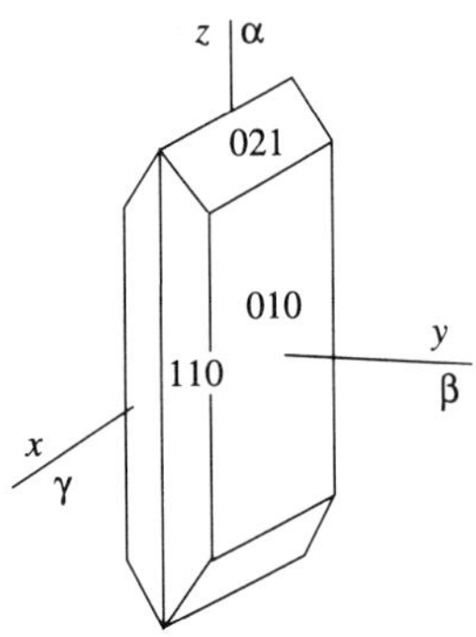

α	1.803
β	2.074
γ	2.076
δ	0.273
$2V_\alpha$	9°
Orientation: $\alpha = z$, $\beta = y$, $\gamma = x$, O.A.P. (010)	
Dispersion:	$r > v$ strong.
D	6.55
H	$3\text{–}3\frac{1}{2}$
Cleavage:	{110} and {021} distinct, {010} and {012} poor.
Twinning:	Repeated twins on {110} common, yielding pseudohexagonal crystal forms or polysynthetic lamellae.
Colour:	Typically white or grey; colourless in thin section or grain mount.
Unit cell:	*a* 5.180, *b* 8.492, *c* 6.134 Å, Z = 4. Space group *Pmcn*

Soluble in warm dilute HNO_3 with effervescence.

Cerussite is a common mineral in the upper oxidized zone of ore deposits containing galena. It usually occurs with smithsonite, anglesite and other secondary minerals. Fine crystals of cerussite are common, and are usually twinned. The name is from the Latin *cerussa*, for Ceruse, a white lead pigment.

Structure

The crystal structure of cerussite was determined by Colby and LaCoste (1933), and refinements were made by Sahl (1974). It has the aragonite-type structure with *a* 5.180, *b* 8.492, *c* 6.134 Å, Z = 4 and space group *Pmcn*. The average Pb–O bond length is 2.69 Å, which is longer than the Ca–O bond in aragonite and the Sr–O bond in strontianite, but shorter than the Ba–O bond in witherite. The average C–O bond length of 1.27 Å and O–C–O bond angle of 121.3° are distinct in comparison with those in alkaline earth carbonates of the aragonite-type structure. The CO_3 groups are slightly aplanar, and the distance of the carbon atom from the plane of the oxygen atoms was determined as 0.026 Å by Chevrier *et al.* (1992). This distance is of the same magnitude as in aragonite and strontianite, and the direction of the deviation from planarity is also the same (Jarosch and Heger, 1986, 1988). Cell dimensions of cerussite from neutron diffraction refinement are *a* 5.179(1), *b* 8.492(3) and *c* 6.141(2) Å (Chevrier *et al.*, 1992).

Chemistry

Most cerussite is very close in composition to $PbCO_3$. Minor amounts of Zn and Sr may replace Pb. Cerussite crystals from Broken Hill and Northampton, Australia, have compositions of 99.69 per cent $PbCO_3$ with 0.20 per cent FeO, and 98.09 per cent $PbCO_3$ with 0.40 per cent $ZnCO_3$ and 1.4 per cent SiO_2 (Warne, 1962). The impurity content of cerussite from Tsumeb, Namibia, was analysed for a number of ions and only trace amounts (< 5 p.p.m.) of Mn and Fe were found (Solé *et al.*, 1987). SrO in the range 0.99–1.25 per cent was detected in cerussite from the Kremikovtsi (Hungary) deposit (Atanassov *et al.*, 1989), and from Åland Island, southwest Finland (Lindqvist and Suominen, 1988). The latter has a unit cell of *a* 5.177, *b* 8.490, *c* 6.137 Å. Phase relationships in the system $PbCO_3$–$SrCO_3$ have been studied by Chang and Brice (1972) who have shown the existence of a complete series of solid solution.

Cerussite is not stable at high temperatures (Ball and Casson, 1975; Yamaguchi *et al.*, 1980). At low CO_2 pressure (< 1 atm), cerussite decomposes to $PbCO_3 \cdot 2PbO$ at ~200 °C and then to PbO at ~300 °C. At medium pressures (~1 atm), two intermediates, $PbCO_3 \cdot PbO$ and $PbCO_3 \cdot 2PbO$, were formed at ~300 and 400 °C, respectively. Cerussite decomposes to PbO via three intermediates, $2PbCO_3 \cdot PbO$, $PbCO_3 \cdot PbO$ and $PbCO_3 \cdot 2PbO$, at high pressure (>4 atm). These results are in good agreement with those reported by Criado *et al.* (1987) who examined the reaction $PbCO_3 \rightarrow PbO + CO_2$ using a constant rate thermal analysis. An equilibrium study of the system PbO–CO_2 at high pressures by Grisafe and White (1964) demonstrated the presence of an additional oxycarbonate, $4PbCO_3 \cdot 3PbO$, which is stable only at pressures >300 bar.

The stability of cerussite in solution was investigated by Garrels (1954) and Takahashi (1960) in terms of Eh, pH and P_{CO_2}. In equilibrium with air ($P_{CO_2} = 10^{-3.5}$ bar), the cerussite field extends above the line of water decomposition and occupies the entire region between the galena and plattnerite (PbO_2) fields in neutral and alkaline solutions. Bilinski and Schindler (1982) examined the solubility of $PbCO_3$ in acid solutions over a wide range of P_{CO_2}. The average log *K* obtained for the reaction $PbCO_3(s) + 2H^+ = Pb^{2+} + CO_2(g) + H_2O$ is 5.20 ± 0.03 at 25 °C and in a solution of constant ionic strength (*I*) of 0.3 mol./dm^3. The solubility of cerussite in a series of carbonate–bicarbonate solutions was also examined and the results are illustrated in the form of $-\log [Pb]_{sol}$ versus $-\log [CO_3^{2-}]$ (Fig. 101). Three tangents are drawn on to the solubility curve. Tangent (a) corresponds with the reaction $PbCO_3(s) = Pb^{2+} + CO_3^{2-}$ with $\log K = -12.15 \pm 0.05$, tangent (b) with the reaction $PbCO_3(s) = PbCO_3^{\circ}$ with $\log K = -6.40 \pm 0.10$, and tangent (c) with the reaction $PbCO_3(s) + CO_3^{2-} = Pb(CO_3)_2^{2-}$ with $\log K = -3.10 \pm 0.20$.

Optical and Physical Properties

Cerussite is biaxial and optically negative with a $2V_\alpha$ close to 9°. Birefringence is very extreme ($\gamma - \alpha = 0.273$) and is related to structure as discussed in the section on aragonite, p. 244. High-order white interference colours are found for most orientations in both grain mount and thin section. Basal sections yield centred acute bisectrix figures with numerous isochromes.

Cerussite is usually white or grey, and often brown or black. Luminescence of considerable intensity is exhibited by cerussite, with a broad emission band

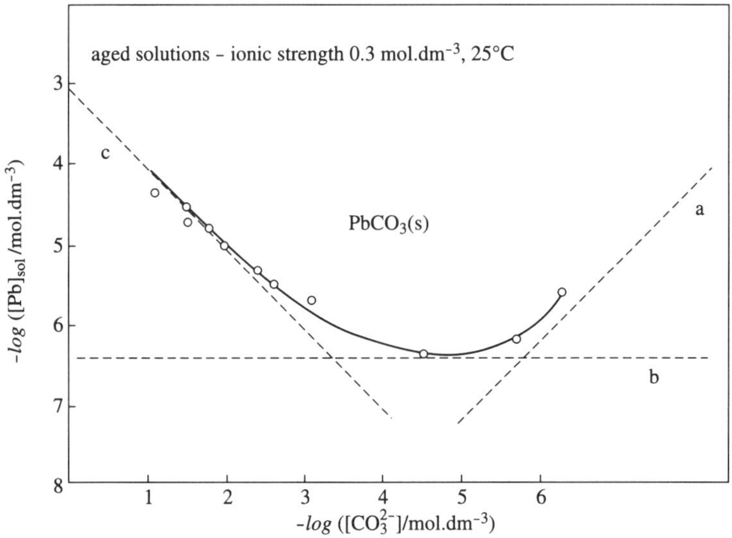

Fig. 101. Solubility of $PbCO_3$ (after Bilinski and Schindler, 1982). The straight lines a, b, c correspond with log $K = -12.15$, log $K = -6.4$ and log $K = -3.1$.

centred at 470 nm (Solé *et al.*, 1987). Under X-ray excitation, cerussite is a highly efficient phosphor at liquid nitrogen temperatures (Lammers and Blasse, 1986).

Cerussite generally occurs in crystals or crystal aggregates. It has a varied habit and shows many forms. Twins are very common, in reticulated networks, in star-like groups or forming sixlings. The morphology of cerussite and its relationship to the conditions of crystal growth have been examined by Franke *et al.* (1981). Cerussite crystals grown from gel have a pseudohexagonal habit resulting from the combination of {111}, {021}, {110} and {010}. Supersaturation of lead ions causes the formation of triplet crystals with the twinning plane (110), and excessive carbonate ions enhance dendritic growth. Fig. 102 illustrates the crystal morphology of cerussite grown by hydrothermal transport under various temperature–pressure conditions. All faces obtained are within the list of the first 12 faces, in ranks of importance for the equilibrium form predicted by the Donnay-Harker law (Donnay and Harker, 1937).

The infrared spectrum of cerussite is shown in Fig. 94 (p. 245). The band width of ν_3 becomes narrower as the cation changes from Ca^{2+} to Pb^{2+} (White, 1974). Like other aragonite-type carbonates, the Raman spectrum of cerussite is characterized by an intense ν_1 (Griffith, 1970).

Distinguishing Features

Cerussite is distinguished from other carbonate minerals by its high specific gravity and adamantine lustre. Twinned crystals and reticulated aggregates are other distinct features. Cerussite is soluble in warm HNO_3 with effervescence, which may be used to distinguish it from anglesite.

Among common carbonate minerals, cerussite, ankerite and strontianite are stained purple in an alizarin red S solution. When boiled in alizarin red S and

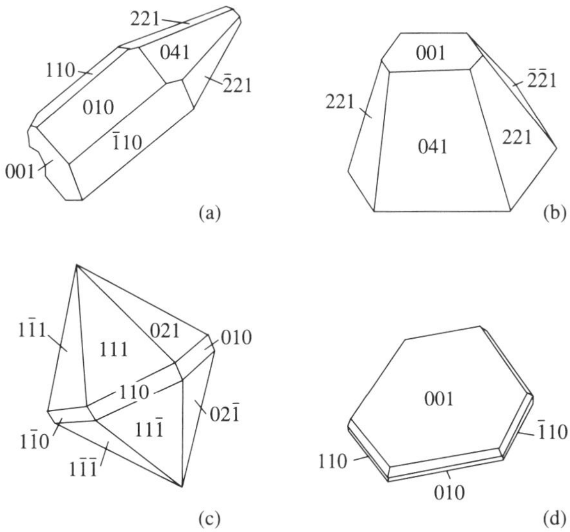

Fig. 102. $PbCO_3$ crystals grown by hydrothermal transport (a) 200 °C, 200 bar; (b) 200 °C, 300 bar; (c) 200 °C, 500 bar; (d) 200 °C, 1000 bar (after Franke *et al.*, 1981).

30 per cent NaOH solution, cerussite reacts rapidly and produces a dark red-brown stain, while strontianite and ankerite are unstained and stained dark purple, respectively (Warne, 1962).

Paragenesis

Cerussite is a common mineral in the upper oxidized zone of ore deposits containing galena. It is usually associated with anglesite, smithsonite, malachite and other secondary minerals. The Broken Hill lead–silver deposits, New South Wales, Australia, contain great masses of cerussite at a depth below about 90 m from the surface (Smith, 1926). Throughout the Leadhills lead and zinc district of southern Scotland cerussite occurs in a great variety of forms with twinning common; it was once worked as 'white lead ore' (Temple, 1956). In the Ag–Mn–Pb–Zn veins and skarn deposits of Uchucchacua, Peru (Bussell *et al.*, 1990), supergene oxidation and mineralization were found in the upper 30–150 m of most veins and ore bodies, characterized by goethite, Mn-oxides and cerussite. In some lead- and zinc-bearing deposits of Tunisia, such as that at Djebel Hallout (Sainfeld, 1956), the oxidized zone reaches at least 100 m below the present groundwater table. The mineral assemblage is composed essentially of cerussite and smithsonite.

In the east Tintic district, Utah, USA, supergene processes have greatly modified the results of hypogene mineralization (Bush *et al.*, 1960). Cerussite and other secondary minerals occur chiefly above the water-table at a depth of around 300 m. The occurrence of cerussite as a supergene product was reported in other deposits of the southwestern United States (Kiersch, 1949; Thomas, 1949; Jicha, 1954). In most of them, oxidation in the deposits is fairly deep.

Cerussite has been found as one of the lead-bearing minerals formed in soils in the vicinity of an abandoned Pb battery processing facility (Whitten and Stratton, 1990).

References

Atanassov, V., Vassileva, M. and **Gorunova, R.,** 1989. [Carbonates with aragonite-type structure (aragonite, witherite and cerussite) in the Kremikovtsi deposit.] *Annual of Higher Institute of Mining & Geology, Sofia, Part I: Geology,* **35**, 73–86. (in Hungarian with English abstract).

Ball, M. and **Casson, M. J.,** 1975. Thermal studies on lead(II) salts. I. Stoichiometry of lead carbonate decomposition at 1 atm. pressure. *J. Inorg. Nucl. Chem.,* **37**, 2253–5.

Bilinski, H. and **Schindler, P.,** 1982. Solubility and equilibrium constants of lead in carbonate solutions. *Geochim. Cosmochim. Acta,* **46**, 921–8.

Bush, J. B., Cook, D. R., Lovering, T. S. and **Morris, H. T.,** 1960. The Chief Oxide–Burgin area discoveries, East Tintic district, Utah; a case history. *Econ. Geol.,* **55**, 1507–40.

Bussell, M. A., Alpers, C. N., Petersen, U., Shepherd, T. J., Bermudez, C. and **Baxter, A. N.,** 1990. The Ag–Mn–Pb–Zn veins, replacement, and skarn deposits of Uchucchacua, Peru: studies of structure, mineralogy, metal zoning, Sr isotopes, and fluid inclusions. *Econ. Geol.,* **85**, 1348–83.

Chang, L. L. Y. and **Brice, W. R.,** 1972. Subsolidus phase relations in the aragonite-type carbonates. II. The systems $CaCO_3$–$SrCO_3$–$PbCO_3$ and $CaCO_3$–$BaCO_3$–$PbCO_3$. *Amer. Min.,* **57**, 155–68.

Chevrier, G., Giester, G., Heger, G., Jarosch, D., Wildner, M. and **Zemann, J.,** 1992. Neutron single-crystal refinement of cerussite, $PbCO_3$, and comparison with other aragonite-type carbonates. *Z. Krist.,* **199**, 67–74.

Colby, M. Y. and **LaCoste, L. J. B.,** 1993. The crystal structure of cerussite. *Z. Krist.,* **84**, 299–309.

Criado, J. M., González, M. and **Macías, M.,** 1987. Influence of CO_2 pressure on the kinetics of thermal decomposition of $PbCO_3$. *Thermochim. Acta,* **113**, 39–47.

Donnay, J. D. H. and **Harker, D.,** 1937. A new law of crystal morphology extending the law of Bravais. *Amer. Min.,* **22**, 446–67.

Franke, W., Lenk, K., Ittyachen, M. A. and **Pillai, K. M.,** 1981. The morphology of cerussite $PbCO_3$ grown in silica gel and on hydrothermal conditions. *J. Cryst. Growth,* **51**, 309–13.

Garrels, R. L., 1954. Mineral species as function of pH and oxidation potentials, with special reference to the zone of oxidation and secondary enrichment of sulfide ore deposits. *Geochim. Cosmochim. Acta,* **6**, 645–70.

Griffith, W. P., 1970. Raman studies on rock-forming minerals. Part II. Minerals containing MO_3, MO_4 and MO_6 groups. *J. Chem. Soc., A,* 286–91.

Grisafe, D. A. and **White, W. B.,** 1964. Phase relations in the system PbO–CO_2 and the decomposition of cerussite. *Amer. Min.,* **49**, 1184–98.

Jarosch, D. and **Heger, G.,** 1986. Neutron diffraction refinement of the crystal structure of aragonite. *Tschermaks Min. Petr. Mitt.,* **35**, 127–31.

Jarosch, D. and **Heger, G.,** 1988. Neutron diffraction investigation of strontianite, $SrCO_3$. *Bull. Min.,* **111**, 139–42.

Jicha, H. L., 1954. Paragenesis of the ores of the Palomas (Hermosa) district, southwestern New Mexico. *Econ. Geol.,* **49**, 759–78.

Kiersch, G. A., 1949. Structural control and mineralization at the Seventynine mine, Gila County, Arizona. *Econ. Geol.,* **44**, 24–39.

Lammers, M. J. J. and **Blasse, G.,** 1986. Luminescence properties of $PbCO_3$ and $Sr_{1-x}Pb_xCO_3$ with the aragonite structure. *Mater. Res. Bull.,* **21**, 529–34.

Lindqvist, K. and **Suominen, V.,** 1988. Cerussite from even-grained rapakivi, Saltvik, Åland Islands, SW Finland. *Bull. Geol. Soc. Finland,* **60**, 75–8.

Sahl, K., 1974. Verfeinerung der Kristallstruktur von Cerussit $PbCO_3$. *Z. Krist.,* **139**, 215–22.

Sainfeld, P., 1956. The lead–zinc bearing deposits of Tunisia. *Econ. Geol.,* **51**, 150–77.

Smith, G., 1926. A contribution to the mineralogy of New South Wales. *Geol. Surv. N. S. W. Min. Res.,* **34**, 145 pp.

Solé, J. G., Rueda, F. A., López, C., Vergara, G. and **Meseguer, F.,** 1987. Optical properties of natural $PbCO_3$ single crystal. *Cryst. Latt. Def. Amorph. Mater.,* **16**, 365–70.

Takahashi, T., 1960. Supergene alteration of zinc and lead deposits in limestone. *Econ. Geol.,* **55**, 1084–115.

Temple, A. K., 1956. The Leadhills–Wanlockhead lead and zinc deposits. *Trans. R. Soc. Edinburgh,* **63**, 85–113.

Thomas, D. E., 1949. Ore deposits of the Wallapai district, Arizona. *Econ. Geol.,* **44**, 663–705.

Warne, S. St J., 1962. A quick field or laboratory staining scheme for the differentiation of the major carbonate minerals. *J. Sed. Petr.,* **32**, 29–38.

White, W. B., 1974. The carbonate minerals. In *The Infrared Spectra of Minerals:* Ed. V. C. Farmer, Min. Soc. Monograph **4**, 227–84.

Whitten, C. W. and **Stratton, M. A.,** 1990. Characterization of minerals within lead contaminated soils. In *Process Mineralogy IX: Applications to Mineral Beneficiation, Metallurgy, Gold, Diamonds, Ceramics, Environment and Health:* Ed. W. Petruk, *Min. Metall. Mater. Soc.,* 547–55.

Yamaguchi, J., Sawada, Y., Sakarai, O., Uematsu, K., Mizutani, N. and **Kato, M.,** 1980. Thermal decomposition of cerussite ($PbCO_3$) in carbon dioxide atmosphere (0–50 atm). *Thermochim. Acta,* **35**, 307–13.

Huntite $Mg_3Ca(CO_3)_4$

Trigonal (−)

ε or ω	~1.615
δ	Very strong
D	2.696
Colour:	White, colourless in thin section.
Unit cell:	a_{hex} 9.503, c_{hex} 7.821 Å, $Z_{hex}=3$. a_{rh} 6.075 Å, α_{rh} 102.93°, $Z_{rh}=1$. Space group *R*32

Soluble with effervescence in cold HCl (1:1).

Huntite occurs typically as an alteration or weathering product of dolomite- or magnesite-bearing rocks. It was first described by Faust (1953) and named in honour of Professor W. F. Hunt of the University of Michigan, USA.

Structure

The crystal structure of huntite was determined by Graf and Bradley (1962) and refined by Dollase and Reeder (1986). It is hexagonal with *a* 9.5027, *c* 7.8212 Å, Z=3 and space group *R*32. Both Ca and Mg atoms exist in the same plane but are ordered, with Ca atoms in a trigonal prism position and Mg atoms in an edge-sharing octahedral position. The major difference between huntite and dolomite is the configuration of the Ca polyhedron. The trigonal prism is not common for the six-coordinated Ca in carbonates but, as pointed out by Dollase and Reeder (1986), the arrangement is not unusual for the *REE* in a series of borates that are isostructural with huntite.

A projection of the huntite structure viewed along the *z* axis is shown in Fig. 103. As in the dolomite structure, there are layers of Ca- and Mg-polyhedra normal to the *z* axis. These polyhedra are linked within the layers by corner sharing with CO_3 groups. A quarter of these are the $C(1)O_3$ groups, which have symmetry 32 and are planar, equilateral triangular groups orientated parallel to (0001). The other three-quarters are the $C(2)O_3$ which have symmetry 2 and are planar but not equilateral triangular groups and not parallel to (0001).

Chemistry

Most huntite is very close in composition to $Mg_3Ca(CO_3)_4$, as shown by chemical analyses listed in Table 37. A Sr-bearing huntite (anal. 8) was reported from Geelong, Victoria, Australia, by Shayan (1984).

When huntite is heated, it dissociates in two steps, as illustrated by the differential thermal analysis (DTA) curve shown in Fig. 104 (Cole and Lancucki, 1982). The dissociation of $MgCO_3$ in huntite takes place at about 570 °C, and that of $CaCO_3$ at about 850 °C. A total weight loss of 45 per cent was recorded. DTAs by Faust (1953) and Baron *et al.* (1957) gave endothermic peaks at 644 and 901 °C, and 500

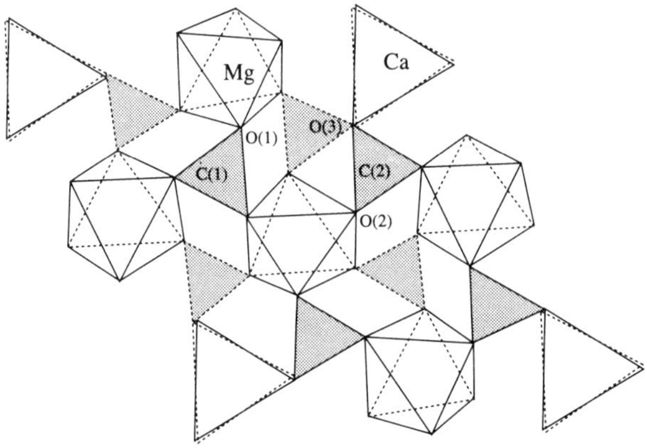

Fig. 103. The structure of huntite viewed along the z axis (after Dollase and Reeder, 1986). Solid and dashed lines represent polyhedral edges at $z = 16/$ to 1/3 and $z = 1/2$, respectively.

and 900 °C, respectively. The difference in the temperature recorded is apparently due to experimental variables. For the Sr-bearing huntite, the endothermic peaks at 537 and 760 °C, and a total weight loss of 44 per cent were observed (Shayan, 1984).

In aqueous solutions at 25 °C and 1 atm total pressure, huntite is metastable with respect to either dolomite or magnesite, dependent upon the activities of Ca^{2+} and Mg^{2+} and P_{CO_2}. Using the free energies of formation of calcium and magnesium carbonates and hydroxides, and aqueous species, Morse and Mackenzie (1990) constructed equilibrium relationships for stable and metastable mineral assemblages. These relationships are shown in Fig. 105. The Gibbs energy of formation of huntite at 298 K and 1 bar has been determined by Walling *et al.*, 1995.

Optical and Physical Properties

Huntite is white and massive with an earthy lustre; in thin section it is colourless and shows strong birefringence. Satisfactory determinations of the optical properties are not possible owing to the grain size of $<2\ \mu m$. An attempt to determine the refractive index by oil immersion gave a value of 1.615 (Vitaliano and Beck, 1963).

Huntite has an even to subconchoidal fracture, and is easily scratched with the fingernail. It is porous, and when placed in water it crumbles into smaller fragments; the displacement of air bubbles may then cause a hissing sound. The specific gravity was found to be 2.696 (Faust, 1953).

Infrared (IR) and Raman (R) spectra of huntite have been analysed by Sheetz and White (1977). The presence of two crystallographically distinct carbonate ions results in two independent internal-mode spectra with different frequencies and selection rules. A splitting of one set of ν_3 and ν_4 modes results from the low site symmetry of the tilted CO_3 ions. A complete assignment of the internal modes of huntite is as follows (in cm^{-1}): ν_1 1105 (IR) and 1122 (R); ν_2 858, 880 (IR) and 887 (R); ν_3 1435, 1460, 1510, 1535 (IR) and 1459 (R); and ν_4 705, 744 (IR) and 703, 724, 743 (R).

Table 37. Huntite Analyses

	1	2	3	4
MgO	34.09	34.4	36.00	34.19
CaO	15.42	16.0	13.53	15.40
H_2O^+	0.86	—	1.14	0.56 (H_2O)
H_2O^-	0.47	—	—	—
	0.06 (SiO_2)			0.01(P_2O_5)
CO_2	48.85(c)	50.4(c)	49.72(c)	49.59(c)
Total	99.75	100.8	100.39	99.75
	Numbers of ions on the basis of six (O)			
Mg	1.506	1.491	1.575	1.511
Ca	0.491	0.501	0.425	0.489
Sr	—	—	—	—
Mn	—	—	—	—
Fe	—	—	—	—
C	2.000	2.000	2.000	2.000

(c) Calculated from stoichiometry.

1 Huntite in dolomite and magnesite rocks from Currant Creek, Nevada, USA (Faust, 1953).
2 Huntite nodules, weathering product in dolomite, Tea Tree Gully, South Australia (Skinner, 1958).
3 Huntite in lightly serpentinized dolomite, Kurgashinkan Pb–Zn deposit, Uzbekistan (Golovanov, 1959).
4 Huntite from fractured and weathered carbonate, Nevada, USA (Vitaliano and Beck, 1963).

Table 37. *Continued*

	5	6	7	8
MgO	32.36	34.26	33.0	33.27
CaO	14.82	15.53	15.0	13.24
H_2O^+	—	1.03	1.30	—
H_2O^-	1.10	—	1.00	0.67
	0.35 (SO_3)		0.01 (MnO)	1.86 (SrO)
	5.91 (ins.)		1.13 (oth.)	1.90 (ins.)
CO_2	46.06(c)	48.19(c)	47.85(c)	48.18
Total	100.60	99.01	99.24	99.17
	Numbers of ions on the basis of six (O)			
Mg	1.505	1.508	1.507	1.528
Ca	0.495	0.491	0.492	0.437
Sr	—	—	—	0.033
Mn	—	—	—	0.001
Fe	—	—	—	0.001
C	2.000	2.000	2.000	2.000

5 Huntite in dolomitized marbles, central Rhodopes, Bulgaria (Maleev, 1974).
6 Huntite in weathered non-tronitized serpentine, Kempirsaya, southern Urals (Ivanov and Pal'gneva, 1976).
7 Huntite in weathered basalts at Deer Park near Melbourne, Australia (Cole and Lancucki, 1975).
8 Huntite in weathered basalts, Geelong, Victoria, Australia (Shayan, 1984). (Includes MnO 0.03, FeO 0.026%.)

Distinguishing Features

Its easy solubility and brisk effervescence in cold (1:1) HCl, HNO_3 or H_2SO_4 establish huntite as a carbonate, and its common occurrence as white chalk-like masses of high porosity may be used to distinguish huntite from crystalline carbonates. Further characterization of huntite requires X-ray diffraction or DTA evidence.

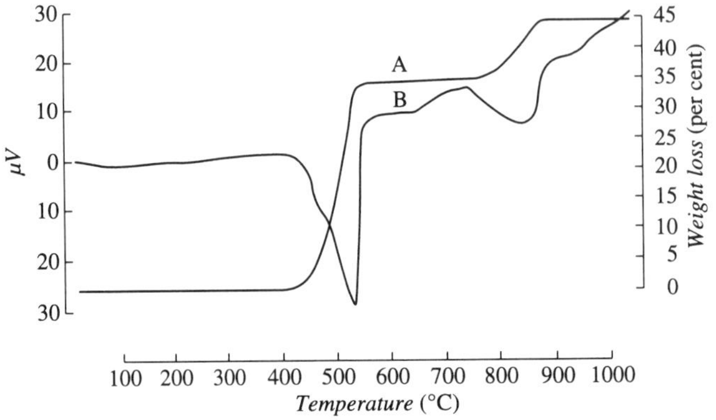

Fig. 104. Combined weight loss (A) and differential thermal analysis (B) curves for huntite from Deer Park, Victoria, Australia (after Cole and Lancucki, 1982).

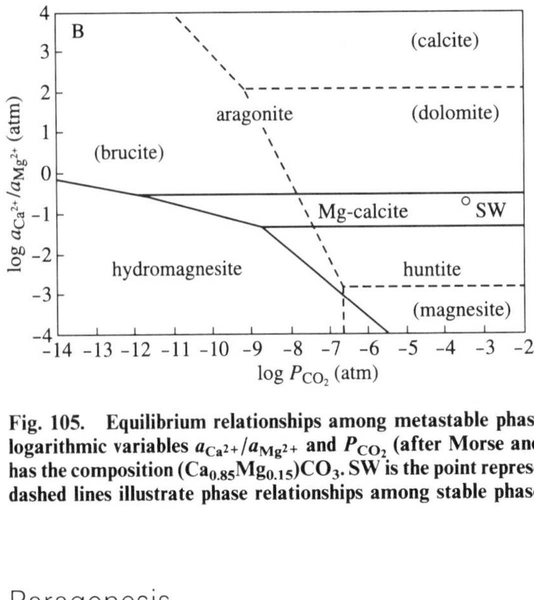

Fig. 105. Equilibrium relationships among metastable phases at 25 °C and 1 atm as a function of the logarithmic variables $a_{Ca^{2+}}/a_{Mg^{2+}}$ and P_{CO_2} (after Morse and Mackenzie, 1990). The magnesian calcite has the composition $(Ca_{0.85}Mg_{0.15})CO_3$. SW is the point representing the composition of average sea-water; dashed lines illustrate phase relationships among stable phases (shown in parentheses).

Paragenesis

The huntite described by Faust (1953) from Currant Creek, Nevada, USA, occurs in vugs and cavities in a bedded tuff formation with magnesite, dolomite and deweylite. Huntite was found to be the last mineral to fill the vugs and cavities. Huntite forms in caves in dolomite as fine-grained 'mountain-milk' in the Herault region of France (Baron *et al.*, 1957). Skinner (1958) reported huntite as a weathering product in the Torrens dolomite near Tea Tree Gully, South Australia. The huntite described by Golovanov (1959) was found as fracture filling in the weathered crust of lightly serpentinized dolomite in the Pb–Zn deposits of Kurgashinkan, Uzbekistan. At Crestmore, California, USA, Carpenter (1961) described an association of huntite, magnesium calcite and aragonite. Huntite is restricted to veinlets which do not continue below the zone of weathering in the carbonate rocks at Gabbs, Nevada, USA (Vitaliano and Beck, 1963). It appears to have been formed from solution percolating downwards from the surface. Diagenetic huntite was reported by Kinsman (1967) from a Recent supratidal evaporite environment along the Trucial Coast, Arabian Gulf. Huntite also occurs in weathered basalt at Deer Park, Melbourne, Victoria, Australia (Cole and Lancucki, 1975). A Sr-bearing huntite was reported from Geelong and Deer Park, Victoria, Australia by Shayan (1984). Kinsman (1967) summarized the environments of huntite formation as of three types: (a) as a cave carbonate in areas where the country rocks such as dolomite, are Mg-rich, (b) as a surface weathering product of Mg-rich rocks, such as magnesite, dolomite or serpentine, and (c) as a diagenetic mineral in Recent sediments where pore solutions are Mg-rich.

In the basal thrust zone of the Oman ophiolite, the weathering serpentine leads to highly magnesian groundwaters, which in turn result in the precipitation of huntite (Stanger and Neal, 1994). In this environment unusually high concentrations of Sr are found in the huntite.

References

Baron, C. S., Caillére, R., Lagrange, R. and **Pobeguin, T.**, 1957. Sur la presence de huntite dans une grotte de l'Herault (La Clamouse). *C. R. Acad. Sci. Paris*, **245**, 92–4.

Baron, G. S., Caillére, R., Lagrange, R. and **Pobeguin, T.**, 1959. Étude du Mondmilch de la grotte de la Clamouse et de quelques carbonates et hydrocarbonates alcalinoterreux. *Bull. Soc. Franç. Min. Crist.*, **182**, 150–8.

Carpenter, A. R., 1961. Mineral assemblages magnesium calcite–aragonite–huntite at Crestmore, California (abs.). *Geol. Soc. Amer.*, 1961 meeting, **22A.**

Cole, W. F. and **Lancucki, C. J.**, 1975. Huntite from Deer Park, Victoria, Australia. *Amer. Min.*, **60**, 1130–1.

Cole, W. F. and **Lancucki, C. J.**, 1982. Huntite from Deer Park, Victoria, Australia: a correction. *Amer. Min.*, **67**, 1290.

Dollase, W. A. and **Reeder, R. J.**, 1986. Crystal structure refinements of huntite, $CaMg_3(CO_3)_4$, with X-ray powder data. *Amer. Min.*, **71**, 163–6.

Faust, G. F., 1953. Huntite, $Mg_3Ca(CO_3)_4$, a new mineral. *Amer. Min.*, **38**, 4–23.

Golovanov, I. M., 1959. Huntite in the Kurgashinkan deposits. *Dokl. Akad. Nauk SSSR*, **124**, 398–401.

Graf, D. L. and **Bradley, W. F.**, 1962. The crystal structure of huntite, $Mg_3Ca(CO_3)_4$. *Acta Cryst.*, **15**, 238–42.

Ivanov, O. K. and **Pal'gneva, G. V.**, 1976. [Huntite from the weathering crust of serpentines of Kempirsaya (southern Urals).] *Tr. Inst. Geol. Geokhim., Ural. Nanchin. Tsentr., Akad. Nauk SSSR.*, **118**, 84–7 (in Russian).

Kinsman, D. J. J., 1967. Huntite from a carbonate evaporite environment. *Amer. Min.*, **52**, 1332–40.

Maleev, M. N., 1974. [Huntite from the area of the Mikhalkovo fluorite deposit, central Rhodopes.] *Mineragenesis*, Sofia, 243–51 (in Bulgarian with English summary).

Morse, J. W. and **Mackenzie, F. T.**, 1990. *Geochemistry of Sedimentary Carbonates* (Developments in Sedimentology, **48**), Amsterdam, Oxford, New York and Tokyo (Elsevier) 707 pp.

Scheetz, B. E. and **White, W. B.**, 1977. Vibrational spectra of the alkaline earth double carbonates. *Amer. Min.*, **62**, 36–50.

Shayan, A., 1984. Strontium in huntite from Geelong and Deer Park, Victoria, Australia. *Amer. Min.*, **69**, 528–30.

Skinner, B. J., 1958. Huntite from Tea Tree Gully, South Australia. *Amer. Min.*, **43**, 159–62.

Stanger, G. and **Neal, C.**, 1994. The occurrence and chemistry of huntite from Oman. *Chem. Geol.*, **112**, 247–54.

Vitaliano, C. J. and **Beck, C. W.**, 1963. Huntite, Gabbs, Nevada. *Amer. Min.*, **48**, 1158–63.

Walling, E. M., Rock, P. A. and **Casey, W. H.**, 1995. The Gibbs energy of formation of huntite, $CaMg_3(CO_3)_4$, at 298 K and 1 bar from electrochemical measurements. *Amer. Min.*, **80**, 355–60.

Nyerereite $Na_2Ca(CO_3)_2$

Orthorhombic (−)

α	1.5112
β	1.5333
γ	1.5345
δ	0.0233
$2V_\alpha$	29°
Orientation:	
$\alpha = z$, $\beta = x$, $\gamma = y$	
D	2.541
H	$\sim 2\frac{1}{2}$
Twinning:	Invariably twinned.
Colour:	Colourless.
Unit cell:	*a* 5.044, *b* 8.809, *c* 12.743 Å Z=4. Space group $Cmc2_1$

High solubility in warm water.

Nyerereite was discovered by Dawson (1962) in the carbonate lavas erupted by the Oldoninyo volcano in Tanzania, and has not, to date, been found elsewhere. Its ideal composition is $(Na_{0.82}K_{0.18})_2Ca(CO_3)_2$, representing the Na-rich end-member of the $Na_2Ca(CO_3)_2$–$K_2Ca(CO_3)_2$ series. The name is after Mr Julius Nyerere, a former president of Tanzania.

Fairchildite, and its low-temperature form bütschliite, are the potassium-rich end-members. Both have the composition $K_2Ca(CO_3)_2$, and their occurrences are restricted.

Structure

Nyerereite is orthorhombic. and has *a* 5.044, *b* 8.809, *c* 12.743 Å, Z=4 and space group $Cmc2_1$. Non-Bragg reflections have $h = n + m\delta$, where *n* and *m* are integers and $\delta = 0.383$, and indicate a structural modulation (McKie and Frankis, 1977). The orthorhombic phase converts to a hexagonal phase at 292 °C, with the disappearance of non-Bragg reflections. On further heating, this intermediate hexagonal phase changes to another hexagonal phase above 340 °C. Cell dimensions for the intermediate phase are *a* 10.10, *c* 12.75 Å, and for the high-temperature hexagonal phase are *a* 5.05 and *c* 12.85 Å. According to McKie and Frankis (1977), neither hexagonal phase is quenchable, and the naturally occurring nyerereite is an inversion product of the high-temperature hexagonal polymorph. Zemkorite, a kimberlite mineral found by Yegorov *et al.* (1988), however, may be a natural analogue of the high-temperature phase. It has a composition richer in sodium (Na/K 6.2) than nyerereite (Na/K=4.5), and its cell dimensions are *a* 10.06, *c* 12.72 Å.

All compositions between $Na_2Ca(CO_3)_2$ and $K_2Ca(CO_3)_2$ have structures which are derived from an aristotype of a hexagonal cell with $a \sim 5$, $c \sim 13$ Å. The aristotype is exemplified by the high-temperature phase of nyerereite and by the structure of the potassium-rich end-member, fairchildite (McKie and Frankis, 1977).

Pertlik (1981) examined the structure of fairchildite synthesized from solid-state reactions according to Mrose *et al.*'s procedure (Mrose *et al.*, 1966). Fairchildite is hexagonal and has *a* 5.294, *c* 13.355 Å and space group $P6_3/mmc$, one of three space groups suggested by Mrose *et al.* In the structure, there are two crystallographically different CO_3 groups. The $C(1)O_3$ group is orientated parallel to (001), and the plane of the $C(2)O_3$ group makes an angle of 69° with (001). The $C(2)O_3$ group exhibits orientational disorder around a common C(2) atom at 0,0,0. Two cation positions exist in the structure, *Me*(1) and *Me*(2), coordinated by seven and eight oxygen atoms, respectively. An ordering arrangement of Ca and K atoms to occupy these positions was not established, and a statistical distribution is possible.

Bütschliite is the low-temperature form of $K_2Ca(CO_3)_2$ with respect to fairchildite (Mrose *et al.*, 1966). A structural analysis performed by Pabst (1974) and refinement by Effenberger and Langhof (1984) assigned a hexagonal unit cell to it with *a* 5.3822, *c* 18.156 Å, Z = 3 and space group $R\bar{3}m$. The structure consists of a charged $[Ca(CO_3)_2]^{2-}$ sheet normal to the *z* axis. in which each Ca is coordinated by six oxygen atoms from six CO_3 groups, which are on the surrounding three-fold axes. K^+ ions are at two different levels and have a total of [6+3] coordination. The K^+ ions situated on the same three-fold axis as the Ca^{2+} ions of a particular sheet are displaced towards the opposite sheet; consequently each K^+ has a markedly polar coordination. The carbonate group was found by Knobloch *et al.* (1980) and Effenberger and Langhof (1984) to be slightly pyramidal in bütschliite. The deviation of the carbon atom from the plane defined by the three oxygen atoms of the CO_3 group is 0.014 Å towards the Ca^{2+} sheet, which is similar to the deviation found in ankerite and dolomite.

Chemistry

The chemical composition of nyerereite from Oldoinyo Lengai, Tanzania, is given in Table 38, which yields an idealized formula of $(Na_{0.2}K_{0.18})_2Ca(CO_3)_2$. Also listed are analytical results of a rounded brown-coloured mineral in association with nyerereite and a matrix consisting of both nyerereite and the brown mineral. Gittins and McKie (1980) assigned a formula of $(Na_{0.78}K_{0.05})_2Ca_{0.17}(CO_3)$ to the brown-coloured mineral, and proposed the name gregoryite. The chemical composition of zemkorite is also given in Table 38.

Fairchildite was synthesized by Mrose *et al.* (1966) from a solid-state reaction between K_2CO_3 and $CaCO_3$ at atmospheric pressure and in the temperature range 704–970 °C, whereas Pabst (1974) successfully prepared bütschliite from three different procedures: (a) spontaneous transformation from fairchildite, (b) solid-state reaction between K_2CO_3 and $CaCO_3$ in the temperature range 420–505 °C, and (c) reaction of a saturated solution of either $KHCO_3$ or $K_2CO_3 \cdot 1.5H_2O$ on $CaCO_3$ in the temperature range 55–60 °C.

Subsolidus phase relationships in the system Na_2CO_3–$CaCO_3$–H_2O are shown in Fig. 106. Two reactions are illustrated: (a) shortite + Na_2CO_3 → nyerereite, and (b) shortite → nyerereite + $CaCO_3$ (Frankis and McKie, 1973). The first reaction indicates that nyerereite has a lower temperature limit of stability, and the second

Table 38. Chemical Analyses of Nyerereite, Gregoryite and Zemkorite

	1	2	3
CaO	24.47	9.10	28.39
MnO	0.07	—	—
MgO	0.12	—	—
SrO	2.20	0.66	—
BaO	0.59	0.24	—
Na_2O	23.94	44.87	25.72
K_2O	7.90	3.95	6.40
CO_2	38.10(c)	36.22(c)	39.20(c)
SO_3	1.50	4.28	—
P_2O_5	0.36	1.92	—
F	0.54	0.40	—
Cl	0.26	0.60	—
H_2O	0.26	—	—
			0.05 (Al_2O_3)
Total	100.31	102.24	99.76

(c) Calculated from stoichiometry.

1 Nyerereite from carbonate lavas, Oldoinyo Lengai volcano, Tanzania (March 1960–September 1965) (McKie and Frankis, 1977).
2 Gregoryite in association with nyerereite, Oldoinyo, Lengai volcano, Tanzania (Gittins and McKie, 1980).
3 Zemkorite from Yakutian kimberlites, eastern Siberia (Yegorov *et al.*, 1988).

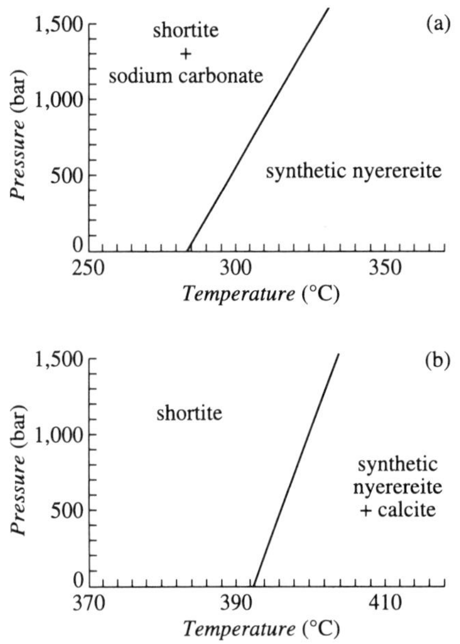

Fig. 106. Pressure–temperature diagrams for reactions (a) shortite + sodium carbonate → synthetic nyerereite and (b) shortite → synthetic nyerereite + calcium carbonate (after Frankis and McKie, 1973).

reaction shows that shortite has an upper temperature limit of stability at any pressure. In the system Na_2CO_3–K_2CO_3–$CaCO_3$, phase relationships at 1 kbar were investigated by Cooper *et al.* (1975). As shown in Fig. 107, there are four primary phase fields characterized by the initial crystallization of (Na_2CO_3–K_2CO_3–$CaCO_3$) solid solution, $(Na,K)_2Ca(CO_3)_2$ nyerereite–fairchildite solid

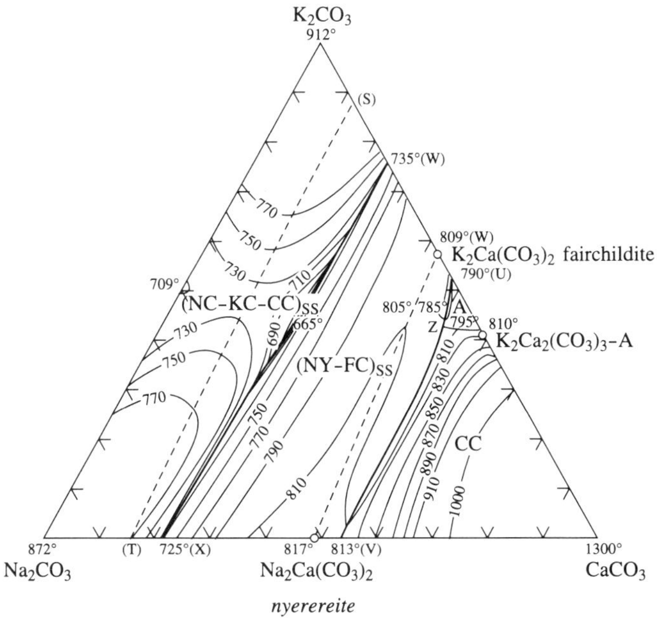

Fig. 107. Isobaric equilibrium diagram for the system Na_2CO_3–K_2CO_3–$CaCO_3$ at 1 kbar, showing the cotectic curves UV and WX and ternary reaction point Z; $(NC–KC–CC)_{ss}$ = ternary solid solutions, $(NY–FC)_{ss}$ = nyerereite–fairchildite solid solutions, A = $K_2Ca_2(CO_3)_3$, CC = $CaCO_3$, NC = Na_2CO_3, KC = K_2CO_3 (after Cooper *et al.*, 1975).

solution, calcite and $K_2Ca_2(CO_3)_3$, a phase that has no natural equivalent and is a compositional equivalent to shortite. The nyerereite–fairchildite join is a complete series of solid solution, with a minimum at 805 °C and a composition of $(K_{0.79}Na_{0.21})Ca(CO_3)_2$, which forms a thermal barrier to trends of liquid fractionation in the system. Compositions of $(NC–KC–CC)_{ss}$ lie along ST, which is based on the maximum solubility of $CaCO_3$ in NC_{ss} determined in the binary system.

Optical and Physical Properties

Nyerereite is biaxial and optically negative with a 2V value of 29°. The refractive indices are α 1.5112 and γ 1.5345. The intermediate refractive index β was evaluated as 1.5333 by measurement of the birefringence γ–β in sections of known thickness exhibiting a centred acute bisectrix figure. Nyerereite is colourless, tabular in habit and invariably twinned.

Zemkorite, the hexagonal form of $Na_2Ca(CO_3)_2$, is uniaxial and optically negative, with refractive indices ω 1.535 and ε 1.513. It is also colourless and has a pearly to glassy lustre. Measured specific gravity is 2.46 and hardness is 2. Zemkorite is readily soluble in warm water.

Both fairchildite and bütschliite are uniaxial and optically negative. Measured refractive indices are ω 1.533 and 1.605, and ε 1.498 and 1.453, respectively (Pabst, 1974). Fairchildite and bütschliite have a fused appearance on their surface, with colour varying from yellowish grey to white to greying black. In thin section both

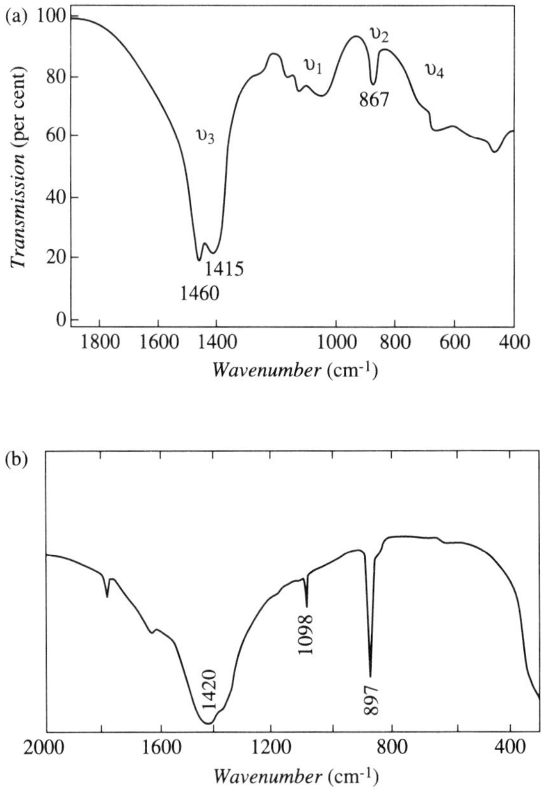

Fig. 108. (a) Infrared spectrum of fairchildite with probable assignments for the carbonate group vibrations (after Pertlik, 1981), and (b) infrared spectrum of bütschliite (after White, 1974).

reveal a texture of aggregates of radiating acicular crystals in association with scattered grains of charcoal and calcite.

Infrared spectra of minerals with compositions in the $Na_2Ca(CO_3)_2$–$K_2Ca(CO_3)_2$ series have not been investigated in detail. Only those of fairchildite and bütschliite are known, which are shown in Fig. 108 (White, 1974; Pertlik, 1981; Balascio and White, 1972). The broad absorption band of the ν_1 symmetric stretching vibrations is caused by the orientational disorder of the $C(2)O_3$ group.

Distinguishing Features

Nyerereite is invariably twinned, which can be detected under a binocular microscope. Crystals are tabular with {001} faces well developed. Its high solubility in warm water may be used to distinguish it from other rock-forming minerals. Both fairchildite and bütschliite occur in fused wood ash stones and always have a fused appearance on their surface.

Paragenesis

The occurrence of nyerereite is restricted to the carbonate lavas erupted by the Oldoinyo Lengai volcano, Tanzania. The lava is porphyritic, containing large, clear plates of nyerereite and rounded brown-coloured grains of gregoryite set in a matrix of the same minerals with an opaque sulphide mineral rich in Mn (Dawson, 1989). The temperatures of the lavas erupted in November 1988 were measured in flow channels and a lava pool. They are consistently in the range 580–590 °C.

References

Balascio, J. R. and **White, W. B.**, 1972. Hydrothermal growth of calcite from alkali carbonate solutions. *Materials Res. Bull.*, **7**, 1461–71.

Cooper, A. F., **Gittins, J.** and **Tuttle, O. F.**, 1975. The system Na_2CO_3–K_2CO_3–$CaCO_3$ at 1 kbar and its significance in carbonatite petrogenesis. *Amer. J. Sci.*, **275**, 534–60.

Dawson, J. B., 1962. Sodium carbonate lavas from Oldoinyo Lengai, Tanzania. *Nature*, **195**, 1075–6.

Dawson, J. B., 1989. Sodium carbonatite extrusions from Oldoinyo Lengai, Tanzania: implications for carbonate complex genesis. In *Carbonatites, Genesis and Evolution:* Ed. K. Bell, London (Unwin Hyman), pp. 255–77.

Effenberger, H. and **Langhof, H.**, 1984. On the aplanarity of the CO_3 group in buetschliite, dipotassium calcium dicarbonate, $K_2Ca(CO_3)_2$: a further refinement of the atomic arrangement. *Acta Cryst.*, **C40**, 1299–300.

Frankis, E. J. and **McKie, D.**, 1973. Subsolidus relations in the system Na_2CO_3–$CaCO_3$–H_2O. *Nature, Phys. Sci.*, **246**, 124–6.

Gittins, J. and **McKie, D.**, 1980. Alkali carbonate magma: Oldoinyo Lengai and its wider applicability. *Lithos*, **13**, 213–5.

Knobloch, D., **Pertlich, F.** and **Zemann, J.**, 1980. Crystal structure refinement of buetschliite and eitelite: a contribution to the stereochemistry of trigonal carbonate minerals. *Neues Jahrb. Min., Monat.*, 230–6.

McKie, D. and **Frankis, E. J.**, 1977. Nyerereite: a new volcanic carbonate mineral from Oldoinyo Lengai, Tanzania. *Z. Krist.*, **145**, 73–95.

Mrose, M. E., **Rose, H. J.** and **Marineko, J. W.**, 1966. Synthesis and properties of fairchildite and buetschliite: their relations to wood-ash stone formation. *Proc. Geol. Soc. Amer.*, San Francisco, 146 (abstr.).

Pabst, A., 1974. Synthesis, properties and structure of $K_2Ca(CO_3)_2$, buetschliite. *Amer. Min.*, **59**, 353–8.

Pertlik, F., 1981. Structural investigation of synthetic fairchildite, $K_2Ca(CO_3)_2$. *Z. Krist.*, **157**, 199–205.

White, W. B., 1974. The carbonate minerals. In *The Infrared Spectra of Minerals:* Ed. V. C. Farmer, Min. Soc. Monograph **4**, 227–84.

Yegorov, N. K., **Ushchapovskaya, Z. F.**, **Kashayev, A. A.**, **Bogdanov, G. V.** and **Sizykh, Ya. I.** Zemkorite, $Na_2Ca(CO_3)_2$, a new carbonate from Yakutian kimberlites. *Dokl. Akad. Nauk SSSR*, **301**, 188–93.

Malachite–Azurite

Malachite $Cu_2(OH)_2CO_3$

Monoclinic (–)

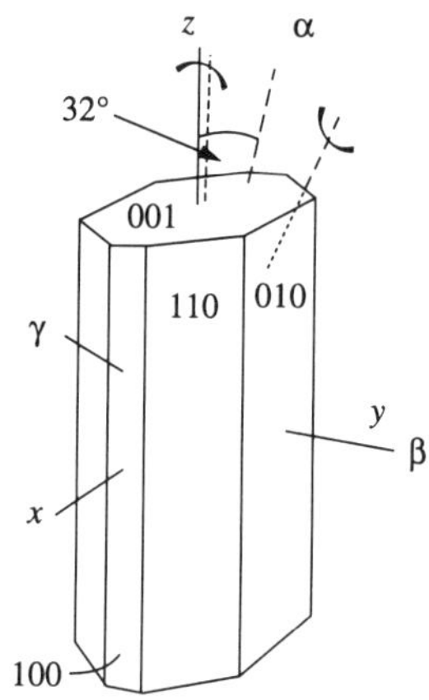

α	1.655
β	1.875
γ	1.909
δ	0.254
$2V_\alpha$	43°

Orientation
$\beta = y$, γ: $x = 23.5°$, α: $z = 32°$, O.A.P. {010}

Dispersion:	$v > r$ distinct.
D	3.7–4.1
H	$3\frac{1}{2}$–4
Cleavage:	($\bar{2}$01) perfect.
Twinning:	Simple twinning on {100}.
Colour:	Green and yellowish green in thin section.
Unit cell:	*a* 9.502, *b* 11.9674, *c* 3.240 Å, β 98.74° Z = 4. Space group $P2_1/a$

Soluble in acid with effervescence.

Azurite $Cu_3(OH)(CO_3)_2$

Monoclinic (+)

α	1.730
β	1.756
γ	1.836
δ	0.106
$2V_\gamma$	68°

Orientation
$\beta = x$, γ: $z = 12°30'$, α: $y = 15°$, O.A.P. $\perp\{010\}$

Dispersion:	$r > v$ distinct.
D	3.77–3.89
H	$3\frac{1}{2}$–4
Cleavage:	(011) perfect; (100) distinct.
Twinning:	Uncommon.
Colour:	Azure blue and blue in thin section.
Unit cell:	*a* 5.011, *b* 5.848, *c* 10.355 Å, β 92.43° Z = 2. Space group $P2_1/c$

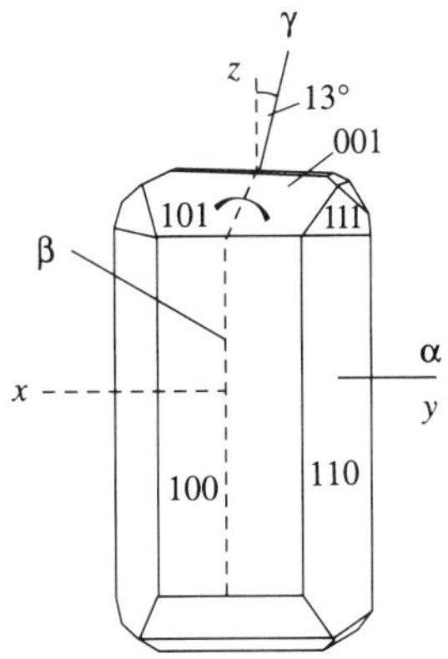

Soluble in acid with effervescence.

Malachite and azurite are two common minerals found in the oxidized zone of copper deposits, especially in regions where limestone is present. The most characteristic physical features of these two carbonate minerals are their colours: the banded grass green of malachite and the deep blue of azurite. Compact malachite may be cut and polished to a high lustre for ornamental pruposes.

The name malachite is derived from the Greek word *malache*, which means mallow, in reference to its green colour, and the name azurite is in allusion to its colour.

Structure

The crystal structure of malachite was determined by Wells (1951) and refined by Süsse (1967) and Zigan *et al.* (1977). It has a monoclinic unit cell with *a* 9.502, *b* 11.974, *c* 3.240 Å, β 98.74°, Z = 2 and space group $P2_1/a$. In the structure, copper atoms have two different environments: $Cu(O_2(OH)_4)$ and $Cu(O_4(OH)_2)$ octahedra. In each case, the spatial arrangement of the six coordinated neighbours is the same, i.e. four approximately coplanar with the copper atom at the corners of a square, and two more at a greater distance (Fig. 109). These octahedra share edges to form chains parallel to the *z* axis. The linked octahedera are also connected by the CO_3 groups. The cleavage $(20\bar{1})$ is parallel to the planes of the carbonate groups. In azurite, all copper atoms are in square planar coordination by 2(O) + 2(OH); further neighbouring oxygen and hydroxyls are in rather irregular positions. The square groups are linked into chains parallel to the *y* axis. Each (OH) group is shared by three Cu^{2+} and each oxygen of the CO_3 group is bonded to one copper atom (Fig. 110) (Gattow and Zemann, 1968; Zigan and Schuster,

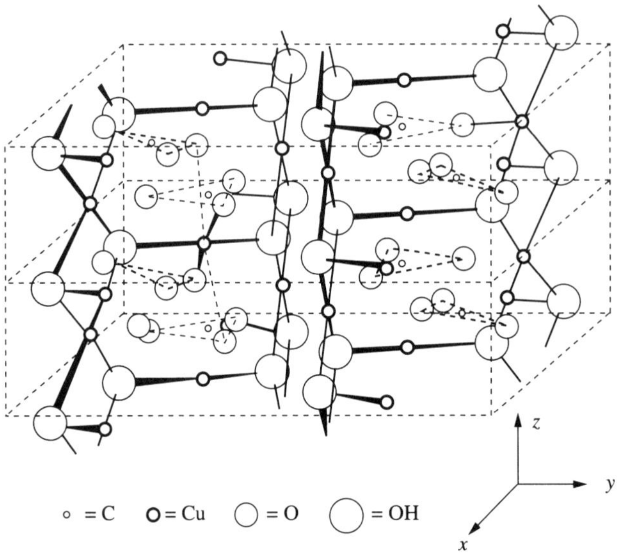

Fig. 109. The crystal structure of malachite (two unit cells) (after Wells, 1951).

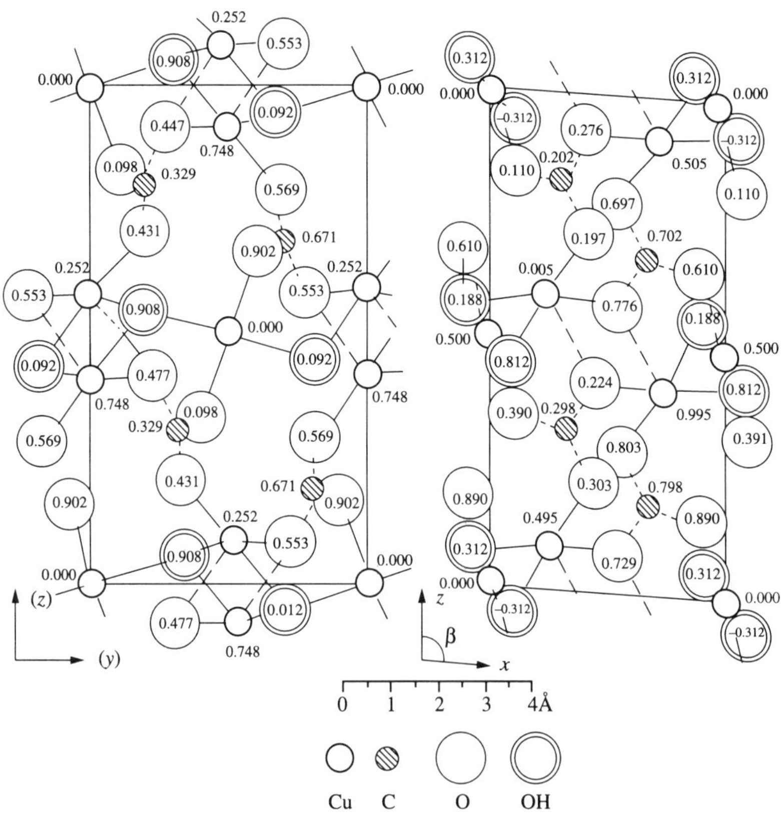

Fig. 110. Projection of the structure of azurite on (100) and (010) (after Gattow and Zemann, 1958).

1972). Azurite has a monoclinic unit cell with *a* 5.0109, *b* 5.8485, *c* 10.345 Å, β 92.43°, Z=4 and space group $P2_1/c$.

Chemistry

Most malachites have compositions close to the ideal $Cu_2(OH)_2CO_3$, with 71.9 per cent CuO, 19.9 per cent CO_2 and 8.2 per cent H_2O. Malachite from old dumps in the form of crystalline crust on the verrucano rocks from Piesky near Banská Bystrica, Czech Republic (Řídkošil, 1981) has a composition of 69.82 per cent CuO, 19.40 per cent CO_2 and 7.43 per cent H_2O, with 1.70 per cent insolubles. Zinc may replace Cu in Cu^{2+}, and Zn-rich varieties are known as rosasite. Analyses of rosasite from Rosas, Sardinia, Durango, Mexico, and Bakara, Bulgaria, give Cu/Zn ratios of 41.5/12.7, 31.6/22.2 and 37.50/31.09, respectively (Nickel and Berry, 1981). Cobaltiferrous malachite from Shaba, Zaïre, was described by Deliens *et al.* (1973) to have a cobalt content ranging from 0.25 to 18.46 per cent. Two groups are distinguished: the cobaltiferous malachite, which has a low cobalt content of 0.25–4.22 per cent, and the rosasite-type minerals, which have the higher Co content (16.05–18.46 per cent). The Co-rich member of the series has been named kolwézite by Deliens and Piret (1980). Glaukosphaerite, $(Cu,Ni)_2(OH)_2CO_3$, and mcquinnessite, $(Mg,Cu)_2(OH)_2CO_3$, are Ni- and Mg-equivalents, respectively, of rosasite (Pryce and Just, 1974; Erd *et al.*, 1981). The composition of azurite appears to vary little from its ideal $Cu_3(OH)_2(CO_3)_2$. An analysis of Piesky azurite gave 66.44 per cent CuO, 24.42 per cent CO_2, 5.55 per cent H_2O and 3.00 per cent insolubles. The compositions of these two minerals differ only in the ratio of CuO, CO_2 and H_2O—in azurite it is 3:2:1 and in malachite 2:1:1. The occurrence of malachite pseudomorphs after azurite is due to the reaction 2 azurite + H_2O = 3 malachite + CO_2.

The dissociation of malachite begins at 310 °C and ends at 420 °C with a differential thermal analysis peak at 375 °C; for azurite dissociation occurs in the temperature range 290–395 °C, with a peak at 350 °C. In both cases, water is released at about 100 °C (Simpson *et al.*, 1964; Seguin, 1975). Stabilities of malachite and azurite in the system Cu–O–H–CO_2 have been calculated by Vink (1986), who illustrated their relationships in Eh–pH diagrams (Fig. 111). In terms of CO_2 partial pressure, the malachite–azurite transition is very sensitive because the equilibrium CO_2 pressure is $10^{-3.45}$ bar, which is close to the normal partial pressure of CO_2 in the atmosphere. Azurite forms at pH values between 7 and 8 at $P_{CO_2} < 10^{-3.45}$ bar, and the corresponding pH value to $P_{CO_2} = 10^{-3.45}$ bar is 6.95. From his calculation, Vink (1986) concludes that azurite can be formed only under acidic conditions at relatively high carbonate activities. Such conditions are rare in nature.

Malachite dissociation and precipitation may be expressed by the equation (Symes and Kester, 1984):

$$Cu_2(OH)_2CO_{3(s)} \rightleftharpoons 2Cu^{2+}_{(aq)} + 2OH^-_{(aq)} + CO^{2-}_{3(aq)},$$

and the equilibrium constant at 25 °C and 1 atm for infinite dilution is $K = 3.5 \pm 0.6 \times 10^{-34}$. For a solution of 0.72 ionic strength, $K = 1.0 \pm 0.2 \times 10^{-32}$, and for 36.9‰ salinity sea-water, $K = 1.3 \pm 0.1 \times 10^{-28}$. The stability fields of malachite and azurite were established by Magalhães (1991) on the basis of their Gibbs energies of formation and experimental determinations of their thermodynamic constants.

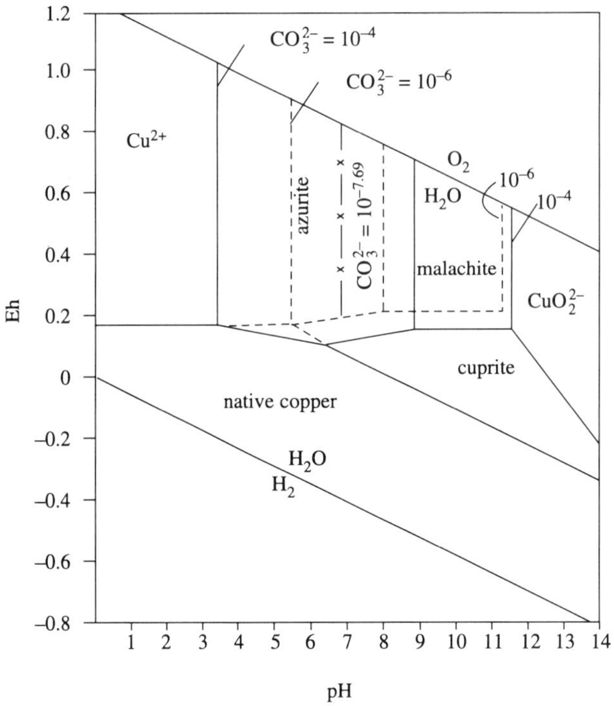

Fig. 111. Eh–pH diagram of the system Cu–O–H–CO_3^{2-} at 25 °C and 1 atm total pressure (after Vink, 1986). Solid and dashed lines are log (molality) $CO_3^{2-} = -4$ and log (molality) $CO_3^{2-} = -6$, respectively; total dissolved copper ion is 10^{-6} molal.

Single crystals of both malachite and azurite of millimetre size have been produced hydrothermally from bicarbonate solutions at 300 °C and 3.4 kbar (Ruszala and Kastiner, 1974).

Optical and Physical Properties

Malachite is biaxial and optically negative, with a 2V of 43°; azurite is biaxial and optically positive, with a 2V of 68°. The maximum birefringence of malachite is very extreme (0.244), seen in the {010} section; that of azurite is extreme (0.107), seen in the {100} section. Interference colours for both minerals are high-order greys in sections or fragments. However, they are usually masked by the green colour in malachite and the blue colour in azurite. Cleavage fragments and sections of malachite cut parallel to $\{\bar{2}01\}$ yield nearly centred acute bisectrix figures showing myriads of isochromes on a blue field. Refractive indices for both malachite and azurite do not vary significantly, because there is relatively little compositional variation. With Co substitution for Cu in malachite, Deliens *et al.* (1973) observed an increase in α with cobalt content. Per cent Co values of 0.25, 0.90, 1.79, 4.22, 16.05 and 18.46, give corresponding α refractive index values of 1.656, 1.656, 1.675, 1.684, 1.688 and 1.692, respectively.

The most characteristic physical features of these two carbonate minerals are their colours: the banded green of malachite and the deep blue of azurite. The

Cu–(O,OH) interatomic distances range from 1.93 to 1.99 Å in azurite and 1.92 to 2.12 Å in malachite. There are additional near-neighbour oxygens and/or hydroxyls bonded to Cu^{2+} cations in azurite (at 2.36–2.98 Å) and in malachite (2.35–2.63 Å), and these seemingly small variations in bonding produce significant differences in the ways in which the *d* electrons of Cu^{2+} absorb energy from a beam of light (Ribbe and Erikson, 1991); thus the idiochromatic blue and green colours are due to a 'crystal field' effect. The hardness of malachite is $3\frac{1}{2}$–4, and its specific gravity ranges from 3.7 to 4.1. Azurite has a hardness of $3\frac{1}{2}$–4 and a specific gravity of 3.77–3.89.

Malachite usually occurs in fibrous, radiate, stalactitic, granular or earthy masses, or as small druse crystals covering other copper minerals. Well-defined crystals are usually twinned on {100}. Some crusts are coated with minute crystals resembling velvet. Azurite is more often found in crystal form than is malachite, and its crystals are usually tabular, prismatic, or wedge-shaped. Radiated aggregates of columnar crystals may assume a stalactitic form.

Goldsmith and Ross (1968) measured the far infrared spectra of malachite and azurite and performed a normal coordinate analysis assuming a square-planar model. The spectrum of malachite is more complex than that of azurite which, as suggested by White (1974), indicates both the increased size of the unit cell and a stronger interaction between the square-planar group and the remainder of the structure.

Distinguishing Features

Malachite is characterized by its bright green colour and botryoidal form, and is distinguished from other green copper minerals by its effervescence with HCl. Azurite is chiefly recognized by its deep blue colour, and its intimate association with malachite.

Paragenesis

Malachite is widespread as a secondary mineral in the oxidized zone of copper deposits, especially in regions where limestone is present. Associated minerals are limonite, calcite, chalcedony and chrysocolla. Azurite forms under the same geological conditions as malachite but is considerably less abundant.

Malachite is the most important mineral in the thick oxidized zone of copper deposits in Katanga, Congo (Gray, 1930; Oosterbosch, 1951). In association with chrysocolla, it is contained in dolomite, dolomitic sandstone and cherty rocks. At Tsumeb, Namibia, malachite and azurite occur as secondary copper minerals in folded limestones and silicified dolomites. Both hypogene and supergene chalcocite are abundant in the sulphide ores (Schneiderhöhn, 1929).

The Bisbee mining district of Arizona, USA, is known to be a producer of oxidized ores contained in limestones. Portions of the orebodies of the Copper Queen mine consist almost entirely of malachite and azurite, with small amounts of cuprite and native copper (Knopf, 1933).Malachite and azurite are also present in the rich oxide ores from contact zones and replacement deposits in limestone of the Morenci–Metcalf district in Arizona, USA (Knopf, 1933). In the Triassic sandstone of Sierra Nacimiente, New Mexico, USA, the stratabound copper deposits contain malachite and azurite as oxidation products of chalcocite, covellite, bornite and chalcopyrite (Woodward *et al.*, 1974).

References

Deliens, M. and **Piret, P.**, 1980. La kolwézite, un hydroxycarbonate de cuivre et de cobalt analogue à la glaukosphaerite et à la rosasite. *Bull. Min.*, **103**, 179–84.

Deliens, M., **Oosterbosch, R.** and **Verbeck, T.**, 1973. Les malachites cobaltifères du Shaba méridional (Zaïre). *Bull. Soc. Franç. Min. Crist.*, **96**, 371–7.

Erd, R. C., **Goff, F. E.**, **Cesbron, F. P.** and **Clark, J. R.**, 1981. Mcquinnessite, a new carbonate from California. *Min. Rec.*, **12**, 143–7.

Gattow, G. von and **Zemann, J.**, 1958. Neubestimmung der Kristallstruktur von Azurit, $Cu_3(OH)_2(CO_3)_2$. *Acta Cryst.*, **11**, 866–72.

Goldsmith, J. A. and **Ross, S. D.**, 1968. The infrared spectra of azurite and malachite. *Spectrochim. Acta*, **24A**, 2131–7.

Gray, A., 1930. The correlation of the ore-bearing sediments of the Katanga and Rhodesian Copper belt. *Econ. Geol.*, **25**, 783–804.

Knopf, A., 1033. *Ore Deposits of the Western States*. New York (Amer. Inst. Mining Metall. Engr.), 797 pp.

Magalhães, M. C. F., 1991. Análise termodinamica de associacões de carbonatos, fosfatos le sulfatos de cobre (II). *Revista de Universidade de Aveiro Geociências*, **6**, 123–31.

Nickel, E. H. and **Berry, L. G.**, 1981. The new mineral, nullaginite and additional data on the related minerals rosasite and glaukosphaerite. *Can. Min.*, **19**, 315–24.

Oosterbosch, R., 1951. Copper mineralization in the Fungurumu region, Katanga. *Econ. Geol.*, **46**, 121–48.

Pryce, M. W. and **Just, J.**, 1974. Glaukosphaerite, a new nickel analogue of rosasite. *Min. Mag.*, **39**, 737–43.

Ribbe, P. H. and **Erikson, S. C.**, 1991. Azurite and malachite: chemically and structurally related minerals. *Min. Rec.*, **22**, 65–6.

Řídkošil, T., 1981. [Azurite and malachite from Piesky near Banská Bystrica.] *Časopis Pro Min. Geol.*, **26**, 263–71 (in Czech with English summary).

Ruszala, F. and **Kastiner, E.**, 1974. The hydrothermal synthesis of single crystals of azurite and malachite. *J. Cryst. Growth*, **26**, 155–6.

Schneiderhöhn, H., 1929. Das Otabivergland und seine Erzlagerstätten. *Z. Prakt. Geol.*, **37**, 85–116.

Seguin, M. K., 1975. Thermogravimetric and differential thermal analysis of malachite and azurite in inert atmospheres and in air. *Can. Min.*, **13**, 127–32.

Simpson, D., **Fisher, R.** and **Libsch, K.**, 1964. Thermal stability of azurite and malachite. *Amer. Min.*, **49**, 1111–4.

Süsse, P. Von., 1967. Verfeinerung der Kristallstruktur des Malachits, $Cu_2(OH)_2CO_3$. *Acta Cryst.*, **22**, 146–51.

Symes, J. L. and **Kester, D. R.**, 1984. Thermodynamic stability studies of the basic copper carbonate mineral, malachite. *Geochim. Cosmochim. Acta*, **48**, 2219–29.

Vink, B. W., 1986. Stability relations of malachite and azurite. *Min. Mag.*, **50**, 41–7.

Wells, A. F., 1951. Malachite: re-examination of crystal structure. *Acta Cryst.*, **4**, 200–4.

White, W. B., 1974. The carbonate minerals. In *The Infrared Spectra of Minerals:* Ed. V. C. Farmer, Min. Soc. Monograph, **4**, 227–84.

Woodward, L. A., **Kaufman, W. H.**, **Schumacher, O. L.** and **Talbutt, L. W.**, 1974. Stratabound copper deposits in Triassic sandstone of Sierra Nacimiento, New Mexico. *Econ. Geol.*, **69**, 108–20.

Zigan, F. and **Schuster, H. D.**, 1972. Verfeinerung der Struktur von Azurit, $Cu_3(OH)_2(CO_3)_2$ durch Neutronenbengung. *Z. Krist.*, **135**, 416–36.

Zigan, F., **Joswig, W.**, **Schuster, H. D.** and **Mason, S. A.**, 1977. Verfeinerung der Struktur von Malachit, $Cu_2(OH)_2CO_3$ durch Neutronenburgung. *Z. Krist.*, **145**, 412–26.

Phosphates

Apatite $Ca_5(PO_4)_3(OH,F,Cl)$
Monazite $(Ce,La,Nd,Th)PO_4$

Apatite $Ca_5(PO_4)_3(OH,F,Cl)$

Hexagonal (–)

ε	1.624–1.665
ω	1.629–1.667
δ	0.001–0.007
Dispersion:	Moderate
D	3.1–3.35
H	5
Cleavage:	$\{0001\}$ and $\{10\bar{1}0\}$, poor.
Twinning:	Twin plane $\{11\bar{2}1\}$ or $\{10\bar{1}3\}$, rare.
Colour:	Green, white, yellow, blue, etc; generally colourless in thin section.
Pleochroism:	Coloured varieties may show weak to moderate pleochroism with absorption $\varepsilon > \omega$.
Unit cell:	*a* 9.36–9.61, *c* 6.76–6.91 Å Z=2. Space group $P6_3/m$

Soluble in HNO_3 or in HCl.

Members of the apatite group are common accessory minerals in almost all igneous rocks and are also found in sedimentary and metamorphic rocks. They are the most abundant phosphorus-bearing minerals, the commonest varieties being represented in the isomorphous series with end-members:

Fluorapatite	$Ca_5(PO_4)_3F$
Chlorapatite	$Ca_5(PO_4)_3Cl$
Hydroxylapatite	$Ca_5(PO_4)_3OH$
Carbonate-apatite	$Ca_5[(PO_4)_{3-x}(CO_3,F)_x](F,OH,Cl)$

Of these, fluorapatite is by far the commonest, and the term apatite is sometimes used synonymously with fluorapatite. The name is from the Greek *apate*, deceit, the mineral having been confused with beryl, olivine, tourmaline, etc. Dahllite is a name given to a variety of carbonate-hydroxylapatite, after the brothers T. and J. Dahll, Norwegian geologists, whereas francolite, carbonate-fluorapatite $Ca_5(PO_4,CO_3,OH)_3F$, was named after its occurrence at Wheal Franco, Devon, England.

Structure

The crystal structure of fluorapatite was determined by Náray-Szabó (1930), Mehmel (1930) and Hendricks *et al.* (1932), and was re-examined in detail by Beevers and McIntyre (1946) (see also Hayashi, 1960). Ca atoms occupy two distinct sites: Ca(1) in seven-fold and Ca(2) in nine-fold coordination. Each F atom is surrounded by three Ca atoms at one level, and in addition Ca–O columns are linked with PO_4 groups forming a hexagonal network (Fig. 112). This arrangement

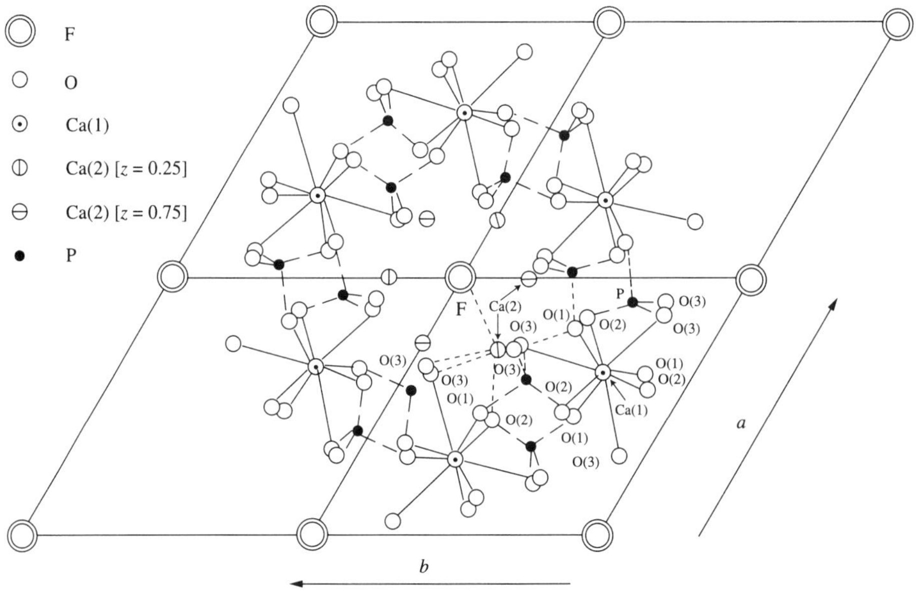

Fig. 112. Schematic depiction of a portion of four unit cells of the fluorapatite structure projected onto the (0001) plane (after Bragg and Claringbull, 1965). Solid, dashed and dotted lines indicate bonds in the different polyhedra (Hughes *et al.*, 1989).

gives a very stable structure to fluorapatite; the structure of hydroxylapatite is slightly expanded compared with that of fluorapatite, and hydroxylapatite is less stable.

More recently much work has been reported on synthetic apatites, and the structures of synthetic chlorapatite and synthetic hydroxylapatite were described by Mackie *et al.* (1972) and Elliott *et al.* (1973), both of these being referred to the monoclinic space group $P2_1/b$. The hexagonal $P6_3/m$ structure of both synthetic and natural fluorapatite was confirmed by Sudarsanan *et al.* (1972).

Three new structure refinements on natural $P6_3/m$ fluorapatite, chlorapatite and hydroxylapatite, all of near end-member composition, have been described in detail by Hughes *et al.* (1989). The hydroxylapatite examined was the well-known greenish yellow material from Durango, Mexico (Young *et al.*, 1969) with $(F_{0.94}Cl_{0.02})$ as shown by microprobe and instrumental neutron activation analyses. Hydroxylapatite was represented by a specimen from Holly Springs, Georgia, USA, with an (F,Cl,OH)$=X$ group formula $(F_{0.06}Cl_{0.03}OH_{0.91})$. The chlorapatite was from Kragero, Norway, with $(F_{0.09}Cl_{0.88})$ in the X position. Changes induced in natural apatites as a result of different anions are propagated throughout the crystal structure; although the effects on the PO_4 tetrahedra and $Ca(1)O_9$ polyhedra are minor, those on the $Ca(2)O_5X(O)$ polyhedra are significant. Thus, the Ca(2)–X bond length is 2.311 Å for fluorapatite, 2.385 Å for hydroxylapatite and 2.759 Å for chlorapatite (Hughes *et al.*, 1989). The (F,Cl,OH) ions in these three natural apatites lie on [0001] hexads. The F anions lie in the special position at the intersection of the hexads with mirror planes at $z=0.25$ and 0.75; each F is bonded to three Ca(2) atoms, which form a triangle within the

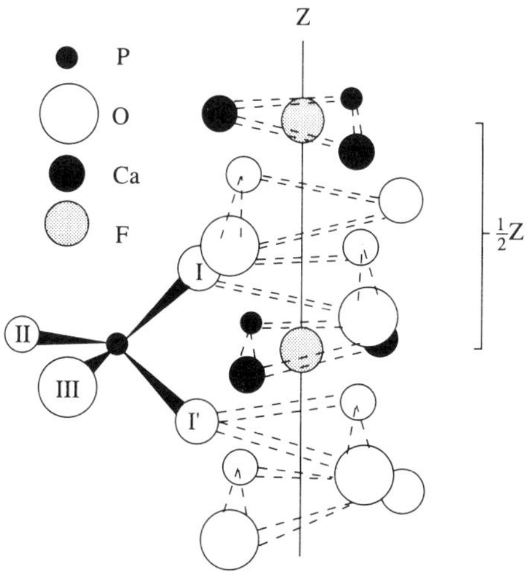

Fig. 113. **A section of the fluorapatite structure showing Ca- and O-triangle arrangements on the screw axis (X-ion column) and an attached PO_4 group (after Binder and Troll, 1989).**

mirror plane (Fig. 113). The OH ions are disordered 0.35 Å above or below the mirror planes, and Cl ions are disordered 1.2 Å above or below the planes.

In pure synthetic chlorapatite, Cl ions are all ordered above the plane in one anion column and below the plane in adjacent columns along *b*, giving a monoclinic structure ($P2_1/b$) with $b_{mon} = 2b_{hex}$ (Mackie *et al.*, 1972). A natural monoclinic chlorapatite (Table 40, anal. 2) was described by Hounslow and Chao (1970). In the near end-member chlorapatite hexagonal apatite, whose structure was refined by Hughes *et al.* (1989), no ordering was observed, though it is possible that the sense of ordering produced by clusters of Cl anions may be reversed by F ion 'impurities'. Thus the structure may contain regions where the anions are locally ordered, while overall the average apatite structure has $P6_3/m$ symmetry (Fig. 114); see also Sänger and Kuhs (1992) who suggest that the $P6_3/m$ space group describes only a time/space average. The transition in chlorapatite from the monoclinic ferroelastic to the hexagonal paraelastic phase was investigated experimentally by Bauer and Klee (1993) who considered that the transition occurs near 360 °C and is weakly discontinuous.

A study of a fluorhydroxylapatite showing sectoral texture and having optically biaxial properties led Akizuki *et al.* (1994) to suggest that although the F/OH sites in hexagonal apatite are symmetrically equivalent in the solid crystal, this equivalence may be lost at a growth surface resulting in a reduction in symmetry. Heating of this apatite to 850 °C caused the disappearance of the optical anomalies due to disordering, which may be related to the loss of OH from the crystal surface.

In further work on the structure of ternary apatites, Hughes *et al.* (1990) describe the structure of natural hexagonal ternary apatite with the *X* position formula ($F_{0.39}Cl_{0.33}OH_{0.28}$), and of a previously unrecognized monoclinic variant in space group $P2_1/b$ with *X* group formula ($F_{0.29}OH_{0.24}Cl_{0.47}$). This monoclinic ternary apatite differs from the chlorapatite considered by Hounslow and Chao (1970) in that all three anion species are accommodated; the reduction in symmetry results from ordering of the *X* group anions into one of the two symmetry-equivalent column-anion sites present in hexagonal ternary apatite. The correlation between

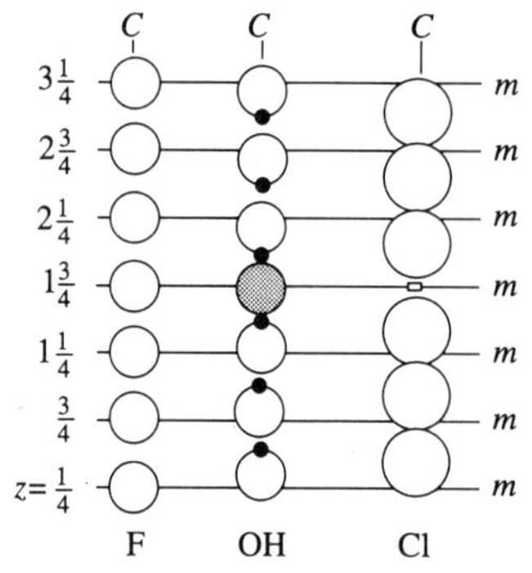

Fig. 114. Anion columns in hexagonal fluorapatite, hydroxylapatite and chlorapatite. Column 'F' shows F anions in fluorapatite located on mirror planes at $z = \frac{1}{4}, \frac{3}{4}$ in successive unit cells. Column 'OH' shows three successive hydroxyls in hydroxylapatite disordered 0.35 Å above the mirror planes (m) and three successive hydroxyls disordered below the mirror planes, with the sense of disordering reversed by an F anion (stippled) impurity at $z = 1\frac{3}{4}$. Column 'Cl' shows three successive Cl anions in chlorapatite disordered above the planes. The vacancy (□) at $z = 1\frac{3}{4}$ in chlorapatite must exist to reverse the sense of ordering as F and OH species are prohibited (after Hughes *et al.*, 1989).

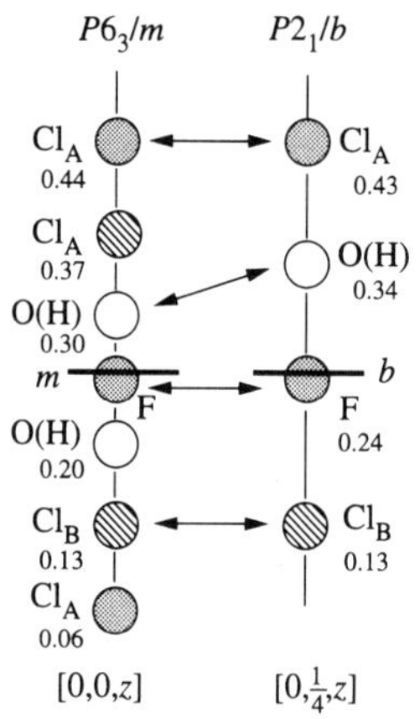

Fig. 115. Correlation of anion sites in $P6_3/m$ and $P2_1/b$ ternary apatites. Atoms represent possible anion sites associated with the mirror plane ($P6_3/m$) or the *b* glide plane ($P2_1/b$) at $z = \frac{1}{4}$ (the *z* positional parameter is given). The shift of the O(H) position between structures is largely a result of the difference in Cl content (after Hughes *et al.*, 1990).

the column-anion sites is shown in Fig. 115. Solid solution in ternary apatite is made possible by a 0.4 Å shift of the Cl atom relative to its position in end-member chlorapatite, enabling an OH neighbour to occupy the position at the next column-anion site at a distance of 2.96 Å.

The structure of an apatite with >1.2 wt. per cent FeO has been found to be monoclinic ($P2_1/m$), due to small shifts of the three subsymmetry equivalents of the $P6_3/m$ site (Hughes *et al.*, 1993). It is suggested that Fe^{2+} preferentially enters these Ca(2) sites in contrast to Mn^{2+}, an ion geochemically similar to Fe^{2+} but which prefers the Ca(1) site. The preference of Fe^{2+} for more distorted sites with a lower coordination number results in ordering into the apatite Ca(2) site; the iron ion is too small for extensive substitution in this site and thus is generally excluded from the apatite structure.

The structural differences discussed so far have been controlled by the variation in content of the column-anion sites, but other structural changes are produced by larger cations partially substituting for calcium. The entry of the smaller Mn^{2+} ion does not cause degeneration of the symmetry from $P6_3/m$. The Mn^{2+} ion preferentially occupies the larger apatite Ca(1) site; the nine O atoms coordinating this site more effectively satisfy the Mn bond valence than the seven ligands of the Ca(2) site (Hughes *et al.*, 1991). This is in contrast to the conclusions of Suitch *et al.* (1985). The Sr^{2+} ion, however, which is larger than Ca^{2+}, is ordered almost completely into the Ca(2) site in the apatite structure. Watson *et al.* (1985) reported that Sr diffusivity rates in apatite were slightly greater perpendicular to the axis than parallel to the axis; if this is so, and as movement through interstitial sites is not a viable diffusion mechanism in the apatite structure, diffusion must be aided by vacancies or structural defects (Hughes *et al.*, 1991). Structural refinements on synthetic lead-substituted calcium hydroxylapatites produced at 900 °C show that Pb also preferentially replaces the Ca(2) atoms in the structure.

Carbonate-apatites which contain appreciable amounts of CO_2 as an essential constituent of the mineral are of widespread occurrence. It has been established beyond doubt that the CO_2 is not present as carbonate impurities or adsorbed on the surface of the crystallites; the carbonate-apatites have distinctive X-ray patterns and rather small *a* parameters. An empirical relationship between the CO_2 content of apatite and the separation Δ of the 211 and 112 X-ray powder diffraction lines (in Å) has been given by O'Brien *et al.* (1990):

$$\text{wt. per cent } CO_2 \text{ in apatite} = 17.335 - (615.524 \times \Delta).$$

The main remaining problem is the exact role of carbon in the apatite structure, i.e. whether the C substitutes for P and/or Ca, and whether there is any effect on the *X* group anions. Little detailed structural work has been carried out on these minerals, though there is convincing evidence from chemical analyses that there is substitution of the CO_3 ion for the PO_4 ion; thus for the carbonate-bearing apatite of Table 42, anal. 4, Gulbrandsen *et al.* (1966) concluded that not only C replaced P but that some excess F was also in the P position, while there is a balanced substitution of Na for the rare earth ions, giving the formula for this apatite as $(Ca_{9.84}Mn_{0.02}Sr_{0.02}Na_{(0.06}REE_{0.06)})(P_{5.87}C_{0.13}O_{23.87}F_{0.13})F_2$. Further work, mainly on biological material, has been reported by McConnell (1970) and Montel (1971); the problem is also discussed in the Chemistry section (p. 314). A review of coupled anion substitution in natural carbon-bearing apatites has been given by Binder and Troll (1989), who concluded that $CO_3^{2-} \cdot F^-$ replaces PO_4^{3-}, with tetravalent carbon substituting for pentavalent phosphorus and a halogen in excess replacing one O^{2-} to give $Ca_{10}[(PO_4)_{6-x}(CO_3,F)_x](F,OH,Cl)_2$. Thus, the tetrahedral coordination of the cations involved is retained and the electrical neutrality of the structure is maintained. Partial replacement of P^{5+} by the smaller C^{4+} ion reduces the lateral distance between the Ca–O columns, leading to a smaller *a* dimension. The cell dimension in the *z* direction is mainly determined by the distance apart of Ca- and O-triangles from the channel anion; hence the *c* dimension remains relatively uniform, with minor amounts of substitution of P for C. That the fundamental crystal chemical substitution in francolite is CO_3^{2-} for PO_4^{3-} has been confirmed statistically for 110 samples by McClellan and Kauwenberg (1990), who also proposed the excess fluorine in the $(CO_3,F)^{3-}$ group. No evidence for direct bonding between CO_3^{2-} and F was found, however, by Regnier *et al.* (1994); quantum mechanical calculation also showed that the hypothetical CO_3F^{3-} ion is unstable in an apatite environment.

The structural refinements reported by Liu and Comodi (1993) yield further information on the $CO_3 = PO_4$ substitution. The tetrahedral size decreases and the distortion angle increases with C content. It is suggested that the triangular planar CO_3 group is disordered on the four faces of the PO_4^{3+} tetrahedron.

X-ray studies on numerous natural and synthetic apatites have been reported, indicating the general trend for the cell parameters of the end-member compositions. The unit cell dimensions of numerous synthetic apatites, with Ca being substituted by Sr, Cd, Pb, Ba, Mn or Ni and P being substituted by As or V, were reported by Kreidler and Hummel (1970). The size limits for substitutions in fluorapatite were found to be $0.29 \leqslant R_p \leqslant 0.60$ Å and $0.95 \leqslant R_c < 1.35$ Å, where R_p and R_c represent the radii of ions occupying the phosphorus and calcium positions, respectively. The size limits for substitution in chlorapatite were similar but with the R_c value as low as 0.80 Å. The detailed structural refinements of Hughes *et al.* (1989, 1990) have demonstrated that the structural configurations of the *X* group anions (F,Cl,OH) are not completely miscible, giving rise to the monoclinic variants discussed above. Nevertheless, in terms of the hexagonal unit cell of most natural apatites, the cell dimensions of the major end-member compositions can be given as:

	a (Å)	*c* (Å)	*c/a*
Fluorapatite	9.39	6.88	0.73
Chlorapatite	9.60	6.78	0.71
Hydroxylapatite	9.42	6.87	0.73
Carbonate-apatite	9.36–9.32	6.89	0.74

though the entry of appreciable strontium or manganese will respectively increase or decrease the cell dimensions of fluorapatite (Hughes *et al.*, 1991; see also Table 39). The cell dimensions of 21 analysed apatites were determined by Liu and Comodi (1993) who found that increasing OH increases both *a* and *c* parameters and leaves the *c/a* ratio almost unchanged; CO_3 and Cl are the most important factors controlling this ratio and have opposite effects. In fluorapatite the *a* parameter is influenced by the CO_3 substitution for PO_4, the value of *a* increasing while *c* remains unchanged; the entry of Cl, however, increases *a* but greatly decreases *c*, giving a sharp decrease in the *c/a* ratio.

The infrared spectra of apatites are distinctive and were reviewed by Ross (1974). For hydroxylapatite, Baddiel and Berry (1966) interpreted the spectra according to C_1 symmetry. The OH stretch of hydroxylapatite lies at 3573 cm^{-1}, but in hydroxylfluorapatites the reaction between OH and F shifts the OH stretch to 3546 cm^{-1}. In heated samples, a higher energy site is partially occupied by OH, giving an absorption at 3535 cm^{-1} (Cant *et al.*, 1971). The CO_3^{2-} absorption bands in francolite and dahllite were tabulated by Carlström (1974).

Chemistry

In the formula $Ca_5(PO_4)_3(F,Cl,OH)$, fluorine, chlorine and the hydroxyl ion can mutually replace each other to form the almost pure end-members fluorapatite, chlorapatite and hydroxylapatite. Although complete solid solution has been obtained in synthetic preparations, it appears that there is not complete immiscibility in natural apatites, there being structural constraints which apply to the anion or *X* position, with some chlorapatite having monoclinic symmetry. In the calcium position, natural apatites show some replacement of Ca by Sr, Mn, Na and the *REE*. The advent of the electron microprobe in petrological studies has meant that analyses of apatites occurring in rocks in accessory amounts are no longer restricted to pegmatitic examples. Some of the wide range of compositions shown in natural apatites are illustrated in Tables 39–42, where the analyses have been recalculated on the basis of 26 (O,OH,F,Cl). The fluorapatites are the commonest by far, many of them containing appreciable quantities of *REE* (Table 39). Wide-ranging reviews of the compositions of natural apatites include those presented by McClellan and Lehr (1969), Lee *et al.* (1973a), McConnell (1973), Vasilieva (1985), Povondra (1992) and Stoppa and Liu (1995); the geochemistry of apatites was considered in detail by Nash (1984), and analyses of 33 apatites from the Stillwater layered basic complex, Montana, USA, were reported by Boudreau and McCallum (1989). Care must be taken with electron microprobe analyses of apatite because F and Cl may diffuse out irregularly under increasing duration of exposure to the electron beam (Stormer *et al.*, 1993).

Many apatites contain minor amounts of strontium, but Sr as a major constituent has been found in fluorapatite from a syenite dyke (Table 39, anal. 13). Edgar (1989) has reported the widespread occurrence of strontium apatites in lamproites, a phlogopite–leucite lamproite from West Kimberley, Western Australia, having SrO 5.78 wt. per cent. Strontium apatite with SrO 46.1 wt. per cent was reported by Efimov *et al.* (1962) and apatite with 92 mol. per cent Sr is reported in serpentinite from Ohmi, Japan (Sakai and Akai, 1994). Some of these strontium apatites also contain appreciable barium (e.g. Table 39, anal. 14), and an extremely Ba-enriched apatite, named alforsite, has been described from Californian evaporites (Newberry *et al.*, 1981), with BaO 67.7 wt. per cent and with Cl the dominant anion. A chalky white hydroxylapatite rich in lead (PbO 21.29 per cent) occurring on vanadinite has also been reported (Livingstone, 1994).

As mentioned above, *REE* are common in many fluorapatites, and in augite syenite, sodalite foyaite and a peralkaline pegmatite from the Ilímaussaq intrusion, south Greenland, the maximum REE_2O_3 content in the apatite exceeds 16 per cent (Rønsbo, 1989). These *REE*-rich apatites also contain appreciable Si and Na (e.g. Table 39, anal. 16), and the substitutions envisaged by Rønsbo are: (a) $REE^3 + Si^{4+} \rightleftharpoons Ca^{2+} + P^{5+}$ and (b) $Na^+ + RE^3 \rightleftharpoons 2Ca^{2+}$. In the Kangerdlugssuaq intrusion, east Greenland, the cumulus apatite is particularly rich in *REE*, with up to 3.6 per cent Ce (Henderson, 1980); the light *REE* concentrate in the apatite rather than sphene. In fluorapatites from a zoned pegmatite in the Black Hills of South Dakota, USA, there is a large range of *REE* concentrations often within 1 mm, with changes from positive to negative Eu anomalies. Joliff *et al.* (1989) found that such changes are not accompanied by variations in major elements, but in fact indicate small-scale disequilibrium due to rapid growth from a heterogeneous melt or one which changed rapidly as volatile minerals crystallized. The *REE* patterns in apatite from the outer parts of the pegmatite show large variations in *REE* concentrations and large positive Eu anomalies, with a strong inverse correlation between these *REE*s and the size of the Eu anomalies. In the apatites from the core of the pegmatite, low *REE* concentration and systematic Ce depletion may indicate a sharp increase in f_{O_2} during core crystallization.

In alkaline igneous rocks, apatite is one of the main carriers of *REE*, giving an obvious negative Eu anomaly and enrichment in light *REE* (Zhou and Wang, 1989). The trace element content of multi-zoned apatite crystals from Panasqueira, Portugal, has been studied optically, mainly using polarospectrography, by Lagerwey (1977). The *REE* and other trace elements in carefully purified transparent crystals of yellowish apatite from Durango, Mexico (Table 39, anal.

Table 39. Fluorapatite Analyses

	1	2	3	4
SiO_2	—	—	0.55	—
Al_2O_3	0.04	—	0.04	—
Fe_2O_3	0.10	0.07	0.14	0.08
FeO	0.04	—	0.24	0.11
MnO	0.11	0.17	0.08	7.59
MgO	—	0.31	—	—
CaO	53.89	54.79	53.91	46.52
SrO	0.04	0.04	—	1.37
Na_2O	0.17	0.04	0.10	0.65
K_2O	0.01	—	0.01	—
P_2O_5	41.00	42.11	41.13	42.07
SO_3	0.11	0.10	—	—
CO_2	0.03	—	—	—
F	2.40	2.44	1.93	3.01
Cl	0.91	—	0.71	—
H_2O^+	0.12	0.61	—	0.34
H_2O^-	0.00	—	—	0.05
	101.14	101.13	101.66	101.79
O=F	1.22	1.02	1.39	1.27
Total	99.92	100.11	100.27	100.52
ε	1.6392	1.627	1.634	1.638
ω	1.6432	1.631	1.640	1.643
D	3.2050	3.190	3.226	3.246
a (Å)	9.4130	—	9.407	—
c (Å)	6.8820	—	6.876	—

Numbers of ions on the basis of 26 (O,OH,F,Cl)

	1		2		3		4	
P	5.964	5.98	5.986	6.00	5.929	6.02	6.060	
Si	—		—		0.094		—	
C	0.007		—		—		—	
S	0.014		0.013		—		—	
Fe^{3+}	0.012	10.14[a]	0.008	10.00[b]	0.018	10.05[c]	0.010	9.95
Mg	—		0.078		—		—	
Fe^{2+}	0.006		—		0.034		0.016	
Mn	0.016		0.024		0.011		1.094	
Na	0.056		0.013		0.033		0.214	
Ca	9.921		9.856		9.837		8.480	
Sr	0.004		0.004		—		0.135	
K	0.002		—		0.002		—	
F	1.304	1.70	1.295	1.98	1.578	1.78	1.620	2.01
OH	0.135		0.683		—		0.386	
Cl	0.265		—		0.205		—	

[a]Includes Al 0.008, *REE* 0.111.
[b]Includes *REE* 0.017.
[c]Includes Al 0.008, *REE* 0.105.

1 Greenish yellow fluor-chlor-oxyapatite, Crystal Lode pegmatite, Devils Canyon, near Eagle, Colorado, USA (Young and Munson, 1966). Anal. E. L. Munson (includes REE_2O_3 1.79, acid insol. 0.38).
2 Apatite, crystal-bearing cavities, South Urals, Russia (Petrukha, 1982) (includes RE_2O_3 0.29, insol. 0.16).
3 Apatite, granitic rock, Cougar Canyon, Dolly Varden Mountains, Elko Co., Nevada, USA (Lee *et al.*, 1973a). Anal. E. L. Munson (includes REE_2O_3 1.69, insol. 0.13).
4 Blue manganoan apatite, lithium-bearing pegmatite, Viitaniemi, Eräjärvi, central Finland (Volborth, 1954). Anal. A. Volborth.

Table 39. *Continued*

	5		6		7		8	
SiO_2	—		—		0.18		—	
Al_2O_3	—		0.08		0.08		—	
Fe_2O_3	—		0.05		0.06		—	
FeO	—		—		0.04		—	
MnO	—		1.82		0.13		—	
MgO	—		0.01		—		—	
CaO	55.36		53.13		54.87		56.03	
SrO	—		0.97		—		0.01	
Na_2O	—		0.00		0.06		—	
K_2O	—		0.00		0.01		—	
P_2O_5	42.04		41.68		41.97		42.18	
SO_3	—		0.01		—		—	
CO_2	—		0.04		—		—	
F	3.14		3.25		3.36		3.41	
Cl	0.09		0.02		0.02		0.01	
H_2O^+	0.27		0.04		—		—	
H_2O^-	—		—		—		—	
	100.90		100.15		101.46		101.82	
O=F	1.34		1.37		1.41		1.44	
Total	99.56		99.78		100.05		100.38	
ε	1.6336		1.6343		1.631		—	
ω	1.6363		1.6367		1.633		—	
D	—		3.228		3.215		—	
a (Å)	9.379		9.3759		9.373		—	
c (Å)	6.881		6.880		6.881		—	

Numbers of ions on the basis of 26 (O,OH,F,Cl)

	5	Σ	6	Σ	7	Σ	8	Σ
P	6.000		6.002		6.001		5.996	
Si	—		—	6.01	0.030	6.03	—	
C	—		0.009		—		—	
S	—		0.001		—		—	
Fe^{3+}	—		0.006		0.008		—	
Mg	—		0.002		—		—	
Fe^{2+}	—		—		0.006		—	
Mn	—		0.262	10.07[d]	0.019		—	10.10[f]
Na	—		—		0.020	10.03[e]	—	
Ca	9.998		9.683		9.928		10.079	
Sr	—		0.096		—		0.001	
K	—		—		0.002		—	
F	1.674		1.748		1.795		1.811	
OH	0.304	2.00	0.002	1.76	—	1.80	—	1.81
Cl	0.026		0.006		0.006		0.002	

[d]Includes Al 0.023.
[e]Includes Al 0.016, *REE* 0.026.
[f]Includes Y 0.016.

5 Apatite, upper zone of Kiglapait layered intrusion, north Labrador, Canada (Huntington, 1979).
6 Colourless to very pale pink apatite, Peerless pegmatite, South Dakota, USA (Young *et al.*, 1966). Anal. E. L. Munson (includes acid insol. 0.05).
7 Acicular apatite, hybrid granitic rock, Snake Creek–Williams Canyon area, Nevada, USA (Lee *et al.*, 1973b). Anal. E. L. Munson (includes REE_2O_3 0.45, insol. 0.23).
8 Apatite, staurolite–sillimanite schist, eastern Connemara, Ireland (Yardley, 1985) (includes Y_2O_3 0.18).

Table 39. Fluorapatite Analyses

	9		10		11		12	
SiO_2	0.34		0.18		—		—	
Al_2O_3	0.07		—		—		—	
Fe_2O_3	0.06		—		—		—	
FeO	—		0.07		—		—	
MnO	0.01		0.47		—		—	
MgO	0.01		0.06		—		—	
CaO	54.02		55.40		54.96		55.52	
SrO	0.07		0.32		—		—	
Na_2O	0.23		0.01		—		—	
K_2O	0.01		0.02		—		—	
P_2O_5	40.78		41.00		42.10		42.01	
SO_3	0.37		0.22		—		—	
CO_2	0.05		—		—		—	
F	3.53		3.72		3.74		3.89	
Cl	0.41		0.04		0.11		—	
H_2O^+	0.01		0.00		0.30		—	
H_2O^-	—		—		0.07		—	
	101.52		101.64		101.28		101.70	
O≡F	1.58		1.58		1.59		1.63	
Total	99.94		100.06		99.69		100.07	
ε	1.6326		—		1.6391		—	
ω	1.6362		—		1.6431		—	
D	3.22		—		3.205		—	
a (Å)	9.391		—		9.41		—	
c (Å)	6.878		—		6.88		—	
	Numbers of ions on the basis of 26 (O,OH,F,Cl)							
P	5.868		5.870		5.979		5.976	
Si	0.058	5.99	0.031	5.93	—		—	
C	0.012		—		—		—	
S	0.047		0.028		—		—	
Fe^{3+}	—		—		—		—	
Mg	—		0.015		—		—	
Fe^{2+}	—		0.010		—		—	
Mn	0.001	10.02[g]	0.067	10.18[h]	—		—	
Na	0.076		0.003		—		—	10.02[i]
Ca	9.837		10.038		9.878		9.996	
Sr	0.007		0.031		—		—	
K	0.002		0.004		—		—	
F	1.897		1.990		1.984		2.067	
OH	0.011	2.03	—	2.00	0.035	2.05	—	
Cl	0.118		0.011		0.031		—	

*Calculated on the basis of 24 (O).
[g]Includes *REE* 0.090, Th 0.001, As 0.009, V 0.001.
[h]Includes La 0.001, Ce 0.003, Nd 0.002, Y 0.001.

9 Yellow apatite, magnetite–apatite deposit, Cerro de Mercado, Durango, Mexico (Young *et al.*, 1969) (includes ThO_2 0.02, REE_2O_3 1.43, As_2O_5 0.09, V_2O_5 0.01 (Al_2O_3, Fe_2O_3, MgO are due to impurities)).
10 Apatite, forsterite–scapolite skarn, north of Gatineau, Quebec, Canada (Hogarth, 1988) (includes La_2O_3 0.02, Ce_2O_3 0.05, Nd_2O_3 0.03, Y_2O_3 0.03).
11 Pale bluish apatite, calc-silicate skarn, Gondivalsa, Orissa, India (Rao *et al.*, 1979).
12 Apatite, adjacent to xenotime and monazite, Gloserheia pegmatite, Froland, southern Norway (Åmli, 1975) (includes Y_2O_3 0.10, Nd_2O_3 0.06, Gd_2O_3 0.06, Yb_2O_3 0.06).

Table 39. *Continued*

	13		14		15		16	
SiO_2	0.14		0.05		0.09		2.49	
Al_2O_3	0.38		0.00		—		—	
Fe_2O_3	0.27		—		—		—	
FeO	—		0.33		0.08		—	
MnO	0.02		—		0.07		—	
MgO	0.25		0.00		0.02		—	
CaO	46.30		48.05		52.40		43.97	
SrO	11.60		2.07		—		—	
Na_2O	0.10		0.00		2.22		1.63	
K_2O	0.02		0.11		0.00		—	
P_2O_5	39.20		36.50		36.64		36.11	
SO_3	—		—		6.56		—	
CO_2	—		—		—		—	
F	2.57		0.00		3.29		3.07	
Cl	tr.		—		0.01		0.02	
H_2O^+	—		—		0.25		—	
H_2O^-	—		—		—		—	
	101.09		100.01		101.63		101.05	
O=F	1.08		—		1.39		1.29	
Total	100.01		—		100.24		99.76	
ε	1.634		—		—		—	
ω	1.638		—		—		—	
D	3.350		—		—		—	
a (Å)	9.410		—		—		—	
c (Å)	6.910		—		—		—	
	Numbers of ions on the basis of 26 (O,OH,F,Cl)							
P	5.943		5.445*		5.137		5.617	
Si	0.025	5.97	0.008	5.45	0.015	5.97	0.457	6.07
C	—		—		—		—	
S	—		—		0.815		—	
Fe^{3+}	0.036		—		—		—	
Mg	0.067		—		0.005		—	
Fe^{2+}	—		0.049		0.011		—	
Mn	0.003	10.32[j]	—	10.29[k]	0.010	10.04	—	10.17[l]
Na	0.034		—		0.712		0.580	
Ca	8.883		9.072		9.298		8.655	
Sr	1.204		0.211		—		—	
K	0.004		0.025		—		—	
F	1.455		—		1.723		1.704	
OH	—		—		0.138	1.86	—	1.71
Cl	—		—		0.003		0.006	

[j]Includes Ba 0.004, Ti 0.005, Al 0.080.
[k]Includes Ti 0.091, Ba 0.842.
[l]Includes La 0.220, Ce 0.434, Pr 0.041, Nd 0.151, Sm 0.042, Y 0.045.

13 Strontian fluorapatite, syenite dyke, near Libby, Montana, USA (Larsen *et al.*, 1952). Anal. M. H. Fletcher (includes BaO 0.06, TiO_2 0.04, ign. loss 0.14).

14 Barian fluorapatite, diopside–leucite lamproite, West Kimberley, Western Australia (Edgar, 1989) (includes TiO_2 0.69, BaO 12.20).

15 Sodian sulphatian fluorapatite, epithermal calcite–quartz vein, Kushikino mine, Kagoshima Prefecture, Japan (Shiga and Urashima, 1987).

16 Sodian *REE*-bearing apatite, sodalite fayaite, Ilimaussaq, south Greenland (Rønsbo, 1989) (includes La_2O_3 3.27, Ce_2O_3 6.47, Pr_2O_3 0.60, Nd_2O_3 2.30, Sm_2O_3 0.66, Y_2O_3 0.46).

Table 40. Chlorapatite Analyses

	1	2	3	4	5
SiO_2	0.12	—	0.14	1.16	—
Al_2O_3	—	—	0.52	—	—
Fe_2O_3	0.04	tr.	0.36	0.18	—
FeO	—	—	—	—	0.21
MnO	—	—	tr.	—	1.52
MgO	—	—	0.20	0.29	0.54
CaO	53.81	53.40	54.01	52.97	52.40
SrO	—	—	—	—	—
Na_2O	0.19	—	0.37	0.22	—
K_2O	—	—	0.15	0.10	—
P_2O_5	40.50	41.20	40.15	40.50	40.98
F	0.00	0.13	1.32	0.17	1.15
Cl	6.24	6.20	4.16	4.13	3.74
H_2O^+	0.25	0.09	—	—	0.06
H_2O^-	—	—	0.63	—	—
	101.23	101.02	102.01	99.72	100.60
O=F	1.37	1.45	1.49	—	1.33
Total	99.86	99.57	100.52	99.72	99.27
ε	—	1.665 (α)	1.651	—	1.653
ω	—	1.667 (β)	1.657	—	1.658
D	—	3.181	3.176	—	—
a (Å)	—	9.606	—	—	—
c (Å)	—	6.785	—	—	—

Numbers of ions on the basis of 26 (O,OH,F,Cl)

	1	Σ	2	Σ	3	Σ	4	Σ	5	Σ
P	5.936	5.98[a]	6.039		5.860	5.88	5.964	6.17	6.000	
Si	0.021		—		0.024		0.202		—	
Fe^{3+}	0.005	10.05	—		0.047	10.34[b]	0.023	10.07	—	10.10
Mg	—		—		0.057		0.075		0.139	
Fe^{2+}	—		—		—		—		0.030	
Mn	—		—		—		—		0.223	
Na	0.064		—		0.124		0.074		—	
Ca	9.981		9.905		9.976		9.872		9.710	
Sr	—		—		—		—		—	
K	—		—		0.033		0.022		—	
F	—	2.12	0.071	1.99	0.720	1.94	0.094	1.31	0.630	1.80
OH	0.288		0.104		—		—		0.068	
Cl	1.831		1.819		1.216		1.218		1.097	

[a]Includes C 0.019.
[b]Includes Al 0.106.

1 Chlorapatite, Ødegårdens verk, Norway (Morton and Catanzaro, 1964) (includes CO_2 0.08).
2 Pinkish white, monoclinic chlorapatite, actinolite–diopside–calcite marble, Bob's Lake, southeast Ontario, Canada (Hounslow and Chao, 1970). Anal. A. W. Hounslow.
3 Chlorapatite, Kurokua, Japan (Seto, 1932).
4 Chlorapatite, Ødegardens verk, Feset, Norway (Bhatnagar, 1968).
5 Greenish yellow inner portion of chlorapatite crystals, cavities in quartz diorite, Kurokura, Kanagawa Prefecture, Japan (Harada, 1938). Anal. A. Kan-nari.

Table 41. Hydroxylapatite Analyses

	1	2	3	4	5
SiO_2	—	—	—	—	0.27
Al_2O_3	—	—	—	—	0.16
Fe_2O_3	—	0.88	—	—	0.71
FeO	—	—	0.32	0.10	—
MnO	0.07	0.19	0.02	0.09	0.15
MgO	0.10	—	0.05	0.11	0.23
CaO	55.84	52.24	55.08	55.69	32.92
SrO	—	—	0.03	—	14.70
Na_2O	—	—	0.04	0.10	0.34
K_2O	—	—	0.01	—	0.13
P_2O_5	42.05	42.52	42.40	41.50	36.25
F	0.16	tr.	1.63	0.45	0.68
Cl	—	tr.	0.20	1.31	—
H_2O^+	1.86	1.84	0.98	0.73	0.60
H_2O^-	—	—	0.08	0.02	—
	100.23	100.47	100.84	100.86	100.17
O=F	0.07	—	0.72	0.48	0.28
Total	100.16	100.47	100.12	100.38	99.89
ε	1.644	1.645	—	1.642	1.648
ω	1.651	1.650	1.642	1.649	1.652
D	3.21	3.20	3.14	3.17	—
a (Å)	9.420	9.414	9.382	9.598	9.843
c (Å)	9.890	6.877	6.887	6.739	6.980
	Numbers of ions on the basis of 26 (O,OH,F,Cl)				
P	5.953	6.026	6.014	6.111	5.989 } 6.04
Si	—	—	—	—	0.053 }
Fe^{3+}	— }	0.111 }	— }	— }	0.104 }
Mg	0.025 }	— }	0.012 }	0.028 }	0.067 }
Fe^{2+}	— }	— }	0.045 }	0.014 }	— }
Mn	0.010 } 10.04	0.027 } 9.85[a]	0.003 } 9.96	0.013 } 10.47	0.024 } 9.87[b]
Na	— }	— }	0.012 }	0.033 }	0.129 }
Ca	10.004 }	9.369 }	9.887 }	10.378 }	6.883 }
Sr	— }	— }	0.003 }	— }	1.663 }
K	— }	— }	0.002 }	— }	0.033 }
F	0.084 }	—	0.868 }	0.248 }	0.420 }
OH	2.074 } 2.16	2.054	1.098 } 2.02	0.846 } 1.48	0.781 } 1.20
Cl	— }	—	0.056 }	0.386 }	— }

[a]Includes Zn 0.346.
[b]Includes Al 0.036, *REE* 0.926.

1 Yellow hydroxylapatite, talc schist, Holly Springs, Cherokee County, Georgia, USA (Mitchell *et al.*, 1943). Anal. D. S. Reynolds (includes insol. 0.15).

2 Chestnut-coloured hydroxylapatite, associated with calcite and brushite in guano deposits near contact with calcareous rock, Castellana Caves, Bari, Italy (Balenzano *et al.*, 1974) (includes ZnO 2.80).

3 Fluorian hydroxylapatite, two-pyroxene granulite, Hitterö, southwest Norway (Howie, 1964). Anal. R. A. Howie.

4 Yellowish green chlor-hydroxylapatite ('asparagus stone'), contact of schists and granite pegmatite, emerald mix, Rajagarh village, Ajmer, India (Babu, 1972) (includes 0.76 per cent, SiO_2, Al_2O_3 and K_2O as sericite).

5 Yellowish green strontian, *REE*-bearing, chlorian hydroxylapatite, pegmatitic alkali syenite, Burpala intrusion, North Baikal region, Russia (Portnov *et al.*, 1970) (includes REE_2O_3 13.03; *REE* percentage composition is La 9.5, Ce 48.2, Pr 5.3, Nd 23.1, Sm 5.0, Eu 0.4, Gd 2.0, Ho 0.2, Er 0.2, Y 6.2).

Table 42. Carbonate-apatite analyses

	1	2	3
SiO_2	0.63	—	0.05
Al_2O_3	—	0.58	—
Fe_2O_3	0.35	0.35	—
FeO	—	—	—
MnO	0.30	—	—
MgO	—	—	—
CaO	54.81	55.01	55.06
SrO	0.08	—	0.47
Na_2O	0.20	—	0.19
K_2O	0.12	—	—
P_2O_5	40.61	41.96	40.99
CO_2	0.23	0.41	0.82
SO_3	0.29	—	—
F	3.90	3.75	3.54
Cl	0.12	0.10	0.06
H_2O^+	0.00	—	0.19
H_2O^-	0.08	—	—
	101.72	102.21	101.78
O=F	1.67	1.59	1.59
Total	100.05	100.62	100.19
ε	1.631	1.631	—
ω	1.634	1.634	—
D	3.147	3.270	—
a (Å)	9.363	—	—
c (Å)	6.883	—	—
	Numbers of ions on the basis of ten (Ca,Mg, etc.)		
P	5.747 } 5.94	5.932 } 6.02	5.805 } 6.00
Si	0.106	—	0.008
C	0.052	0.092	0.187
S	0.036	—	—
Fe^{3+}	0.044 } 10.00	0.044 } 10.00[a]	— } 10.00[b]
Mg	—	—	—
Fe^{2+}	—	—	—
Mn	0.042	—	—
Na	0.065	—	0.062
Ca	9.815	9.841	9.868
Sr	0.008	—	0.046
K	0.026	—	—
F	2.061 } 2.09	1.980 } 2.01	1.872 } 2.10
OH	—	—	0.212
Cl	0.034	0.028	0.017
O,OH,F,Cl	25.90	26.19	25.98

[a]Includes Al 0.114.
[b]Includes La 0.004, Ce 0.012, Nd 0.006, Y 0.002.

1 Pale bluish brown carbonate-apatite associated with bustamite and spessartine in skarn, Broken Hill mine, New South Wales, Australia (Harada *et al.*, 1974). Anal. H. Sekino.
2 Colourless apatite, pegmatite dyke, Mansjö Mountain, Hälsingland, Sweden (Eckermann, 1922). Anal. H. von Eckermann and J. Petrén (including ign. loss 0.05).
3 Core of *REE*-bearing apatite, sövite, Qaqarssuk carbonatite complex, southwest Greenland (Knudsen, 1991) (includes La_2O_3 0.09, Ce_2O_3 0.20, Nd_2O_3 0.10, Y_2O_3 0.02).

Table 42. *Continued*

	4		5		6	
SiO_2	—		1.76		—	
Al_2O_3	—		—		1.30	
Fe_2O_3—	—		0.13		—	
FeO	—		—		—	
MnO	0.15		—		0.15	
MgO	—		—		0.34	
CaO	54.70		53.91		53.42	
SrO	0.18		0.28		—	
Na_2O	0.18		0.28		0.06	
K_2O	—		—		0.10	
P_2O_5	40.90		38.26		39.41	
CO_2	0.57		0.88		1.72	
SO_3	—		0.72		—	
F	4.00		3.92		2.32	
Cl	—		—		0.01	
H_2O^+	—		0.29		1.19	
H_2O^-	—		—		—	
	101.44		102.25		100.02	
O=F	1.68		1.65		0.98	
Total	99.76		100.50		99.04	
ε	—		1.6330		1.637	
ω	—		1.6380		1.640	
D	3.22		3.280		3.19	
a (Å)	9.368		9.3796		9.370	
c (Å)	6.890		6.8924		6.900	
	Numbers of ions on the basis of ten (Ca,Mg, etc.)					
P	5.825	5.96	5.467	6.06	5.595	5.99
Si	—		0.297		—	
C	0.132		0.202		0.394	
S	—		0.091		—	
Fe^{3+}	—	10.00[c]	0.016	10.00[d]	—	10.00[e]
Mg	—		—		0.084	
Fe^{2+}	—		—		—	
Mn	0.021		—		0.021	
Na	0.059		0.091		0.019	
Ca	9.854		9.750		9.598	
Sr	0.018		0.027		—	
K	—		—		0.021	
F	2.128		2.091	2.42	1.231	2.57
OH	—		0.326		1.332	
Cl	—		—		0.003	
O,OH,F,Cl	25.92		25.63		26.01	

[c] Includes *REE* 0.048.
[d] Includes La 0.027, Ce 0.076, Y 0.013.
[e] Includes Al 0.257.

4 Green apatite, Faraday Township, Ontario, Canada (Gulbrandsen *et al.*, 1966) (includes REE_2O_3 0.76 (av. atomic wt. of 139.11 used in recalculation)).
5 Deep green to reddish brown *REE*-bearing apatite, in diopside–fluorite–calcite skarn, Huddersfield Township, Quebec, Canada (Trzcienski *et al.*, 1974) (includes La_2O_3 0.44, Ce_2O_3 1.23, Y_2O_3 0.15).
6 White apatite in andesitic ignimbrite, China (Qu, 1985).

Table 42. *Continued*

	7	8	9
SiO_2	0.67	—	—
Al_2O_3	0.46	tr.	—
Fe_2O_3	0.22	tr.	0.34
FeO	0.00	0.00	—
MnO	—	0.00	—
MgO	0.39	0.00	0.10
CaO	53.86	54.64	53.94
SrO	—	0.13	—
Na_2O	0.16	0.00	—
K_2O	0.04	0.00	—
P_2O_5	42.05	38.25	38.13
CO_2	0.66	1.98	3.40
SO_3	—	—	—
F	2.81	2.80	3.71
Cl	—	0.00	—
H_2O^+	0.06	0.66	0.46
H_2O^-	—	0.06	0.01
	101.38	100.71	101.92
O≡F	1.18	1.18	1.56
Total	100.20	99.53	100.36
ε	1.6385	†	1.624
ω	1.6420	—	1.629
D	3.20	3.21	3.14
a (Å)	—	—	—
c (Å)	—	—	—

Numbers of ions on the basis of 10 (Ca,Mg, etc.)

	7		8		9	
P	5.921	6.18	5.524	5.99	5.571	6.37
Si	0.111		—		—	
C	0.150		0.462		0.801	
S	—		—		—	
Fe^{3+}	0.028	9.9	—	10.00^f	—[g]	10.00
Mg	0.097		—		0.026	
Fe^{2+}	—		—		—	
Mn	—		—		—	
Na	0.052		—		—	
Ca	9.720		9.987		9.974	
Sr	—		0.013		—	
K	0.008		—		—	
F	1.478	1.54	1.512	2.26	2.025	2.55
OH	0.066		0.748		0.529	
Cl	—		—		—	
O,OH,F,Cl	26.75		25.86		26.80	

† α 1.620, β 1.627, γ 1.628, $2V_\alpha \sim 37°$

7 Pale green apatite, carbonatite, Kudangulam, Tamil Nadu, India (Ramaswamy, 1992).
8 Francolite, oolitic ironstone, Robin Hood quarry, Thorpe-on-the-Hill, Leeds, Yorkshire, England (Deans, 1938). Anal. H. G. C. Vincent (includes insol. 2.10).
9 Francolite, Wheal Franco, Buckland Monachorum, Tavistock, Devon, England (Sandell *et al.*, 1939). Anal. E. B. Sandell (includes insol. 1.83).

Table 42. *Continued*

	10		11		12	
SiO_2	—		0.60		—	
Al_2O_3	—		—		—	
Fe_2O_3	0.25		3.61		0.03	
FeO	—		—		—	
MnO	—		0.16		0.02	
MgO	0.53		0.40		—	
CaO	54.84		52.22		55.24	
SrO	—		—		—	
Na_2O	0.20		0.15		—	
K_2O	—		0.08		—	
P_2O_5	35.01		34.93		40.91	
CO_2	4.43		5.42		1.18	
SO_3			—		—	
F	5.60		1.65		1.05	
Cl	0.03		—		2.31	
H_2O^+	1.51		0.48		0.68	
H_2O^-	0.16		—		—	
	102.66		99.70		101.42	
O=F	2.36		0.69		1.17	
Total	100.30		99.01		100.25	
ε	*		1.619		1.646	
ω	—		1.624		1.651	
D	3.116		3.02		3.18	
a (Å)	—		—		—	
c (Å)	—		—		—	
	Numbers of ions on the basis of 10 (Ca,Mg, etc.)					
P	4.930	5.94	4.947	6.28	5.84	6.110
Si	—		0.099		—	
C	1.006		1.237		0.27	
S	—		—		—	
Fe^{3+}	0.031	10.00	0.455	10.00	0.004	10.00
Mg	0.131		0.100		—	
Fe^{2+}	—		—		—	
Mn	—		0.022		0.003	
Na	0.064		0.048		—	
Ca	9.774		9.358		9.993	
Sr	—		—		—	
K	—		0.017		—	
F	2.946	4.63	0.873	1.41	0.561	1.99
OH	1.671		0.535		0.661	
Cl	0.008		—		0.766	
O,OH,F,Cl	26.66		25.872		26.151	

* α 1.614 – 1.617, β 1.627, γ 1.627 – 1.630, $2V_\alpha$ 0–36°
[f] Includes Al 0.090.
[g] $Fe3^+$ omitted as it is reportedly present as hematite.

10 Francolite, altered lava, Annisfontein Farm, Richtersveld, Namaqualand, South Africa (Villiers, 1942). Anal. C. J. Liebenberg (includes insol. 0.10).
11 Francolite, cryptocrystalline aggregate, Gylinskogo Massif, Russia (Egorov, 1975). Anal. Yu. S. Nebelitskaya.
12 Blue chloro-fluoro-hydroxyl carbonate-apatite, serpentinized marble quarry, Matale, north-central Sri Lanka (Cooray, 1970).

9), were reported by Young *et al.* (1969); this fluorapatite has Ce 4700, La 4200, Nd 2000, Sm 300, Eu < 100, Gd 200 and Y 760 p.p.m. In a review of the relationship of the *REE* composition of apatites to geological environments, apatite of marine sedimentary origin was found to show marked depletion in Ce; apatites of siliceous igneous rocks seem favourable to heavy lanthanides and Y (Fleischer and Altschuler, 1969). The apatite in the apatite-rich iron ores of the Avnik region of Turkey all show large Eu anomalies, but Ce is not depleted and a magmatic origin is suggested (Helvaci, 1987). Other substitutions reported in natural apatites include As substituting for P (Ishida, 1985; Çelebi, 1992) and S for P (Shiga and Urashima, 1987); in the latter variety the charge decrease of $(SO_4)^{2-}$ for $(PO_4)^{3-}$ is considered to be electrostatically compensated by the concomitant substitutions of Na^+ for Ca^{2+} and minor $(SiO_4)^{4-}$ for $(PO_4)^{3-}$ (e.g. Table 39, anal. 15; see also Vasilieva *et al.*, 1958). A detailed study of the geochemistry of uranium in apatite was made by Altschuler *et al.* (1958) who gave values of 10–100 p.p.m. U for primary igneous apatite and 50–200 p.p.m. in sedimentary marine apatite.

Minor and trace elements in 47 apatites from igneous and metamorphic rocks were studied by Cruft (1966). The stable chlorine isotopes in apatite were studied by Morton and Catanzaro (1964) who found that the $^{35}Cl/^{37}Cl$ ratio for three apatites ranged from 3 to 129. The occurrence of 25 p.p.m. of I in carbonate-apatite was reported by Walters and Luth (1969). The trace element content, in particular the *REE*, in apatite from a Scottish metabentonite horizon allows the correlation of two separate outcrops and provides a chemical fingerprint with which to correlate this Silurian horizon with similar occurrences elsewhere (Batchelor and Clarkson, 1993). The *REE* patterns of apatite crystals in two distinct bentonite beds in Gotland show light *REE* enrichment, and have a Ce/Y ratio of 9.0, interpreted as indicating a transitional alkaline igneous origin (Batchelor and Jeppsson, 1994).

The analyses of certain members of the apatite group show them to contain appreciable CO_2 (as much as 5 or 6 per cent), and separate names have been proposed for these carbonate-apatities. Francolite has been used for an apatite containing appreciable CO_2 and also >1 per cent fluorine, i.e. a carbonate-fluorapatite. Less frequently used is the name dahllite for an apatite with essential CO_2 but with OH dominant over F, i.e. a carbonate-hydroxylapatite. Analyses of carbonate-apatites are given in Table 42, where because of the replacement of PO_4 by CO_3 or by CO_3OH the formulae have been calculated not on the usual basis of 26 (O,OH,F,Cl) but on the basis of ten atoms in the Ca position.

It is now accepted that the CO_2 reported is not present as impurities of calcite or aragonite, nor is it present adsorbed on the surface of the crystallites, and although some may occur in fluid inclusions, these would only contribute a minor amount. Infrared absorption studies show that the absorption peaks characteristic of simple carbonate minerals are absent (Tuddenham and Lyon, 1960). Carbonate-fluorapatite has a distinctive X-ray pattern and smaller *a* dimension than normal fluorapatite or hydroxylapatite. It also shows a decrease in refractive indices with increasing CO_2.

Following a review of earlier proposals for the incorporation of the carbonate group into apatite, Binder and Troll (1989) reported an extensive investigation of 18 carbonate-apatites from carbonatites, pegmatites, marbles and volcanic rocks, using high-grade chemical techniques and X-ray diffraction data. In several of these samples the fluorine content exceeded the theoretical maximum value of 3.767 wt. per cent, and these samples contained less P than the near-stoichiometric apatites. Binder and Troll (1989) found a significant positive correlation between CO_2 content and the excess of the sum of X anions (F + Cl + OH) in a number of samples, and offered an explanation involving a partial substitution expressed as $PO_4^{3-} \rightarrow CO_3^{2-}F^-$. They suggest that one carbon replaces one phosphorus and that the charge balance is maintained by the substitution of O^{2-} by F^-, thus allowing the anion composition of carbonate-apatites to be depicted by the chemical formula $Ca_{10}[(PO_4)_{6-x}(CO_3,F)_x](F,OH,Cl)_2$ and explaining the apparent excess of F over the accepted formula.

McClellan and Kauwenbergh (1990) consider that the maximum substitution of CO_2 in francolite is between 6 and 7 per cent, or ~25 per cent of the PO_4^{3-} sites. An apatite with 11.30 per cent CO_2, however, has been described (Eckermann, 1974) which, even after the deduction of 3.92 per cent present in gas-filled vesicles, gives a formula with C 1.65, i.e. $(P_{4.03}C_{1.65}Si_{0.24}S_{0.07})$, and with low refractive indices (ω 1.590, ε 1.593).

A detailed study by McClellan and Kauwenbergh (1991) has revealed that there is a systematic decrease in the CO_2 content and an increase in unit-cell *a* values (i.e. decarbonation of francolite) with increasing geological age (see also McClellan, 1980).

Experimental Work

The first synthesis of apatite was that by Daubrée (1851) who obtained crystals of apatite by passing phosphorus trichloride vapour over red-hot lime. Fluorapatite may be synthesized by the fusion of $Ca_3(PO_4)_2$ with CaF_2 (Montel and Chaudron, 1951), or with NH_4F. Pure fluorapatite has also been prepared by melting CaF_2 with $Ca_3(PO_4)_2$ at 1660 °C. Following the reaction $CaF_2 + 3Ca_4(PO_4)_2 = 2Ca_5(PO_4)_3F$, Franz (1983) reported two eutectic points: one between CaF_2 and $CaFPO_4$ at 1210 ± 5 °C, and the other between $Ca_5(PO_4)_3F$ and $Ca_3(PO_4)_2$ at 1550 ± 30 °C. This fluorapatite was found to be more resistant to acids than hydroxylapatite and able to activate the remineralization of dental enamel. Due to a different microstructure, the synthetic apatite is, however, less resistant to mechanical stress. The growth of fluorapatite from modified sea-water solutions has been studied by Cappellen and Berner (1991), to determine the dependence of the growth rate on solution supersaturation, pH, temperature and dissolved Mg. The growth kinetics of fluorapatite in these conditions are controlled by surface processes and are affected by temperature; adsorbed Mg ions retard the crystal growth of fluorapatite by blocking surface growth sites. The crystallites grow to 0.1–10.0 μm in size, comparable with those formed during burial of modern phosphatic sediment. Hydroxylapatite is most readily obtained by precipitation from solutions of calcium salts with the addition of ammoniacal phosphate solutions, e.g. Arnold (1950) prepared hydroxylapatite by the simultaneous addition of calcium acetate and ammonium phosphate to CO_2-free ammonium acetate. Pure hydroxylapatite crystals up to 0.3 mm in length were synthesized hydrothermally by the hydrolysis of monetite ($CaHPO_4$) at 300 °C and 1250 lb/in^2 saturated steam pressure (National Bureau of Standards, 1957). By varying the pH at which apatite was precipitated, 'defect' hydroxylapatites were prepared with atomic ratios of Ca:P between 10:6 and 8.2:6, the latter value being close to that commonly found for bone, teeth, etc. Hayek *et al.* (1958) prepared crystalline hydroxylapatite by hydrothermal synthesis at 400 °C, in the pH range 5–11: they reported that the complete substitution of OH by F can be achieved in the presence of KF. Very pure stoichiometric single crystals of hydroxylapatite were prepared by Arends *et al.* (1979) using a three-stage hydrothermal process, with final recrystallization at 2 kbar and 430–500 °C. Vacuum-dried amorphous tricalcium phosphate can be converted into apatite in the presence of normal moist atmosphere; after two years, long needles with a diameter of ~10 nm are produced (Puech *et al.*, 1982). Wallaeys and Chaudron (1950) heated synthetic chlorapatite at 300 °C with varying amounts of CaF_2 and converted it to fluorapatite: similarly hydroxylapatite heated with $CaCl_2$ was changed to chlorapatite, or heated with CaF_2 was changed to fluorapatite. Chlorapatite may be synthesized by fusing calcium phosphate with $CaCl_2$ (Borgström, 1932); with the addition of NaCl, a continuous series of mixed crystals from chlorapatite to $3Ca_3P_2O_5 \cdot NaCl$ is obtained (see also Prener, 1967).

The hydrothermal synthesis of chlorapatite from an aqueous saturated solution of calcium chloride and orthophosphoric acid, in the range 200–850 °C and 1–3000 bar, was reported by Argiolas and Baumer (1978); acicular crystals in particular were obtained at >750 °C and 1000 bar. Earlier, Wyllie *et al.* (1962) had noted that apatite crystals coexisting with liquid or vapour in the system $CaO–CaF_2–P_2O_5–H_2O–CO_2$ are equant, whereas those precipitated from a liquid during a quench are greatly elongated parallel to *z*, thus suggesting that acicular apatites might be a criterion for the former existence of a liquid phase in rocks whose origin is debatable. Fluorapatite, chlorapatite and five intermediate members were synthesized at 640 (8 kbar), 750 (10 kbar) and 900 °C (12 kbar), and showed a complete range of solid solution between the two end-members (Shaw *et al.*, 1993).

The hydrothermal synthesis of all three of the normal apatite end-members was reported by Baumer and Argiolas (1981), who obtained crystallites 50–500 μm in size and determined the cell dimensions. The reactions involved were, e.g. for chlorapatite:

$$10CaCl_2 + 6H_3PO_4 \rightleftharpoons Ca_{10}(PO_4)_6Cl_2 + 18HCl$$

at 400 °C and <3 kbar. The synthesis and stability of carbonate-fluorapatite was examined by Jahnke (1984), who demonstrated that it is the stable apatite phase in solutions relatively rich in carbonate such as sea-water. Loss of carbonate from the apatite should occur only when the crystals are exposed to low-carbonate solutions. The compression of fluorapatite and hydroxylapatite to 15 GPa at ⩽2300 °C shows that apatites decompose to an assemblage with $Ca_3(PO_4)_2$ above 12 GPa (Murayama *et al.*, 1986); this phase could be the host for LIL (light-ion-lithophile) incompatible elements in the deep upper mantle. Thermodynamic modelling of the partitioning of F–Cl–OH between biotite and apatite has led to a revised apatite–biotite geothermometer (Zhu and Sverjensky, 1992). Phase relationships in the phlogopite–apatite system at 20 kbar with an F/OH ratio of 1 (Vukadinovic and Edgar, 1993) have indicated that F will tend to remain in the solid phases rather than the melt during partial melting in the mantle.

Elements and radicals which have been substituted in synthetic apatites include Ba, Pb, Mg, Ni, Sr, Zn, Cr^{3+}, Cr^{6+}, Al, Fe, Si, Na, Ce, Y, OH, F, Cl, CO_3 and SO_4: details of many syntheses, the methods used, and physical properties of the products were given by Jaffe (1951). Apatite containing a probable maximum of 2 per cent Al_2O_3 was synthesized by Pieruccini (1947), who also produced voilet-coloured crystals containing about the same amount of Fe_2O_3: in these examples a possible replacement is $Ca_2^{2+}P^{5+} \rightleftharpoons 3(Al,Fe)^{3+}$. Green chromium-bearing chlorapatite (Cr_2O_3 0.75 per cent) was synthesized by Minguzzi (1941) from a mixture of $Ca_3(PO_4)_2 + CaCrO_4 + CaCl_2$ fused at 1100 °C, and a pure chromium analogue of hydrous apatite was synthesized by Johnson (1960) from CaO and Cr_2O_3 heated in air at 950 °C: in these compounds the substitution which occurs is probably $C^{3+} + 2Cr^{6+} \rightleftharpoons 3P^{5+}$. Various synthetic products with the apatite structure were obtained by Dihn and Klement (1942) by sintering mixtures of $Ca_3(PO_4)_2$, Ca_2SiO_4, $CaSO_4$, CaF_2 and Na_2SO_4 at 200 °C; in some of the products $Si^{4+} + S^{6+}$ replace $2P^{5+}$, while in others the partial replacement $P^{5+} \rightleftharpoons S^{6+}$ is balanced by $Ca^{2+} \rightleftharpoons Na^{+}$. Strontian hydroxylapatite was synthesized by Klement (1939) by heating at 1150 °C $Sr_3(PO_4)_2$ with $SrCO_3$ in a current of steam; the hydrolysis of $SrHPO_4$ with NaOH and precipitation of Sr salts with alkali phosphate solutions gave only impure material. The synthetic $Sr_5(PO_4)_3(OH)$ has *a* 9.76, *c* 7.21 Å. Strontian chlorapatite was synthesized by Carobbi (1950) by fusing $Sr_3(PO_4)_2$, $SrCl_2$ and $HgCl_2$; the ionic radii of Hg^{2+}(1.10 Å) and Sr^{2+}(1.12 Å) are similar, but the product contained only 0.001 per cent Hg. A study of strontium diffusion in apatite was reported by Cherniak and Ryerson (1993) using Rutherford back-scattering spectroscopy and ion implantation; the results indicated that Sr in the cores of apatites in felsic magmas may be unaffected by the Sr compositions of their surroundings for the temperature and duration of later events.

Early reports of the direct synthesis of carbonate-apatite must be discounted because the products were not checked chemically or by X-ray methods, and may have represented a mixture of hydroxylapatite and $CaCO_3$. There is in fact no evidence that an apatite containing structural CO_3^{2-} can be synthesized except by phosphate replacement of a carbonate (Ames, 1959). The replacement of $CaCO_3$ by alkaline phosphate solutions to yield carbonate-apatite was first demonstrated by Irvine and Anderson (1891). Part of the system Na_3PO_4–$CaCO_3$–H_2O was investigated at low temperatures by Ames (1959), who used ^{14}C tracer techniques and equilibrium data to study the replacement of $CaCO_3$ by carbonate-apatite. The carbonate-apatite thus formed is due to incomplete replacement of CO_3^{2-} by PO_4^{3-}. The equilibrium data and structural formulae indicate that this carbonate can be entirely present in the apatite structure when the apatite contains <10 wt. per cent CO_3^{2-}; with a greater amount of CO_3^{2-} the carbonate is at least partly present as a separate phase.

In an experimental study of the substitution of Ca by Sr and Ba in synthetic apatites, the Ca–Sr substitution was found to be continuous whereas that of Ba is limited, $BaHPO_4$ forming instead (Lucas *et al.*, 1990). The synthesis of trivalent-cation substituted calcium oxyhydroxylapatites was studied by Tamishita *et al.* (1986), who showed that in the formula $Ca_{10-x}M_x(PO_4)_6(OH)_{2-x}O_x$, where $M = Al^{3+}$ or Y^{3+}, the limit of substitution for both cations is $x = 2$, i.e. $Ca_8Al_2(PO_4)_6O_2$. Simple substitutions of Co^{2+} and Ni^{2+}, charge-coupled substitutions of alkali REE^{3+}, $REE^{3+}Si^{4+}$ and $REE^{3+}O^{2-}$ (with $REE =$ La, Pr and Nd), and types and defect varieties in the apatite structure were investigated by Grisafe and Hummel (1970); the solubility limits of Co^{2+} in fluorapatite and chlorapatite were found to be 15 and 25 mol. per cent, respectively.

The ion exchange of OH^-, F^- and Cl^- in apatites was studied hydrothermally by Latil and Maury (1977) at 300–600 °C and 200–1000 bar. They found that temperature controls the fixation of F, whereas pH controls the fixation of Cl^-; F^- displaces both Cl^- and OH^- but Cl^- has no influence on F^- and OH^-. A thermodynamic study of the halogens in melt–vapour–apatite equilibria has been made by Candela (1986). The solubility of apatite in supercritical fluids was investigated by Ayers and Watson (1991). They found that for fluorapatite the solubility in pure water is low, at <0.4 per cent, but it increases strongly with increasing pH. In peraluminous liquids studied over the range 750–1000 °C and 2–5 kbar, a dramatic increase occurs in the solubility of apatite and in the P content of the melt compared with the solubility of apatite in metaluminous or peralkaline compositions (Pichavant *et al.*, 1992). High-pressure apatite solubility in carbonate-rich liquids in the range 1120–1430 °C at 30 kbar was determined by Baker and Wyllie (1992); the dissolution of apatite into haplogranitic melts at 750°C and 200 MPa (P_{H_2O}) was studied by Wolf and London (1994).

Fission tracks. Microscopic defects and tracks that occur in some apatites represent radiation damage caused primarily by the fission of ^{238}U impurities; their study and analysis is a widely used method for dating (Fleischer *et al.*, 1965, 1975; Wagner and Haute, 1992). Particular studies of fission tracks in apatite have been reported and reviewed by Green *et al.* (1986) and Paul and Fitzgerald (1992). Their lengths have been used to study the degree of annealing of induced fission tracks and hence to establish 'closure temperatures', nominal temperatures below which the radiogenic fission tracks are effectively retained. A standard fission-track reference age of 31.4 ± 0.5 m.y. for the Durango apatite was obtained after analysis by four analysts using three different fission-track methods (Jonckheere *et al.*, 1993). Detailed analysis of apatite fission-track data from 47 samples from Oligocene strata in Texas, USA, yielded track lengths which agree well with those for a similar data set from the Lower Cretaceous rocks of the Otway basin, southeast Australia (Corrigan, 1993). Model equations were presented by Crowley *et al.* (1991) describing the reduction in etched length as a function of time and temperature; they provide first-order predictions of annealing of etched fission tracks on geological time scales.

Estimates of closure temperatures for fission-track retention in apatite are usually in the range 75–125 °C for cooling rates between 1 and 100 °C/m.y. (Wagner and Reimer, 1972; Gleadow and Lovering, 1978). This unique sensitivity of the apatite fission-track system is now of considerable economic importance due to the coincidence between the temperature range over which annealing occurs and that over which liquid hydrocarbons are generated. Other applications include the determination of the timing of emplacement and the thermal history of ore deposits. There is abundant literature on both fission-track dating and its use in evaluating the tectonic and thermal history of rocks. It must suffice here to indicate the nature of the techniques used; see, for example, Green *et al.* (1989) who have quantitatively modelled the thermal annealing of fission tracks in apatite. Accuracy of predictions is limited principally by the effect of apatite compositions on annealing kinetics. However, from tests of the validity of the method to known geological situations, where geological age and thermal history is known with some confidence, Green *et al.* concluded that such a quantitative understanding of apatite fission-track analysis could provide the basis for rigorous palaeotemperature estimation (see also Grivet *et al.*, 1993). For apatites from basement granites in Antarctica, a break in slope in the apatite age profile is considered to represent the base of an uplifted partial annealing zone, using the effective closure temperature for fission tracks in apatite as ~100 °C together with the dating of the ~60 °C cooling temperature (Wagner *et al.*, 1989).

Using transmission electron microscopy giving magnifications of ×100 000 to ×500 000, it is now possible to observe directly the fission tracks in apatite; annealing, however, remains a very complicated process to decipher (Green *et al.*, 1988; Paul and Fitzgerald, 1992). The anisotropy of track annealing means that fission-track ages should be determined on the $(10\bar{1}0)$ plane in apatite, the plane most resistant to annealing (Sandhu *et al.*, 1987).

Isotopic composition. Although the isotopic exchange between aqueous inorganic solutions and PO_4^{3-} ions is so slow as to be negligible, even over geological time scales at low temperatures, it has been demonstrated that in enzyme-catalysed reactions (such as those suggested for the biological formation of Ca-phosphates in bones and teeth) the exchange of ^{18}O between water and PO_4^{3-} is extremely rapid. Thus biogenic apatites respond sensitively to the aqueous environment during their formation and preserve their record after death of the organism

(Kolodny *et al.*, 1983). The oxygen isotope record is better preserved in phosphates than in carbonates or cherts; the $\delta^{18}O$ values of phosphate oxygen in ancient phosphorites decrease with increasing age, probably due to variations in temperature (Shemesh *et al.*, 1983). The oxygen isotope compositions of phosphate ($\delta^{18}O_p$) in modern guano apatite and in ancient apatite on Pacific oceanic islands were studied by Ayliffe *et al.* (1992); the ancient apatites have $\delta^{18}O_p$ values which can best be explained by re-equilibration with circulating meteoric waters.

The isotopic composition of the structural carbonate in apatite of bones and teeth ($\delta^{18}O_c$ and $\delta^{13}C$) has also been measured. In sedimentary francolites McArthur *et al.* (1980) showed that $\delta^{13}C$ values allow a distinction to be made between authigenic francolite precipitated within anoxic sediments and diagenetic varieties produced by phosphatization of carbonates. The relationship between $\delta^{18}O_p$ and $\delta^{18}O_c$ suggests early diagenetic replacement of an originally phosphatic phase by cabonate-fluorapatite (Kolodny and Luz, 1991). The isotopic composition ($\delta^{18}O_p$) in francolite from a Lower Pliocene phosphate deposit was also altered by diagenesis, and any palaeoenvironmental interpretation must allow for this effect (McArthur and Herczeg, 1990). The isotopic composition of strontium in apatite was measured by Landa *et al.* (1984) who reported $^{87}Sr/^{86}Sr$ ratios mainly in the range 0.734–0.706. Creaser and Gray (1992) reported that the initial $^{87}Sr/^{86}Sr$ ratio is better preserved in apatite than in feldspar or biotite when Proterozoic rocks have been modified by igneous cooling or hydrothermal alteration.

The diffusion of Sr in apatite is small, but at 900–1250 °C it was found to be slightly greater in the direction perpendicular to *z* than in a direction parallel to *z* (Watson *et al.*, 1985). Measurement of O and Sr diffusion in apatite at 550–1200 °C for O and 640–1200 °C for Sr with a P_{H_2O} of 1 kbar showed that the diffusion coefficient for O transport parallel to *z* is about three orders of magnitude greater than that perpendicular to the *z* axis (Farver and Giletti, 1989), though in these experiments no anisotropy was observed for Sr. The data show that closure temperatures for Sr diffusion in apatite are lower than high-temperature studies predict, and that apatite would remain open to Sr diffusion well below the solidus temperature of most igneous rocks, which could result in higher $^{87}Sr/^{86}Sr$ ratios in the apatite than the whole-rock initial value.

The partition coefficients between apatite and liquid for the *REE* and Sr were studied experimentally by Watson and Green (1981). $D_{REE}^{ap/liq}$ is much higher for silicic melts than for basic ones; Sr (and Eu^{2+}) is less concentrated by apatite than are the trivalent *REE*. Partitioning of Sr is insensitive to changes in melt composition, temperature and pressure. The partitioning of F and Cl between apatite and non-silicate fluids at high pressures (1–2 GPa) and high temperatures (950–1050 °C) was found to be dependent on fluid composition (Brenan, 1993a) and suggests that chlorine-rich aqueous fluids or brines may play an important role in mass transport in the upper mantle. Further experimental work has shown that, arising from changes in pressure, temperature or fluid/melt chemistry, apatite may rapidly develop measurable F–Cl–OH zoning profiles via diffusion along the *z* axis, but with no gradient perpendicular to this axis (Brenan, 1993b). The differential thermal analysis curves of apatites are generally smooth and featureless (Silverman *et al.*, 1952), but may have a minor endothermic peak at about 800 °C, corresponding to loss of hydroxyl. The solubility of the various members of the apatite group depends on the degree of crystallinity rather than on the exact chemistry of the specimen (Visse, 1953). Pseudomorphs of serpentine or kaolinite after apatite have been recorded, and apatite may also be replaced by wavellite, florencite, rhabdophane, svanbergite or woodhouseite.

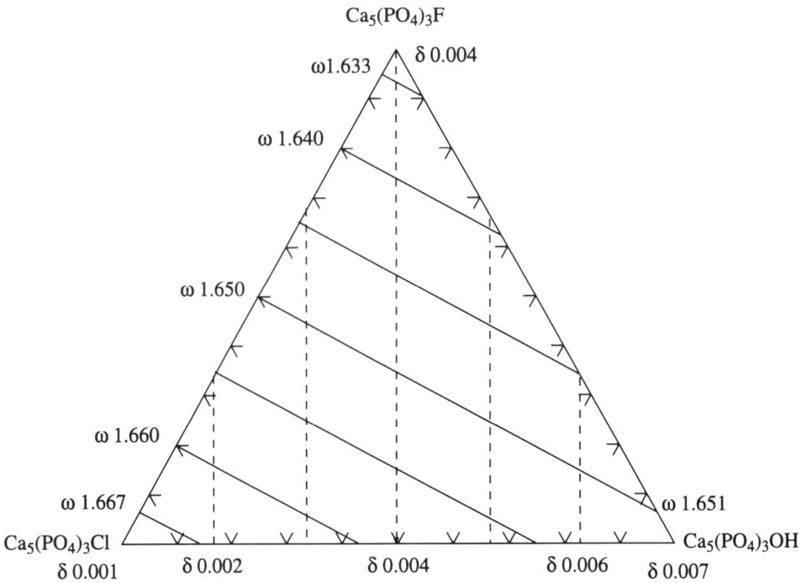

Fig. 116. Optical properties of the fluorapatite, chlorapatite and hydroxylapatite series, based on end-member values.

Optical and Physical Properties

Apatite is optically negative and is normally uniaxial, though biaxial varieties with an optic angle of up to 20° are known. The carbonate-bearing apatites in particular may have anomalous optics. e.g. the francolite of Table 42, anal. 8, has a 2V value of 25° at the centre of the crystal and of 40° at the hexagonal edge (Deans, 1938). Twinning may occur, giving six triangular sectors in the basal plane, the extinction position of these sectors being at about 40° to the hexagonal edge.

The large variations in composition within the apatite group make the accurate correlation of optical data difficult. In general, however, the refractive indices are highest for chlorapatite (for synthetic chlorapatite values of indices are ω 1.6684, $\omega-\varepsilon$ 0.0009; Borgström, 1932), and are reduced by the substitution of OH and still more by the substitution of F (Fig. 116). The partial replacement of P by C also brings about a general reduction in the values of the refractive indices (see Table 42); values down to ω 1.607 have been recorded. The incorporation of Mn increases the refractive indices and also increases the specific gravity. The substitution of Sr for Ca does not appreciably alter the refractive indices, though the specific gravity is considerably increased (Table 39, anal. 13; Table 42, anal. 5). For fluorapatite from Knappenwand, Sulzbachtal, Austria, Baumhauer (1908) determined the refractive indices at various wavelengths, as follows:

λ (μm)	ε	ω
447	1.6462	1.6497
502	1.6413	1.6445
588	1.6357	1.6388
688	1.6326	1.6356

The birefringence also varies with substitution in the (OH,F,Cl) group, being lowest for chlorapatite (0.001), intermediate for fluorapatite (around 0.004) and highest for hydroxylapatite (0.007). The birefringence of some carbonate-apatites may be even stronger, e.g. Deans (1938) recorded a δ value of 0.008 for francolite (Table 42, anal. 8), while that of anal. 10 (Table 42) is 0.013. The cryptocrystalline variety of apatite, collophane, may appear isotropic or show only weak birefringence. This is due to the aggregate polarization effect produced by a mass of superimposed submicroscopic crystallites. The chlorapatite of Table 40, anal. 2, is biaxial with $2V_\alpha \sim 10°$. This is in line with observations for some synthetic chlorapatites (Prener, 1967; Young *et al.*, 1968). A natural fluorhydroxylapatite from the Asio mine, Japan, shows sector growth and has $2V_\alpha \leqslant 40°$; Akizuki *et al.* (1994) suggest that these unusual optical properties were produced during non-equilibrium crystal growth. Heating of this apatite to 850 °C results in the virtually complete disappearance of the optical anomalies. Apatite crystals from Llallagua, Bolivia, and Minas Gerais, Brazil, showing heterogeneous distribution of trace elements that correspond with sectoral zoning and optical anomalies have been reported by Rakovan and Reeder (1994).

The colour of apatite is extremely variable. Vasilieva (1958) has shown that the intensity of the colour increases with an increase in Mn content of the apatite, but that the colour itself depends on the state of oxidation of the manganese. Thus, Mn^{2+} produces pale pink and blue tones, Mn^{3+} blue and Mn^{7+} violet. Hogarth (1957) has described orange, red, brown and green apatites from veins, with associated radioactive minerals; the red and brown colours were due to finely dispersed hematite. Ferric and ferrous iron together produce a green colour in both synthetic and natural apatites (Hoffmann, 1938); some other colours may be due to *REE*. Synthetic apatites containing cobalt, nickel, praseodymium or neodymium substituting for calcium have violet, pale yellow-green, light green and blue to blue-violet colours, respectively (Grisafe and Hummel, 1970). Some apatites have been reported to show a yellow or a pale violet fluorescence. Iwase (1935) recorded strong broad fluorescence bands from tabular apatites of metalliferous veins, compared with the weak narrow bands given by prismatic apatite from basic igneous rocks. The luminescence and colour of zoned apatite crystals from Panasqueira, Portugal, were studied by Knutson *et al.* (1985), but see also the earlier study of zoning in apatite from this locality by Lagerwey (1977) and comments by Leckebusch (1979). The latter author attributed a blue colour to Eu^{2+} and yellow to Mn^{2+}, and considered that a zoned bright yellow/dark green crystal owed its colour to fluctuations in the Mn content. The infrared luminescence of several specimens of apatite was recorded by Barnes (1958). The infrared absorption spectra of apatites show large absorption bands between 9 and 10.5 μm due to the PO_4^{3-} ion (Omori and Seki, 1960).

Zonal violet and yellow cathodoluminescence in apatite of igneous rocks (Wenzel and Ramseyer, 1992) has been attributed to the incorporation of the *REE* (Eu^{2+}, Sm^{3+}, Dy^{3+}, Ce^{3+}) and Mn^{2+}, respectively. Under cathodoluminescence, apatite crystals from diorite and monzodiorite of the Meissen massif, Germany, show growth and corrosion zoning, with 5 per cent of the crystals containing well-rounded, probably relict, cores (Wenzel and Ramseyer, 1992). The violet and yellow luminescing zones are caused by different abundances of Sm^{3+}, Dy^{3+}, Eu^{2+}, Ce^{3+} and Mn^{2+}. These trace elements are taken to suggest alkaline affinities for the parental melt.

The coloured varieties of apatite may, in thick sections, show moderate pleochroism with absorption $\varepsilon > \omega$; blue apatite, for example, may have ε greenish blue, ω light blue. Zonal distribution of the colour is also known, particularly in the blue pegmatitic manganapatites. Portnov *et al.* (1977) concluded that the dominant process in apatite pleochroism was probably the diffusion of iron, accompanied by the reduction of trivalent elements and the creation of colour centres. A greenish yellow to reddish brown apatite from Tanzania has been described showing the cat's eye effect; this is due to fine, fibrous parallel inclusions of goethite (Gübelin and Schmetzer, 1983).

Distinguishing Features

The high relief and low birefringence of apatite serve to distinguish it from most light-coloured minerals. Its straight extinction and the absence of a good cleavage help to distinguish it from sillimanite and melilite. Topaz is optically positive, as well as being biaxial, while eudialyte has a more distinct cleavage and, in general, lower refractive indices. Vesuvianite and zoisite have considerably higher relief and may show anomalous interference colours. The coloured slightly pleochroic varieties may be distinguished from tourmaline by having absorption $\varepsilon > \omega$, with tourmaline having $\omega \geqslant \varepsilon$. The addition of a drop of nitric acid followed by some ammonium molybdate will confirm the presence of a phosphate by giving a yellow precipitate of ammonium phosphomolybdate.

Paragenesis

Apatite is a common accessory mineral in many types of rock and is the most abundant phosphorus-bearing mineral. A useful compilation of the many different types of occurrence of apatite was given by McConnell (1973).

Igneous rocks. Apatite occurs as an accessory mineral in almost all igneous rocks from basic to acid; in some it may amount to as much as 5 per cent by volume, though 0.1 to 1.0 per cent is the more normal range. In most igneous apatite the fluorapatite molecule is generally dominant, often with appreciable OH in addition, giving a hydroxyl-fluorapatite. Apatite is a fairly common mineral of granitic pegmatites, where fluorapatite is particularly common (e.g. Table 39, anals. 1, 4, 6 and 12) and where the mineral is often a manganoan variety (anals. 4 and 6). These pegmatitic apatites may be associated with lithium minerals such as lepidolite, spodumene, amblygonite or lithophilite, and with triplite, beryl, elbaitic tourmaline, etc. The apatites are often zoned and may also occur as a hydrothermal or different type of alteration product of other phosphate minerals. Some pegmatite minerals contain minor amounts of CO_2 (e.g. Table 39, anal. 6). The apatite-bearing veins of the Nisikkatch Lake area of northern Saskatchewan, Canada, occur in zoned pegmatitic bodies; associated minerals include hornblende, apatite, calcite, baryte, hyalophane and allanite, one or more of which may be present in any individual vein (Hogarth, 1957). Apatite also occurs in hydrothermal veins and cavities, and is found in Alpine-type veins associated with quartz, adularia, sphene, etc. Fluorapatite has also been detected in granitic rocks (Table 39, anals. 3 and 7). The use of *REE* in apatite to distinguish between two different series of granitic rocks has been described by Zhang *et al.* (1985). Minor fluorapatite in granular xenoliths (ejecta) from the ultramafic rocks of the Alban Hills, central Italy, invariably contains appreciable amounts of Si, F and SO_3 and can be referred to as F-bearing silico-sulphate apatite (Federico *et al.*, 1994).

Fluorapatite is a common accessory mineral in syenites. The strontian apatite of Table 39 (anal. 13) occurs in a pyroxene–sphene–apatite patch in a syenitic dyke cutting apatite pyroxenite; the apatites from elsewhere in this dyke contain almost no strontium (Larsen *et al.*, 1952). A strontian, *REE*-bearing chlorian hydroxylapatite (Table 41, anal. 5) has been described from pegmatitic alkali syenite of the Burpala intrusion, North Baikal, Russia (Portnov *et al.*, 1970). Alkali syenites with mafic layers produced by *in situ* accumulation of clinopyroxene and apatite have been described from Bahia, Brazil (Conceição *et al.*, 1989). A carbonate-fluorapatite (Table 42, anal. 6) has been described from an andesitic ignimbrite, China (Qu, 1985).

The role of apatite in the petrogenesis of some alkali basalt suites was studied by Exley and Smith (1982), who considered that apatite-rich xenoliths in alkali basalt represent a probable source of positive phosphorus anomalies in mantle enrichment processes. In the Kiglapait intrusion, Labrador, the fractional crystallization of basaltic magma produced cumulus apatite starting at the 94 per cent solidification level, when P_2O_5 reached saturation in the liquid (Huntington, 1979); the earliest apatites are fluorine-rich, but those from the outer border zone have a higher Cl content. A detailed investigation of the varying composition of apatites in the Stillwater basic layered complex, Montana, USA, has been reported by Boudreau and McCallum (1989); the F, Cl and OH contents were not constrained by crystal-chemical factors (in contrast to micas and amphiboles), and thus changes in the abundance of these components provide an indicator of halogen fugacity variations and give insights into the degassing history of these rocks. In the intercumulus apatites of Stillwater, Cl-rich apatites, typically with <0.4 wt. per cent F and >6.0 wt. per cent Cl, occur throughout the lower third of the complex, excluding the basal series. A change to more F-rich apatites occurs within the olivine-bearing zone of the banded series; in the platiniferous J–M reef in this series, the apatites typically contain >1.2 wt. per cent F and <3.0 wt. per cent Cl. The most F-rich apatites occur in the uppermost exposed cumulates. In the chlorapatites, the *REE* concentrations show a marked peak in the olivine-rich rocks of the J–M reef and contain up to 2 wt. per cent $Ce_2O_3 + La_2O_3$. Boudreau and McCallum (1989) predict that a similar stratigraphic change from chlorapatite to more F-rich apatite should also occur in the Bushveld complex at the stratigraphic transition from the critical zone to the main zone. Cawthorn (1994) has presented evidence to show that the Bushveld complex formed from a nearly anhydrous magma and did not release a hydrous fluid before apatite began to crystallize; it is suggested that apatite in the earliest cumulates equilibrated with trapped interstitial liquid, which converted it from the typical F-rich composition of cumulate apatite to a Cl-rich composition.

The fluorapatites of lamproites from West Kimberley, Western Australia, are notably rich in barium (Table 39, anal. 14) and also contain appreciable TiO_2 (Edgar, 1989). Sodian *REE*-bearing apatites have been reported from sodalite foyaite of the Ilímaussaq intrusion, south Greenland (anal. 16), and here the *REE* oxides amount to >14 wt. per cent (Rønsbo, 1989). The association of magnetite and apatite in cross-cutting bodies of apparent igneous origin has often been reported; in a garnetiferous pegmatitic rock within a Precambrian cordierite–garnet migmatite in Western Australia, a faintly pleochroic yellow F–Cl apatite has FeO 2.2 wt. per cent, its unusually high iron content being considered due to the high temperature and low oxygen fugacity found in the water-poor melt (Fransolet and Schreyer, 1981). Dendritic magnetite fringes around apatite in miniature diapir-like bodies have been noted from the Kirunavaara iron ore, Sweden (Nyström and Henriquez, 1989), where they have been attributed to differentiation of the iron ore melt. The yellow fluorapatite from the magnetite–apatite deposit of Cerro de Mercado, Durango, Mexico (Table 39, anal. 9), has only minor iron (Fe_2O_3 0.06) but has appreciable *REE* (Paulick and Newesely, 1968; Young *et al.*, 1969). Apatites from lamproites of the Jharia coalfied, eastern India, have been reported to have up to 20 per cent SrO and complex *REE* distributions (Rock *et al.*, 1992).

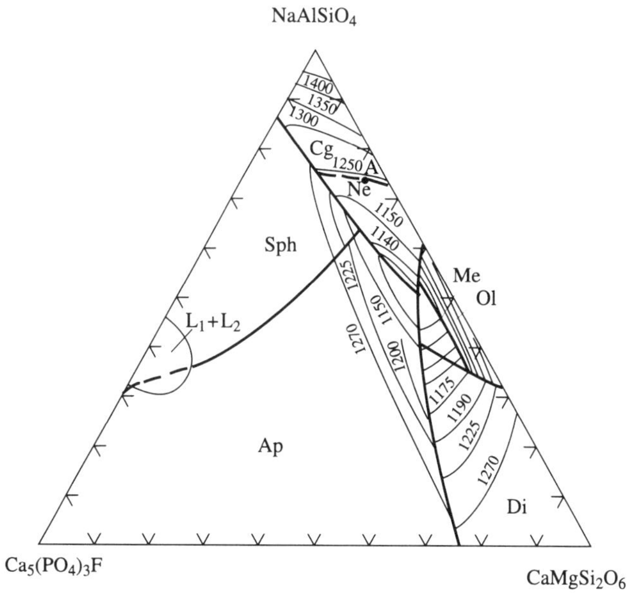

Fig. 117. Liquidus phase boundaries in the system nepheline–fluorapatite–diopside at 1 atm: Ol, olivine; Me, melilite; Sph, silicophosphate; Ap, apatite; L1 + L2, immiscibility field. Point A represents a weighted average composition of the apatite-bearing rocks of Khibina, Russia (after Kogarko *et al.*, 1976).

In the Munni Munni complex and associated intrusions of the West Pilbara block, Western Australia, the mol. fraction of fluorapatite increases from 60 to 70 per cent in cumulates of the ultramafic zone, to >90 per cent near the top of the gabbroic zone (Boudreau *et al.*, 1993); the percentage of chlorapatite decreases from 30 to 10 per cent. This is attributed to the separation of a volatile fluid phase during crystallization, with loss of Cl and OH relative to F.

Carbonatites generally contain appreciable apatite (e.g. Hogarth, 1989). The several apatite deposits of the Khibina tundra, Kola Peninsula, Russia, have been investigated by many authors. The Khibina massif hosts the world's largest apatite deposit and is of considerable economic importance as a source of fertilizer. It is a layered igneous body ~1 km thick. The phase equilibria of the apatite-bearing ijolite–urtite rocks of Khibina can be approximated by the system $NaAlSiO_4$–$CaMgSi_2O_6$–$Ca_5(PO_4)_3F$ (Fig. 117); nepheline crystallizes first, followed by nepheline + apatite as the temperature falls. Kogarko and Khapaev (1986) have reviewed the petrogenesis of the complex and present two alternative models, one based on sedimentation in steady-state convection and the other on non-structural convection taking account of mass and heat transfer. A model for apatite ore formation has been proposed in which the main condition is that the alkaline magma must contain ~2 per cent P_2O_5. Geothermometry on inclusions indicates that the apatite formed during magmatism at 1000–750 °C; the magma crystallized at 10–15 per cent nepheline, then an apatite–nepheline–diopside cotectic was reached. Apatite separation is caused by sedimentation and convection-stirring in the magma chamber (Kogarko, 1991). The constant occurrence of magnetite and fertilizer-type apatite in the alkaline ultramafic Kovdor complex has been ascribed by Zimin *et al.* (1984) to immiscibility of initial phosphides and acid–base reactions in the magma; they considered that the generation of such igneous apatite occurred at the top of the lower mantle by the reaction of H_3PO_4 with peridotite.

In the Maimecha–Kotuisky province, the *REE* and F contents of apatite in rocks of the phoscorite–carbonatite group in ijolite–carbonatite complexes indicate four successively injected differentiation products from the non-silicate magma (Yegorov, 1984). The petrology of the Magan apatite deposit in this Russian province was described by Kravchenko *et al.* (1987); the main mass of apatite occurs in ijolite as a result of differentiation of the initial phosphorus-rich nephelinite magma. In the Amba Dongar carbonatites, India, the early cumulate apatite is rich in F (1.68–3.75 per cent) and SrO (0.93–2.02 per cent); Ce_2O_3 and La_2O_3 rise to 0.83 and 0.69 per cent, respectively, in apatites of this complex (Viladkar and Wimmenauer, 1992).

Apatite pyroxenite and related rocks are important in the Phalabora complex and at Loolekop, both in the Transvaal, South Africa. In the Phalabora shonkinite, apatite forms about 3 per cent of the rock, and the apatite–diopside rock and apatite rock of this complex has up to 96 per cent phosphate. Similar apatite-bearing carbonatites are known from Zimbabwe, Malawi, Uganda and Tanzania. In nephelinite lavas from Shombole volcano, southern Kenya, fluorapatite occurs associated with fluorite, strontianite, Sr–Ca and K–Ba zeolites and aegirine (Kjarsgaard and Peterson, 1991). Carbonate-apatite also occurs in the calcitic carbonatites (alvikites) of the Alnö alkaline complex, Sweden, and may amount to 13 per cent in the sövites (Eckermann, 1948, 1974). It is also reported in sövite of the Qaqarssuk carbonatite complex, southwest Greenland (Table 42, anal. 3), where it contains minor amounts of *REE* (Knudsen, 1991). In the lenticular ultramafic pyroxenite of the Loch Borralan igneous complex in the northwest Highlands of Scotland, apatite amounts to 1–10 per cent, and it has been estimated that the complex contains a total of some five million tons of apatite; Notholt *et al.* (1985) have calculated that an apatite concentrate with 20 per cent P_2O_5 could be obtained, for possible application as a slow-acting fertilizer.

Metamorphic rocks. Apatite occurs in both contact and regionally metamorphosed rocks. Fluorapatite is a frequent associate of chondrodite and phlogopite in metasomatized calc-silicate rocks and impure limestones, whereas chlorapatite ('asparagus stone') (Table 41, anal. 4) has been described from the contact of a granulite pegmatite and schist at the Rajagarh emerald mine, Ajmer, India. Alternation during metamorphism involved the replacement of the chlorapatite at Ødegårdens, southern Norway, (e.g. Table 40, anal. 1), by hydroxylapatite and rare-earth phosphates (Liefrink *et al.*, 1994). Fluorapatite from a fosterite–scapolite skarn (Table 39, anal. 10) has been described from north of Gatineau, Quebec, Canada (Hogarth, 1988), but carbonate-fluorapatite is also known from skarn deposits, e.g. Table 42, anals. 1 and 5. Carbonate-apatite has been reported from contact metamorphosed rocks, as at Magnet Cove, Arkansas, USA (McConnell and Gruner, 1940); the *REE*-bearing apatite from Mineville, New York, USA, occurs in association with contact metamorphic magnetite ore (McKeown and Klemic, 1957). Fine acicular apatite may be considered to be an indicator of contact metamorphism or rapid crystallization of magma. Thus it may be found at the outer contact of the younger of two intrusions (Fenogenov, 1977). The higher the temperature, the more elongate are the apatite crystals (see also Wyllie *et al.*, 1962).

In regionally metamorphosed rocks, hydroxylapatite (Table 41, anal. 1) in the talc schist and fluor-hydroxylapatite in chlorite schist, both from a serpentinite near Holly Springs, Georgia, USA, are considered to owe their formation to the hydrous environment (Mitchell *et al.*, 1943). The chloro-fluoro-hydroxyl carbonate-apatite of Table 42, anal. 12, occurs in a serpentinized marble quarry in Sri Lanka (Cooray, 1970). The monoclinic chlorapatite of Table 40, anal. 2, occurs in an actinolite–diopside–calcite marble at Bob's Lake, Ontario, Canada

(Hounslow and Chao, 1970). In higher grade metamorphic rocks, fluorapatite (Table 39, anal. 8) has been reported from a staurolite–sillimanite schist in southwestern Ireland, and fluorian hydroxylapatite (Table 41, anal. 3) occurs in a two-pyroxene granulite from southwest Norway. From a study of the compositional variation of apatites in regionally metamorphosed rocks, Kapustin (1986) showed that in general terms those of the greenschist facies had low contents of Sr, Ba, Mn and Na, but higher Cl and SO_3 compared with apatites from amphibolite-facies rocks, which had higher Sr, Mn and Na; these elements reached a maximum in apatites of granulite-facies rocks. The metamorphism of sedimentary phosphorite deposits may give rise to finely crystalline apatite rocks; indeed it has been suggested that many apatite schists in metamorphic rocks represent metamorphosed syngenetic deposits (Serdyuchenko, 1956). The intrusion of igneous sills and dykes into the Permian phosphate-bearing beds of Montana, USA, has converted collophane and francolite to colourless fine-grained apatite, sometimes giving a quartz–apatite assemblage (Lowell, 1955). Apatite–magnetite mineralization occurs along the Singhbhum shear zone of Bihar, India, where it is generically related to the Dhanjori volcanic rocks and associated granites (Dasgupta and Bhattacharya, 1992). Oriented monazite inclusions occur in the cores of apatite porphyroblasts in schist at the Hemlo gold deposit, Ontario, Canada, depleting the surrounding apatite in the light *REE* (Pan *et al.*, 1993).

Sedimentary rocks. Apatite is not uncommon in sedimentary rocks, where it may be found both as a detrital mineral and as a primary deposit. The francolite (Table 42, anal. 8) found in oolitic ironstone near Leeds, England, occurs in freshwater sediments (Deans, 1938); here the phosphate crystallized within kaolinite ooliths during diagenesis. Deposits containing appreciable vertebrate remains are often highly phosphatic, and fish scales or bones may act as nuclei for the secondary concentrations of calcium phosphate to form nodules. Konta (1956) recorded Permian fish scales with a core of francolite and a thin outer layer of hydroxyl-fluorapatite, whereas Osmond and Sawin (1959) noted an approach to carbonate-hydroxylapatite in terrestrial mammals, though shark and fish apatite is usually carbonate-fluorapatite. A study of francolite compositions was made by Egorov (1975), and a detailed investigation of francolite from bones and coprolites has been reported by Kulikova *et al.* (1991), including thermal effects, cell dimensions and chemical analyses for 14 francolites. In Lower Zechstein argillaceous dolomites in Poland, francolite replaces fossil remnants, algae, etc., and was formed from the phosphorus released from organic compounds during auto-oxidation of the organic matter (Majer and Piestrzynski, 1986). The francolite in many unaltered Tertiary phosphorites is highly substituted, with ~ 1.2 per cent Na, 0.25 per cent Sr, 0.36 per cent Mg and 2.7 per cent SO_4, substituting for Ca and PO_4 (McArthur, 1985). In the francolite of the Permian Phosphoria formation of western North America, the Na/P_2O_5 ratios and CO_2 contents show areal trends related to depth of burial and degree of thermal metamorphism. These two processes, together with weathering and freshwater diagenesis, explain all observed variations in francolite composition. Francolites are metastable with respect to fluorapatite, and those of various geological ages show systematic trends in composition and cell parameters with time. Francolites in rocks of Palaeozoic or older ages are generally close to fluorapatite in composition, whereas Mesozoic and younger francolites have a range of compositions that vary with the type and intensity of post-depositional alteration (McClellan and Kauwenbergh, 1991).

In the Phosphoria formation, beds containing up to 80 per cent phosphate are interbedded with phosphatic shales and impure limestones, such phosphate-rich strata being termed phosphorites. The mineralogy of phosphorites is complex, the phosphate occurring as a cryptocrystalline, often concretionary, virtually isotropic material. The name collophane has been applied to the cryptocrystalline mineral component, and is a term analogous to 'limonite'; it is used only when the apatite-like phase cannot be definitely identified. Many of the remaining problems with the study of phosphorites have been reviewed by Notholt and Jarvis (1990). A study of 177 Israeli phosphorites of Santonian to Eocene age showed that apatites from the carbonate-rich Santonian–Campanian Menuha formation contain on average 5.5 per cent CO_2, whereas the average CO_2 content of the apatites from the phosphatic Campanian–Maastrichtian Mishash formation is 3.7 per cent, such changes reflecting the early diagenetic environment and, more specifically, the varying composition of the interstitial waters (Nathan *et al.*, 1990). The physical and mineralogical characteristics of the phosphorites of the southeastern United States and of Togo in West Africa are the result of both primary depositional factors and secondary diagenetic processes. Thus, the North Carolina (USA) phosphorites appear to be largely unaltered, whereas both the Florida and Togo phosphorites show systematic decarbonatization of the francolite, the development of Fe and Al phosphates and clay mineral alteration profiles (Kauwenbergh and McClellan, 1990). A mathematical model for the early diagenesis of P and F in marine sediments has been proposed by Cappellen and Berner (1988), involving the decomposition of organic matter and biogenic mineral debris, the reduction of hydrous ferric oxides, the precipitation of authigenic fluorapatite and the reversible adsorption of dissolved phosphorus.

In the Jurassic sandstones in the oilfields of the North Sea, the dissolution of apatite has been found to be a function of depositional environment rather than burial depth. Thus, in shallow marine Upper Jurassic sands of the Claymore, Clyde and Tartan fields, apatite is ubiquitous even down to a burial depth of 3800 m; however, in fluviodeltaic sands of the Beatrice and Murchison fields apatite has dissolved at above 3300 m depth (Morton, 1986). The association of apatite with glauconite formed within the anoxic zone within sediments from the east Australian continental margin has been ascribed to an interaction between two processes of iron-cycling between the oxic and anoxic sediment zones and sediment mixing (O'Brien *et al.*, 1990). Carbonate-apatite is the chief inorganic material found in various human calculi and in concretions accompanying certain pathological conditions (Frondel and Prien, 1946); hydroxylapatite has also been recognized, but in comparison with material of bones and teeth, the apatites of calculi have lower refractive indices.

Apatite occurs as a minor accessory in some chondrite and achondrite meteorites, and may contain appreciable *REE* (Reed and Smith, 1985).

References

Akizuki, M., Nisidoh, H., Kudoh, Y., Watanabe, T. and **Kurata, K.,** 1994. Sector growth and symmetry of (F,OH) apatite from the Asio mine, Japan. *Min. Mag.*, **58**, 307–14.

Altschuler, Z. S., Clark, R. S. and **Young, E. J.,** 1958. Geochemistry of uranium in apatite and phosphorite. *U. S. Geol. Surv., Prof. Paper*, **314D**, 45–90.

Ames, L. L., Jr., 1959. The genesis of carbonate apatites. *Econ. Geol.*, **54**, 829–41.

Åmli, R., 1975. Mineralogy and rare earth geochemistry of apatite and xenotime from the Gloserheia granite pegmatite, Froland, southern Norway. *Amer. Min.*, **60**, 607–20.

Arends, J., Schuthof, J., Linden, W. H. Van der, Bennema, P. and **Van den Bergh, P. J.,** 1979. Preparation of pure hydroxyapatite single crystals by hydrothermal recrystallization. *J. Cryst. Growth*, **46**, 213–20.

Argiolas, R. and **Baumer, A.,** 1978. Synthèse de chlorapatite par voie hydrothermale: étude de l'influence de la sursaturation sur l'evolution des faciès des cristaux. *Can. Min.*, **16**, 285–90.

Arnold, P. W., 1950. The nature of precipitated calcium phosphates. *Trans. Faraday Soc.*, **46**, 1061.

Ayers, J. C. and **Watson, E. B.**, 1991. Solubility of apatite, monazite, zircon, and rutile in supercritical aqueous fluids with implications for subduction zone geochemistry. *Phil. Trans. R. Soc. A*, **335**, 365–75.

Ayliffe, L. K., **Vech, H. H.** and **Chivas, A. R.**, 1992. Oxygen isotopes of phosphate and the origin of island apatite deposits. *Earth Planet. Sci. Lett.*, **108**, 119–29.

Babu, S. K., 1972. On the occurrence of chlorapatite at the emerald mine, Rajagarh village, *Amer. Min. Mag.*, **38**, 972–4.

Baddiel, C. B. and **Berry, E. E.**, 1966. Spectra–structure correlations in hydroxyapatite and fluorapatite. *Spectrochim. Acta*, **22**, 1407–16.

Baker, M. B. and **Wyllie, P. J.**, 1992. High-pressure apatite solubility in carbonate-rich liquids: implications for mantle metasomatism. *Geochim. Cosmochim. Acta*, **56**, 3409–22.

Balenzano, F., **Dell'Anna, L.** and **DiPierro, M.** 1974. Richerche mineralogiche su alcuni fosfati Rinvenuti nelle Grotte di Castellana (Bari): strengite alluminfera, vivianite, taranakite, brushite e idrossiapatite. *Soc. Min. Ital. Min. Petr.*, **30**, 543–73.

Barnes, D. F., 1958. Infrared luminiscence of minerals. *U.S. Geol. Surv., Bull.*, **1052C**, 71–157.

Batchelor, R. A. and **Clarkson, E. N. K.**, 1993. Geochemistry of a Silurian metabentonite and associated apatite from the North Esk Inlier, Pentland Hills. *Scott. J. Geol.*, **29**, 123–30.

Bauer, M. and **Klee, W. E.**, 1993. The monoclinic–hexagonal phase transition in chlorapatite. *Eur. J. Min.*, **5**, 307–16.

Batchelor, R. A. and **Jeppsson, L.**, 1994. Late Llandovery bentonites from Gotland, Sweden, as chemostratigraphic markers. *J. Geol. Soc.*, 151, 000.

Baumer, A. and **Argiolas, R.**, 1981. Synthèses hydrothermales et déterminations RX d'apatites chlorée, fluorée ou hydroxylée. *Neues Jahrb. Min., Monat.*, 344–8.

Baumhauer, H., 1908. Über die Brechungsexponenten und die Doppelbrechung des Apatits von verschiedenen Fundorten. *Z. Krist.*, **45**, 555–62.

Beevers, C. A. and **McIntyre, D. B.**, 1946. The atomic structure of fluorapatite and its relation to that of tooth and bone material. *Min. Mag.*, **27**, 254–7.

Bhatnagar, V. M., 1068. Infrared spectra of mineral apatites. *Bull. Soc. Franç. Min. Crist.*, **91**, 90–1.

Binder, G. and **Troll, G.**, 1989. Coupled anion substitution in natural carbon-bearing apatites. *Contrib. Min. Petr.*, **101**, 394–401.

Borgström, L. H., 1932. Syntetisk NaCl–apatit. *Finska Kemi. Meddel.*, **2**, 51–4.

Boundreau, A. E. and **McCallum, I. S.**, 1989. Investigations of the Stillwater complex: part V. Apatites as indicators of evolving fluid composition. *Contrib. Min. Petr.*, **102**, 138–53.

Boundreau, A. E., **Love. C.** and **Hoatson, D. M.**, 1993. Variations in the composition of apatite in the Munni Munni complex and associated intrusions of the West Pilbara block, Western Australia. *Geochim. Cosmochim. Acta*, **57**, 1467–77.

Bragg, W. L. and **Claringbull, G. F.**, 1965. *Crystal Structures of Minerals*. London (G. Bell and Sons, Ltd), 409 pp.

Brenan, J., 1993a. Partitioning of fluorine and chlorine between apatite and aqueous fluids at high pressure and temperature: implications for the F and Cl content of high *P–T* fluids. *Earth Planet. Sci. Lett.*, **117**, 251–63.

Brenan, J., 1993b. Kinetics of fluorine, chlorine and hydroxyl exchange in fluorapatite. *Chem. Geol. inc. Isotope Geol.*, **110**, 195–210.

Candela, P. A., 1986. Towards a thermodynamic model for the halogens in magmatic systems: an application to the melt–vapor–apatite equilibria. *Chem. Geol.*, **57**, 289–301.

Cant, N. W., **Bett, J. A. S.**, **Wilson, G. R.** and **Hall, W. H.**, 1971. Vibrational spectra of hydroxyl groups in hydroxy-apatites. *Spectrochim. Acta*, **20**, 937–47.

Cappellen, P. Van and **Berner, R. A.**, 1988. A mathematical model for the early diagenesis of phosphorus and fluorine in marine sediments: apatite precipitations. *Amer. J. Sci.*, **228**, 289–333.

Cappellen, P. Van and **Berner, R. A.**, 1991. Fluorapatite crystal growth from modified seawater solutions. *Geochim. Cosmochim. Acta*, **55**, 1219–34.

Carlström, D., 1974. Mineralogical carbonate-containing apatites. In *International Symposium on Structural Properties of Hydroxyapatite and Related Compounds:* Eds. W. E. Brown and R. A. Young.

Carobbi, G., 1950. Celestina e apatite di stronzio sintetiche contenenti piccole quantità di mercurio. *Atti (Rend). Accad. Naz. Lincei, Cl. Sci. Fis. Mat. Nat. Ser. 8*, **8**, 87–93.

Cawthorn, R. G., 1994. Formation of chlor- and fluor-apatite in layered intrusions. *Min. Mag.*, **58**, 299–306.

Çelebi, H., 1992. Geochemie des metamorphosen Fluorapatits von Avnik/Ost-Türkei. *Chem. der Erde*, **52**, 115–29.

Cherniak, D. J. and **Ryerson, F. J.**, 1993. A study of strontium diffusion in apatite using Rutherford back-scattering spectroscopy and ion implantation. *Geochim. Cosmochim. Acta*, **57**, 4653–62.

Conceição, H., **Bonin, B.**, **Sabaté, P.** and **Lameyre, J.**, 1989. Le massif syénitique protérozoique d'Itiúba (Bahia Brésil): contexte géologique et pétrologie. *C. R. Acad. Sci. Paris, Sér. II*, **308**, 45–50.

Cooray, P. G., 1970. A carbonate-bearing fluor–chlor–hydroxyapatite from Matale, Ceylon. *Amer. Min.*, **55**, 2038–41.
Corrigan, J. D., 1993. Apatite fission-track analysis of Oligocene strata in south Texas, U.S.A.: testing annealing models. *Chem. Geol. (Isotope Geosci. Sect.)*, **104**, 227–49.
Creaser, R. A. and **Gray, C. M.**, 1992. Preserved initial $^{87}Sr/^{86}Sr$ in apatite from altered felsic igneous rocks: a case study from the Middle Proterozoic of South Australia. *Geochim. Cosmochim. Acta*, **56**, 2789–95.
Crowley, K. D., Cameron, M. and **Schaefer, R. L.**, 1991. Experimental studies of annealing of etched fission tracks in apatite. *Geochim. Cosmochim. Acta*, **55**, 1449–65.
Cruft, E. F., 1966. Minor elements in igneous and metamorphic apatite. *Geochim. Cosmochim. Acta*, **30**, 375–98.
Dasgupta, S. and **Bhattacharya, D. K.**, 1992. Apatite mineralisation along Singhbhum and Purulia–Bankura shear zones: their nature and physico-chemical characteristics. *Ind. Min.*, **46**, 123–32.
Daubrée, A., 1851. Expériences sur la production artificielle de l'apatite, de la topaze, et de quelques autres métaux fluorifères. *C. R. Acad. Sci. Paris*, **32**, 625.
Deans, T., 1938. Francolite from sedimentary ironstones of the Coal Measures. *Min. Mag.*, **25**, 135–9.
Dihn, P. and **Klement, R.**, 1942. Isomorphe Apartitarten. *Z. Electrochem. Angew. Physik. Chem.*, **48**, 331–3.
Eckermann, H. von, 1922. The rocks and contact minerals of the Mansjö Mountain. *Geol. För. Förh., Stockholm*, **44**, 203–410.
Eckermann, H. von, 1948. The alkaline district of Alnö Island. *Sveriges Geol. Undersök., Ser. Ca.*, **36**, 176 pp.
Eckermann, H. von, 1974. The chemistry and optical properties of some minerals of the Alnö alkaline rocks. *Arkiv. Min. Geol.*, **5**, 94–210.
Edgar, A. D., 1989. Barium- and strontium-enriched apatites in lamproites from West Kimberley, Western Australia. *Amer. Min.*, **74**, 889–95.
Efimov, A. S., Kravchenko, S. M. and **Vasih'eva, Z. V.**, 1962. Strontium apatite: a new mineral. *Dokl. Akad. Nauk USSR*, **142**, 439–42.
Egorov, L. S., 1975. [Nelsonites, francolite breccia and the problem of apatite content in carbonatite complexes.] *Zap. Vses. Min. Obshch.*, **104**, 272–88 (in Russian).
Elliott, J. C., Mackie, P. E. and **Young, R. A.**, 1973. Monoclinic hydroxylapatite. *Science*, **180**, 1055–7.
Exley, R. A. and **Smith, J. V.**, 1982. The role of apatite in mantle enrichment processes and in the petrogenesis of some alkali basalt suites. *Geochim. Cosmochim. Acta*, **46**, 1375–84.
Farver, J. R. and **Giletti, B. J.**, 1989. Oxygen and strontium diffusion kinetics in apatite and potential applications to thermal history determinations. *Geochim. Cosmochim. Acta*, **53**, 1621–31.
Federico, M., Peccerillo, A., Barbieri, M. and **Wu, T. W.**, 1994. Mineralogical and geochemical study of granular xenoliths from the Alban Hills volcano, Central Italy: bearing on evolutionary processes in potassic magma chambers. *Contrib. Min. Petr.*, **115**, 384–401.
Fenogenov, A. N., 1977. Fine acicular apatite, an indicator of contact metamorphism and rapid crystallization of magma. *Dokl. Acad. Sci. USSR, Earth Sci. Sect.*, **235**, 148–9.
Fleischer, M. and **Altschuler, Z. S.**, 1969. The relationship of the rare-earth composition of minerals to geological environment. *Geochim. Cosmochim. Acta*, **33**, 725–32.
Fleischer, R. L., Price, P. B. and **Walker, R. M.**, 1965. Effects of temperature, pressure and ionisation on the formation and stability of fission tracks in minerals and glasses. *J. Geophys. Res.*, **70**, 1497–582.
Fleischer, R. L., Price, P. B. and **Walker, R. M.**, 1975. *Nuclear Tracks in Solids: Principles and Applications*. Berkeley, California (Univ. of California Press), 548 pp.
Fransolet, A.-M. and **Schreyer, W.**, 1981. Unusual iron-bearing apatite from a garnetiferous pegmatoid, Northampton block, Western Australia. *Neues Jahrb. Min., Monat.*, 317–27.
Franz, E.-D., 1983. Fluorapatite $Ca_5F(PO_4)_3$—Ein Modell zur Synthese der Zahnhart-substanz im System CaF_2–$Ca_3(PO_4)_2$. *Z. Naturforschung*, **38b**, 1037–40.
Frondel, C. and **Prien, E. L.**, 1946. Deposition of calcium phosphates accompanying senile degeneration and disease. *Science*, **103**, 326.
Gleadow, A. J. W. and **Lovering, J. F.**, 1978. Thermal history of granitic rocks from western Victoria: a fission track dating study. *J. Geol. Soc. Aust.*, **25**, 323–40.
Green, P. F., Duddy, I. R., Gleadow, A. J. W., Tingate, P. R. and **Laslett, G. M.**, 1986. Thermal annealing of fission tracks in apatite: 1. A quantitative description. *Chem. Geol.*, **59**, 237–53.
Green, P. F., Duddy, I. R. and **Laslett, G. M.**, 1988. Can fission track annealing in apatite be described by first-order kinetics? *Earth Planet. Sci. Lett.*, **87**, 216–28.
Green, P. F., Duddy, I. R., Laslett, G. M., Hegarty, K. A., Gleadow, A. J. W. and **Lovering, J. F.**, 1989. Thermal annealing of fission tracks in apatite: 4. Quantitative modelling techniques and extension to geological time. *Chem. Geol. (Isotope Geosci. Sect.)*, **79**, 155–82.
Grisafe, D. A. and **Hummel, F. A.**, 1970. Crystal chemistry and color in apatites containing cobalt, nickel, and rare-earth ions. *Amer. Min.*, **55**, 1131–45.

Grivet, M., Rebetez, M., Ben Ghouma, N., Chambaudet, A., Jonkheere, R. and **Mars, M.,** 1993. Apatite fission-track age correction and thermal history analysis from projected track distribution. *Chem. Geol.*, **103**, 157–69.

Gübelin, E. and **Schmetzer, K.,** 1983. A new gemstone variety from Tanzania. *J. Gemmol.*, **18**, 592–5.

Gulbrandsen, R., Kramer, J. R., Beatty, L. B. and **Mays, R. E.,** 1966. Carbonate-bearing apatite from Faraday Township, Ontario, Canada. *Amer. Min.*, **51**, 819–24.

Harada, K., Sekino, H., Nagashima, K., Watanabe, T. and **Momoi, H.,** 1974. High-iron bustamite and fluorapatite from the Broken Hill mine, New South Wales, Australia. *Min. Mag.*, **39**, 601–4.

Harada, Z., 1938. Beiträge zur Kenntnis der optischen und chemischen Eigenschaften des Apatis von Kurokura, Kanagawa Präfektur. *J. Fac. Sci. Hokkaido Univ., Ser. 4*, **4**, 11–16.

Hayashi, S., 1960. Analysis of crystal structure of fluorapatite by nuclear magnetic resonance. *J. Chem. Soc. Japan*, **81**, 540–2.

Hayek, E., Böhler, W., Lechleitner, J. and **Petter, H.,** 1958. Hydrothermalsynthese von Calcium–Apatiten. *Z. Anorg. Chem.*, **295**, 241–6.

Helvaci, C., 1987. Rare earth elements in apatite-rich iron deposits and associated rocks of the Avnik (Bingöl) region, Turkey. *Schweiz Min. Petr. Mitt.*, **67**, 307–19.

Henderson, P., 1980. Rare earth element partition between sphene, apatite and other coexisting minerals of the Kangerdlugssuaq intrusion. E. Greenland. *Contrib. Min. Petr.*, **72**, 81–5.

Hendricks, S. B., Jefferson, M. E. and **Mosley, V. M.,** 1932. The crystal structures of some natural and synthetic apatite-like substances. *Z. Krist.*, **81**, 352–69.

Hoffmann, J., 1938. Über Ionen- und Atomfärbungen künstlich hergestellter und natürlicher Apatite. *Chem. Erde*, **11**, 552–75.

Hogarth, D. D., 1957. The apatite-bearing veins of Nisikkatch Lake, Saskatchewan. *Can. Min.*, **6**, 140–50.

Hogarth, D. D., 1988. Chemical composition of fluorapatite and associated minerals from skarn near Gatineau, Quebec. *Min. Mag.*, **52**, 347–58.

Hogarth, D. D., 1989. Pyrochlore, apatite and amphibole: distinctive minerals in carbonatite. In: *Carbonatites: Genesis and Evolution:* Ed. K. Bell, London (Unwin Hyman), pp. 105–48.

Hounslow, A. M. and **Chao, G. Y.,** 1970. Monoclinic chlorapatite from Ontario. *Can. Min.*, **10**, 252–9.

Howie, R. A., 1964. A pyroxene granulite from Hittero, southwest Norway. In *Advancing Frontiers in Geology and Geophysics* Ed. Krishan, M. S., Hyderabad (Indian Geophysical Union): pp. 297–307.

Hughes, J. M., Cameron, M. and **Crowley, K. D.,** 1989. Structural variations in natural F, OH and Cl apatites. *Amer. Min.*, **74**, 870–6.

Hughes, J. M., Cameron, M. and **Crowley, K. D.,** 1990. Crystal structures of natural ternary apatites: solid solution in the $Ca_5(PO_4)_3X$ (X = F,OH,Cl) system. *Amer. Min.*, **75**, 295–304.

Hughes, J. M., Cameron, M. and **Crowley, K. D.,** 1991. Ordering of divalent cations in the apatite structure: crystal structure refinements of natural Mn- and Sr-bearing apatite. *Amer. Min.*, **76**, 1857–62.

Hughes, J. M., Fransolet, A.-M. and **Schreyer, W.,** 1993. The atomic arrangement of iron-bearing apatite. *Neues Jahrb. Min., Monat.*, 504–10.

Huntington, H. D., 1979. Kiglapait mineralogy I: apatite, biotite and volatiles. *J. Petr.*, **20**, 625–52.

Irvine, R. and **Anderson, W. S.,** 1891. On the action of metallic and other salts on carbonate of lime. *Proc. R. Soc. Edinburgh*, **17**, 42.

Ishida, K., 1985. [Arsenian apatite from the Munakata mine, Fukuoka Prefecture.] *Dept. Earth Sci., Coll. Gen. Educ., Kyushu Univ.*, **24**, 7–12 (M.A. **90M**–3376) (in Japanese).

Iwase, E., 1935. Über das Fluoresczenzspektrum des Apatites im ultravioletten Lichte. *Sci. Papers Inst. Phys. Chem. Res., Tokyo*, **27**, no. 567, 1–9.

Jaffe, E. B., 1951. Abstracts of the literature on syntheses of apatites and some related phosphates. *U. S. Geol. Surv., Circ.* **135**, 78 pp.

Jahnke, R. A., 1984. The synthesis and solubility of carbonate fluorapatite. *Amer. J. Sci.*, **284**, 58–78.

Johnson, W., 1960. Two synthetic compounds containing chromium in different valency states. *Min. Mag.*, **32**, 408–11.

Joliff, B. L., Papike, J. J., Shearer, C. J. and **Shimizu, N.,** 1989. Inter- and intra-crystal REE variations in apatite from the Bob Ingersoll pegmatite, Black Hills, South Dakota. *Geochim. Cosmochim. Acta*, **53**, 429–41.

Jonckheere, R., Mars, M., Van den Haute, P. Rebetez, M. and **Chambaudet, A.,** 1993. L'apatite de Durango (Mexique): analyse d'un minéral standard pour la datation par traces de fission. *Chem. Geol.*, **103**, 141–54.

Kapustin, Yu. L., 1986. The composition of apatite from metamorphic rocks. *Geochem. Int.*, **24(4),** 45–51.

Kauwenbergh, S. J. Van and **McClellan, G. H.,** 1990. Comparative geology and mineralogy of the southeastern United States and Togo phosphorites. In *Phosphorite Research and Development:* Eds. A. J. G. Notholt and I. Jarvis, *Geol. Soc. Spec. Publ.*, **52**, 139–55.

Kjarsgaard, B. and **Peterson, T.,** 1991. Nephelinite–carbonatite liquid immiscibility at Shombole volcano, East Africa: petrographic and experimental evidence. *Min. Petr.*, **43**, 293–314.

Klement, R., 1939. Basische Phosphate zweiwertiger Metalle. IV. Strontium–hydroxylapatit. *Z. Anorg. Chem.*, **242**, 215–21.

Knudsen, C., 1991. Petrology, geochemistry and economic geology of the Qaquarssuk carbonatite complex, southern West Greenland. *Mon. Ser. Min. Deposits*, **29**, 110 pp.

Knutson, C., Peacor, D. R. and **Kelly, W. C.**, 1985. Luminescence, color and fission-track zoning in apatite crystals from the Panasqueira tin–tungsten deposit, Beira-Baixa, Portugal. *Amer. Min.*, **70**, 829–37.

Kogarko, L. N., 1991. *PTX* and mechanisms of formation of apatite and rare-metal deposits related with alkaline rocks. In *Source, Transport and Deposition of Metals:* Eds. M. Pagel and J. L. Leroy (A. A. Balkema), pp. 759–61.

Kogarko, L. N. and **Khapaev, V. V.**, 1986. The modelling of formation of apatite deposits of the Khibina massif (Kola Peninsula). In *Origins of Igneous Layering:* Ed. I. Parsons Dordrecht, (Reidel: NATO ASI Series C, **196**), pp. 589–611.

Kogarko, L. N., Krigman, L. D., Petrova, Ye. N. and **Solov'eva, I. P.**, 1976. Phase equilibria in the system nepheline–diopside–fluor-apatite in connection with genesis of apatite deposits of the Khibina. *Geochemistry*, **12**, 14–17.

Kolodny, Y. and **Luz, B.**, 1991. Oxygen isotopes in phosphates of fossil fish—Devonian and Recent. In *Stable Isotope Geochemistry: A Tribute to Samuel Epstein. Geochim. Soc., Spec. Publ.*, **3**, 105–19.

Kolodny, Y., Luz, B. and **Navon, O.**, 1983. Oxygen isotope variations in phosphate of biogenic apatites. I. Fish bone apatite—rechecking the rules of the game. *Earth Planet. Sci. Lett.*, **64**, 398–404.

Konta, J., 1956. Regularly-orientated growth of two phosphates in the scales of Permian fishes (*Palaeoniscidae*). *Casopis pro Min. Geol.*, **1**, 15–23.

Kravchenko, S. M., Kolenko, Yu. A. and **Rass, I. T.**, 1987. [Petrology of Magan apatite deposit in the Maimecha–Kotuiskaya province.] *Izvest. Akad. Nauk USSR, Ser. Geol.*, **4**, 88–100 (in Russian).

Kreidler, E. R. and **Hummel, F. A.**, 1970. The crystal chemistry of apatite: structure fields of fluor- and chlorapatite. *Amer. Min.*, **55**, 170–84.

Kulikova, G. V., Lipatova, E. A., Rozinova, E. L., Sosedko, T. A. and **Furmakova, L. N.**, 1991. [Results and methodical aspects of diagnostics for francolites from fossils and coprolites.] *Zap. Vses. Min. Obshch.*, **120(3)**, 116–33 (in Russian).

Lagerwey, A. A. F., 1977. Detection of trace elements in apatite crystals from Panasqueira, Portugal, by non-destructive optical methods, especially polarospectrography. *Scripta Geol.*, **42**, 52 pp.

Landa, E. A., Murina, G. A., Sheragina, Yu. P. and **Krasnova, N. I.**, 1984. Isotopic compositions of strontium in apatite and apatite-bearing rocks of carbonatite complexes. *Doklady Acad. Sci. USSR, Earth Sci. Sect.*, **264**, 214–6.

Larsen, E. S., Jr., Fletcher, M. H. and **Cisney, E. A.**, 1952. Strontian apatite. *Amer. Min.*, **37**, 656–8.

Latil, C. and **Maury, R.**, 1977. Contribution à létude des échanges d'ions OH^-, Cl^- et F^- et de leur fixation dans les apatites hydrothermales. *Bull. Soc. Franç. Min. Crist.*, **100**, 246–50.

Leckebusch, R., 1979. Comments on the luminescence of apatites from Panasqueira (Portugal). *Neues Jahrb. Min., Monat.*, 17–21.

Lee, D. E., Brandt, E. L. M., Loenen, R. E. Van and **Rose, H. J., Jr.**, 1973a. The chemistry of five accessory rock-forming apatites. *J. Research U. S. Geol. Surv.*, **1**, 267–72.

Lee, D. E., Leonen, R. E. Van and **Mays, R. E.**, 1973b. Accessory apatite from hybrid granitic rocks of the southern Snake Range, Nevada. *J. Research. U.S. Geol. Surv.*, **1**, 89–98.

Lieftink, D. J., Nijland, T. G. and **Majer, C.** 1994. The behavior of rare-earth elements in high-temperature Cl-bearing aqueous fluids: results from the Ødegårdens Verk natural laboratory. *Can. Min.*, **32**, 149–58.

Liu, Y. and **Comodi, P.**, 1993. Some aspects of the crystal chemistry of apatites. *Min. Mag.*, **57**, 709–19.

Livingstone, A., 1994. An apatite high in lead from Wanlockhead, Strathclyde Region, Scotland. *Min. Mag.*, **58**, 159–63.

Lowell, W. R., 1955. Igneous intrusions and metamorphism in some phosphatic rocks of southwestern Montana. *Econ. Geol.*, **50**, 715–37.

McArthur, J. M., 1985. Francolite geochemistry—compositional controls during formation, diagenesis, metamorphism and weathering. *Geochim. Cosmochim. Acta*, **49**, 23–35.

McArthur, J. M. and **Herczeg, A.**, 1990. Diagenetic stability of the isotopic composition of phosphate–oxygen: palaeoenvironmental implications. In *Phosphorite Research and Development:* Eds. A. J. G. Notholt and I. Jarvis, *Geol. Soc. Spec. Publ.*, **52**, 119–24.

McArthur, J. M., Coleman, M. L. and **Bremner, J. M.**, 1980. Carbon and oxygen isotope composition of structural carbonate in sedimentary francolite. *J. Geol. Soc.*, **137**, 669–73.

McClellan, G. H., 1980. Mineralogy of carbonate fluorapatite. *J. Geol. Soc.*, **137**, 675–82.

McClellan, G. H. and **Kauwenbergh, S. J. Van**, 1990. Mineralogy of sedimentary apatites. In *Phosphorite Research and Development:* Eds. A. J. G. Notholt and I. Jarvis, *Geol. Soc. Spec. Publ.*, **52**, 23–31.

McClellan, G. H. and **Van Kauwenbergh, S. J.** 1991. Mineralogical and chemical variation of francolites with geological time. *J. Geol. Soc.*, **148**, 809–12.

McClellan, G. H. and **Lehr, J. R.**, 1969. Crystal chemical investigation of natural apatites. *Amer. Min.*, **54**, 1374–91.

McConnell, D., 1970. Crystal chemistry of bone mineral: hydrated carbonate apatites. *Amer. Min.*, **55**, 1659–69.

McConnell, D., 1973. *Apatite—its Crystal Chemistry, Mineralogy, Utilization and Geologic and Biologic Occurrences*. Vienna, Heidelberg and New York (Springer-Verlag), 111 pp.
McConnell, D. and **Gruner, J. W.**, 1940. The problem of the carbonate-apatites. III. Carbonate-apatite from Magnet Cove, Arkansas. *Amer. Min.*, **25**, 157–67.
McKeown, F. A. and **Klemic, H.**, 1957. Rare-earth-bearing apatite, at Mineville, Essex County, New York. *Bull. U.S. Geol. Surv.*, **1046B**, 9–23.
Mackie, P. E., **Elliott, J. C.** and **Young, R. A.**, 1972. Monoclinic structure of synthetic $Ca_5(PO_4)_3Cl$, chlorapatite. *Acta Cryst.*, **B28**, 1840–948.
Majer, W. and **Piestrzynski, A.**, 1986. Francolite from the Lower Zechstein sediment, Rudna mine, Fore–Sudetic monocline. *Min. Polonica*, **17**, 77–84.
Mehmel, M., 1930. Über die Struktur des Apatits. *Z. Krist.*, **75**, 323–31.
Minguzzi, C., 1941. Apatiti sintetiche con cromo trivalente ed esavalente. *Periodico Min. Roma*, **12**, 343–80.
Mitchell, L., **Faust, G. T.**, **Hendricks, S. B.** and **Reynolds, D. S.**, 1943. The mineralogy and genesis of hydroxylapatite. *Amer. Min.*, **28**, 356–71.
Montel, G., 1971. Sur les structures de quelques apatites d'intérêt biologique et leurs imperfections. *Bull. Soc. Franç. Min. Crist.*, **194**, 300–13.
Montel, G. and **Chaudron, G.**, 1951. Synthèse de la fluorapatite calcique par voir sèche à partir de différents phosphates de calcium. *Acad. Sci. Paris*, **233**, 318–9.
Morton, A. C., 1986. Dissolution of apatite in North Sea Jurassic sandstones: implications for the generation of secondary porosity. *Clay Min.*, **21**, 711–33.
Morton, R. D. and **Catanzaro, E.**, 1964. Stable chlorine isotope abundances in apatites from Ødegårdens verk, Norway. *Norsk. Geol. Tidsskr.*, **44**, 307–13.
Murayama, J. K., **Nakai, S.**, **Kato, M.** and **Kumazawa, M.**, 1986. A dense polymorph of $Ca_3(PO_4)_2$: a high pressure phase of apatite decomposition and its geochemical significance. *Phys. Earth Planet. Interiors*, **44**, 293–303.
Náray-Szabó, S., 1930. The structure of apatite. *Z. Krist.*, **75**, 387–98.
Nash, W. P., 1984. Phosphate minerals in terrestrial igneous and metamorphic rocks. In *Phosphate Minerals:* Eds. J. O. Nriagu and P. B. Moore, Berlin and New York (Springer-Verlag), pp. 215–41.
Nathan, Y., **Soudry, D.** and **Avigour, A.**, 1990. Geological significance of carbonate substitution in apatites: Israeli phosphorites as an example. In *Phosphorite Research and Development:* Eds. A. J. G. Notholt and I. Jarvis, *Geol. Soc. Spec. Publ.*, **52**, 179–91.
Natio.,al Bureau of Standards, 1957. Crystal chemistry of tooth and bone material. *J. Washington Acad. Sci.*, **47**, 202–4 (M.A. **14**–17).
Newberry, N. G., **Essene, J.** and **Peacor, D. R.**, 1981. Alforsite, a new member of the apatite group: the barium analogue of chlorapatite. *Amer. Min.*, **66**, 1050–3.
Notholt, A. J. G. and **Jarvis, I.**, Eds., 1990. *Phosphorite Research and Development. Geol. Soc. Spec. Publ.*, **52**, 326 pp.
Notholt, A. J. G., **Highley, D. E.** and **Harding, R. R.**, 1985. Investigation of phosphate (apatite) potential of Loch Borralan igneous complex, northwest Highlands, Scotland. *Trans. Inst. Mining Metall. (Sect. B: Appl. Earth Sci.)*, **94**, 58–65.
Nyström, J. O. and **Henriquez, F.**, 1989. Dendritic magnetite and miniature diapir-like concentrations of apatite: two magmatic features of the Kirunavaara iron ore. *Geol. För. Förh., Stockholm*, **111**, 53–64.
O'Brien, G. W., **Milnes, A. R.**, **Veeh, H. H.**, **Heggie, D. J.**, **Riggs, S. R.**, **Cullen, D. J.**, **Marshall, J. F.** and **Cook, P. J.**, 1990. Sedimentation dynamics and redox iron-cycling: controlling factors for the apatite–glauconite association on the East Australian continental margin. In *Phosphorite Research and Development:* Eds. A. J. G. Notholt and I. Jarvis, *Geol. Soc. Spec. Publ.*, **52**, 61–86.
Omori, K. and **Seki, T.**, 1960. Infrared studies of some phosphate minerals. *J. Jap. Assoc. Min. Econ. Geol.*, **44**, 7–13.
Osmond, J. K. and **Sawin, H. J.**, 1959. Unit-cell dimensions of recent and fossil tooth apatites. *Bull. Geol. Soc. Amer.*, **70**, 1963 (abstr.).
Pan, Y., **Fleet, M. E.** and **McRae, N. D.**, 1993. Oriented monazite inclusions in apatite porphyroblasts from the Hemlo gold deposit, Ontario, Canada. *Min. Mag.*, **57**, 697–707.
Paul, T. A. and **Fitzgerald, P. G.**, 1992. Transmission electron microscope investigation of fission tracks in fluorapatite. *Amer. Min.*, **77**, 336–44.
Paulick, J. and **Newesely, H.**, 1968. Zur Kenntnis der Apatite der Cerro de Mercado, Durango, Mexiko. *Neues Jahrb. Min., Monat.*, 224–35.
Petrukha, L. M., 1982. [Apatite from crystal-bearing cavities of South Ural.] *Zap. Vses. Min. Obshch.*, **111**, 593–7 (in Russian).
Pichavant, M., **Montel, J.-M.** and **Richard, L. R.**, 1992. Apatite solubility in peraluminous liquids: experimental data and an extension of the Harrison–Watson model. *Geochim. Cosmochim. Acta*, **56**, 3855–61.

Pieruccini, R., 1947. Sulla possibilità di introdurre alluminio, ferro, piombo o molibdeno nel reticolo delle apatiti. *Atti. Soc. Toscana Sci. Nat.*, **54**, 295–308.

Portnov, A. M., Dubinshuk, V. T. and **Stolyarova, T. I.**, 1970. A natural rare-earth oxy-apatite. *Dokl. Acad. Sci. USSR, Earth Sci. Sect.*, **192**, 114–6.

Portnov, A. M., Litvintsev, E. G., Filippova, V. N. and **Rakov, L. T.**, 1977. Pleochroism of apatite. *Dokl. Acad. Sci. USSR, Earth Sci. Sect.*, **228**, 121–2.

Povondra, P., 1992. Crystal chemistry of rock-forming apatites from the Bohemian massif. *Acta Univ. Carol., Geol.*, 197–224.

Prener, J. S., 1967. The growth and crystallographic properties of calcium fluor- and chlorapatite crystals. *J. Electrochem. Soc.*, **114**, 77–83.

Puech, J., Heughebart, J.-C. and **Montel, G.**, 1982. A new mode of growing apatite crystallites. *J. Cryst. Growth*, **56**, 20–4.

Qu, G., 1985. [Discovery of hydroxyl-fluorapatite in a certain locality of China.] *Acta Min. Sinica*, **5**, 158–63 (in Chinese).

Rakovan, J. and **Reeder, R. J.**, 1994. Differential incorporation of trace elements and disymmetrization in apatite: the role of surface structure during growth. *Amer. Min.* **79**, 892–903.

Ramaswamy, R., 1992. Carbonate-apatite from carbonatites of Kudangulam near Cape Comorin, Tamilnadu. *Ind. Min.*, **46**, 91–4.

Rao, A. T., Raman, C. V. and **Prakasa Rao, Ch. S.**, 1979. Fluor-apatites from calc-silicate skarn vein contacts, Gondivalsa, Orissa, India. *Min. Mag.*, **43**, 439.

Reed, S. J. B. and **Smith, D. G. W.**, 1985. Ion probe determination of rare earth elements in merrillite and apatite in chondrites. *Earth Planet. Sci. Lett.*, **72**, 238–44.

Regnier, P., Lasaga, A. C., Berner, R. A., Han, O. H. and **Zilm, K. W.** Mechanism of CO_3^{2-} substitution in carbonate-fluorapatite: evidence from FTIR spectroscopy, ^{13}C NMR, and quantum mechanical calculations. *Amer. Min.*, **79**, 809–18.

Rock, N. M. S., Griffin, B. J. and **Edgar, A. D.**, 1992. A spectrum of potentially diamondiferous lamproites and minettes from the Jharia coalfield, eastern India. *J. Volc. Geotherm. Research*, **50**, 55–83.

Rønsbo, J. G., 1989. Coupled substitutions involving REEs and Na and Si in apatites in alkaline rocks from the Ilímaussaq intrusion, South Greenland, and its petrological implications. *Amer. Min.*, **74**, 896–901.

Ross, S. D., 1974. Phosphates and other oxy-anions of group V. In *The Infrared Spectra of Minerals:* Ed. V. C. Farmer, London (Min. Soc. Monograph **4**), pp. 383–422.

Sakai, M. and **Akai, J.**, 1994. Strontium, barium and titanium-bearing minerals and their host rocks from Ohmi, Japan. *Sci. Rept. Niigata Univ., Ser. E (Geol. & Min.)*, **9**, 97–118.

Sandell, E. B., Hey, M. H. and **McConnell, D.**, 1939. The composition of francolite. *Min. Mag.*, **25**, 395–410.

Sandhu, A. S., Singh, S. and **Virk, H. S.**, 1987. Anisotropic track annealing in apatite. *Min. J.*, **13**, 307–13.

Sänger, A. T. and **Kuhs, W. F.**, 1992. Structural disorder in hydroxyapatite. *Z. Krist.*, **199**, 123–48.

Serdyuchenko, D. P., 1956. On certain types of sedimentary–metamorphic mineral formations. *Problems Min. Sediment Formations. Books 3 and 4,* Publ. Lvov. Univ., 486–515 (M.A. **13**–590).

Seto, K., 1932. Chemical composition of chlorine–fluorine–apatite from Kurokura. *J. Jap. Assoc. Min. Petr. Econ. Geol.*, **7**, 180–1.

Shaw, R. K., Venkatesh, T. L. and **Gupta, A. K.**, 1993. Experimental study of the system fluorapatite–chlorapatite under 10 and 12 kbar at 640, 750 and 900°C. *Natl Acad. Sci. Lett.*, **16**, 27–35.

Shemesh, A., Kolodny, Y. and **Luz, B.**, 1983. Oxygen isotope variations in phosphate of biogenic apatites. II. Phosphorite rocks. *Earth Planet. Sci. Lett.*, **64**, 405–16.

Shiga, Y. and **Urashima, Y.**, 1987. A sodian sulfatian fluorapatite from an epithermal calcite–quartz vein of the Kushikino mine, Kagoshima Prefecture, Japan. *Can. Min.*, **25**, 673–81.

Silverman, S. R., Fuyat, R. K. and **Weiser, J. D.**, 1952. Quantitative determination of calcite associated with carbonate-bearing apatites. *Amer. Min.*, **37**, 211–22.

Stoppa, F. and **Liu, Y.**, 1995. Chemical composition and petrogenetic implications of apatites from some ultra-alkaline Italian rocks. *Eur. J. Min.*, **7**, 391–402.

Stormer, J. C., Pierson, M. L. and **Tacker, R. C.**, 1993. Variation of F and Cl X-ray intensity due to anisotropic diffusion in apatite during electron microprobe analysis. *Amer. Min.*, **78**, 641–8.

Suedersanan, K., Mackie, P. E. and **Young, R. A.**, 1972. Comparison of synthetic and mineral fluorapatite, $Ca_5(PO_4)_3F$, in crystallographic detail. *Mat. Res. Bull.*, **7**, 1331–8.

Suitch, P. R., LaCout, J. L., Hewat, H. and **Young, R. W.**, 1985. The structural location and role of Mn^{2+} partially substituted for Ca^{2+} in fluorapatite. *Acta. Cryst.*, **B41**, 173–9.

Tamishita, K., Owada, H., Nakagawa, H., Umegaki, T. and **Kanazawa, T.**, 1986. Trivalent-cation-substituted calcium oxyhydroxyapatite. *J. Amer. Ceram. Soc.*, **69**, 590–4.

Trzcienski, W. E., Jr., Perrault, G. and **Hebert, P.**, 1974. A note on apatite from Huddersfield Township, Quebec. *Can. Min.*, **12**, 289–91.

Tuddenham, W. M. and **Lyon, R. J. P.**, 1960. Infrared techniques in the identification and measurement of minerals. *Anal. Chem.*, **32**, 1630–4.

Vasilieva, Z. V., 1958. On the role of manganese in apatites. *Mem. All-Union Min. Soc.*, **87**, 455–68.

Vasilieva, Z. V., 1985. [Typomorphism of apatite of alkaline volcano-plutonic rocks of South Mongolia.]. *Zap. Vses. Min. Obshch.*, **114**, 348–60 (in Russian).

Vasilieva, Z. V., **Litzarev, M. A.** and **Organova, N. I.**, 1958. On a natural sulphate-apatite. *Dokl. Acad. Sci. USSR*, **118**, 577–80.

Viladkar, S. G. and **Wimmenauer, W.**, 1992. Geochemical and petrological studies on the Amba Dongar carbonatites (Gujarat, India). *Chem. Erde*, **52**, 277–91.

Villiers, J. E. de, 1942. The carbonate-apatites: francolite from the Richtersveld, South Africa. *Amer. J. Sci.*, **240**, 443–47.

Visse, L., 1953. Minéralogie des temperatures calciques. Facteurs de solubilité. *C. R. Acad. Sci. Paris*, **236**, 1583–5.

Volborth, A., 1954. Phosphatminerale aus dem Lithiumpegmatit von Viitaniemi, Eräjärvi, Zentral-Finnland. *Ann. Acad. Sci. Fennicae, Ser. A*, **3**, 39.

Vukadinovic, D. and **Edgar, A. D.**, 1993. Phase relations in the phlogopite–apatite system at 20 kbar; implications for the role of fluorine in mantle melting. *Contrib. Min. Petr.*, **114**, 247–54.

Wagner, G. and **Van den Haute, P.** 1992. *Fission-Track Dating*. Dordrecht, The Netherlands (Kluwer Acad. Publ.), 285 pp.

Wagner, G. A. and **Reimer, G. M.**, 1972. Fission track tectonics: the tectonic interpretation of fission track ages. *Earth Planet. Sci. Lett.*, **14**, 263–8.

Wagner, G., **Gleadow, A. J. W.** and **Fitzgerald, P. G.**, 1989. The significance of the partial annealing in apatite fission track analysis: projected track length measurements and uplift chronology of the Transantarctic Mountains. *Chem. Geol.* (*Isotope Geosci. Sect.*), **79**, 295–305.

Wallaeys, R. and **Chaudron, G.**, 1950. Sur la préparation de certaines apatites mixtes. *C. R. Acad. Sci. Paris*, **231**, 355–7.

Walters, L. J., Jr. and **Luth, W. C.**, 1969. Unit-cell dimensions, optical properties, halogen concentrations in several apatites. *Amer. Min.*, **54**, 156–62.

Watson, E. B. and **Green, T. H.**, 1981. Apatite/liquid partition coefficients for the rare earth elements and strontium. *Earth Planet. Sci. Lett.*, **56**,405–21.

Watson, E. B., **Harrison, T. M.** and **Ryerson, F. J.**, 1985. Diffusion of Sm, Sr and Pb in fluorapatite. *Geochim. Cosmochim. Acta*, **49**, 1813–23.

Wenzel, T. and **Ramseyer, K.**, 1992. Mineralogical and mineral–chemical changes in a fractionation-dominated diorite–monzodiorite–monzonite sequence: evidence from cathodoluminescence. *Eur. J. Min.*, **4**, 1391–9.

Wolf, M. B. and **London, D.**, 1994. Apatite dissolution into peraluminous haplogranitic melts: an experimental study of solubilities and mechanisms. *Geochim. Cosmochim Acta*, **58**, 4127–45.

Wyllie, P. J., **Cox, K. G.** and **Biggar, G. M.**, 1962. The habit of apatite in synthetic systems and igneous rocks. *J. Petr.*, **3**, 238–43.

Yardley, B. W. D., 1985. Apatite composition and the fugacities of HF and HCl in metamorphic fluids. *Min. Mag.*, **49**, 77–9.

Yegorov, L. S., 1984. Rare-earth element and fluorine contents of apatite as reflecting formation conditions, alteration, and potential mineralization for rocks of the phoscorite–carbonatite group in ijolite–carbonatite complexes. *Geochem. Int.*, **21(1)**, 93–107.

Young, E. J. and **Munson, E. L.**, 1966. Fluor-chlor-oxy-apatite and sphene from Crystal Lode pegmatite near Eagle, Colorado. *Amer. Min.*, **51**, 1476–93.

Young, E. J., **Sheridan, D. M.** and **Munson, E. L.**, 1966. Manganese- and strontium-bearing apatite from the Peerless pegmatite, South Dakota. *Amer. Min.*, **51**, 1516–24.

Young, E. J., **Myers, A. T.**, **Munson, E. L.** and **Conklin, N. M.**, 1969. Mineralogy and geochemistry of fluorapatite from Cerro de Mercado, Durango, Mexico. *U.S. Geol. Surv., Prof. Paper*, **650D**, 84–93.

Young, R. A., **Sudarsanan, K.** and **Mackie, P. E.**, 1968. Structural origin of some physical property differences among three apatites. In *Colloque Intern. Phosphates Mineraux Solides* (*Toulouse*), 114–7.

Zhang, S., **Wang, L.**, **Zhu, W.** and **Yang, W.**, 1985. [Use of *REE* in apatites to distinguish the petrogeno-mineralization series of granitic rocks.] *Geochimica*, 45–56 (in Chinese with English abstract).

Zhou, L. and **Wang, Y.**, 1989. REE geochemical characteristics of apatite, sphene and zircon from alkaline rocks. *Chin. J. Geochem.*, **8**, 245–53.

Zhu, C. and **Sverjensky, D. A.**, 1992. F–Cl–OH partitioning between biotite and apatite. *Geochim. Cosmochim. Acta*, **56**, 3435–67.

Zimin, S. S., **Lennikov, A. M.**, **Oktyabr'skiy, R. A.** and **Solyanik, A. N.**, 1984. Phosphide model of apatite-bearing igneous complexes and apatite. *Dokl. Acad. Sci. USSR, Earth Sci. Sect.*, **267**, 127–30.

Monazite (Ce,La,Nd,Th)PO_4

Monoclinic (+)

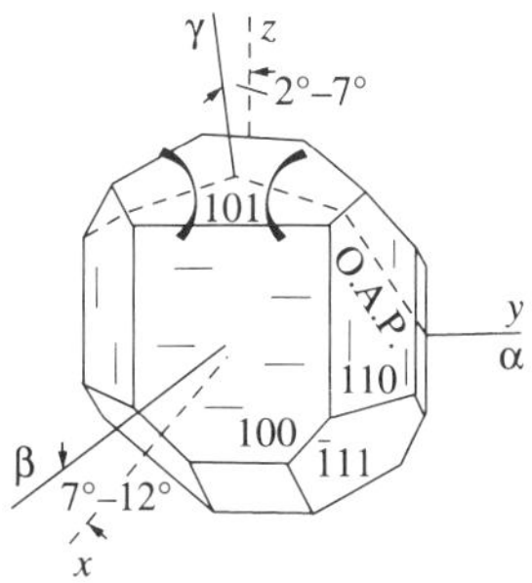

α	1.722–1.793
β	1.778–1.800
γ	1.823–1.860
δ	0.047–0.067
$2V_\gamma$	10–26°
γ: z 2–7°, O.A.P. ⊥(010)	
Dispersion:	Usually $r < v$, weak
D	4.98–5.43
H	5
Cleavage:	{100} moderate, {001} variable.
Twinning:	Twin plane {100} common; also rare lamellar twinning on {001}.
Colour:	Yellow, brown, reddish brown, grey; yellow or colourless in thin section or opaque.
Pleochroism:	Very weak, light yellow to very light yellow.
Unit cell:	a 6.75–6.84, b 6.96–7.04, c 6.44–6.58 Å, β 103.54–104.00° Z=4. Space group $P2_1/n$

Slowly decomposed by acids. The grey variety may be metamict.

Monazite occurs as an accessory mineral in acidic igneous rocks and gneisses, and may be found as relatively large crystals in pegmatites. It also occurs as a heavy detrital mineral derived from the weathering of such rocks. It is rich in *REE*, and although the cerium-bearing variety, monazite-(Ce), is by far the commonest, monazite-(La) and monazite-(Nd) also occur, where La and Nd, respectively, are the dominant *REE*. The name is from the Greek *monazein*, to be solitary, in allusion to its rather rare occurrence.

Structure

The structure of monazite was investigated by Gliszczysnki (1939), by Parrish (1939) who determined the cell parameters and space group, and by Kokkoros (1942) who showed that the structure is made up of PO_4 tetrahedra with each metal atom almost equidistant from eight oxygen atoms. A full structural determination was made on a thorium-bearing monazite from Japan by Ueda (1953), who reported that the distorted PO_4 tetrahedra have P–O distances of 1.62–1.66 Å, and that the Ce atom is surrounded by nine O atoms, which form irregular polyhedra. Each O atom is shared by one P and two Ce atoms. The arrangement of PO_4 groups is clearly responsible for the distinct {100} cleavage (Fig. 118(a)) Two recent structural refinements giving more reasonable P–O bond lengths of ~1.54 Å, however, have been reported (Mi and Shen, 1994; Ni *et al.*, 1995); the REO_9 polyhedron is particularly suitable for accomodation of the larger light *REE* (La–Gd).

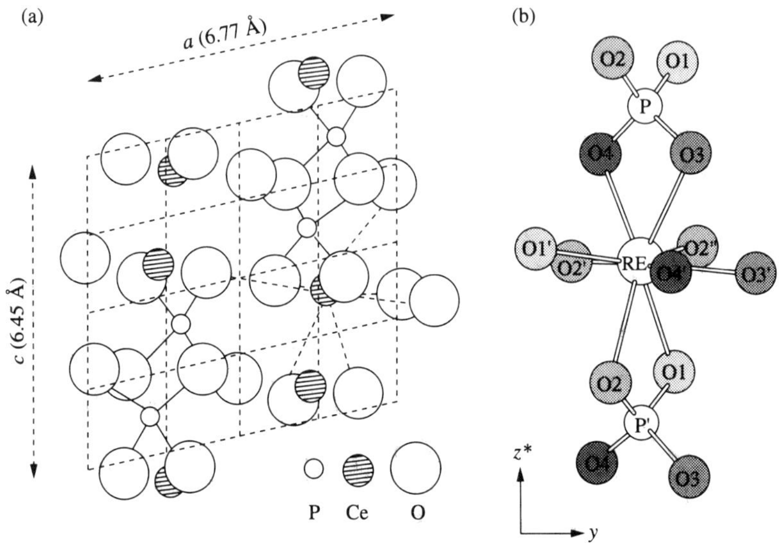

Fig. 118. (a) The crystal structure of monazite projected on the *xz* plane (after Ueda, 1953). (b) Chain of alternating PO_4 tetrahedra and REO_9 polyhedra parallel to *z* in monazite (after Ni *et al.*, 1995).

The X-ray powder pattern of monazite was indexed by Gliszczynski (1939) and Pabst (1951), and a detailed comparative study of the X-ray powder patterns of ten monazites was made by Molloy (1959). Further single-crystal X-ray work was reported by Shankar and Khubchandani (1956) on thorium-bearing monazite. Insufficient data are available to allow a correlation of the unit cell parameters with the dominant *REE*, but there are indications that monazite-(La) may have a larger unit cell than monazite-(Ce).

The calcium-bearing, thorium-rich member of the monazite group, cheralite, has the idealized composition (Th,Ca,Ce)(P,Si,)O_4 and is isostructural with monazite (Bowie and Horne, 1953). There is a systematic decrease in the cell dimensions with increasing substitution of the smaller Ca and Th ions for Ce and La. The cheralite from the type locality at Kutlakuzhi, Kerala, India, was re-examined by Bowles *et al.* (1980) who determined its cell parameters as *a* 6.751, *b* 6.962, *c* 6.468 Å, *β* 103.9°; see also Finney and Nagaraja Rao (1967). Huttonite, ideally $ThSiO_4$ (Pabst, 1951), is also a member of the group; ordered domain structures and continuity in the monazite–huttonite series was demonstrated by Kucha (1980). A huttonite with a (*REE*+Ca):P ratio of 0.93 and a *REE*:P ratio of 1.55 has *a* 6.61, *b* 7.19, *c* 6.32 Å, *β* 101.8°.

Chemistry

Monazite-(Ce), the commonest species, can be expressed via the formula (Ce,La,Nd,Th)PO_4; 14 representative analyses are given in Table 43. The thorium content is very variable, ranging from virtually 0.0 to 19.5 per cent ThO_2 (anal. 13). Frondel (1958) reported contents of up to at least 27 wt. per cent, probably in solid solution with huttonite (see below). No apparent correlation exists between Th and the *REE* values; see also Murata *et al.* (1953). Calcium may also occur in the rare-earth position, substituting for the slightly larger cerium ion, often simultaneously with the replacement of Ce by Th; minor amounts of U, Al and

Fe^{3+} may also be found. Electron microprobe analyses of 18 monazite-(Ce) grains from detrital sands in Victoria, Australia, have been reported by Henry (1993); these have fairly low values of U ($\leqslant$0.92 wt. per cent) with high Th contents (ThO_2 4.0–13.6 wt. per cent) and are thought to be of pegmatitic origin. The dark, often nodular, monazite-(Ce) being increasingly recognized in Lower Palaeozoic rocks is often richer in Eu than average monazite compositions, with Eu_2O_3 up to 1.40 per cent (e.g. Rosenblum and Mozier, 1983; Read *et al.*, 1987).

Cheralite $(REE,Th,Ca,U)(P,Si)O_4$ is isostructural with monazite (Bowie and Horne, 1953; Bowles *et al.*, 1980), and huttonite $ThSiO_4$ forms the silicate end-member (Pabst, 1951; Kucha, 1980). There is thus a ternary system $REEPO_4$–$CaTh(PO_4)_2$–$ThSiO_4$. In monazite or huttonite when *REE* or Th are substituted, the resultant valency differences are compensated by the entrance of Ca and minor F and $(OH)_4$, with monazite containing up to 27 wt. per cent Th and huttonite having up to 20 wt. per cent *REE*. Earlier work recording appreciable Si in monazites includes that of Johnstone (1914), Starynkevitch (1922), Wylie (1948) and Jaffe (1955); see also Nekrasov (1972).

In addition to analyses reported in Table 43, *REE* in monazite were determined by Borovsky and Gerasimovsky (1945), Wylie (1950), Murata *et al.* (1959) and Heinrich *et al.* (1960). The latter workers found the distribution of the *REE* for 11 monazites to be Ce > Nd > La > Sm > Pr, though for three monazites from the Brown Derby pegmatite, Colorado, USA, it was Ce > Nd > Sm > La > Pr. The average *REE* compositions of monazites from granitic pegmatites, granitic rocks and alkaline igneous rocks and carbonatites were given by Fleischer and Altschuler (1969). Monazite-(Ce) inclusions in apatite porphyroblasts in a muscovite schist are enriched in *LREE* but markedly depleted in *HREE* compared with monazite grains in the rock matrix and cross-cutting veins (Pan *et al.*, 1993). In a study of the Sweetwater Wash granite suite, southeast California, USA, Wark and Miller (1993) found that monazite in the more differentiated rocks (aplites) was relatively depleted in the *LREE* and had higher U and Th. Further analytical work on the *REE* in monazites has been reported by Montel (1993) and Kingsbury *et al.* (1993).

Monazite-(La), whose generalized formula is $(La,Ce,Nd)PO_4$, was probably first recognized by Borovsky and Gerasimovsky (1945). It is now known to be fairly common (Table 44) and includes the thorium-free monazite (anal. 4) from Llallagua, Brazil (Gordon, 1939), and several samples analysed by Kato (1958) from Japan.

Monazite-(Nd), with formula $(Nd,La,Ce)PO_4$, has been reported by Maksimović and Pantó (1980) and Graser and Schwander (1987). This mineral (Table 44, anal. 8) differs from other monazites in forming long prismatic crystals, which are mostly translucent or milky but always bright rose in colour.

The isotope ratios $^{238}U/^{206}Pb$, $^{235}U/^{207}Pb$, $^{207}Pb/^{206}Pb$ and $^{232}Th/^{208}Pb$ in monazite have been extensively used for determining the absolute ages of monazite-bearing pegmatites, etc., e.g. Holmes (1954) and Ahrens (1955). Tilton and Nicolaysen (1957), however, showed that there is a heterogeneous distribution of U, Th and radiogenic Pb in some monazite crystals. Monazite from Norway, on decomposition with H_2SO_4, has yielded as much as 5.14 ml of helium per ml of monazite (Erichsen, 1951); see also Prasad *et al.* (1957).

Monazite is particularly useful in the dating of peraluminous rocks, where zircon inheritance often precludes a precise U–Pb age determination for magmatic zircon. Monazite favours the incorporation of Th relative to U and can therefore contain considerable amounts of excess ^{206}Pb derived from initially incorporated ^{236}Th, an intermediate decay product of ^{238}U; although corrections can be made using the Th/U ratio of the host rock, they may not always be valid. Monazite may lose Pb during episodic prolonged heating events in the uppermost amphibolite and

Table 43. Monazite-(Ce) analyses

	1	2	3	4	5	6	7
SiO_2	0.70	0.34	0.2	0.81	0.68	—	0.30
ThO_2	3.01	—	tr.	2.28	—	3.6	—
U_3O_8	—	—	—	—	—	0.7	—
Ce_2O_3	36.19	34.22	34.0	33.28	32.19	30.6	30.43
La_2O_3	19.65	23.88	15.5	16.88	11.05	14.3	14.17
Nd_2O_3	8.20	7.25	14.4	9.90	17.73	12.8	14.58
Pr_2O_3	2.94	1.99	1.0	4.59	5.71	—	4.67
Sm_2O_3	0.85	—	2.6	1.31	2.61	2.5	3.11
Gd_2O_3	—	—	1.5	2.61	0.75	4.1	1.96
Y_2O_3	—	—	0.5	0.16	0.96	1.4	—
Dy_2O_3	—	—	—	0.01	0.27	—	—
Al_2O_3	—	—	—	—	—	—	—
Fe_2O_3	—	—	—	—	—	0.1	—
MgO	—	—	—	—	—	—	—
CaO	—	0.57	0.3	0.16	—	0.5	—
PbO	0.12	—	—	—	—	—	—
P_2O_5	29.23	30.27	29.1	27.33	28.66	29.7	29.93
Total	100.89	100.00	99.8	100.20	100.94	100.3	99.55
α	1.779	—	—	—	—	—	1.783
β	1.781	—	—	—	—	—	1.784
γ	1.833	—	—	—	—	—	1.830
$2V_\gamma$	26°	—	—	—	—	—	15–17°
D	4.980	—	—	—	—	—	5.340 (calc.)
a (Å)	—	—	—	—	—	—	6.760
b (Å)	—	—	—	—	—	—	7.000
c (Å)	—	—	—	—	—	—	6.450
β	—	—	—	—	—	—	104.0°

	Numbers of ions on the basis of 16 (O)													
P	3.888	4.00	3.966	4.02	3.921	3.95	3.751	3.88	3.850	3.96	3.958		3.980	4.03
Si	0.108		0.053		0.032		0.132		0.108		—		0.047	
Ce	2.081		1.939		1.981		1.975		1.868		1.764		1.750	
La	1.138		1.363		0.910		1.009		0.646		0.830		0.821	
Nd	0.460		0.401		0.819		0.573		1.005		0.726		0.818	
Pr	0.168		0.114		0.059		0.275		0.300		—		0.271	
Sm	0.046		—		0.143		0.073		0.041		0.221		0.106	
Gd	—		—		0.082		0.145		0.041		0.221		0.168	
Dy	—		—		—		0.001		0.014		—		—	
Y	—	4.01	—	4.07[a]	0.042	4.11[b]	0.014	4.21[c]	0.081	4.11[d]	0.117	4.04	—	3.99[e]
Th	0.108		—		—		0.084		—		0.129		—	
U	—		—		—		—		—		0.024		—	
Al	—		—		—		—		—		—		—	
Fe^{3+}	—		—		—		—		—		0.012		—	
Mg	—		—		—		—		—		—		—	
Ca	—		0.092		0.051		0.028		—		0.084		—	
Pb	0.005		—		—		—		—		—		—	

[a]Includes Fe^{2+} 0.092, Sr 0.069.
[b]Includes Lu 0.019.
[c]Includes Eu 0.017, Er 0.001, U 0.019.
[d]Includes Ho 0.008, Er 0.004, Yb 0.005.
[e]Includes Mn 0.053.

1 Monazite-(Ce), baryte–carbonate rock near alkali syenite, Mountain Pass, San Bernardino County, California, USA (Jaffe, 1955). Anal. M. Sherwood and H. J. Rose.
2 Monazite-(Ce) intergrown with allanite, carbonatite, Uyaynah area, United Arab Emirates (Woolley *et al.*, 1991) (includes FeO 0.71, SrO 0.77).
3 Central part of yellow-orange monazite-(Ce) showing alexandrite effect, quartz vein, Uwharrie National Forest, 8.4 km east of Badin, Montgomery County, North Carolina, USA (Bernstein, 1982) (includes Lu_2O_3 0.4, F 0.5, less O = F 0.2).
4 Monazite-(Ce), sheared charnockitic rock, Airport Hill, Visakhapatnam, India (Kamineni *et al.*, 1991) (includes Eu_2O_3 0.32, Er_2O_3 0.03, UO_2 0.53).
5 Yellow monazite-(Ce), inclusion in apatite, associated with xenotime, granite pegmatite, Gloserheia, 9 km north of Arendal, Norway (Åmli, 1975) (includes Ho_2O_3 0.16, Er_2O_3 0.07, Yb_2O_3 0.10).
6 Monazite-(Ce), wolframite concentrate, Mount Pleasant deposit, New Brunswick, Canada (Petruk and Owens, 1975) (average of six probe analyses).
7 Rose monazite-(Ce), Blausee, Binntal, Switzerland (Graeser and Schwander, 1987) (average of six analyses; includes MnO 0.40).

Table 43. *Continued*

	8	9	10	11	12	13	14
SiO_2	0.2	0.4	1.37	—	—	0.8	0.16
ThO_2	7.7	11.6	5.7	7.69	3.52	19.5	11.34
U_3O_8	—	0.4	0.1	—	0.87	1.6	—
Ce_2O_3	28.5	28.7	25.8	25.58	24.57	19.5	16.31
La_2O_3	13.4	21.7	14.4	12.83	19.62	9.6	13.89
Nd_2O_3	9.6	6.0	14.1	9.15	10.89	4.5	2.34
Pr_2O_3	3.4	1.9	3.7	3.07	3.85	2.5	1.64
Sm_2O_3	3.3	0.6	1.9	2.69	1.78	5.4	—
Gd_2O_3	1.0	0.09	1.3	1.48	0.30	2.0	—
Y_2O_3	0.3	—	0.4	2.30	3.22	2.7	1.01
Dy_2O_3	—	0.2	—	1.08	0.23	—	0.05
Al_2O_3	—	0.02	0.15	—	—	—	—
Fe_2O_3	—	—	0.4	—	0.28	—	—
MgO	—	—	0.05	—	—	—	—
CaO	1.4	1.0	1.25	1.61	—	3.3	4.45
PbO	—	—	—	0.66	—	—	—
P_2O_5	31.0	27.4	27.83	31.0	29.63	30.0	31.02
Total	100.2	100.16	99.23	99.85	99.40	101.5	98.56
α	—	1.795	1.778	—	—	—	1.772
β	—	—	1.780	—	—	—	1.778
γ	—	1.845	1.831	—	—	—	1.823
$2V_\gamma$	—	—	12.14°	13.5°	12.75°	—	14.5°
D	—	5.247	5.15	—	5.10	—	5.150 (calc.)
a (Å)	6.788	6.761	6.778	—	6.811	6.771	6.78
b (Å)	7.009	6.966	7.003	—	7.039	6.982	6.96
c (Å)	6.464	6.478	6.479	—	6.501	6.471	6.48
β	103.57°	103.58°	103.57°	—	103.54°	103.81°	103.90°

	Numbers of ions on the basis of 16 (O)						
P	4.048	3.790	3.757	3.946	3.965	3.902	4.088
Si	0.030	0.065	0.218	—	—	0.123	0.024
	4.08	3.85	3.97			4.03	4.11
Ce	1.609	1.716	1.506	1.408	1.422	1.102	0.930
La	0.762	1.307	0.847	0.711	1.144	0.544	0.797
Nd	0.529	0.350	0.803	0.491	0.587	0.256	0.130
Pr	0.194	0.115	0.218	0.171	0.225	0.142	0.094
Sm	0.175	0.034	0.104	0.139	0.097	0.286	—
Gd	0.053	0.004	0.071	0.076	0.016	0.105	—
Dy	—	0.010	—	0.052	0.012	—	0.002
Y	0.025	—	0.034	0.184	0.271	0.221	0.084
Th	0.270	0.431	0.207	0.263	0.127	0.682	0.402
U	—	0.014	0.003	0.184	0.030	0.054	0.542
Al	—	0.004	0.029	—	—	—	—
Fe^{3+}	—	—	0.048	—	0.034	—	—
Mg	—	—	0.012	—	—	—	—
Ca	0.231	0.175	0.215	0.259	—	0.543	0.742
Pb	—	0.004	—	0.027	—	—	—
	3.86[f]	4.18[g]	4.11[h]	3.97	4.00[i]	3.94	3.86[j]

[f]Includes U 0.014.
[g]Includes Ho 0.002.
[h]Includes Ti 0.017.
[i]Includes Eu 0.103, Er 0.025, Tb 0.005.
[j]Includes K 0.138.

8 Yellow monazite-(Ce) associated with zircon, Nibbio, Val d'Ossola, Italy (Mannucci *et al.*, 1986) (average of seven analyses; includes UO_2 0.4).
9 Reddish orange, gem-quality monazite-(Ce), gem gravels, Sri Lanka (Jobbins *et al.*, 1977) (includes Ho_2O_5 0.05, Pb_3O_4 0.1).
10 Monazite-(Ce), biotite-rich veins in migmatitic gneiss, Puumala, eastern Finland (Haapala *et al.*, 1969) (includes TiO_2 0.14, MnO 0.004, H_2O^+ 0.610, H_2O^- 0.030).
11 Monazite-(Ce) within polymerized hydrocarbon envelope, Permo-Triassic sandstone, northern Perth Basin, Western Australia (Rasmussen and Glover, 1990) (includes UO_2 0.71).
12 Monazite-(Ce), hydrothermal veins, cutting syenite, Crni Kamen ridge, Alinci, Yugoslavia (Bermanec *et al.*, 1988) (includes Eu_2O_3 0.05, Er_2O_3 0.49, Tb_2O_3 0.10).
13 Greenish thorian monazite-(Ce), pegmatite, Rio Graia, Val Vigezzo, Italy (Mannucci *et al.*, 1986).
14 Greenish thorian, uranian monazite, pegmatite, Piona, northern Italy (Gramaccioli and Segalstad, 1978) (includes UO_2 15.64, K_2O 0.71).

Table 44. Monazite-(La) and monazite-(Nd) analyses

	1	2	3	4	5	6	7	8
SiO_2	2.24	0.18	1.54	0.27	1.18	2.46	2.46	0.22
ThO_2	4.51	—	7.32	0.00	5.37	14.52	5.32	—
U_3O_8	0.21	—	0.32	—	—	0.32	1.23	—
Ce_2O_3	24.58	23.83	22.63	31.41	28.43	22.95	22.95	19.85
La_2O_3	38.51	35.77	34.63	33.19	32.49	30.56	28.28	2.48
Nd_2O_3	—	—	—	—	—	—	—	30.32
Pr_2O_3	—	—	—	—	—	—	—	6.47
Sm_2O_3	—	—	—	—	—	—	—	8.81
Gd_2O_3	—	—	—	—	—	—	—	2.26
Y_2O_3	—	6.09	4.66	5.08	—	—	4.28	—
Al_2O_3	0.36	0.29	0.10	—	1.08	0.43	—	—
Fe_2O_3	0.48	1.73	0.08	—	2.88	0.93	2.88	—
PbO	0.026	—	0.33	—	0.09	0.20	—	—
CaO	0.86	0.92	0.35	0.34	0.02	0.82	—	—
P_2O_5	26.81	29.92	27.89	29.29	28.57	26.84	27.12	28.65
H_2O^+	0.91	0.66	0.34	—	0.46	0.35	—	—
H_2O^-	—	0.61	—	—	—	—	—	—
Total	99.50	100.00	100.21	99.80	100.57	100.38	100.00	99.83
α	1.790	—	—	1.785	—	—	1.791	1.793
β	1.791	—	—	1.787	—	—	1.792	1.795
γ	1.840	—	—	1.840	—	—	1.844	1.860
$2V_\gamma$	10°	—	—	—	—	—	12–14°	18–20°
D	5.200	—	5.27	5.173	—	—	5.26	5.43 (calc.)
a (Å)	6.820	—	—	—	6.782	6.843	—	6.745
b (Å)	7.003	—	—	—	6.920	7.041	—	6.964
c (Å)	6.480	—	—	—	6.580	6.506	—	6.435
β	103.65°	—	—	—	103.75°	103.72°	—	103.65°

	1		2		3		4		5		6		7		8	
	Numbers of ions on the basis of 16 (O)															
P	3.621	3.98	3.865	3.89	3.714	3.96	3.856	3.90	3.719	3.90	3.608	4.00	3.536	3.92	3.895	3.93
Si	0.358		0.027		0.243		0.042		0.181		0.391		0.379		0.035	
La	2.266		2.013		2.011		1.903		1.842		1.790		1.606		0.147	
Ce	1.435		1.331		1.308		1.788		1.600		1.334		1.294		1.167	
Nd	—		—		—		—		—		—		—		1.739	
Pr	—		—		—		—		—		—		—		0.384	
Sm	—		—		—		—		—		—		—		0.487	
Gd	—		—		—		—		—		—		—		0.124	
Y	—	4.14	0.494	4.24	0.390	4.09[a]	0.420	4.22[b]	—	4.17	—	4.00	0.350	4.03[c]	—	4.11[d]
Th	0.164		—		0.262		—		0.187		0.525		0.186		—	
U	0.007		—		0.011		—		—		0.011		0.032		—	
Pb	0.001		—		0.014		—		0.004		0.008		—		—	
Al	0.068		0.052		0.019		—		0.196		0.080		—		—	
Fe^{3+}	0.057		0.199		0.009		—		0.333		0.111		0.334		—	
Ca	0.146		0.150		0.059		0.057		0.004		0.139		—		—	

[a]Includes Mg 0.004.
[b]Includes Mg 0.051.
[c]Includes (Ta,Nb) 0.229.
[d]Includes Zr 0.060.

1 Yellowish to dark brown monazite-(La), quartz vein, Ebisie mine, Wada, Hirukawa village, Ena County, Gifu Prefecture, Japan (Kato, 1958). Anal. T. Kato.
2 Monazite-(La), mixed gneiss, Poiana Marului, Muntele Mic Mountains, Romania (Pavelescu and Pavelescu, 1972).
3 Monazite-(La), Dickens Township, Nipissing district, Ontario, Canada (Ellsworth, 1932). Anal. H. V. Ellsworth (includes MgO 0.02).
4 Flesh-pink, thorium-free monazite-(La), Llallagua, Brazil (Gordon, 1939) (includes MgO 0.22).
5 Monazite-(La), Minas Gerais, Brazil (Kato, 1958). Anal. T. Kato.
6 Monazite-(La), Ratunapura, Sri Lanka (Kato, 1958). Anal. T. Kato.
7 Pale yellow monazite-(La), stream sand, Ngahere dredge concentrate, Grey River, Westland, New Zealand (Hutton, 1950). Anal. F. T. Seelye (recalc. after eliminating 15 per cent of zircon tantalian cassiterite and ilmenite; includes $(Ta,Nb)_2O_5$ 5.48).
8 Bright rose monazite-(Nd), aplitic–pegmatitic veins with rutile, xenotime, allanite, gadolinite and bastnä–site in micaceous gneiss, Pta. Glogstafel, Val Formazza, Italy (Graeser and Schwander, 1987) (includes ZrO_2 0.77).

granulite facies; its closure temperature is $\sim 725 \pm 25\,^{\circ}C$. Several examples of U–Pb systematics from most of these situations have been presented by Parrish (1990) to illustrate the use and complexity of monazite in geochronological studies. Cliff *et al.* (1991) found monazites very suitable for single-grain isotopic analysis, yielding concordant results more often than zircon; most of the monazite in the Carboniferous sandstones from the Pennine basin of northern England gave Caledonian ages of 410–430 m.y., placing constraints on possible sourcelands.

Lead isotope ratios in monazites determined by ion microprobe allow $^{207}Pb/^{206}Pb$ dating with precisions as good as 4–5 m.y. Thus DeWolff *et al.* (1993), for a single large (500 μm) monazite grain with a recrystallized core and resorbed rim, obtained age estimates for these three regions of 2788 ± 9, 2663 ± 4 and 2523 ± 6 m.y., respectively, showing the preservation of a 250 m.y. history, even within a single grain. Monazite inclusions in garnet that are fully armoured may provide estimates for the time of garnet growth, even in high-grade terrains where most chronometers are reset.

Experimental Work

The synthesis of monazite was first accomplished by Radominski (1875) by the fusion of $CePO_4$ with excess $CeCl_3$. Karkhanavala (1956) obtained monoclinic $CePO_4$ by heating $Ce(NO_3)_2 \cdot 4H_2O$ and $(NH_4)_2HPO_4$ to about 200 °C, followed by sintering at 700 °C: a mixture of this monoclinic $CePO_4$ and huttonite, after heating to 1400 °C for 120 hours, gave a product whose X-ray pattern agreed with that of natural monazite. Single crystals of monazite were synthesized hydrothermally in bombs at a temperature as low as 200 °C by Anthony (1957), using cerium hydroxide gel together with excess phosphoric acid. Runs of 48 hours at 300 °C produced crystals up to 0.7 mm along the *z* axis.

The synthesis of monazite at 200 °C in air was reported by Hikichi and Hukuo (1977) who heated mixtures of Ce^{3+} salts and various phosphates or phosphoric acids with a P/Ce ratio of 1 for 5 hours; platy crystals were obtained up to 8 μm in length. Further such work has been reported by Hikichi (1991) who synthesized a single monazite-(La, Ce, Nd or Sm) phase by heating mixtures of rare-earth chlorides, oxides or nitrates with phosphoric acid at 400–1000 °C for 1 day.

On the basis of experimental data on monazite solubility and the fractionation of the *REE* by monazite in obsidian from Macusani, Peru, a geothermometer using *REE* fractionation between monazite and the parent rock has been developed and used to decipher the genesis and differentiation pattern of peraluminous granitic magmas in the continental crust (Montel, 1993). The capacity for substitution of *REE* in monazite has also been examined by Nekrasov and Novilov (1993).

The differential thermal analysis curve of monazite is featureless, the exothermic peaks sometimes obtained being caused by iron-bearing alteration products (Molloy, 1959).

Much of the more recent work has been concerned with surface reactions of synthetic monazite and the leaching of U and Th from natural samples. A series of monazite dissolution experiments was conducted in a hydrous (1–6 wt. per cent) granitic melt at 8 kbar over the range of 1000–1400 °C using a polished cube of monazite (Rapp and Watson, 1986). The *REE* diffusing away from the monazite are charge-compensated by concurrent P diffusion, the dissolved water being of significant importance. Calculations showed that monazite grains $< 30\,\mu$m diameter would probably dissolve geologically instantaneously under most reasonable conditions of crustal anatexis, whereas grains $>50\ \mu m$ would be residual. Ayers and Watson (1991) found that the solubility of monazite in pure water for 24 hours at 1000 °C and 1–2 GPa was <0.2 wt. per cent but increased with increasing pressure and temperature and decreasing pH. The *REE* composition of monazite may reflect changes in the relative stability of the light *REE* complexes as a function of melt composition, with the lightest *REE* (La + Ce + Pr) being more stable in a less-evolved, less-polymerized melt. Selective ion-exchange reactions of synthetic monazite were studied by Jonasson *et al.* (1988) using aqueous solutions at pH 6; Sm- and Nd-monazite showed surface-exchange reactions within 5 days at 65 °C.

Incongruent leaching of the various members of the ^{238}U and ^{232}Th decay series from natural monazite was investigated by leaching finely divided powdered samples in bicarbonate–carbonate solutions for up to 6.8 years (Eyal and Olander, 1990). The effect of prior high-temperature annealing on the preferential loss of ^{238}Th relative to ^{232}Th was also studied. This leaching was modelled by Olander and Eyal (1990a,b) who concluded that the penetration depth of the concentration perturbation caused by these processes is 5 to ~ 50 nm. For Th in the annealed sample, no penetration occurred; here most of the ^{228}Th in solution stemmed from the decay of ^{228}Ra rejected from the matrix during annealing.

Optical and Physical Properties

Monazite is typically light yellow or yellowish brown in thin section or in grains, but rarely shows any appreciable pleochroism. The refractive indices are high and are further increased by the entry of thorium or the thorium silicate components (the isostructural monoclinic thorium silicate, huttonite, has α 1.898, β 1.900, γ 1.922, $2V_\gamma$ 25°, D 7.1; Pabst, 1951). With the replacement of Ce and La by Ca and Th, as in cheralite, the increase in the refractivity due to Th appears to be balanced by the effect of Ca. Partly metamic monazite has been reported with a low birefringence and with *n* 1.79; such material gives a diffuse X-ray powder pattern, but after heating for several hours at 1130 °C the birefringence becomes strong (with γ 1.83), a sharp X-ray pattern is obtained and the specific gravity increases from 5.33 to 5.62 (Karkhanavala and Shankar, 1954). Weak horizontal dispersion may be recognized.

The infrared absorption spectrum of monazite was given by Omori and Seki (1960): as for many phosphates, a large absorption band occurs at 9–10 μm (see also Hunt *et al.*, 1972; Bowles *et al.*, 1980). Strong absorption bands that occur in the 5252 and 570–80 nm region have been ascribed to *REE*, notably neodymium. The absorption spectrum of the red gem monazite from Sri Lanka (Table 43, anal. 9) has been recorded by Jobbins *et al.* (1977). A monazite has been described with the alexandrite effect (reddish orange under incandescent light, yellow orange in daylight), which results from several narrow Nd^{3+} absorption bands rather than a smaller number of broad bands (Bernstein, 1982); this specimen is unusual in being virtually free of Th, U and iron-group elements (Table 43, anal. 3), which can cause deep colours that mask the pastel colours due to the *REE*. Brown, yellow and grey monazites from high-grade regionally metamorphosed paragneiss in Antarctica each have a distinctive U–Pb isotopic composition which appears to result from different proportions of loss of radiogenic lead (Black *et al.*, 1984). The colour type is apparently related to the capacity of different minerals to shield included monazite grains from fluids circulating in the rock system.

Distinguishing Features

The light yellow colour, in conjunction with the high relief, high birefringence and the absence of appreciable pleochroism, are distinctive. Zircon has higher refractive indices and no good cleavage and is uniaxial, while sphene has a higher birefringence. Olivine and epidote have larger optic axial angles. A simple spectroscopic eye-piece has been described which can be used in conjunction with a normal microscope for the identification of monazite by its characteristic absorption band in the yellow region of the spectrum (Smithson, 1959).

Paragenesis

Monazite occurs as a rather rare accessory mineral in granitic rocks and as larger crystals in syenitic and granitic pegmatites. Monazite-bearing pegmatites occur at numerous localities in Norway, as at Arendal, and they are also found in Madagascar, Zambia, India, New Mexico and Brazil. Monazite is often associated with zircon, xenotime, cassiterite, columbite, apatite, wolframite or magnetite (e.g. Table 43, anals. 5, 6). In the eastern Minas Gerais, Brazil, Murata *et al.* (1959) found that xenotime and monazite were deposited together during the crystallization of certain pegmatites, but that the ratio of monazite to xenotime decreased as the crystallization proceeded from the wall to the core of the pegmatite. Monazite occurs also in Alpine-type veins associated with sphene, albite, rutile, etc., as in the Swiss Alps, and in quartz veins (e.g. Table 44, anal. 1). The calcium-rich thorium-bearing member of the monazite group, cheralite, occurs in a kaolinized pegmatite, associated with tourmaline, chrysoberyl, zircon and smoky quartz. The pleochroic haloes in the biotite of a New Zealand granite in some cases enclose nuclei of monazite (Hutton, 1947). Rose *et al.* (1958) have described an occurrence of earthy monazite from Magnet Cove, Arkansas, USA, which is believed to be the weathering product of rare-earth apatite from an apatite–pyrite vein in carbonatite.

In the Kola Peninsula, Russia, monazite-(Ce) occurs associated with yttrofluorite and fluocerite in amazonite pegmatites at Mount Ploskaya (Vetoshkina *et al.*, 1980). The occurrence of thorian monazite-(Ce) in granite pegmatites in the Alps (Mannucci *et al.*, 1986) is exemplified by anals. 13 and 14 (Table 43). The size, morphology and colour of 1500 samples of detrital monazite derived from pegmatites in the Calabrian region of Italy were studied by Brondi *et al.* (1972), who record both euhedral and opaque monazites. In granitic rocks of the Malá Fatra Mountains of eastern Europe, monazite was considered to be related to low pressure–temperature conditions in the later magmatic differentiates (Broska, 1986); it is reported as being typically associated with zircon rich in Hf, but to be antagonistic to allanite. In some carbonatites, however, e.g. Table 43, anal. 2, monazite is found associated with allanite. A similar alkaline paragenesis is represented by Table 43, anal. 1, for monazite-(Ce) from a baryte–carbonate rock near alkali syenite Mountain Pass, California, USA (Jaffe, 1955). Monazite-(Ce) is also reported in hydrothermal veins in syenite (Table 43, anal. 12), and two generations of hydrothermal monazite-(Ce) have been reported from albitized rocks of gold deposits in the east of the Sudbury complex, Ontario, Canada; their U/Pb geochronology has enabled the albitization to be dated at 1700 ± 2 m.y. (Schandl *et al.*, 1994). The general geological occurrence of monazite was surveyed by Overstreet (1967), who concluded that its thorium content is controlled by three factors: (a) pressure–temperature conditions of crystallization in magmatic rocks, (b) the regional metamorphic facies and (c) the composition of the host rocks. The widespread occurrence of monazite in S-type granites has been related to fixing during the weathering of mobile *REE* in hydrous phosphates such as florencite and rhabdophane as ~3 μm particles; subsequent dehydration during anatexis gives monazite (Sawka *et al.*, 1986). The low solubility of P in S-type granite melts inhibits the dissolution of monazite, leading to it being transported in such magmas.

In metamorphic rocks an unusual monazite deposit of metasomatic origin has been described from the Namib Desert, Namibia (Knorring and Clifford, 1960), where large reddish-brown crystals occur in an iron-stained dolomitic marble. Monazite may also occur in gneissic rocks, e.g. Table 43, anal. 10, and has been reported from charnockite (anal. 4) and other granulite-facies rocks (Kamineni *et al.*, 1991), as well as in quartz veins and pegmatites within gneissic rocks (e.g. Fortson and Navarre, 1959). Monazite is of widespread occurrence in Precambrian pelitic schists metamorphosed to staurolite–kyanite grade in the Great Smoky Mountains of North Carolina, USA (Mohr, 1984), where it forms zoned euhedral porphyroblasts with the charge-coupled substitution $Th^{4+} + Ca^{2+} = 2REE^{3+}$. An occurrence of grains of monazite within hydrocarbon envelopes in Permo-Triasic sandstone has also been examined (Rasmussen and Glover, 1990); the monazite (Table 43, anal. 11) is considered to have been derived from local gneissic basement rocks. In the eastern Mojave Desert, California, USA, monazite is absent in shale and low-grade schists but is abundant in schists at staurolite and higher grades. The lower-grade schists instead include numerous minute oxides of Th and Ce and unidentified Ce-poor light *REE* phosphates that are apparently the low-temperature precursors to monazite, the latter forming when the schist passes through lower-amphibolite-facies conditions (Kingsbury *et al.*, 1993).

Monazite-(La) is less common than monazite-(Ce) but has been described from a quartz vein in Japan (Table 44, anal. 1) and gneissic rocks (anal. 2). The bright rose-coloured monazite-(Nd) is the rarest variety and has been recognized in aplitic–pegmatitic veins in micaceous gneiss (Table 44, anal. 8) from northern Italy (Mattioli, 1985; Graeser and Schwander, 1987). A description of monazite-(Nd) as inclusions in brindleyite in rare-earth-rich Marmara bauxite, Greece, has been given by Maksimović and Pantó (1980).

A grey, opaque variety of monazite occurring as ellipsoidal granules in Palaeozoic shales was first recognized by Donnot *et al.* (1973) from black shales in Brittany, France. These nodules, 0.1–1.8 mm in diameter, have a specific gravity of 4.65 and are either polycrystalline or have an internal radiating mosaic structure; they are notably high in Eu_2O_3 (0.05 per cent) and Th. In Brittany they occur surrounded by phyllitic pressure shadows scattered across the schistosity of weakly metamorphosed shales, with some chloritoid and minor graphite. Donnot *et al.* (1973) suggested that they originated as an early diagenetic gel of rare-earth phosphate in marine silts, followed by transformation into fibrous spherules of rhabdophane before eventual conversion into monazite. Pellet-like black monazite from Taiwan was considered to owe its colour to cryptocrystallinity and not just to inclusions; this monazite lacks Th but is high in Eu (Matzko and Overstreet, 1977). It was considered to represent normal monazite-(Ce) as detrital grains from igneous rocks which later underwent low-grade metamorphism and incorporated Eu from clay minerals during diagenesis.

Such europium-rich dark monazite is now being more widely reported, for example from 60 alluvial samples in Alaska, USA (Rosenblum and Mozier, 1983); here the turbid grains have inclusions of graphite and sagenitic rutile, D 4.25–4.70, and an average value of 0.3 per cent Eu. In Britain it has been found in panned stream sediment concentrates (Cooper *et al.*, 1983; Cooper and Read, 1984; Smith *et al.*, 1994), both in Lower Palaeozoic rocks of Wales and in southwest England (where they were possibly derived from a Welsh source). Further work by Read *et al.* (1987) revealed a pronounced zonation of light and heavy *REE* within individual nodules, a low concentration of Th, and a high concentration of Eu in these monazite nodules compared with igneous monazite; their uniform grain size and low-grade metamorphic inclusion fabric indistinguishable from the host rock were considered indicative of *in situ* formation. Nodules of monazite 'grise' have also been reported from the Palaeozoic rocks of Belgium (Nonnon, 1984, 1989; Limbourg, 1986), and the dark ellipsoidal authigenic monazite relatively rich in lanthanides and termed kularite by Nekrasova and Nekrasov (1984) is clearly a similar variety; this monazite from the Kular Ridge, Russia, is host to small inclusions of sulphides of Fe and As.

Composite grains of detrital monazite and solid hydrocarbons, ranging up to 0.5 mm in diameter, have been found in Phanerozoic sedimentary rocks from three localities in Western Australia (Table 43, anal. 11). The most common form is a core of monazite-(Ce) surrounded by a hydrocarbon envelope, which apparently congealed when hydrocarbons were polymerized and crosslinked by irradiation from the monazite (Rasmussen and Glover, 1990). This monazite is a normal transparent variety, but most grains have a slightly patchy extinction, perhaps related to crystal damage caused by the radioactivity (these monazites have 7.0–11.5 per cent ThO_2). In the northern Perth Basin, Western Australia, these composite grains, with cores of monazite derived from local gneissic basement rocks, formed after the growth of siderite cement but before extensive silicification.

From lunar material, inclusions of Th-poor monazite have been found in a ferrohedenbergite grain in Apollo 11 mare basalt; Lovering *et al.* (1974) considered this to be late-stage phase (see also Smith and Steele, 1974).

Monazite is moderately resistant to weathering and is frequently concentrated as a detrital mineral in stream and beach sands in areas where granitic pegmatites, etc., are undergoing erosion. In many parts of the world monazite concentrates in sands are worked as a source of cerium and other *REE*. In the monazite-bearing sediments of the southeastern United States, ilmenite and leucoxene commonly make up half or more of the heavy mineral suite, followed in order of abundance by zircon, rutile, monazite, staurolite and kyanite. A detailed investigation of monazite from New Zealand stream sands (Table 44, anal. 7) was made by Hutton (1950). In Liberia, monazite constitutes up to 10.4 per cent of heavy mineral deposits (Rosenblum, 1974), and Australian resources in such sands have been given as 229 000 tonnes (Towner, 1986). Detrital monazite is important in heavy mineral sands in North Carolina, USA, and here is relatively rich in thorium and uranium, with ThO_2 4.9–6.1 per cent and U_3O_8 0.71–1.9 per cent, the highest values recorded in the USA (Overstreet *et al.*, 1971). Three belts of monazite-rich deposits were located by panning in this area and can be used to indicate the occurrence of granitic rocks beneath the overburden (Mertie, 1979).

References

Ahrens, L. H., 1955. The convergent lead ages of the oldest monazites and uraninites (Rhodesia, Manitoba, Madagascar and Transvaal). *Geochim. Cosmochim. Acta*, **7**, 294–302.

Åmli, R., 1975. Mineralogy and rare earth geochemistry of apatite and xenotime from the Gloserheia granite pegmatite, Froland, southern Norway. *Amer. Min.*, **60**, 607–20.

Anthony, J. W., 1957. Hydrothermal synthesis of monazite. *Amer. Min.*, **42**, 904.

Ayers, J. C. and **Watson, E. B.**, 1991. Solubility of apatite, monazite, zircon and rutile in supercritical aqueous fluids with implications for subduction zone geochemistry. *Phil. Trans. R. Soc. A*, **335**, 365–75.

Bermanec, V., Tibljaš, D., Gessner, M. and **Kniewald, G.**, 1988. Monazite in hydrothermal veins from Alinici, Yugoslavia. *Min. Petr.*, **38**, 139–50.

Bernstein, L. R., 1982. Monazite from North Carolina having the alexandrite effect. *Amer. Min.*, **67**, 356–9.

Black, L. P., Fitzgerald, J. D. and **Harley, S. L.**, 1984. Pb isotopic composition, colour, and microstructure of monazites from a polymetamorphic rock in Antarctica. *Contrib. Min. Petr.*, **85**, 141–8.

Borovsky, I. B. and **Gerasimovsky, V. I.**, 1945. Rare earths in minerals. *Doklady Acad. Sci. USSR*, **49**, 353–6.

Bowie, S. H. U. and **Horne, J. E. T.**, 1953. Cheralite, a new mineral of the monazite group. *Min. Mag.*, **30**, 93–9.

Bowles, J. F. W., Jobbins, E. A. and **Young, B. R.**, 1980. A re-examination of cheralite. *Min. Mag.*, **43**, 885–8.

Brondi, A., Ferretti, O. and **Locardi, E.**, 1972. Sulla provienza dei diversi dipi di zircone e monazite contenuti nei sedimenti fluviali della Calabria. *Rend. Soc. Ital. Min. Petr.*, **28**, 31–52.

Broska, I., 1986. [Typomorphic accessory minerals in the granitoids of the crystalline Malà Fatra Mountains.] *Geol. Zbornik, Geol. Carpathica (Bratislava)*, **37**, 693–707 (in Russian).

Cliff, R. A., Drewery, S. E. and **Leeder, M. R.**, 1991. Sourcelands for the Carboniferous Pennine river system: constraints from sedimentary evidence and U–Pb geochronology using zircon and monazite. In *Developments in Sedimentary Provenance Studies:* Eds. A. C. Morton *et al.*, *Geol. Soc. Spec. Publ.*, **57**, 137–59.

Cooper, D. C. and **Read, D.**, 1984. The distribution of monazite in panned stream sediment concentrates from Britain: palaeogeographic implications. *Rep. Br. Geol. Surv.*, **84/1**, 1–4.

Cooper, D. C., Basham, I. R. and **Smith, T. K.**, 1983. On the occurrence of an unusual form of monazite in panned stream sediments in Wales. *Geol. J.*, **18**, 121–7.

DeWolff, C. P., Belshaw, N. and **O'Nions, R. K.**, 1993. A metamorphic history from micron-scale $^{207}Pb/^{206}Pb$ chronometry of Archean monazite. *Earth Planet. Sci. Lett.*, **120**, 207–20.

Donnot, M., Guigues, J., Lulzac, Y., Magnien, A., Parfenoff, A. and **Picot, P.**, 1973. Un nouveau type de gisement d'europium; la monazite grise à europium en nodules dans les schistes paléozoïques de Bretagne. *Min. Deposita*, **8**, 7–18.

Ellsworth, H. V., 1932. Monazite coloured by carbon from Dickens Township, Nipissing, Ontario. *Amer. Min.*, **17**, 19–28.

Erichsen, L. von, 1951. Über die Heliumabgabe von Monazit in Abhängkeit von Gasphase, Druck und Temperatur. *Neues Jahrb, Min., Monat.*, 25–33.

Eyal, Y. and **Olander, D. R.**, 1990. Leaching of uranium and thorium from monazite. I. Initial leaching. *Geochim. Cosmochim. Acta*, **54**, 1867–77.

Finney, J. J. and **Nagaraja Rao, N.**, 1967. The crystal structure of cheralite. *Amer. Min.*, **52**, 13–9.

Fleischer, M. and **Altschuler, Z. S.**, 1969. The relationship of the rare-earth composition of minerals to geological environment. *Geochim. Cosmochim. Acta*, **33**, 725–32.

Fortson, C. W. and **Navarre, A. T.**, 1959. Monazite-bearing pegmatites in the south Georgia Piedmont. *Econ. Geol.*, **54**, 1309–15.

Frondel, C., 1958. *Systematic Mineralogy of Uranium and Thorium.* Washington, DC (US Govt. Printing Office), 400 pp.

Gliszczynski, S. von, 1939. Beitrag zur 'Isomorphie' von Monazit und Krokoit. *Z. Krist.*, **101**, 1–16.

Gordon, S. G., 1939. Thorium-free monazite from Llallagua, Brazil. *Notulae Naturae, Acad. Nat. Sci. Philadelphia*, **2**, 7 pp. (M.A. **7**–508).

Graeser, S. and **Schwander, H.**, 1987. Gasparite-(Ce) and monazite-(Nd): two new minerals to the monazite group from the Alps. *Schweiz Min. Petr. Mitt.*, **67**, 103–13.

Gramaccioli, C. M. and **Segalstad, T. V.**, 1978. A uranium- and thorium-rich monazite from a south-alpine pegmatite at Piona, Italy. *Amer. Min.*, **63**, 757–61.

Haapala, I., Ervamaa, P., Löfgren, A. and **Ojanperä, P.**, 1969. An occurrence of monazite in Puumala, eastern Finland. *Bull. Geol. Soc. Finland*, **41**, 117–24.

Heinrich, E. W., Borup, R. A. and **Levinson, A. A.**, 1960. Relationships between geology and composition of some pegmatitic monazites. *Geochim. Cosmochim. Acta*, **19**, 222–31.

Henry, D. A., 1993. Phosphate minerals in detrital deposits. In *Phosphate Minerals of Victoria:* Eds. W. D. Birch and D. A. Henry, Melbourne (Min. Soc. Victoria, Spec. Publ. 3), pp. 111–20.

Hikichi, Y., 1991. Synthesis of monazite (RPO_4, R = La, Ce, Nd or Sm) by solid state reaction. *Min. J.*, **15**, 268–75.

Hikichi, Y. and **Hukuo, K.**, 1977. [Synthesis of monazite ($CePO_4$) by dry method.] *Bull. Nagoya Inst.*

Techn., **29**, 203–8 (in Japanese with English summary).
Holmes, A., 1954. The oldest dated minerals of the Rhodesian shield. *Nature*, **173**, 612–4.
Hunt, G. R., Salisbury, J. W. and **Lenhoff, C. J.,** 1972. Visible and near-infrared spectra of minerals and rocks. V. Halides, phosphates, arsenates, vanadates and borates. *Modern Geol.*, **3**, 121–32.
Hutton, C. O., 1947. The nuclei of pleochroic haloes. *Amer. J. Sci.*, **245**, 154–7.
Hutton, C. O., 1950. Studies of heavy detrital minerals. *Bull. Geol. Soc. Amer.*, **61**, 635–716.
Jaffe, A. W., 1955. Precambrian monazite and zircon from the Mountain Pass rare-earth district, San Bernardino County, California. *Bull. Geol. Soc. Amer.*, **66**, 1247–56.
Jobbins, E. A., Tresham, A. E. and **Young, B. R.,** 1977. Gem monazite from Sri Lanka. *J. Gemmol.*, **15**, 295–9.
Johnstone, S. J., 1914. Monazites from some new localities. *J. Soc. Chem. Industr.*, **33**, 55.
Jonasson, R. G., Bancroft, G. M. and **Boatner, L. A.,** 1988. Surface reactions of synthetic, end-member analogues of monazite, xenotime and rhabdophane, the evolution of natural waters. *Geochim. Cosmochim. Acta*, **52**, 767–70.
Kamineni, D. C., Rao, A. T. and **Bonardi, M.,** 1991. The geochemistry of monazite types from the Eastern Ghats granulite terrain, India. *Min. Petr.*, **45**, 119–30.
Karkhanavala, M. D., 1956. The synthesis of huttonite and monazite. *Curr. Sci.*, **25**, 166–7.
Karkhanavala, M. D. and **Shankar, J.,** 1954. An X-ray study of natural monazite. I. *Proc. Ind. Acad. Sci.*, **40A**, 67.
Kato, T., 1958. A study on monazite from the Ebisu mine, Gifu prefecture. *Min. J. (Japan)*, **2**, 224–35.
Kingsbury, J. A., Miller, C. F., Wooden, J. L. and **Harrison, T. M.,** 1993. Monazite paragenesis and U–Pb systematics in rocks of the eastern Mojave Desert, California, U.S.A.: implications for thermochronometry. *Chem. Geol. incl. Isotope Geol.*, **110**, 147–67.
Knorring, O. von and **Clifford, T. N.,** 1960. On a skarn monazite from the Namib Desert near Usakos South-West Africa. *Min. Mag.*, **32**, 650–3.
Kokkoros, P., 1942. La structure de la monazite. *Acta Acad. Athens*, **17**, 163–5.
Kucha, H., 1980. Continuity in the monazite–huttonite series. *Min. Mag.*, **43**, 1031–4.
Limbourg, Y., 1986. Découverte de monazite grise en nodules dans l'Eodévonien du Synclinorium de Neufchâteau (Belgique). *Bull. Soc. Belge Géologie*, **95(4),** 281–5.
Lovering, J. F., Wark, D. A., Gleadow, A. J. W. and **Britten, R.,** 1974. Lunar monazite: a late-stage (mesostasis) phase in mare basalt. *Earth Planet. Sci. Lett*, **21**, 164–8.
Maksimović, Z. and **Pantó, G.,** 1980. Bastnäsite-(La) and monazite-(Nd), a new variety of monazite from the Marmara bauxite deposit (Greece). *Bull. Acad. Serbe Sci. Arts, Cl. Sci. Naturelles Math., Sci. Naturelle, Nouvelle Sér.*, **20**, 35–42 (M.A., **83M**–1933).
Mannucci, G., Diella, V., Gramaccioli, C. M. and **Pilati, T.,** 1986. A comparative study of some pegmatitic and fissure monazite from the Alps. *Can. Min.*, **24**, 469–74.
Mattioli, V., 1985. Un nuovo minerale alpino nell'Ossola (NO). *Revista Min. Ital.*, **8**, 129–34.
Matzko, J. J. and **Overstreet, W. C.,** 1977. Black monazite from Taiwan. *Proc. Geol. Soc. China (Taiwan)*, **20**, 16–35.
Mertie, J. B., Jr., 1979. Monazite in the granitic rocks of the southeastern Atlantic States—an example of the use of heavy minerals in geologic exploration. *U.S. Geol. Surv. Prof. Paper*, **1094,** 79 pp.
Mi, J. and **Shen, J.,** 1994. Structural refinement of monazite-(Ce). *Abstracts vol., I.M.A. Meeting, Pisa*, 276–7.
Mohr, D. W., 1984. Zoned porphyroblasts of metamorphic monazite on the Anakeesta Formation, Great Smoky Mountains, North Carolina. *Amer. Min.*, **69**, 98–103.
Molloy, M. W., 1959. A comparative study of ten monazites. *Amer. Min.*, **44**, 510–32.
Montel, J. M., 1993. A model for monazite/melt equilibrium and application to the generation of granitic magmas. *Chem. Geol. incl. Isotope Geol.*, **110**, 127–46.
Murata, K. J., Rose, H. J. and **Carron, M. K.,** 1953. Systematic variation of rare earths in monazite. *Geochim. Cosmochim. Acta*, **4**, 292–300.
Murata, K. J., Dutra, C. V., de Costa, M. T. and **Branco, J. J. R.,** 1959. Composition of monazites from pegmatites in eastern Minas Gerais, Brazil. *Geochim. Cosmochim. Acta.* **16**, 1–14.
Nekrasov, I. Y., 1972. New data on a mineral of the monazite–cheralite–huttonite group. *Dokl. Acad. Sci. USSR, Earth Sci. Sect.*, **204**, 134–6.
Nekrasov, I. Y. and **Novikov, M. P.,** 1993. Isomorphic substitution capacity of monazite for rare earth elements. *Dokl. Akad. Nauk USSR, Earth Sci. Sect.*, **321**, 167–70 (M.A. **94M**–1883).
Nekrasova, R. A. and **Nekrasov, I. Y.,** 1984. Kularite, an authigenic monazite variety. *Dokl. Acad. Sci. USSR, Earth Sci. Sect.*, **268**, 151–5.
Ni, Y., Hughes, J. M. and **Mariano, A. N.,** 1995. Crystal chemistry of monazite and xenotime structures. *Amer. Min.*, **80**, 21–6.
Nonnon, M., 1984. Découverte de monazite grise en nodules et d'or alluvionaire dans le Massif de la Croix-Scaille. *Bull. Soc. Belge Géol.*, **3**, 307–14.

Nonnon, M., 1989. Une nouvelle occurrence de monazite grise en Belgique. *Bull. Soc. Belge Géol.*, **98**, 73–6.

Olander, D. R. and **Eyal, Y.**, 1990a. Leaching of uranium and thorium from monazite. II. Elemental leaching. *Geochim. Cosmochim. Acta*, **54**, 1879–87.

Olander, E. R. and **Eyal, Y.**, 1990b. Leaching of uranium and thorium from monazite. III. Leaching of radiogenic daughters. *Geochim. Cosmochim. Acta*, **54**, 1889–96.

Omori, K. and **Seki, T.**, 1960. Infra-red studies of some phosphate minerals. *J. Jap. Assoc. Min. Petr. Econ. Geol.*, **44**, 7–13.

Overstreet, W. C., 1967. The geologic occurrence of monazite. *U. S. Geol. Surv. Prof. Paper*, **530**, 327 pp.

Overstreet, W. C., Warr, J. J., Jr. and **White, A. M.**, 1971. Possible petrogenetic relations of thorium, uranium, and Ce/(Nd+Y) in detrital monazite from Surry and Stokes counties, North Carolina. *Southeastern Geol.*, **13**, 99–125.

Pabst, A., 1951. Huttonite, a new monoclinic thorium silicate. *Amer. Min.*, **36**, 60–9.

Pan, Y., Fleet, M. E. and **McFae, N. D.**, 1993. Oriented monazite inclusions in apatite porphyroblasts from the Hemlo gold deposit, Ontario, Canada. *Min. Mag.*, **57**, 697–707.

Parrish, R. R., 1990. U–Pb dating of monazite and its application to geological problems. *Can. J. Earth Sci.*, **27**, 1431–50.

Parrish, W., 1939. Unit cell and space group of monazite, (La,Ce,Y)PO_4. *Amer. Min.*, **24**, 651–2.

Pavelescu, L. and **Pavelescu, M.**, 1972. Study of some allanites and monazites from the south Carpathians (Romania). *Tschermaks Min. Petr. Mitt., Ser. 3*, **17**, 208–14.

Petruk, W. and **Owens, D.**, 1975. Monazite from the Mount Pleasant deposit, New Brunswick. *Can. Min.*, **13**, 298–9.

Prasad, N. S. K., Kamat, K. D., Fareeduddin, S. and **Sethna, H. N.**, 1957. Helium from monazite sand. *Curr. Sci.*, **26**, 247–8.

Radominski, M. F., 1875. Reproduction artificielle de la monazite et de la xenotime. *C. R. Acad. Sci. Paris*, **80**, 304–7.

Rapp, R. P. and **Watson, E. B.**, 1986. Monazite solubility and dissolution kinetics: implication for the thorium and light rare earth chemistry of felsic magmas. *Contrib. Min. Petr.*, **94**, 304–16.

Rasmussen, B. and **Glover, J. E.**, 1990. The diagenetic and economic significance of composite grains of monazite and hydrocarbon in Western Australian arenites. *J. Geol. Soc.*, **148**, 809–16.

Read, D., Cooper, D. C. and **McArthur, J. M.**, 1987. The composition and distribution of nodular monazite in the Lower Palaeozoic rocks of Great Britain. *Min. Mag.*, **51**, 271–80.

Rose, H. J., Blade, L. V. and **Ross, M.**, 1958. Earthy monazite at Magnet Cove, Arkansas. *Amer. Min.*, **43**, 995–7.

Rosenblum, S., 1974. Analyses and economic potential of monazite in Liberia. *J. Res. U. S. Geol. Surv.*, **2**, 689–92.

Rosenblum, S. and **Mozier, E. L.**, 1983. Mineralogy and occurrence of europium-rich dark monazite. *US Geol. Surv. Prof. Paper*, **1181**, 67 pp.

Sawka, W. N., Banfield, J. F. and **Chappell, B. W.**, 1986. A weathering-related origin of widespread monazite in S-type granites. *Geochim. Cosmochim. Acta*, **50**, 171–5.

Schandl, E. S., Gorton, M. P. and **Davis, D. W.**, 1994. Albitization at 1700±2 Ma in the Sudbury–Wanapitei Lake area, Ontario: implications for deep-seated alkalic magmatism in the Southern province. *Can. J. Earth Sci.*, **31**, 597–607.

Shankar, J. and **Khubchandani, P. G.**, 1956. An X-ray study of natural monazite. II. Single crystal data on Indian monazite mineral. *Proc. Ind. Acad. Sci.*, **44A**, 130–3.

Smith, J. V. and **Steele, I. M.**, 1974. Lunar mineralogy: a heavenly detective story. Part II. *Amer. Min.*, **61**, 1059–116.

Smith, R. T., Cooper, D. C. and **Bland, D. J.**, 1994. The occurrence and economic potential of nodular monazite in south-central Wales. *British Geological Survey Technical Rept.*, **WF/94/1**, 45 pp.

Smithson, F., 1959. A simple spectroscopic eye-piece for testing monazite under the microscope. *Min. Mag.*, **32**, 176.

Starynkevitch, I., 1992. [The chemical formula of monazite and several analyses of Russian monazites.] *Acad. Sci. USSR, Comptes Rendus*, **31**, 28–30 (in Russian).

Tilton, G. R. and **Nicolaysen, L. O.**, 1957. The use of monazites for age determination. *Geochim. Cosmochim. Acta*, **11**, 28–39.

Towner, R. R., 1986. Resources of mineral sands. What of their future. In *Australia: a World Source of Ilmenite, Rutile, Monazite and Zircon:* Perth (Australasian Inst. Mining Metall.), pp. 35–50.

Ueda, T., 1953. The crystal structure of monazite ($CePO_4$). *Mem. Coll. Sci., Univ. Kyoto, Ser. B*, **20**, 225–46.

Vetoshkina, A. M., Gordinenko, V. V., Elina, N. A. and **Polezhaeva, L. I.**, 1980. [Yttrofluorite and accompanying rare earth minerals from amazonite pegmatites at Mt. Ploskaya (Kola Peninsula).] *Min. Zhurn.*, **2**, 51–8 (in Russian).

Wark, D. A. and **Miller, C. F.**, 1993. Accessory mineral behavior during differentiation of a granite suite: monazite, xenotime and zircon in the Sweetwater Wash pluton, southeastern California, U.S.A. *Chem. Geol. incl. Isotope Geol.*, **110**, 49–67.

Woolley, A. R., Barr, M. W. C., Din, V. K., Jones, G. C., Wall, F. and **Williams, C. T.**, 1991. Extrusive carbonatites from the Uyaynah area, United Arab Emirates. *J. Petr.*, **32**, 1143–67.

Wylie, A. W., 1948. Composition of monazite. *Nature*, **161**, 97.

Wylie, A. W., 1950. Composition of some Australian monazites. *Aust. J. Appl. Sci.*, **1**, 164–71.

Halides

Fluorite CaF_2
Halite NaCl

Fluorite CaF_2

Cubic

n	1.433–1.435
Dispersion:	Weak.
D	3.18
H	4
Cleavage:	{111} perfect.
Twinning:	{111}, commonly as interpenetrant cubes.
Colour:	Extremely variable; colourless, white, yellow, green, blue and purple varieties are the most common. Colourless, pale green or pale violet in thin section.
Unit cell:	*a* 5.463 Å Z = 4. Space group *Fm*3*m*

Soluble in H_2SO_4 with liberation of HF; slightly soluble in HCl.

Fluorite commonly occurs as a vein mineral often associated with quartz, calcite and ores of lead and zinc. It can also be found in igneous rocks, chiefly in granites and pneumatolytic deposits, and in sedimentary rocks, such as limestones and dolomites, as a result of metasomatism. The name is derived from the Latin *fluo*, meaning I flow, in allusion to its readily fusible nature and its use as a flux in smelting. The name fluorspar is used for material containing sufficient fluorite to make it of commercial interest. The variety of fluorite known as 'Blue John' occurs in spheroidal masses with a radiating crystalline structure; it contains blue bands of varying intensity arranged concentrically, parallel to the nodular surface, and between these, colourless, yellow or paler blue bands. This variety is virtually restricted to an area near Castleton, Derbyshire, UK. The name may be derived from a comparison with Black Jack (sphalerite) or more probably represents a corruption of the French description *bleu et jaune*.

Structure

The crystal structure of fluorite was one of the earliest to be investigated by X-rays. It was determined by Bragg (1914), who demonstrated that the calcium ions are arranged on a cubic face-centred lattice, while each fluorine ion is at the centre of one of the smaller cubes obtained by dividing the unit cube into eight parts. Each calcium is therefore coordinated by eight fluorine ions, while each fluorine ion is surrounded by four calcium ions arranged at the corners of a regular tetrahedron (Fig. 119). This is the simplest of the structures commonly assumed by AX_2 compounds and represents the highest possible coordination (8:4), the radius condition being that the ratio radius A/radius X >0.732. The work of Goldschmidt and Thomassen (1923) indicated that the cell dimensions may increase slightly with the introduction of appreciable amounts of Y and Ce, as in yttrian fluorite. Chatterjee (1940) showed that natural crystals containing about 14 per

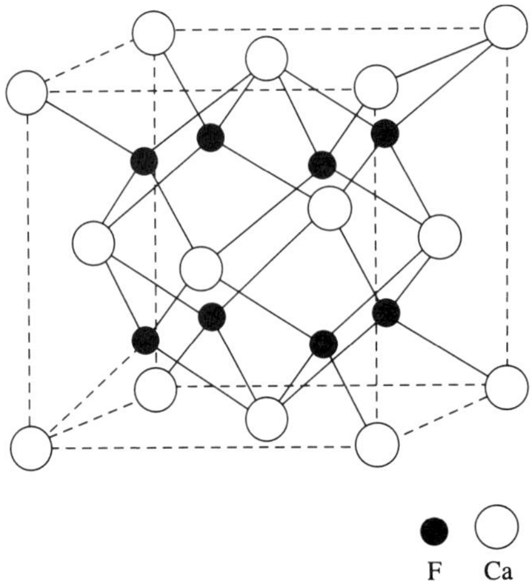

Fig. 119. The crystal structure of fluorite (after Bragg and Claringbull, 1965).

cent YF_3 retain the fluorite structure; the extra F^- is accommodated interstitially in the yttrofluorite series $(Ca,Y)F_{2-3}$ (Zintl and Udgård, 1939). An yttrofluorite with an unusually large content of rare earth elements (RE_2O_3 44.04 per cent) has *a* 5.528 Å (Gerasimovskii and Kataeva, 1984). Indexed X-ray powder data for various fluorides were given by Steyn (1954), together with corresponding cell sizes; specimens with appreciable Sr show a distinct increase in the cell edge, some having *a* 5.4686 Å, an increase of 0.0055 Å. The structural explanation for the frequent occurrence of interpenetration twins in fluorite, to the exclusion of contact (spinel) twins, was given by Gliszczynski (1949). Allen (1952) recorded no significant difference in the cell sizes of 18 specimens of varying colour and paragenesis; the value for optical fluorite is *a* 5.46295 Å at 28 °C.

Ordered arrays of point defect clusters in the (111) planes of natural blue fluorite have been observed by transmission electron microscopy; during continued observation with the 200 kV electron beam, a mosaic feature changed to an ordered array of defect clusters with an accompanying intensification of the blue colour (Murr, 1974), clearly indicating that the defects are colour centres (see p. 360). Selected area electron diffraction normal to (111) showed superlattice reflections due to the ordered arrays, which can be indexed on a cubic cell with A (superlattice)$=\sqrt{3}a$, where $a=5.463$ Å, the cell edge of fluorite, The defect aggregates have a mean diameter of ~100 Å, with a mean linear separation of ~200 Å. These point defects are considered to be due to anion (F^-) vacancies in which electrons are trapped and cluster into ordered arrays.

Crystal structure analysis of fluorite at 7 GPa has been reported (Kudoh, *et al.*, 1993); it has *a* 5.388 Å, a reduction of 4.4%. At somewhere between 7 and 8.3 GPa a pressure-induced phase transition to the α-$PbCl_2$ structure occurs.

Chemistry

Most fluorite is at least 99 per cent CaF_2, and the small amounts of Si, Al and Mg sometimes reported are probably due to impurities or inclusions (Saito, 1950). The chief substitutions which can take place are Y and Ce for Ca, and in the variety yttrofluorite, $(Ca,Y)F_{2-3}$, the YF_3 molecule may amount to 10–20 per cent with minor amounts of CeF_3 (Sverdrup, 1968). Some fluorites have been reported to contain free fluorine and on grinding these specimens may give a strong odour of ozone and HF (Sine, 1925; Hoffmann, 1937); this variety is sometimes known as antozonite.

Some fluorites, particularly the dark portions of the banded variety Blue John from Treak Cliff, Castleton, Derbyshire, contain hydrocarbons; Garnett (1920) obtained up to 0.27 per cent carbon from the combustion of dark Blue John. Blount and Sequeria (1919) analysed the blue and white portions of the same Blue John sample, but the results were very similar and did not account for the colour difference; the same authors extracted small amounts of oily matter from Blue John. Dark purple fluorites may contain appreciable amounts of strontium; Steyn (1954) reported up to 1 per cent Sr in purple fluorite from the Transvaal, South Africa; the dark purple Treak Cliff material is also comparatively rich in this element (Dunham, 1952). In fluorite from the Amba Dongar carbonatites, however, Viladkar and Wimmenauer (1992) report only 140 p.p.m. of Sr in the violet material but 1600–1800 p.p.m. Sr in the white and colourless crystals.

The main economic uses of fluorite are as a flux in the metallurgical industry, in ceramics and in the production of hydrofluoric acid. Some 55 per cent of the world's output is used in metallurgy, for which a content of >67 per cent fluorite (with low sulphur and lead) is required. Purity required in the ceramic industry is 80–96 per cent fluorite, and the acid-grade material must have at least 97 per cent fluorite (with SiO_2 <1.5 per cent and low S). The name fluorspar is used for raw material containing sufficient fluorite to make it of commercial interest. A report on the production of acid-grade fluorspar in the south Pennines has been given by Bramley (1990).

Experimental Work

Fluorite can be synthesized by the evaporation of a solution of CaF_2 in HCl, by the diffusion of ammonium fluoride into $CaCl_2$ solution or by the reaction of calcium fluorosilicate and $CaCl_2$ solution. Large synthetic crystals for optical use have been obtained by fusing precipitated CaF_2 in a conical-bottomed graphite crucible in a vacuum furnace with a temperature gradient (Stockbarger, 1949); better results were reported with crushed natural fluorite and the addition of PbF_2 to act as a scavenger for impurities. The solubility of fluorite was investigated by Thugutt (1936), who found that when heated in water in sealed tubes for 50 hours at 206 °C powdered fluorite yielded a solution containing 0.00296 g CaF_2 for 100 ml.

The solubility equilibria of fluorite in 351 hot springs and surface waters from geothermal areas in the western USA show that the solubility of fluorite controls the dissolved fluorite activity; undersaturated waters underwent dilution by non-thermal waters as shown by decreased conductivity and temperature values (Nordstrom and Jenne, 1977).

As noted above, yttrium and the rare earth elements have long been known to substitute for calcium in fluorite. The distribution coefficient REE_{CaF_2}/REE_{soln} is always high, and thus fluorite records with magnification but without major distortion the *REE* pattern of the magmatic or hydrothermal solutions involved; it also provides a useful indicator of the geochemistry of these solutions (Marchand *et al.*, 1976). Until recently, the determination of the rare earth elements was analytically tedious and subject to error, but the advent of modern techniques, such as neutron activation analysis, mass spectrometry, X-ray fluorescence, the electron microprobe and inductively coupled plasma spectrometry, has led to increased interest in the analysis of the *REE* in fluorite. Early results were summarized by Fleischer (1969) who noted that fluorites generally have Y > *REE* and light *REE* > heavy *REE*. More recent results have been presented by: Marchand *et al.* (1976) for 59 French hydrothermal fluorites; Gundlach *et al.* (1976) for 13 hydrothermal fluorites from the Harz Mountains, Germany; Bellanca *et al.* (1981) for 13 fluorites from Sicily; Chatagnon and Galland (1982) for 40 natural fluorites; Howie *et al.* (1982) for 11 English fluorites; and Naldrett *et al.* (1987), Mitra (1981), Méany *et al.* (1985) and Viladkar and Wimmenauer (1992) for fluorites from the Amba Dongar carbonatite in India. Fayziyev (1990) examined 372 fluorites from Russia and noted that the highest levels of Y occurred in fluorites from granitic pegmatites (mean 749 p.p.m.), with the highest in fluorites from granitic pegmatites in Tadzhikistan (2884 p.p.m.); carbonatites and metasomatic rocks had fluorites with averages of 337 and 343 p.p.m., respectively, the lowest level of 132 p.p.m. being found in accessory fluorite from magmatic rocks (see also Faiziev and Koplus, 1992). The *REE* contents have also been determined by Bastos Neto *et al.* (1991) in fluorites from Santa Catarina, Brazil, by Subias Pérez and Fernandez-Nieto (1991) for samples from Huesca, Spain, and for 94 fluorites from Zhejiang Province, China (Jiang and Li, 1993). Numerous further *REE* analyses are reported in discussions (below) on the causes of colour in fluorite.

An investigation of the uranium content of 160 pegmatitic and hydrothermal fluorites gave U values of 10^{-5}–10^{-4} per cent (Baranov, 1966), i.e. 0.1–1 p.p.m.; values for U + Th of 1–18 p.p.m. have been reported for fluorite from Bayan Obo, Inner Mongolia (Li *et al.*, 1985).

The solubility of fluorite in NaCl solutions at up to 100 °C was studied by Richardson and Holland (1979); the solubility decreases above this temperature. The solubility in NaCl–$CaCl_2$ and NaCl–$CaCl_2$–$MgCl_2$ solutions was found to be controlled by the common ion effect and by the presence of HF^0. In the system CaF_2–$MgCl_2$–H_2O at 150–300 °C and 4.5–85 bar, Schaefer and Strubel (1979) found a maximum solubility of 5.16×10^{-1} g CaF_2/kg H_2O at 150 °C, 4.4 bar, and 8.6 g CaF_2/kg H_2O at 300 °C, 85 bar.

Fluorite is often pseudomorphed by other minerals, principally quartz or chalcedony. Murdoch (1936) described fine granular aggregates of quartz or feathery chalcedony, with an outer zone of chalcedony in which the fibres are normal to the cube faces of the fluorite. Other pseudomorphs after fluorite include calcite, dolomite, sphalerite, pyrite and various oxides of iron and manganese.

Optical and Physical Properties

The low refractive index, small dispersion and isotropic character make fluorite a suitable material for optical use. A detailed study of the variations in the optical and physical properties of fluorite was made by Allen (1952) who tabulated colour, crystal form, mode of occurrence, optical and X-ray data, trace element content and change in colour on exposure to X-rays for 39 specimens from various localities.

Using the method of minimum deviation, Allen (1952) obtained a refractive index of 1.43385 (20 °C) for a perfect prism of optical fluorite, in exact agreement with the earlier work of Merwin (1911). The total range of variation for the refractive index in sodium light for various coloured fluorites was 0.00111 (20 °C), with the lowest value of 1.43349 for green fluorite from Rossie, New York, and the highest value of 1.43460 for brown fluorites from Weardale, northern England, and Clay

Center, Ohio, USA. Fluorites with principal light absorption at long wavelengths are characterized by low refractive indices, unless minor constituents obscure the effect, while fluorites with principal absorption at short wavelengths have abnormally high refractive indices. The abnormally low indices are associated with a fluorine deficiency compensated for by the presence of colloidal calcium; the abnormally high indices are attributed to excess calcium in structural holes (Allen, 1952). For fluorite with substantial substitution of Y for Ca, the refractive index and specific gravity are considerably higher (Vogt, 1911; Zambonini, 1915); an yttrofluorite with RE_2O_3 44 per cent has n 1.470 and D 4.12 (Gerasimovskii and Kataeva, 1984). The dispersion of fluorite is relatively constant, with $n(4359)–n(6550)$ Å ~ 0.007.

The cause of the colour in fluorite has been discussed extensively; early reviews of work on fluorite colour and on the varying luminescence were given by Dunham (1952) and Przibram (1956). Haberlandt and Scheiner (1935) made a detailed study of the crystallographic distribution of the colour and discussed various explanations for the phenomenon. The causes suggested include defect structures, radioactive inclusions or emanations from nearby radioactive material, traces of rare earth elements and the presence of inclusions of carbonaceous material, MnO_2, etc.

Fluorite often demonstrates complex colour-centre phenomena in which defects and aggregates of defects in the crystal structure occur. Such aggregates are found in Blue John fluorite, where the presence of calcium colloids is now recognized as being responsible for the blue colour (Mackenzie and Green, 1971; Braithwaite *et al.*, 1973; Galwey *et al.*, 1979). Here the colloidal calcium is attributed to radiation damage, the Ca being trapped along $\{111\}$ cleavage planes. The radiation may have been derived from the fluorite's proximity to U-bearing collophane deposits. Although Blue John fluorite contains minor amounts of organic material, these cannot be the cause of the colour. Colloidal calcium released by radioactivity has also been given as an explanation for the zonal arrangement of colour in fluorite from Bayan Obo, Inner Mongolia (Li *et al.*, 1985). When colourless fluorite is exposed to irradiation in a cathodoluminescence chamber it becomes purple and may show delicate colour zoning (Dickson, 1980), although the colour fades after a few weeks.

Fluorite which had been irradiated naturally was examined by Berman (1957), who showed that purple samples found in association with radioactive materials became colourless on heating to temperatures $>175\,^{\circ}C$. This purple colour is associated with an increase in refractive index and a decrease in specific gravity, and these effects were believed to be caused by nuclear particles disrupting the structure along their paths. However, it is now known that most fluorites fade or lose their colour completely when heated to around 200 °C. Raising the temperature allows the sample to recrystallize with fewer strains and imperfections. Haloes in dark blue fluorite have been ascribed to radioactive disintegration, the colour being due to the release of colloidal calcium; strongly radioactive fluorite with haloes has been called radiofluorite (Hess, 1931). White fluorite when exposed to radiation was reported by Hoffmann (1937) to become blue, whereas deep purple fluorite in a mine used for storing radium became completely decolourized (Dunham, 1952), although this material later developed a greenish colour.

Thermoluminescence studies of various fluorites have demonstrated a relationship between the presence of rare earth elements and the colour of the fluorite. Thus when a trivalent *REE* ion substitutes for a divalent calcium ion, the charge compensation required is provided by an extra F^- occupying an interstitial position in the fluorite structure, or O^{2-} substituting for a structural F^-. The released electrons reduce the trivalent *REE* to divalent *REE* ions, most of which absorb visible light. The remaining neutral fluorine atoms (F^0) or monovalent oxygen ions (O^-) represent 'holes' that can attract electrons, these being held in various 'traps' in the fluorite structure. As the temperature is raised, traps are successively emptied as their activation energies are reached. This allows the 'hole' to extract an electron from a divalent *REE* ion, leaving a trivalent *REE* ion in an excited state which then reverts to a stable state by releasing energy as light (see, for example, Naldrett *et al.*, 1987). The colour due to the divalent *REE* ion disappears and the absorption of the trivalent *REE* ion does not contribute to the colour.

Although colourless fluorites have lower Σ*REE*, and coloured fluorites from some localities are richer in the heavy *REE* than in the light *REE* (e.g. some brownish rose and brown fluorites; Trinkler *et al.*, 1993), it is rare to be able to associate the abundance of a particular *REE* with a colour. The development of improved analytical techniques has only now made the determination of all 13 *REE* readily available. Among the few attempts to relate fluorite colour with Y, La and each of the 13 naturally occurring *REE* is that of Naldrett *et al.* (1987), who concluded that there does not appear to be any association between the presence of a particular *REE* (even in comparatively large amounts) and the colour of the fluorite. Thus, as noted by Naldrett *et al.* the intensity of colour increases with the *REE* content but no particular colour is favoured. Nevertheless, it is tempting to speculate that Sm and/or Dy may be important chromophoric elements in green fluorite (although the colour fades from the Weardale green fluorite on exposure to sunlight); a yellow coloration may be attributable to O^{2-} or Sr. There is some consensus of opinion that the presence of Y or the YO_2 complex in fluorite is responsible for a rose coloration, e.g. Bill *et al.* (1967), Bill and Calas (1978) and Millán Chagoyen *et al.* (1990).

Direct experiments by Blanchard (1976) and Marchand *et al.* (1976) showed that when fluorite is deposited from a solution of CaF_2 spiked with 0.1 per cent of individual *REE*, it gives thermoluminescence glow curves with peak temperatures and general configurations characteristic of the lanthanide species, whereas pure CaF_2 was found to be thermoluminescent only at low temperatures. This work also showed that the composition of the fluorite deposited parallels the composition of the solution, but with an enhancement of the *REE* relative to CaF_2 (a phenomenon also noted in natural samples (see p. 357)) and with the partition coefficient of crystal to liquid increasing from La to Lu. A CaF_2 crystal containing 0.1 per cent Dy displayed an unexpected, strong original thermoluminescence which disappeared on heating. The salts of Dy are known to have a green colour. Earlier experiments had shown that synthetic fluorite grown from melts containing variable quantities of samarium fluoride is green in colour, leading to the suggestion that the green coloration of some natural fluorites may be due to the presence of Sm^{2+} (Feofilov, 1956).

The differential thermal analysis curves and thermal luminescence spectra of fluorites have been recorded by Mukerji *et al.* (1981), together with the effects of thermal pretreatment, irradiation at different temperatures, ultraviolet radiation and electric polarization. The thermoluminescence (TL) glow peaks were investigated by Ganguly and Kaul (1984). Both studies used fluorite from the Amba Dongar carbonatite, Gujarat, India. The use of TL curves of fluorite to assign the associated ores to particular genetic and mineralogical types of deposit has been advocated by Faiviev and Iskandarov (1990), who found that the change in the TL curve with increasing sample depth could be used to estimate the extent of erosional removal above the deposit and consequently to predict the occurrence of ore mineralization at depth. The fluorites of ore deposits in igneous and metamorphic rocks have stronger TL than those in sedimentary rocks. The monitoring of the emission spectrum during TL, rather than simply recording light emission during heating, is reported to give more details of impurity ions, radiation doses and the thermal history of the fluorite (Rendell *et al.*, 1993). The green and yellow fluorites from Spain were found to show the optical and luminescence peaks characteristic, respectively, of Sm^{2+} and Eu^{2+} for the green fluorite, and O^{3-} substitution for two adjacent F^- ions and photoluminescence spectra of Ce^{3+} and Eu^{2+} for the yellow fluorite. Purple fluorite from the USA showed colloidal calcium and no photoluminescence. All three fluorites, after cooling, irradiation with X-rays and re-heating, showed a series of features at wavelengths characteristic of Dy^{3+} transitions at 50 °C, together with a series of broad peaks in the blue region at higher temperatures.

Early attempts to establish a relationship between individual rare earth elements and a particular colour led to a plethora of somewhat confusing results due partly to the presence of too many variables. It is useful to compare the ratio of a particular *REE* to the total *REE* in a sample with the proportion of that *REE* to the total *REE* of average chondrites. Numerous plots of *REE* in French fluorites versus normalized chondritic abundance values are, for example, given by Marchand *et al.* (1976) and Grappin *et al.* (1979). The plots mostly show small positive Eu anomalies, but the green Weardale fluorite (Fig. 120) is particularly enriched in Eu, whereas the blue and colourless fluorites from southern Derbyshire show a negative Eu anomaly (and, interestingly, are non-fluorescent).

Qualitative determinations have indicated that green fluorite does indeed contain Sm^{2+} (Bill *et al.*, 1967; Recker *et al.*, 1968). In a study of coloured fluorites from Sicily, Bellanca *et al.* (1981) found that the green samples were richer in Sm and Eu than the violet, blue or white varieties, although some of the dark grey and white samples were also relatively rich in Eu (see also Millán Chagoyen *et al.*, 1990; Fanlo *et al.*, 1992). In the fluorites from the Amba Dongar carbonatite, India, the abundance of ΣREE was found to increase in the order colourless < yellow < green < violet. In a study of fluid inclusions and *REE* in fluorite from five Bulgarian mineral deposits, Stolynova (1993) determined the green fluorite to be the highest temperature product and to have higher contents of *REE*, Co, Fe, U and Hf than violet fluorites.

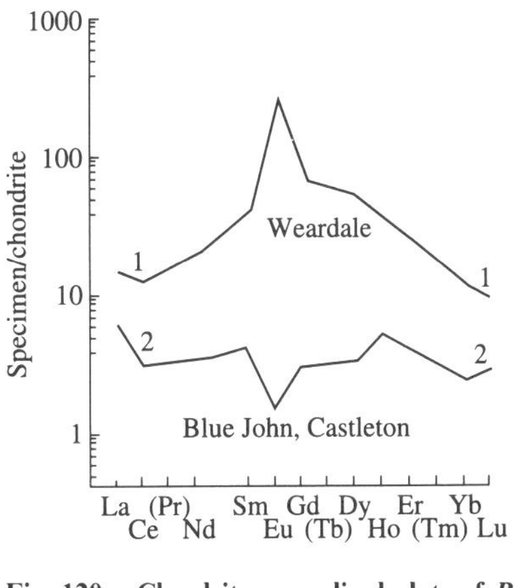

Fig. 120. Chondrite-normalized plots of *REE* in (1) fluorescent green fluorite from Weardale and (2) Blue John fluorite from Castleton, Derbyshire (after Howie *et al.*, 1982).

The fluorescence of many fluorites in ultraviolet light gave its name to the phenomenon, the principal activator for blue-violet fluorescence generally being Eu^{2+} (e.g. Huber-Schausberger and Schroll, 1967; Recker *et al.*, 1968; Bill and Calas, 1978). A yellow-green fluorescence has been ascribed to the presence of Yb^{2+}, and Huber-Schausberger and Schroll reported that for blue-violet fluorescence to occur the Eu:Yb ratio must be >6.

Distinguishing Features

In hand specimen fluorite may be distinguished by its perfect octahedral cleavage, in addition to its vitreous lustre and its cubic habit when this is free to develop. The colour is so variable as to be of little help, but varying shades of purple and violet are common colours, as are green and yellow, and colourless material is not rare. It is relatively soft, does not effervesce with like calcite, but is attacked by H_2SO_4. Under the microscope its isotropic character and very low refractive index are characteristic; purple varieties often have sufficient depth of colour to remain purple or violet in thin section. Cryolite has an even lower relief, is very weakly birefringent, and has a pseudo-cubic {001} cleavage; it is less readily fusible than fluorite. Members of the sodalite group have a slightly higher refractive index and the {110} cleavage is poor. Opal has no cleavage. Halite has a perfect {001} cleavage and a higher refractive index.

Paragenesis

Fluorite occurs most commonly in mineralized sedimentary carbonate rocks because of the latter's susceptibility to replacement by relatively low-temperature hydrothermal fluids. Fluorite is also found in fissure veins and stockworks in granitic rocks, greisens and pegmatites; deposits associated with carbonatites and other alkaline igneous rocks are becoming of increasing commercial importance. Occurrences in metamorphic rocks are also known, and fluorite has been recorded as a volcanic sublimate and as an authigenic mineral in evaporitic sediments. General accounts of the various environments of deposition have been given by Grogan *et al.* (1974), Hodge (1986) and Heyl (1993).

Fluorite mineralization is common in the Carboniferous Limestone of the English Pennines, where it is mainly restricted to the central area of the orefield (Dunham and Wilson, 1985). Studies of fluid inclusions in fluorite from this area have indicated a range of temperatures of homogenization (T_h) of 92–154 °C which, after a pressure correction of 11 °C, gives mean temperatures of formation of 103–165 °C (Rogers, 1977, 1978). Freezing studies on the fluid inclusions indicate that the ore-forming fluids had salinities in the range 18–28 wt. per cent equivalent NaCl. Within individual ore zones, trace element geochemistry and fluid inclusion studies have shown that vein intersections acted as major conduits for ascending solutions, resulting in deposits in the northern Pennines which have yielded 2 000 000 tons of fluorspar (Greenwood and Smith, 1977). In the southern Pennines, the emplacement of a flat at Masson Hill, Matlock, metasomatizing Carboniferous Limestone, was ascribed to mineralizing fluid being trapped above a thick lava flow and below a volcanic clay horizon, down-dip from a clay-filled fault (Ixer, 1978).

The unique character of the concentrically banded Blue John variety has been described by Ford (1955) as being due to the coincidence of mineralization in which fluorite predominated, oil-rich sediments and an anticlinal structure in the limestone of Treak Cliff, Derbyshire, though the Blue John occurs in pipe veins and cavities rather than in the usual mineralized veins of the district. A general account of British fluorite deposits was given by Dunham (1952), but see also that by Ixer and Vaughan (1993).

In the fluorite deposits of Argentolle, Saône-et-Loire, France, the T_h values for green fluorite are 120–150 °C, whereas the yellow fluorites give 80–120 °C (Valette, 1983). Elsewhere in France, the occurrence of green fluorite was attributed to alkaline parent fluids (Marchand *et al.*, 1976). A study of fluid inclusions in fluorite of the Spanish Pyrenees gave T_h values of 111–156 °C, with around 19 wt. per cent equivalent NaCl (Fanlo *et al.*, 1992). In the Middle Trias of Austria, veins of fluorite in schist have a low content of *REE*, leading Götzinger and Weinke (1984) to assign them to a sedimentary origin. In the Alps, veins in Triassic strata have been attributed to the remobilization of Pb, Zn and F in underlying Carboniferous strata, the fluorite here being associated with syndiagenetic dolomitization (Hein *et al.*, 1984). Later fluid inclusion studies on these Alpine deposits were interpreted (Hein *et al.*, 1990) as indicating that early fluorite was deposited from highly saline H_2O–NaCl–$CaCl_2$ solutions containing appreciable H_2S; during vein formation the fluids changed from low salinity ($\sim$7 wt. per cent NaCl equivalent) and medium temperatures ($T_h \sim 230$ °C), corresponding with the precipitation of early quartz, towards high salinity ($\sim$20 wt. per cent NaCl equivalent) and low temperatures ($T_h \sim 170$ °C) during the precipitation of late-stage fluorite. Their *REE* content is relatively low and favours a genetic model for fluorite mineralization involving the leaching of suitable rock units by formation waters. In the Transvaal, South Africa, two large bodies of nearly pure fluorite yielded over 40 000 tons of the mineral (Kupferburger, 1927); here, the fluorite occurred as a replacement deposit in dolomite, with the mineralizing solutions having risen along lines of brecciation, faults and joint planes.

Fluid inclusions in hydrothermal fluorite from the Kulikolon deposit in central Asia were studied by Ermakov (1944) who determined their T_h as 101–136 °C. The fluorite in the Tÿrnÿauz scheelite–fluorite mineralization (Vasil'kova *et al.*, 1980) is recorded to be low in *REE* ($<$10 p.p.m.) but to have a relatively high temperature of crystallization (400–700 °C). The Urtuy fluorite deposit is the largest in Eastern Transbaikalia and is 600 m deep along a length of 2100 m; it is an adularia–quartz–fluorite vein deposit (Balnayev, 1993). The fluorite deposits in the Kalat region of northwest Pakistan are of Mississippi Valley type and contain fluid inclusions rich in water ± hydrocarbons, giving a T_h of 90–140 °C and a composition around 4 wt. per cent NaCl equivalent (Alderton, 1991).

Similar investigations on Illinois (USA) fluorite gave 83–115 °C for 'primary' inclusions and 112–172 °C for 'secondary' inclusions (Grogan and Schrode, 1952). The important and widespread Illinois–Kentucky fluorite deposits are epigenetic and include both vein deposits and bedding replacement deposits; the physical guides were structural and stratigraphical (Weller *et al.*, 1952; Grogan *et al.*, 1974). The dilution of mineralizing fluids by groundwater and the change in temperature gradient are considered to have been the main chemical factors governing the formation of these fluorite deposits. The epithermal fluorite deposits of northeast Idaho have low salinity ($\leqslant$0.6 wt. per cent equivalent NaCl) and give low to moderate temperatures of homogenization of their fluid inclusions (T_h 98–146 °C), indicating a meteoric origin for the fluids. Their $\delta^{18}O$ and δD isotopic values are identical to those of modern meteoric water, i.e. the Eocene climatic conditions were similar to present day, and not warmer as deep-ocean Eocene evidence indicates (Seal and Rye, 1993).

In Eocene carbonate mudstone in Florida, fluorite has been reported within primary gypsum nodules at a depth of 235–240 m below the surface; the survival of this assemblage is attributed to a lack of groundwater circulation (Cook *et al.*, 1985). The fluorite probably formed contemporaneously with gypsum in a sabkha environment from sea-water-concentrated brine. Submicroscopic nearly spherical grains of fluorite have been reported in lacustrine, tuffaceous zeolitized Pliocene mudstones in Oregon (Sheppard and Gude, 1969); the overall fluorite content is around 5 per cent but $\leqslant 16$ per cent in zeolitic tuff. The origin of this fluorite is attributed to the diagenesis of sediments in an alkaline lake. Fluorite is sometimes found as a cementing material in sandstones; violet grains of fluorite are fairly common as a detrital mineral in sands, being derived from acid igneous rocks and hydrothermal deposits.

In igneous rocks fluorite may occur as a late-crystallizing mainly hydrothermal product, especially in granites, syenites and greisens. It is a rather common accessory mineral in some granitic pegmatites; rare subhedral crystals may be as much as 2 m in diameter (Jahns, 1953). The examination of fluid inclusions in pegmatite fluorites has given a T_h range of 450–550 °C. The vein-type fluorite deposits of the northern Bavarian basement occur as veins in the granites themselves, as siderite–baryte–fluorite veins in the wallrock of granites or in sulphide-bearing veins in sediments and volcanic rocks of Variscan age (Dill *et al.*, 1986); they have undergone intensive redeposition during late-stage activity. Calcium-rich fluid inclusions were noted in fluorite from pegmatites associated with the Strezegom granite, Lower Silesia (Kozlowski, 1984); they gave T_h values of 165–180 °C, the compositions of the solutions being H_2O 76, $CaCl_2$ 18 and NaCl 6 per cent. Some Newfoundland deposits in granite consist of 1.5–6.0 m thick veins, containing 75–95 per cent CaF_2.

Fluorite has been reported from nepheline syenite (Jérémine, 1950) and from apatite-rich deposits in alkaline rocks in the Transvaal (Kent *et al.*, 1943), where it occurs in economically important quantities. Pneumatolytic fluorite in a fluorite–apatite rock grading into mariupolite has been found in an alkali granite (Lebedev and Moleva, 1935). In these igneous occurrences, associated minerals include cassiterite, topaz, apatite, lepidolite, etc., for the pneumatolytic deposits, and calcite, pyrite, apatite, etc., for the hydrothermal deposits. Fluorite has also been recorded in the drusy cavities of blocks ejected from volcanoes, and as a volcanic sublimate (e.g. Kvasha, 1954). In the alkaline complex of Alnö Island, Sweden, calcite–fluorite dykes have been described (Eckermann, 1948).

Fluorite is frequently associated with carbonatite complexes. In the Amba Dongar carbonatite, Gujarat, India, for example, fluorite occurs both as an early magmatic mineral and as thick veins and replacement deposits in both sövite and ankeritic carbonatite (Viladkar and Wimmenauer, 1992). A study of fluid inclusions in the Amba Dongar fluorite (presumably from a replacement deposit) indicated a crystallization temperature of 100–150 °C (Roedder, 1973). The total reserves here have been estimated at 11.6 million tons, averaging 30 per cent fluorite. In a study of African fluorspar resources, Van Alstine and Schruben (1980) reported that the main sources are Kenya and South Africa, where most ore is produced from carbonate mantos. Fluorspar is important in 24 of 115 carbonatite localities associated with rift structures; the ores average 22 per cent CaF_2 and amount to 192 million metric tonnes.

In metamorphic rocks fluorite is less common. However, a silicified mica schist breccia with a groundmass of dark violet fluorite was described from Banffshire, northeastern Scotland (Russell, 1936). In Trentino, Italy, a thick vein of fluorite cutting mica schist near a granodiorite contact has inclusions of a yellow organic compound which decomposes at 220–230 °C, indicating a maximum temperature for the deposition of this fluorite (Morgante, 1939). In San Lui Potosi, Mexico, the Cuevas fluorite deposit occurs at the contact between a Lower Cretaceous limestone and a Tertiary rhyolite breccia; it is a limestone pore-filling and replacement deposit (Ruiz *et al.*, 1980). Here, the fluid inclusions have a salinity of 0 per cent equivalent NaCl and give T_h values of 60–130 °C. The pink fluorite found in Huanuco Province, Peru, forms crystals up to 5 cm size in vugs in sulphide ore (Belsher, 1982). A variant from these metasomatic occurrences is the fluorite found in granulite-grade metamorphic rocks of the Adirondacks. Fayalite–ferrohedenbergite–fluorite–quartz$\pm$magnetite and ferrosalite–fluorite–quartz–magnetite assemblages occur in orthogneisses buffered by narrow f_{O_2}–f_{F_2} fields at constant P–T conditions (Bohlen and Essene, 1978). The conditions of crystallization of this fluorite are considered to have been f_{F_2} $10^{-29\pm1}$ bar, f_{O_2}-$10^{-16\pm1}$ bar at 1000 K and 7 kbar.

References

Alderton, D. H. M., 1991. Fluid inclusion evidence for the genesis of fluorite deposits in the Kalat region of Baluchistan Province, Pakistan. *Acta Mineralogica Pakistanica*, **5**, 4–8.

Allen, R. D., 1952. Variations in chemical and physical properties of fluorite. *Amer. Min.*, **37**, 910–30.

Alstine, R. E. Van and **Schruben, P. G.**, 1980. Fluorspar resources of Africa. *Bull. U.S. Geol. Surv.*, **1487**, 25 pp.

Balnayev, K. B., 1993. Vertical zonation of peri-ore metasomatic products in the Urtuy fluorite deposit, Eastern Transbaikalia. *Dokl. Acad. Nauk USSR, Earth Sci. Sect.*, **318**, 138–42.

Baranov, E. N., 1966. Uranium content in fluorite. *Geochem. Internat.*, **3**, 795–7.

Bastos Neto, A., **Touray, J.-C.**, **Dardenne, M.** and **Charvet, J.**, 1991. Chronologie et évolution des fluides hydrothermaux dans le district à fluorine de Santa Catarina, Brésil: données de l'analyse des terres rares et de l'étude des inclusions fluides. *Min. Deposita*, **26**, 211–6.

Bellanca, A., **di Salvo, P.**, **Möller, P.**, **Neri, R.** and **Schley, F.**, 1981. Rare-earth and minor element distribution and petrographic features of fluorites and associated Mesozoic limestones of northwestern Sicily. *Chem. Geol.*, **32**, 255–69.

Belsher, D. O., 1982. Pink octahedral fluorite from Peru. *Min. Rec.*, **13**, 29–30.

Berman, R., 1957. Some physical properties of naturally irradiated fluorite. *Amer. Min.*, **42**, 191–203.

Bill, H. and **Calas, G.**, 1978. Color centres, associated rare-earth ions and the origin of coloration of natural fluorites. *Phys. Chem. Min.*, **3**, 117–31.

Bill, H., **Sierro, J.** and **Lacroix, R.**, 1967. Origin of coloration in some fluorites. *Amer. Min.*, **52**, 1003–8.

Blanchard, F. N., 1967. Thermoluminescence of synethetic fluorite. *Amer. Min.*, **52**, 371–9.

Blount, B. and **Sequeira, J. H.**, 1919. 'Blue John' and other forms of fluorite. *J. Chem. Soc.*, **115**, 705.

Bohlen, S. R. and **Essene, E. J.**, 1978. The significance of metamorphic fluorite in the Adirondacks. *Geochim. Cosmochim. Acta*, **42**, 1669–78.

Bragg, W. L., 1914. The structure of some crystals as indicated by their diffraction of X-rays. *Proc. Roy. Soc.*, *A*, **89**, 468.

Bragg, W. L. and **Claringbull, G. F.**, 1965. *Crystal Structures of Minerals*. London (G. Bell & Sons), p. 37.

Braithwaite, R. S. W., **Flowers, W. T.**, **Hazeldine, R. N.** and **Russell, M.**, 1973. The cause of the colour of Blue John and other purple fluorites. *Min. Mag.*, **39**, 409–11.

Bramley, J. V., 1990. Production of acid-grade fluorspar in the south Pennines. In *Minerals, Materials and Industry:* Ed. M. J. Jones, London (Institution of Mining and Metallurgy), pp. 431–8.

Chatagnon, B. and **Galland, D.**, 1982. R.P.E. de la fluorite naturelle: inventaire des ions lanthanide et autres impurétes paramagnétiques. *Bull. Min.*, **105**, 37–42.

Chatterjee, N., 1940. Struktur-Untersuchungen von näturlichem und kunstlichem Yttrofluorit mit Hilfe der Fluoreszenz und Absorptionsspektren. *Z. Krist.*, **102**, 245–84.

Cook, D. J., **Randazzo, A. F.** and **Sprinkle, C. L.**, 1985. Authigenic fluorite in dolomitic rocks of the Floridan aquifer. *Geology*, **13**, 390–1.

Dickson, J. A. D., 1980. Artificial colouration of fluorite by electron bombardment. *Min. Mag.*, **43**, 820–2.

Dill, H., **Dulski, P.** and **Möller, P.**, 1986. Fluorite mineralization and REE patterns in vein-type deposits from the N Bavarian basement. *Neues Jahrb. Min., Abhdl.*, **154**, 141–51.

Dunham, K. C., 1952. Fluorspar. *Mem. Geol. Surv. Gt Britain; Special Rep. Min. Resources Gt Britain*, **4**.

Dunham, K. C. and **Wilson, A. A.,** 1985. *Geology of the Northern Pennine Orefield: Volume 2, Stainmore to Craven.* London (British Geological Survey), 247 pp.

Eckermann, H. von, 1948. The alkaline district of Alnö Island. *Sveriges Geol. Unders., Ser. Ca,* no. 36.

Ermakov, N. P., 1944. Determination of temperature of formation of hydrothermal minerals by studying liquid inclusions. *Dokl. Acad. Sci. USSR,* **45**, 202–3.

Faiziev, A. R. and **Iskandarov, F. Sh.,** 1990. Thermoluminescence of fluorites of the Pamirs and central Tadzhikistan. *Zapiski Vses. Min. Obshch.,* **119**, 48–60.

Faiziev, A. R. and **Koplus, A. V.,** 1992. Rare earths within the fluorite of different genesis. *Zapiski Vses. Min. Obshch.,* **121**, 79–88 (Russian with English abstract).

Fanlo, I., Subias, I. and **Fernandez-Nieto, C.,** 1992. Inclusiones fluidas y distribucion de elementos de las tierras raras en las fluoritas del yacimiento de Parzan (Bielsa, Huesca). *Bol. Soc. Española Min.,* **15**, 147–51.

Fayziyev, A. R., 1990. Yttrium in fluorite from endogenous shows in USSR. *Geochem. Internat.,* **27**, 114–9.

Feofilov, P. P., 1956. On the nature of the green colour of fluorite. *Mém. Soc. Russe Min.,* **85**, 569–71.

Fleischer, M., 1969. The lanthanide elements in fluorite. *Indian Min.,* **10** 36–9.

Ford, T. D., 1955. Blue John fluorspar. *Proc. Yorks. Geol. Soc.,* **30**, 35–60.

Galwey, A. K., Jones, K. A., Reed, R. and **Dollimore, D.,** 1979. The blue coloration in banded fluorite (Blue John) from Castleton, Derbyshire, England. *Min. Mag.,* **43**, 243–50.

Ganguly, S. and **Kaul, I. S.,** 1984. Analysis of the thermoluminescence glow peaks from natural calcium fluoride. *Modern Geol.,* **8**, 155–61.

Garnett, C. S., 1920. Colouring matters of red and blue fluorite. *J. Chem. Soc.,* **117**, 620.

Gerasimovskiĭ, V. V. and **Kataeva, Z. T.,** 1984. [An yttrofluorite with an unusually large content of rare-earth elements.] *Min. Zhurn.,* **6**, 36–42 (in Russian).

Gliszczynski, S. von, 1949. Notiz über Flußspatzwillinge. *Neues Jahrb. Min. Geol.,* Abt. A, *Monat.,* 45–9.

Goldschmidt, V. M. and **Thomassen, L.,** 1923. Die Krystallstruktur näturlicher und synthetischer Oxyde von Uran, Thorium und Cerium. *Vid. Skrift., I. Mat.-nat. Kl., no. 2* (M.A.2–206).

Götzinger, M. A. and **Weinke, H. H.,** 1984. Spurenelementgehalte und Entstehung von Fluoritmineralisationen in den Gutensteiner Schichte (Anis-Mittelrais), Nördliche Kalkalpen, Österreich. *Tschermaks Min. Petr. Mit.,* **33**, 101–19.

Grappin, C., Trueill, M., Yaman, S. and **Touray, J.-C.,** 1979. Le spèctre des terres rares de la fluorine en tant que marqueur des propriétés du milieu de dépôt et des interactions entre solutions minéralisantes et roches sources. Exemple pris dans le district de la Marche Occidentale (France). *Min. Deposita,* **14**, 297–309.

Greenwood, D. A. and **Smith, F. W.,** 1977. Fluorspar mining in the northern Pennines. *Trans. Instn Min. Metall. (Sect. B: Appl. Earth Sci.),* **86**, 181–90.

Grogan, R. M. and **Shrode, R. S.,** 1952. Formation temperature of southern Illinois bedded fluorite as determined from fluid inclusions. *Amer. Min.,* **37**, 555–66.

Grogan, R. M., Cunningham-Dunlop, P. K., Bartlett, H. F. and **Czel, L. J.,** 1974. The environments of deposition of fluorspar. *Spec. Publ. Kentucky Geol. Surv. Ser. x,* **22**, 4–9. [This Special Publication contains several other papers on regional fluorspar deposits.]

Gundlach, H., Möller, P., Parekh, P. P. and **Stoppel, D.,** 1976. Zur Genese des Fluorits auf den Barytgängen des Südwest-Harzes. *Geol. Jahrb., D,* **20**, 3–22.

Haberlandt, H. and **Scheiner, A.,** 1935. Über Farbverteilungen beim Fluorit in ihrem Zusammenhang mit dem Kristallbau. *Z. Krist.,* **90**, 193–214.

Hein, U. F., Möller, P. and **Schneider, H.-J.,** 1984. Rare earth element fractionation in fluorite as a geochemical indicator of ore deposition in Triassic carbonate rocks of the eastern and southern Alps. *Neues Jahrb. Min., Monat.,* 305–16.

Hein, U. F., Luders, V. and **Dulski, P.,** 1990. The fluorite vein mineralization of the southern Alps: combined application of fluid inclusions and rare-earth element (REE) distribution. *Min. Mag.,* **54**, 325–33.

Hess, F. L., 1931. Radioactive fluorspar from Wilberforce, Ontario. *Amer. J. Sci.,* **22**, 215–21.

Heyl, A. V., 1993. Classification and habit of fluorine deposits in the Rio Grande depression, New Mexico and Colorado. In *Proceedings of the Eighth IAGOD Symposium:* Ed. Y. T. Maurice, Berlin (Springer Verlag), pp. 233–41.

Hodge, B. L., 1986. Occurrence and exploitation of fluorite. In *Geology in the Real World—the Kingsley Dunham Volume*: Eds. R. W. Nesbitt and I. Nichol, London (Inst. Mining Metall.), p. 165.

Hoffmann, J., 1937. Über photolytisch neutralisierte Elementarstoffe der Fluorite. *Chem. der Erde,* **11**, 368–74.

Howie, R. A., Pegram, E. and **Walsh, J. N.,** 1982. The content of rare earth elements in English fluorites. *J. Russell Soc.,* **1**, 22–5.

Huber-Schausberger, I. and **Schroll, E.,** 1967. UV-Lumineszenz und Seltenerdgehalte in Flusspaten. *Geochim. Cosmochim. Acta,* **31**, 1333–41.

Ixer, R. A., 1978. The emplacement of a fluorspar flat at Masson Hill, Matlock, Derbyshire. *Mercian Geol.*, **6**, 245–55.

Ixer, R. A. and **Vaughan, D. J.**, 1993. Lead–zinc–fluorite–baryte deposits of the Pennines, North Wales and the Mendips. In *Mineralization in the British Isles:* Eds. R. A. D. Patrick and D. A. Polya, London (Chapman and Hall), pp. 325–418.

Jahns, R. H., 1953. The genesis of pegmatites. I. Occurrence and origin of giant crystals. *Amer. Min.*, **38**, 563–98.

Jérémine, E., 1950. Sur quelques minéraux des syénites néphéliniques de Bou Agrao (Bou Aougra), Haut Atlas (Maroc). *C. R. Acad. Sci. Paris*, **250**, 110–1.

Jiang, X. and **Li, C.**, 1993. [REE geochemistry of fluorite deposits in Zhejiang Province.] *Min. Deposita*, **12**, 55–66 (in Chinese with English abstract).

Kent, L. E., **Russell, H. D.** and **Rooyen, D. P. van**, 1943. Fluorspar in the Union of South Africa and South West Africa. *Bull. Geol. Surv. South Africa*, **14**, 69 pp.

Kozłowski, A., 1984. Calcium-rich inclusion solutions in fluorite from the Strzegom pegmatite, Lower Silesia. *Acta Geol. Polonica*, **34**, 131–8.

Kudoh, Y., **Nagase, T.**, **Ohta, S.-I.** and **Sasaki, S.**, 1993. Crystal structure of fluorite, CaF_2, at 7 GPa using a diamond anvil and single crystal X-ray diffraction method. *KEK Progress Report*, **93-1**, p. 214.

Kupferburger, W., 1927. The fluorspar, lead and zinc deposits of the western Transvaal. *Trans. Geol. Soc. South Africa*, **30**, 5–56.

Kvasha, L. G., 1954. The study of samples of sublimates from the lava flows and craters of the 'Jubilee' break of the Klyuchevsky volcano. *Trans. Lab. Volcanology, Acad. Sci. USSR*, **8**, 100–4.

Lebedev, P. I. and **Moleva, V. A.**, 1935. Pneumatolytic rock of the Azov sea-shore crystalline mass. *Trav. Inst. Petrogr. Loewinson-Lessing, Acad. Sci. USSR*, **5**, 13–20 (M.A. **6**–315).

Li, X., **Liu, T.** and **Zhao, Y.**, 1985. [A study of colour-causing mechanism of fluorite from Bayan Obo, Nei Monggol.] *Acta Min. Sinica*, **5**, 164–8 (in Chinese with English abstract).

Mackenzie, K. J. D. and **Green, J. M.**, 1971. The cause of coloration in Derbyshire Blue John banded fluorite and other blue banded fluorites. *Min. Mag.*, **38**, 459–70.

Marchand, L., **Joseph, D.**, **Touray, J. C.** and **Treuil, M.**, 1976. Criteres d'analyse géochimique des gisements de fluorine basés sur l'étude de la distribution des lanthanides—application au gite de Maine (71 Cordesse, France). *Min. Deposita*, **11**, 357–79.

Méary, A., **Touray, J. C.**, **Galland, D.** and **Jebrak, M.**, 1985. Interprétation de l'anomalie en europium des fluorines hydrothermales—donnés de la résonance paramagnétique électronique: application au gîte de fluorine de Montroc (Tarn, France). *Chem. Geol.*, **48**, 115–24.

Merwin, H. E., 1:911. Quartz and fluorite as standards of density and refractive index. *Amer. J. Sci.*, Ser. 4, **32**, 429–32.

Millán Chagoyen, A., **Garcia Guinea, J.**, **Arribas Gernandez, J. G.** and **Calderón García, T.**, 1990. Caracterizacíon optica de fluoritas españolas. *Bol. Soc. Española Min.*, **13**, 99–108.

Mitra, S., 1981. Nature and genesis of colour centres in yellow and colourless fluorites from Amba Dongar, Gujarat, India. *Neues Jahrb. Min., Abhdl.*, **141**, 290–308.

Morgante, S., 1939. Il giacimento di Corvara. *Periodico Min. Roma*, **10**, 379–96.

Mukerji, S., **Sengupta, S.** and **Kaul, I. K.**, 1981. Studies on some influencing parameters in the thermoluminescence of natural fluorites. *Modern Geol.*, **8**, 1–11.

Murdoch, J., 1936. Silica–fluorite pseudomorphs. *Amer. Min.*, **21**, 18–32.

Murr, L. E., 1974. Ordered lattice defects in colored fluorite. *Science*, **183**, 206–8.

Naldrett, D. L., **Lachaine, A.** and **Naldrett, S. N.**, 1987. Rare earth elements, thermal history, and the colour of natural fluorites. *Can. J. Earth Sci.*, **24**, 2082–8.

Nordstrom, D. K. and **Jenne, E. A.**, 1977. Fluorite solubility equilibria in selected geothermal waters. *Geochim. Cosmochim. Acta*, **41**, 175–88.

Przibram, K., 1956. *Irradiation Colours and Luminescence: a Contribution to Mineral Physics*. London (Pergamon Press), 332 pp.

Quirk, D. G., 1993. Origin of the Peak District orefield. *UK J. Mines Min.*, **13**, 4–15.

Recker, K., **Neuhaus, A.** and **Leckebusch, R.**, 1968. Vergleichende Untersuchungen der Farb- und Lumineszenzeigenschaften natürlicher und gezüchteter definiert dotierter Fluorite. *IMA Papers Proc. Fifth Gen. Meeting, Cambridge 1966*, London (Min. Soc.), 145–52.

Rendell, H. M., **Khanlary, M.-R.**, **Townsend, P. D.**, **Calderon, T.** and **Luff, B. J.**, 1993. Thermoluminescence spectra of minerals. *Min. Mag.*, **57**, 217–22.

Richardson, C. K. and **Holland, H. D.**, 1979. The solubility of fluorite in hydrothermal solutions, and experimental study. *Geochim. Cosmochim. Acta*, **43**, 1313–25.

Roedder, P. L., 1973. Fluid inclusions from the fluorite deposits associated with carbonatites Amba Dongar, India and Okurusu, South West Africa. *Trans. Instn Mining Metall. (Sect. B: Appl. Earth Sci.)*, **82**, 33–9.

Rogers, P. J., 1977. Fluid inclusion studies in fluorite from the Derbyshire orefield. *Trans. Instn Mining*

Metall. (Sect. B: Appl. Earth Sci.), **86**, 128–32.

Rogers, P. J., 1978. Fluid inclusion studies on fluorite from the Askrigg Block. *Trans. Instn Mining Metall. (Sect. B: Appl. Earth Sci.)*, **87**, 125–31.

Ruiz, J., **Kesler, S. E.**, **Jones, L. M.** and **Sutter, J. F.**, 1980. Geology and geochemistry of the Las Cuevas fluorite deposit, San Luis Potosi, Mexico. *Econ. Geol.*, **75**, 1200–9.

Russell, A., 1936. Notes on the occurrence of fluorite in Aberdeenshire and Banffshire. *Min. Mag.*, **24**, 307–17.

Saito, T., 1950. Chemical composition of fluorites of Japan. *J. Fac. Sci. Hokkaido Univ., Ser. 4, Geol. Min.*, **7**, 307–14.

Schaefer, B. and **Strubel, G.**, 1979. Hydrothermal solubility of fluorite in the system CaF_2–$MgCl_2$–H_2O. *Neues Jahrb. Min., Montat.*, 233–40.

Seal, R. R. III and **Rye, R. O.**, 1993. Stable isotope study of fluid inclusions in fluorite from Idaho: implications for continental climates during the Eocene. *Geology*, **21**, 219–22.

Sheppard, R. A. and **Gude, A. J.**, 1969. Authigenic fluorite in Pliocene lacustrine rocks near Rome, Malheur County, Oregon. *Prof. Paper U.S. Geol. Surv.*, **650-D**, 69–74.

Sine, F. L., 1925. Antozonite from Monteagle township, Hastings County, Ontario. *Univ. Toronto Studs., Geol. Ser.*, **20**, 22–4.

Steyn, J. G. D., 1954. Spectrographic and X-ray data on some fluorites from the Transvaal, South Africa. *Min. Mag.*, **30**, 327–32.

Stockbarger, D. C., 1949. The production of large artificial fluorite crystals. *Discuss. Faraday Soc.*, **5**, 294–9.

Stolynova, M., 1993. [Fluid inclusions and rare earths in the fluorite from some Bulgarian mineral deposits.]. *Ann. Univ. Mining & Geol.*, **39(1)**, 29–32 (Bulgarian with English abstract).

Subias Pérez, I. and **Fernandez-Nieto, C.**, 1991. Geologia y Geoquímica (REE y trazas) del filón de fluorita ana mary (Lanuza, Huesca). *Bol. Soc. Española Min.*, **14**, 151–60.

Sverdrup, T. L., 1968. Yttrofluorite–yttrocerite–cerfluorite in Norwegian pegmatites. *Norsk Geol. Tidsskr.*, **48**, 245–52.

Thugutt, S. J., 1936. O koloidalnym roztworze fluorytu—Sur une solution colloidale de la fluorine. *Arch. Min. Tow. Naui, Warsaw*, **12**, 187–92.

Trinkler, M., **Kempe, U.**, **Plotze, M.** and **Rieser, U.**, 1993. Über rosa und braunen Fluorit aus Sn-W-Lagerstatten. *Chem. der Erde*, **53**, 165–81.

Valette, C.-O., 1983. Karsts et filons à fluorine dans le faisceau synclinal du Morvan: le gisement d'Argentolle (Saône-et-Loire, France). *Documents Bur. Rech. Geol. Min.*, **58**, xx + 299 pp.

Vasil'kova, N. N., **Kulikov, I. V.**, **Lyapunov, S. M.** and **Mineev, D. E.**, 1980. [Fluorite, an indicator of the conditions of formation and zonation of the Tÿrnÿauz scheelite-fluorite mineralization.] *Min. Zhurn.*, **2**, 45–57 (in Russian).

Viladkar, S. G. and **Wimmenauer, W.**, 1992. Geochemical and petrological studies on the Amba Dongar carbonatites (Gujarat, India). *Chem. der Erde*, **52**, 277–91.

Vogt, T., 1911. Vorläufige Mitteilung über Yttrofluorite, eine neue Mineralspezies aus dem nordlichen Norwegen. *Centralbl. Min.*, 373–5.

Weller, J. M., **Grogan, R. M.** and **Tippie, F. E.**, 1952. Geology of the fluorspar deposits of Illinois. *Illinois State Geol. Surv., Bull.*, **76**, 20 pp.

Zambonini, F., 1915. Suelle soluzioni solide de comopositi di calcio, stronzio, bario e piombo conquelli delle 'terre rare' e loro importanza par la mineralogica chemica. *Riv. Min. Crist. Italiana*, **45**, 3–185.

Zintl, E. and **Udgård, A.**, 1939. Über die Mischkrystallbildung zwischen einigen Salzartigen Fluoriden von verschiedenem Formeltypus. *Z. Anorg. Chem.*, **240**, 150–6.

Halite NaCl

Cubic

n	1.544
Dispersion:	Moderate.
D	2.16–2.17.
H	$2\frac{1}{2}$.
Cleavage:	{100} perfect.
Twinning:	On {111} for synthetic crystals.
Colour:	Colourless or white when pure, more typically orange or red due to inclusions of iron compounds; it may also be grey, pink, yellow or blue. Colourless in thin section.
Unit cell:	*a* 5.640 Å Z=4, Space group *Fm*3*m*

Soluble in water. Salty taste. Colours a flame deep yellow.

Halite, or rock salt, is typically found as an evaporite, occurring in sedimentary rocks of all ages, often associated with gypsum, anhydrite and other soluble salts. It is used for culinary purposes and in the chemical industry. The name is derived from the Greek *hals*, salt.

Structure

The structure of halite was the first to be analysed by X-rays. It was shown by Bragg (1914) to consist of sodium and chlorine ions arranged alternately along rows parallel to the edges of a face-centred cubic cell. If ions of one type are taken at the corners and face centres of the cell, those of opposite sign lie at the mid-points of cell edges and at the cube centre (Fig. 121). Each sodium atom is thus surrounded by six chlorine atoms and each chlorine atom by six sodium atoms. This analysis of the structure by Bragg (1914) first established the fact that in typical inorganic compounds there is no grouping of atoms or ions into molecules: each chlorine atom is equally related to all six of the surrounding sodium ions.

Previously, etch figures on the cube faces of halite had been observed to consist of square pits whose edges were slightly inclined to the cube faces, leading to the conclusion that the crystals possessed holoaxial symmetry. The X-ray results indicate that the mineral is holosymmetric, however, and Hettich (1926) showed that under conditions of perfect cleanliness the etch figures substantiated this: the presence of an optically active substance in the etching reagent may also cause asymmetry (see also Bragg and Claringbull, 1965, p. 59).

The cell edge of halite was the first crystal dimension to be measured, all the early measurements of crystal spacings and X-ray wavelengths being based on it. A precision determination by Frondel (1955) gave a value of 5.6404±0.0001 Å at 25 °C.

When a flat rectangular plate of rock salt is bent, the edge faces become broken

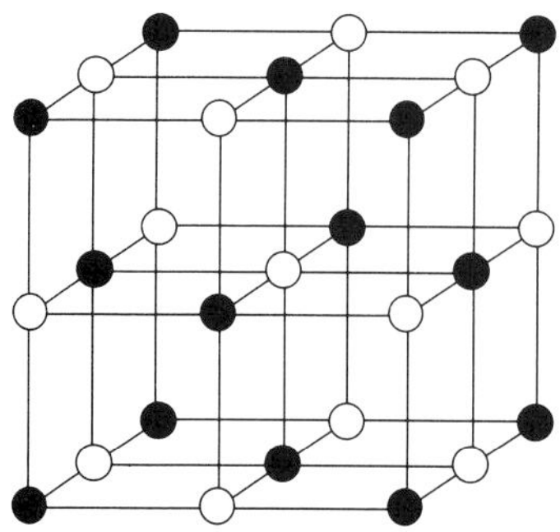

Fig. 121. The structure of halite. The two types of circle represent the two types of atom: it is immaterial which type is taken to represent Na and which Cl (after Bragg and Claringbull, 1965).

into facets at small angles to one another, the plate itself developing a series of birefringent stripes which correspond with the boundaries of these plane facets. The structure of a crystal deformed in this way was discussed in detail by Brilliantow and Obreimow (1937). Preston (1939) recorded diffuse spots which appear on Laue photographs of single crystals of halite and which increase in intensity when the temperature is raised to 500 °C. Their presence was attributed to a breaking up of the crystal into groups of atoms, the thermal vibrations causing the interatomic distances to vary slightly from one group to another so that new diffraction maxima characteristic of each group of atoms are produced. These diffuse reflections have also been studied by Olmer (1942) at constant and varying temperatures, and by Sun and Ch'u (1948).

The crystal structure of hydrohalite, $NaCl{\cdot}2H_2O$, was determined by Klewe and Pedersen (1974); it is monoclinic ($P2_1/c$) with *a* 6.3313, *b* 10.1178, *c* 6.5029 Å, β 114.407° and Z=4.

Chemistry

Some analyses of halite are listed in Table 45. Carefully purified material normally contains over 99 per cent NaCl; massive rock salt may, however, contain admixed clay, iron oxides, gypsum and other evaporite minerals. Detailed analyses show small amounts of Br also present, in the order of 0.023–0.030 per cent (Mir, 1946). In natural material at normal temperatures there does not appear to be much replacement of Na by K, although sylvite (KCl) is isomorphous with halite; Under metamorphic conditions of crystallization, however, an almost complete solid solution series exists (Fig. 122, p. 372).

In the sylvinite mining zone of Esterhazy, Saskatchewan, The K-rich sylvinite locally contains bodies of halite (called salt horses) 20–200 square feet in cross-section; Br contents in the halite are around 102 p.p.m., while the sylvite has 1470 p.p.m. In the 'normal' pores of this deposit halite has 81 and sylvite 1020 p.p.m., showing that the salt horses were formed by the local leaching of sylvite (McIntosh and Wardlaw, 1968). In the lacustrine deposits of the Green River Formation, northwestern Colorado, halite occurs with nahcolite in rhythmic evaporite facies in oil shale; the Br contents range from 15 to 182 p.p.m., and were considered to reflect the palaeosalinities of deposition and the salinity fluctuation of ancient lake waters (Dyni *et al.*, 1970). In boreholes encountering halite deposits

Table 45. Halite Analyses

	1	2	3	4	5	6	7	8
Na	39.00	37.69	36.34	—	—	—	—	—
NaCl	—	—	—	99.53	99.09	98.016	97.53	95.60
K	0.12	0.30	0.19	—	—	—	—	—
Ca	0.08	0.00	0.73	—	—	0.921	—	—
$CaSO_4$	—	—	—	0.17	0.40	—	1.49	1.29
$CaCl_2$	—	—	—	—	0.12	—	—	0.03
Mg	0.03	0.00	0.11	—	—	—	—	—
$MgSO_4$	—	—	—	0.02	—	—	0.23	—
$MgCl_2$	—	—	—	0.22	0.03	—	—	0.08
SO_4	0.27	0.00	0.56	—	—	0.869	—	—
Cl	60.27	54.16	57.43	—	—	—	—	—
CO_3	—	3.57	0.17	—	—	0.013	—	—
H_2O	—	3.83	3.61	0.19	—	—	—	—
insol.	—	—	0.28	0.06	0.36	—	—	3.00
Total	99.77	99.83	99.59	100.19	100.00	100.00	99.68	100.00

1 Salmon-coloured halite, Cardona, Spain (Wells, 1937). Anal. E. T. Erickson.
2 Salt, Makarikari salt pan, Bechuanaland (now Botswana) (Golding and Phayp, 1936). Anal. F. Golding (includes SiO_2 0.24, $Al_2O_3+Fe_2O_3$ 0.04).
3 Salt, bed of White Volta river, Daboiya, Ghana (Junner and James, 1947). Anal. R. Pickup (includes Al_2O_3 0.17).
4 Large coarse white crystals, Comet Vale, Western Australia (Simpson, 1952).
5 Rock salt, Malagash, Nova Scotia (Johnstone, 1954).
6 Salt produced by solar evaporation, Big Salt Plain, Oklahoma, USA (Johnson, 1970) (includes Al 0.05, Fe 0.004, Si 0.10, Mn 0.002, Br 0.020, Bi 0.002, Cu 0.002, Sr 0.001).
7 Rock salt, Stassfurt, Germany (Johnstone, 1954) (includes Na_2SO_4 0.43).
8 Rock salt, Winsford, Cheshire, England (Johnstone, 1954).

in the Mediterranean, the halite has Br values of 0.014–0.254 wt. per cent, with three maxima within the borehole coinciding with three postulated horizons of desiccation (Kühn and Hsü, 1974; see also Holser and Wilgus, 1981).

For NaCl-saturated sea-water in a marine salt works at Secovlje, Yugoslavia, the partition coefficient, wt. per cent Br (halite)/wt. per cent Br (in solution), was found to be 0.14 ± 0.02 at 25 °C (Hermann, 1980). In Jurassic halite from the Gulf of Mexico sedimentary basin, Land *et al.* (1988) reported Br as <60 p.p.m., and attributed this low value to deformation which caused rock–water interactions with loss of Br to the co-existing brines and the incorporation of radiogenic strontium. Massive inclusion-free halite from Permian evaporites in Kansas, USA, has Sr values <5 p.p.m. and no appreciable Rb; the halite has $^{87}Sr/^{86}Sr$ values of 0.7078 ± 0.0005, indicating that these salts remained isolated from external fluids (Brooks and McMillen, 1970).

Fluid inclusions are relatively common in halite, the entrapped liquid being brine; these cavities may take the form of 'negative crystals'. Two- and three-phase inclusions are also found, the gas phase being mainly CO_2 (but occasionally H_2S, air or hydrocarbons) while the solid phases may be sylvite, anhydrite or $MgCl_2 \cdot 6H_2O$. The decrepitating 'Knistersalz' of Werra, Germany, contains spherical gaseous inclusions consisting mainly of air, with only 3.3 per cent CO_2 and 5.8 per cent hydrocarbons; the pressure in the air inclusions is 10–12 kg/cm^2 (Tamman and Seidel, 1932).

In water-poor inclusions in Cambrian halite from Siberia, experimental freezing gave solid CO_2 and H_2S; the H_2S melted at the nearly eutectic temperature of -96.5 °C in CO_2–H_2S, and at -98 to -100 °C in CO_2–H_2S–CH_4 or

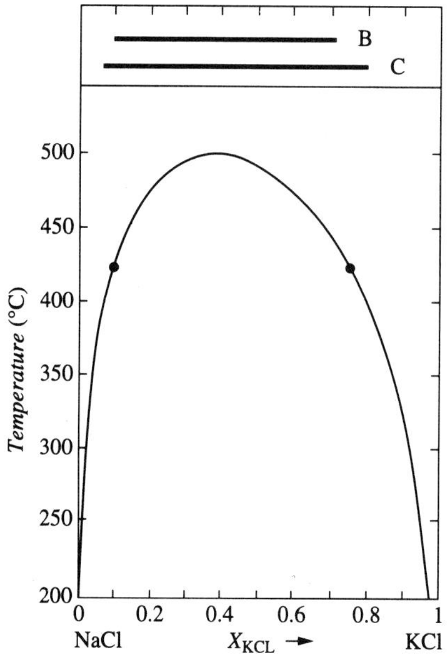

Fig. 122. The two-phase region in the system NaCl–KCl (Waldbaum, 1969). The two points shown on the two-phase curve correspond with analyses of two zones in a dendritic inclusion of halite in quartz from Campolungo, Switzerland (after Trommsdorff *et al.*, 1985). The horizontal bars labelled B and C correspond with the range of compositions found in solid inclusions in quartz crystals (B representing 40 microprobe analyses in specimens from Campolungo and C based on 100 analyses on inclusions in quartz from Cornone di Blumone, Italy).

CO_2–H_2S–N_2 inclusions (Grishina, 1987). Fluid inclusions in large limpid cleavage fragments of recrystallized halite from former brine pockets were found to be saline K–Mg–Fe–Cl solutions, with daughter crystals of bischofite and minor iron giving delicate hexagonal crystals of pyrrhotite (Fabricius, 1989). In the halite from Oligocene salt beds in France, careful examination of the fluid inclusions avoiding contamination revealed the presence of Br, K and Mg, the Br and K being in both the fluid and solid components while the Mg was only in the fluid (Moretto, 1988). A study of the brines from fluid inclusions in Silurian halite from the Salina Group of Michigan, USA, demonstrated that the composition of Silurian sea-water did not necessarily differ from that of modern seas. Six fluid inclusions in halite from the Leine rock salt of the Gorleben salt dome in Lower Saxony were found to contain Na, K, Mg, Cl and SO_4 as major constituents, with Br 3450–3920 and Li 40–321 p.p.m. (Herrmann and Borstel, 1991); concentrated and saturated solutions with high $MgCl_2$ contents are characteristic. These inclusions are considered to be the products of deposition from and re-dissolution of chloride minerals by normal sea-water in the Zechstein basin.

The isotopic analysis of fluid inclusions has been reported by Horita and Matsuo (1986) and Knauth and Beeunas (1986). In Permian halites from Death Valley, Texas and Arizona, USA, the latter authors reported the fluid inclusions to be enriched in ^{18}O relative to meteoric matter but depleted in D relative to ocean water.

The chemical composition of fluid inclusions in halite can be determined accurately from the observed melting temperatures of the solid phases hydrohalite (a well-defined hydrate, $NaCl \cdot 2H_2O$, stable at or below 0 °C), ice and sylvite (see for example Davies *et al.*, 1990). A study of the chemistry of fluid inclusions in halite from Lower Palaeozoic evaporites led to the conclusion that the chemistry and density of Silurian sea-water were not necessarily different from those for modern sea-water (Das *et al.*, 1990).

Experimental Work

The melting point of halite is 800 °C. The thermodynamic mixing properties of NaCl–KCl liquids were investigated by Waldbaum (1969) who calculated solidus and liquidus curves from these results (Fig. 122) and compared them with data from previous studies (see also Green, 1970). Isobaric curves giving the composition of liquids saturated with NaCl in the system $NaCl–H_2O$ were obtained by Akella *et al.* (1969), Bowers and Helgeson (1983) and Trommsdorff *et al.* (1985) (see Fig. 123). A further study by Bischoff *et al.* (1986) showed that at 373 °C the NaCl content of the vapour increases smoothly with pressure from that of the three-phase assemblage (0.0029 per cent NaCl at 135 bar) to just below the vapour pressure of pure water (0.012 per cent NaCl at 184 bar) (see Fig. 124). The solubility of halite in vapour-saturated liquids above 445 °C was investigated by Chou (1987) who redetermined the phase equilibrium properties in the system $NaCl–H_2O$ to 1000 °C and 1500 bar. The solubility of NaCl can be expressed as:

$$\text{NaCl (wt. per cent)} = 19.39 - 0.0364T + 3.553 \times 10^{-4}T^2 - 2.298 \times 10^{-7}T^3,$$

where $447 \leqslant T \leqslant 800$ °C (see also Gunter *et al.*, 1983; Anderko and Pitzer, 1993). The *P–T* position of the halite liquidus for a salinity of 40 wt. per cent has been determined by Bodnar (1994) using synthetic fluid inclusions.

High-temperature data on halite saturation in the system $NaCl–CaCl_2–H_2O$ were determined by Vanko *et al.* (1988). They have been used to calculate halite solubility over a wide range of compositions and temperatures to 330 °C, and hence to construct a temperature-contoured phase diagram for this system

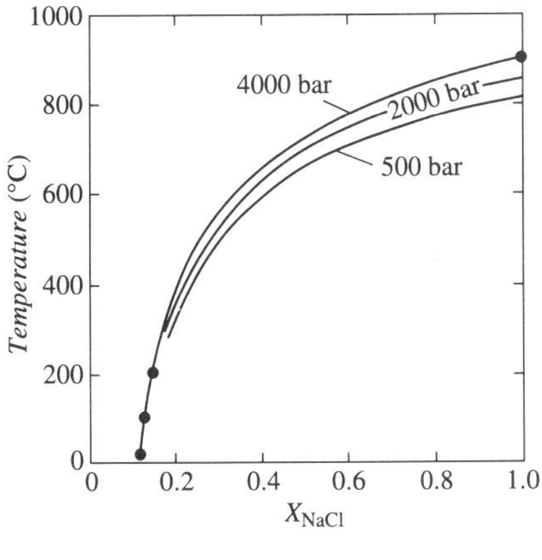

Fig. 123. Isobaric curves giving the composition of liquids saturated with NaCl in the system $NaCl–H_2O$ (after Trommsdorff *et al.*, 1985). The melting point of pure NaCl is from Akella *et al.* (1969). Solid circles represent data points for 4000 bar. Curves at 2000 and 500 bar are from Bowers and Helgeson (1983).

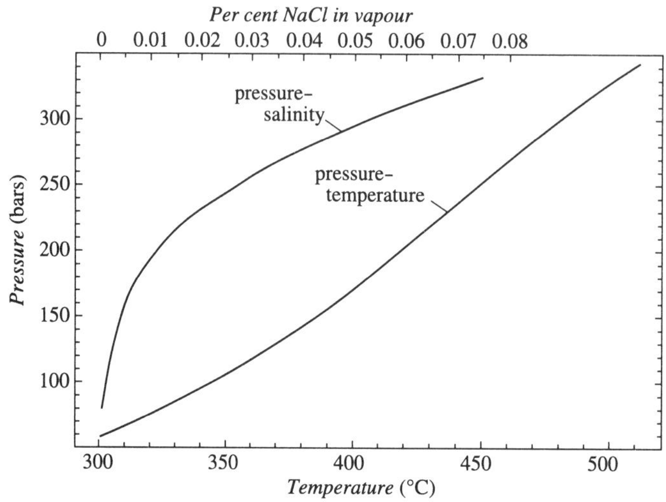

Fig. 124. Experimental determinations of pressure, temperature and NaCl content of vapour for the three-phase assemblage vapour–liquid–halite (after Bischoff *et al.*, 1986).

(Williams-Jones and Samson, 1990). This diagram can be used to make graphical estimates of the NaCl and $CaCl_2$ contents of halite-bearing aqueous fluid inclusions using the final melting temperature of ice and the temperature of halite dissolution.

Crystals of halite are readily obtainable by the evaporation of brine. Large single crystals used for making lenses and prisms for ultraviolet and infrared spectroscopes and cameras are manufactured by the slow cooling of the fused salt over a period of 7–10 days: single crystals 19 cm in diameter and weighing 14 kg are thus produced (Taylor and Kremers, 1944). Egli (1949) obtained the best results in growing halite from solution when a small amount of lead was present in the solution. An experimental study of the conditions of recrystallization of pressure-deformed artificial and natural halite was carried out by Müller (1935), who was able to measure the velocity of recrystallization at various temperatures.

Because of its ready solubility, halite is often replaced and pseudomorphed by other evaporite minerals such as anhydrite, gypsum and polyhalite, as well as by dolomite and pyrite. The solution of cubes of halite may leave cubic cavities in sediments, which are later filled in by clay; similar casts are preserved in limestone. Linck (1946) has, however, suggested that the sand–halite crystals of the Namib Desert, southwest Africa, analogous to the well-known, sand-filled crystals of calcite, were formed in sandy beds deposited in salt water.

Optical and Physical Properties

Halite is normally colourless in thin section, but material which is deeply coloured in hand specimen may show a faint colour in section. The refractive index is fairly constant at about 1.5443 for Na light at 18 °C. The values for wavelengths between 2380 and 11 200 Å have been tabulated by Harting (1948) using an interpolation formula. The dispersion is moderately strong with $v-r = 0.013$. The variation of refractive index with temperature was given by Marbach (1921). Halite and sylvite, rather than forming a continuous series of mixed crystals when crystallizing together, produce a series of 10 or 12 double salts ranging in refractive index from 1.544 for pure halite to 1.490 for pure sylvite. On cooling, these double salts break down submicroscopically and the refractive index of the aggregate is proportional to the relative amounts of NaCl and KCl (Slawson, 1929); thus the amount of NaCl in a mixture of these chlorides can be quantitatively determined by a measurement of the refractive index.

While pure rock salt is normally colourless, the natural material shows a variety of colours. In most cases the colouration is due to impurities: the grey colours are caused by inclusions of clay, or sometimes talc, while the yellow, orange, pink and red colours commonly found are due to various iron compounds, mostly oxides or hydroxides, mechanically enclosed in the crystals. Brown tints may be related to bituminous inclusions. Certain yellow, violet and blue colourations of halite, however, appear to be due to other causes, of which radiation is the most often advocated: extensive work on the subject was summarized by Przibram (1956). The realization that a crystal structure always contains voids led to an interpretation in which the colour-centre (*F*-centre) which produces the yellow colouration in rock salt is an electron held in an anion vacancy, i.e. a structural site where a chlorine ion is missing (see also Gordon, 1959). The blue colour of some halite was attributed to the presence of colloidal sodium particles. It has been shown, however, that some of the blue and violet tints are caused by electrons situated at centres of stronger disturbance in the structure (*M*- and *R*-centres). Under a pressure of about 400 kg/cm^2 halite acquires a yellow colour; when it is exposed to radioactivity and then to light it becomes blue, this colour being dispelled by heat. These observations are proposed to confirm the opinion that the coloration is caused by a permanent disturbance in the structure. Large single crystals grown with a seed crystal dipped into molten NaCl (Ekstein, 1938) were occasionally pink; ultramicroscopic examination revealed the presence of minute particles which were suggested to be free sodium. Deep blue halite from Stassfurt, Germany, had 23 p.p.m. of gold, and Friend and Allchin (1940) suggested that the presence of colloidal gold is related to the colour of the mineral. Vendl and Mándy (1958) reported that the blue colour was not due to iron (total Fe only 0.5 p.p.m.) but could be satisfactorily explained by radioactivity. Aprodov (1945) recorded blue halite with brown and black spots which occurred in proximity to sylvite, especially in crushed zones: the radioactivity of potassium was suggested as the cause of the coloration. However, a study of coloured natural halites by Thomson and Wardle (1954) showed up to 20.4×10^{-1} cc/g of helium in blue and yellow material from Werra, Germany; among the trace elements estimated were U and Th. The matter has been reviewed more recently by Kirchheimer (1976) who considered that radioactivity, particularly radioactive potassium, could not be the only colouring agent for blue and violet halite. After a study of some 3000 published papers, it was concluded that no satisfactory reason is known.

Some specimens of halite are fluorescent. Murata and Smith (1946) found that halites containing minute amounts of Mn and Pb show a bright red fluorescence in ultraviolet rays. NaCl crystallized from a solution containing Mn gave the red fluorescence only if Pb (0.001–0.010 per cent) was also present. A red-orange fluorescence in halite has been attributed to Mn^{2+}.

The visible and infrared spectra of halite were presented by Hunt *et al.* (1972) who also discussed colour centres. When crushed in a mortar, some halite shows a fairly strong blue-white triboluminescence: this property has been discussed by Inoue *et al.* (1939). The dielectric constants of halite were determined by Rao (1949): their thermal expansion (non-linear) has been measured as far as the melting point (Frischmeister, 1956). Supersaturated solutions of NaCl yield twin groups of halite consisting of a central octahedron, with three cubes in twinned position having their face diagonals parallel to [110] of the octahedron (Deicha, 1948).

Many halites show signs of deformation, and it has been shown (Wilkins *et al.*, 1981) that proton irradiation at 150–200 °C reveals deformation microstructures by selective coloration, e.g. glide bands, subgrain boundaries and healed fracture surfaces. When fluid inclusions are heated to beyond their homogenization temperature they migrate along coloured trails. The transition between frictional slip and ductile flow in halite shear zones was examined by Shimamoto (1986), who found a complete transition with decreasing velocity (or slip rate) or increasing confining pressure for a thin layer of halite undergoing large shearing deformation. The frictional–velocity relationship is crucial when considering stability analysis of fault motions (see also Shimamoto, 1989). The effects of grain/interphase boundary structures on pressure solution in halite have been studied by Hickman and Evans (1991) who showed, in experiments with convex lenses of single crystals of halite and flat lenses of halite or fused silica at 50.2 °C and 0.1 MPa fluid pressure, that convergence developed not between two halite lenses but between halite and silica in brine.

The heating of halite crystals at 720–760 °C for several hours develops bright lines on their surface, which conform with the symmetry of the crystals, and makes rough surfaces smooth. These changes were interpreted (Lukirsky, 1945) as indicating a surface migration of material towards areas of higher surface tension, the maximum being in the direction of [111]. In extension at 2 kbar confining pressure, halite deformation was reported to occur by translation gliding by {110} $\langle 1\bar{1}0 \rangle$ slip, with self-diffusion as the probable rate-limiting process (Carter and Heard, 1970).

Distinguishing Features

In hand specimen the perfect cubic cleavage, saline taste, solubility and relative softness are characteristic. In thin section the isotropic nature of the mineral and the low relief are distinctive. Sylvite differs chiefly in having a refractive index which is less than the normal mounting medium and low enough ($n = 1.490$) to give the mineral an appreciable relief.

Paragenesis

Halite occurs chiefly in sedimentary rocks, where it has been deposited by evaporation from sea-water or salt lakes. In wide areas of coastal flats bordering lagoons, the sabhka environment precipitates evaporites dominated by carbonate–sulphate deposits, in addition to halite. This type of environment, first described from the Trucial Coast of the Arabian Gulf (Shearman, 1966), is now recognized to be of widespread occurrence throughout geological time (see, for example, Chaudhuri *et al.*, 1987). Hopper-shaped cubes of halite sometimes occur, and these have been ascribed to deposition under conditions of relatively high temperature, high ionic concentration, rapid crystallization and fluctuating water levels, as in the Dea Sea (Gornitz and Schreiber, 1981). Such hopper-shaped crystals are frequently found pseudomorphed by other minerals or are represented by negative casts on the bottom surface of subsequent deposits of clay or sandstone.

Analyses indicate that NaCl represents 77.6 per cent of the salts formed by evaporation at the present day. Detailed comparisons with similar evaporite deposits elsewhere, e.g. in the Stassfurt Permo-Triassic deposits of Germany or the Permian salt deposits of Texas, USA, reveal no significant variation in this percentage. On evaporation, halite begins to crystallize when the sea-water has been reduced to about 10 per cent of its original volume; it appears after the carbonates of Ca and Mg, after $CaSO_4$, but before the very soluble sulphates and chlorides of Mg and K. A detailed review of the geology and mineralogy of salt deposits was given by Stewart (1954) and Lotze (1957). In most of the well-documented Permo-Triassic evaporite deposits, such as those of northwest Europe, the beds of halite are found associated with carbonates and sulphates (gypsum and/or anhydrite), but in the Triassic strata of Cheshire, England, vast deposits of rock salt occur in Keuper Marl, with little evidence of accompanying sulphates. The two main rock salt horizons in the Winsford area of Cheshire are separated by some 300 m of mudstone; reserves are estimated at 4×10^6 tons. Conflicting ideas on the enormous amounts of salt in the Cheshire basin have been discussed by Wilson (1993). At Winsford, at the famous Stassfurt mines in Germany and at Wieliczka in Poland, the rock salt has been mined directly. However, in many other localities the halite is produced commercially by pumping water down, dissolving the salt, and evaporating the solution in brine pans at the surface. Elsewhere halite is extracted by the direct evaporation of sea-water, a process employed around the Mediterranean and in tropical countries for many centuries.

Thin ($\leqslant 1$ mm) delicate winged crystals of halite $\leqslant 80$ mm across were found in Permian Marl in northeast England and are considered to have formed while the enclosing mud was sufficiently plastic to be moved by the growing crystals (Smith, 1993). The occurrence of halite as a cement in sandstone has been documented from Denmark (Laier and Nielsen, 1989) and attributed to the hyperfiltration of saturated formation brines through low-permeability layers of shale. The involvement of nitrogen gas late in the diagenetic history of the Triassic strata in southwestern Denmark may have been responsible.

Deformation applied to stratified rocks with interbedded halite deposits may give rise to the upthrusting or intrusion of massive salt domes, which are usually circular in cross-section and may be several miles in diameter. These diapiric structures may form structural traps for petroleum and their recognition is thus of economic importance.

Halite in the form of stalactites and stalagmites has been recorded in mines, deposited from the mine waters (Spiroff, 1937). Halite is also found in speleothems in caves in the Nullarbor Plain, Western Australia, where its origin has been ascribed to sea spray. The Nullarbor halite has abundant liquid inclusions and gas bubbles.

Halite also occurs as a sublimation product in volcanic regions, having been recorded, for example, in a white crust in the crater of Vesuvius (Alfani, 1934). It is also found as a surface efflorescence in arid regions.

In the Mir and Udachnaya kimberlite pipes, northeastern Siberia, halite amounts to 1.5 per cent at deep levels (Pavlov and Hupin, 1973). It contains inclusions of serpentine, anhydrite and carbonates. Halite–magnetite ores are well known in the Angara iron-ore province on the Siberian Platform. These are skarn-type deposits in which the surfaces of the magnetites are covered by Mg–Al chlorites; cavities in these brecciated metasomatized rocks are filled with late halite derived from Cambrian salt deposits in the sedimentary cover (Vakhurshev, 1981).

Giant crystalline halite in cleavable masses up to 50 cm or more has been called saltspar (Murzaev, 1941); it has been variously proposed to be caused by dynamometamorphism, recrystallization in sink holes, or original recrystallization in the deeper and stiller waters of a lagoon. The Zayarsk borehole in the Irkutsk basin penetrated salt beds intruded by dolerite; here the halite recrystalllized at the dolerite contact to give coarse-grain (4–6 cm) salt crystals. The dolerite was probably at a temperature of >950 °C, considerably above the 800 °C melting point of halite (Pavlov, 1970). The Tuba borehole penetrated a halite–amphibole–anhydrite–carbonate rock produced by contact metamorophism of halite with clay, carbonate and sulphate impurities.

Solid inclusions of halite have been found in several samples of calc-silicate rocks from Switzerland and northern Italy which have been metamorphosed to lower amphibolite grade, with the development of talc and tremolite. Idiomorphic crystals and dendrites of halite have been analysed (Trommsdorff *et al.*, 1985) and show a range of compositions in the system NaCl–KCl which indicate a temperature maximum of greater than 500 °C and pressures of 2 kbar or higher.

References

Akella, J., Vaidya, S. N. and **Kennedy, G. C.,** 1969. Melting of sodium chloride at pressures of 65 kb. *Physical Rev.*, **185**, 1135–40.

Alfani, M., 1934. Sui sali di ferrodellefumarole vesuviane a cloruri e solfati alcalini del maggio-giugno 1993. *Ann. R. Osservatorio Vesuviano*, ser. **4**, **3**, 81–101 (M.A. **6**–155).

Anderko, A. and **Pitzer, K. S.,** 1993. Equation-of-state representation of phase equilibria and volumetric properties of the system NaCl–H_2O above 573 K. *Geochim. Cosmochim. Acta*, **57**, 1657–89.

Aprodov, V. A., 1945. Blue halite from the Solidamsk potassium deposits. *Doklady Acad. Sci. USSR*, **48**, 346–8.

Bischoff, J. L., Rosenbauer, R. J. and **Pitzer, K. S.,** 1986. The system NaCl–H_2O: relations of vapor–liquid near the critical temperature of water and of vapor–liquid–halite from 300° to 500°C. *Geochim. Cosmochim. Acta*, **50**, 1437–44.

Bodnar, R. J., 1994. Synthetic fluid inclusions: XII. The system H_2O–NaCl. Experimental determination of the halite liquidus and isochores for a 40 wt.% NaCl solution. *Geochim. Cosmochim. Acta*, **58**, 1053–63.

Bowers, T. S. and **Helgeson, H. C.,** 1983. Calculation of the thermodynamic and geochemical consequences of nonideal mixing in the system H_2O–CO_2–NaCl on phase relations in geological systems: equation of state for H_2O–CO_2–NaCl fluids at high pressures and temperatures. *Geochim. Cosmochim. Acta*, **47**, 1247–75.

Bragg, W. L., 1914. The structure of some crystals as indicated by their diffraction of X-rays. *Proc. Roy. Soc. London, A*, **89**, 248–52.

Bragg, W. L. and **Claringbull, G. F.,** 1965. *Crystal Structures of Minerals*. London (G. Bell & Sons), p. 38.

Brilliantow, N. A. and **Obreimow, I. W.,** 1937. On plastic deformation of rock salt, IV. *Phys. Zeit. Soujetunion*, **12**, 7–19 (M.A. 7–400).

Brookins, D. G. and **McMillen, H. O.,** 1970. Isotopic composition of strontium in halite and included brines from Hutchinson, Kansas. *Trans. Kansas Acad. Sci.*, **73**, 277–8.

Carter, N. L. and **Heard, H. C,** 1970. Temperature and rate dependent deformation of halite. *Amer. J. Sci.*, **269**, 193–249.

Chaudhuri, A. K., Sarkar, S. and **Chanda, S. K.,** 1987. Proterozoic coastal sabkha halite pans: an example from the Pranhita–Godavari valley, South India. *Precambrian Res.*, **37**, 305–21.

Chou, I.-M., 1987. Phase relations in the system NaCl–KCl–H_2O. III: solubilities of halite in vapor-saturated liquids above 445°C and redetermination of phase equilibrium properties in the system NaCl–H_2O to 1000°C and 1500 bars. *Geochim. Cosmochim. Acta*, **51**, 1965–75.

Das, N., Horita, J. and **Holland, H. D.,** 1990. Chemistry of fluid inclusions in halite from the Salina Group of the Michigan Basin: implications for late Silurian seawater and the origin of sedimentary brines. *Geochim. Cosmochim. Acta*, **54**, 319–27.

Davies, D. W., Lowenstein, T. K. and **Spencer, R. J.,** 1990. Melting behaviour of fluid inclusions in laboratory-grown halite crystals in the systems NaCl–H_2O, NaCl–KCl–H_2O, NaCl–$MgCl_2$–H_2O and NaCl–$CaCl_2$–H_2O. *Geochim. Cosmochim. Acta*, **54**, 591–601.

Deicha, G., 1948. Macles et déséquilibre cristallo-génétique. *Experimenta, Basel*, **4**, 67–8.

Dyni, J. R., Hite, R. J. and **Raup, O. B.**, 1970. Lacustrine deposits of bromine-bearing halite, Green River formation, northwestern Colorado. *Third Symp. Salt*, **1**, 166–80.

Egli, P. H., 1949. Crystal research. *Sci. Monthly* (*Amer. Assoc. Adv. Sci.*), **68**, 270–8.

Ekstein, H., 1938. Sur la couleur des crsitaux de sel gemme artificiel. *Bull. Soc. Franç. Min.*, **61**, 239–44.

Fabricius, J., 1989. Geochemical investigations of NaCl–KCl–$MgCl_2$–$CaCl_2$–$FeCl_2$ solutions in Zechstein 2 salt, Suldrup Dome, Denmark: microthermometry on fluid inclusions in halite. *Danmarks Geol. Unders., A*, **27**, 33 pp.

Fischmeister, H. F., 1956. The thermal expansion of sodium chloride and some other alkali halides at high temperatures. *Acta Cryst.*, **9**, 416–20.

Friend, J. N. and **Allchin, J. P.**, 1940. Blue rocksalt. *Nature*, **145**, 266–7.

Frondel, C., 1955. A precision X-ray powder camera. *Amer. Min.*, **40**, 876–84.

Golding, E. and **Phaup, A. E.**, 1936. Chemical analyses of the rocks, ores and minerals of Southern Rhodesia. *Bull. Geol. Surv. Southern Rhodesia*, no. 29.

Gordon, R. G., 1959. Colour centers in crystals. *Amer. Scientist*, **47**, 361–75.

Gornitz, V. M. and **Schreiber, B. C.**, 1981. Displacive halite hoppers from the Dead Sea: some implications for ancient evaporative deposits. *J. Sedim. Petr.*, **51**, 787–94.

Green, E. J., 1970. Predictive thermodynamic models for mineral systems. I. Quasi-chemical analysis of the halite–sylvite solvus. *Amer. Min.*, **55**, 1692–1713.

Grishina, S. N., 1987. H_2S-bearing inclusions in recrystallized halite. *Chem. Geol.*, **61**, 91–4.

Gunter, W. D., Chou, I.-M. and **Girsperger, S.**, 1983. Phase relations in the system NaCl–KCl–H_2O. II: Differential thermal analyses of the halite liquidus in the NaCl–H_2O binary above 450°C. *Geochim. Cosmochim. Acta*, **47**, 863–73.

Harting, H., 1948. Die Brechzahlen einiger Halogenidkristalle. *Sitz. Deutsch. Akad. Wiss., Math.-nat. Kl.*, **4**, 25 pp.

Herrmann, A. G., 1980. Bromide distribution between halite and NaCl–saturated seawater. *Chem. Geol.*, **28**, 171–7.

Herrmann, A. G. and **Borstel, L. E. von**, 1991. The composition and origin of fluid inclusions in Zechstein evaporites of Germany. *Neues Jahrb. Min., Monat.*, 263–9.

Hettich, A., 1926. Über die äußere scheinbare unsymmetrie der Alkalihalogenide. *Z. Krist.*, **64**, 265–95.

Hickman, S. H. and **Evans, B.**, 1991. Experimental pressure solution in halite: the effect of grain/interphase boundary structure. *J. Geol. Soc.*, **148**, 549–60.

Holser, W. T. and **Wilgus, C. K.**, 1981. Bromide profiles of the Röt salt, Triassic of northern Europe, as evidence of marine origin. *Neues Jahrb. Min., Monat.*, 267–76.

Horita, J. and **Matsuo, S.**, 1986. Extraction and isotopic analysis of fluid inclusions in halites. *Geochem. J.*, **20**, 261–72.

Hunt, G. R., Salisbury, J. W. and **Lenhoff, C. J.**, 1972. Visible and near-infrared spectra of minerals and rocks: V. Halides, phosphates, arsenates, vanadates and borates. *Modern Geol.*, **3**, 121–32.

Inoue, T., Kunitomi, M. and **Shibata, E.**, 1939. Über Triboluminescenz. *Sci. Hiroshima Univ., A*, **9**, 129–36.

Johnson, K. S., 1970. Salt produced by solar evaporation on Big Salt Plain, Woods County, Oklahoma. *Oklahoma Geol. Notes*, **30**, 47–54.

Johnstone, S., 1954. *Minerals for the Chemical and Allied Industries*. London (Chapman and Hall).

Junner, N. R. and **James, W. T.**, 1947. Chemical analyses of Gold Coast rocks, ores and minerals. *Gold Coast Geol. Surv., Bull.*, **15**.

Kirchheimer, F., 1976. Blaues Steinsalz und sein Vorkommen im Neckarund Oberrheingebeit. *Geol. Jahrb., Ser. D*, **18**, 3–139.

Klewe, B. and **Pederson, B.**, 1974. The crystal structure of sodium chloride dihydrate. *Acta Cryst.*, **B30**, 2363–71.

Knauth, L. P. and **Beeunas, M. A.**, 1986. Isotope geochemistry of fluid inclusions in Permian halite with implications for the isotope history of ocean water and the origin of saline formation waters. *Geochim. Cosmochim. Acta*, **50**, 419–33.

Kühn, R. and **Hsü, K. J.**, 1974. Bromine content of Mediterranean halite. *Geology*, **2**, 213–6.

Laier, T. and **Nielsen, B. L.**, 1989. Cementing halite in Triassic Bunter Sandstone (Tonder, southwest Denmark) as a result of hyperfiltration of brines. *Chem. Geol.*, **76**, 353–63.

Land, L. S., Kupecz, J. A. and **Mack, L. E.**, 1988. Louann salt geochemistry (Gulf of Mexico sedimentary basin, USA): a preliminary synthesis. *Chem. Geol.*, **74**, 22–35.

Linck, O., 1946. Die sogenannten Steinsalz-Pseudomorphosen als Kristall-Relikte. *Abh. Senckenberg Nat. Gesell. Frankfurt a.M.*, **470 , 44 pp.**

Lotze, F., 1957. *Steinsalz und Kalisalz I* Allgemeingeologischer Teil), Zweite Auflage. Berlin (Gebrüder Borntraeger).

Lukirsky, P. U., 1945. Experiments with rock-salt crystals. *Doklady Acad. Sci. USSR*, **46**, 274–6.

Marbach, E., 1921. Brechungsindices von KCl und NaCl. *Z. Krist.*, **56**, 212.

McIntosh, R. A. and **Wardlaw, N. C.**, 1968. Barren halite bodies in the sylvinite mining zone at Esterhazy, Saskatchewan. *Can. J. Earth Sci.*, **5**, 1221–38.

Mir, J., 1946. Sobre el contenido en bromo de las sales potásicas de la cuenca del Llobregat (Barcelona). *Notas y Comunic. Inst. Geol. y Min. de España*, **16**, 269–87 (M.A. **10**–304).
Moretto, R., 1988. Observations on the incorporation of trace elements in halite of Oligocene salt beds, Bourg-en-Bresse Basin, France. *Geochim. Cosmochim. Acta*, **52**, 2809–14.
Müller, H. G. 1935. Zur Natur de Rekristallisationsvorgänge. *Physik*, **96**, 279–327.
Murata, K. J. and **Smith, R. L.**, 1946. Manganese and lead as coactivators of red fluorescence in halite. *Amer. Min.*, **31**, 527–38.
Murzaev, P. M., 1941. On saltspar and its origin. *Doklady Acad. Sci. USSR*, **33**, 306–7.
Olmer, P., 1942. Contribution à l'étude de la diffusion des rayons X èn dehors des réflexions sélectives de Bragg. *Mém. Faculté Sci. Univ. Paris*, 20 (M.A. **9**–220).
Pavlov, D. I., 1970. Halite anatectite and some of its less highly metamorphosed analogs. *Dokl. Acad. Sci. USSR, Earth Sci. Sect.*, **195**, 171–3.
Pavlov, D. I. and **Hupin, I. P.**, 1973. Halite in Yakutian kimberlite, its relation to serpentine and the source of its parent solutions. *Dokl. Acad. Sci. USSR, Earth Sci. Sect.*, **213**, 178–80.
Preston, G. D., 1939. Diffraction of X-rays by crystals at elevated temperatures. *Proc. Roy. Soc. London, A.*, **172**, 116–26.
Przibram, K., 1956. *Irradiation Colours and Luminiscence.* London (Pergamon Press), 332 pp.
Rao, D. A. A. S. N., 1949. Dielectric constants of crystals—III. *Proc. Indian Acad. Sci., A.*, **30**, 82–6.
Shearman, D. J., 1966. Origin of marine evaporites by diagenesis. *Trans. Inst. Min. Metall.*, **B75**, 207–15.
Shimamoto, T., 1986. Transition between frictional slip and ductile flow for halite shear zones at room temperature. *Science*, **231**, 711–4.
Shimamoto, T., 1989. Mechanical behaviours of simulated halite shear zones; implications for the seismicity along subducting plate-boundaries. In *Rheology of Solids and of the Earth:* Eds. S. I. Karato and M. Toirumi, Oxford (OUP), 351–73.
Simpson, E. S., 1952. *Minerals of Western Australia, vol. III.* Perth (Govt. Printer).
Slawson, C. B., 1929. The quantitative optical determination of sodium and potassium chlorides. *Amer. Min.*, **14**, 293–8.
Smith, D. B., 1993. Slate halite in the late Permian Rotten Marl of Teeside, north-east England. *Proc. Yorks. Geol. Soc.*, **49**, 293–4.
Spiroff, K., 1937. An unusual occurrence of halite. *Amer. Min.*, **22**, 931–3.
Stewart, F. H., 1954. Permian evaporites and associated rocks in Texas and New Mexico compared with those of northern England. *Proc. Yorks. Geol. Soc.*, **29**, 185–23.
Sun, T.-Y. and **Ch'u, S.-L.**, 1948. X-ray diffraction pattern of a strained rock-salt crystal. *Acta Cryst.*, **1**, 135–7.
Tammann, G. and **Seidel, K.**, 1932. Zur Kenntnis der Kohlensäureausbrüche in Bergwerken. *Z. anorg. Chem.*, **205**, 209–29.
Taylor, R. L. and **Kremers, H. C.**, 1944. Synthetic optical crystals. *Chem. Industries*, **55**, 906–11.
Thomson, S. J. and **Wardle, G.**, 1954. Coloured natural rocksalts: a study of their helium contents, colours and impurities. *Geochim. Cosmochim. Acta*, **5**, 169–84.
Trommsdorff, V., **Skippen, G.** and **Ulmer, P.**, 1985. Halite and sylvite as solid inclusions in high-grade metamorphic rocks. *Contr. Min. Petr.*, **89**, 24–9.
Vakrushev, V. A., [The halite–magnetite ores of the Siberian Platform.] *Geol. Rudnykh Mestorozhdenii*, **23**, 100–4 (in Russian) (M.A. 83M/2670).
Vanko, D. A., **Bodnar, R. J.** and **Sterner, S. M.**, 1988. Synthetic Fluid Inclusions VIII. Vapor-saturated halite solubility in part of the system $NaCl–CaCl_2–H_2O$ with application to fluid inclusions from oceanic hydrothermal system. *Geochim. Cosmochim. Acta*, **52**, 2351–6.
Vendl, A. and **Mándy, T.**, 1958. Über die blaue Farbe einiger Mineralien. *Acta Min.-Petr. Univ. Szeged*, **11**, 61–7.
Waldbaum, D. R., 1969. Thermodynamic mixing properties of NaCl–KCl liquids. *Geochim. Cosmochim. Acta*, **33**, 1415–27.
Wells, R. C., 1937. Analyses of rocks and minerals from the laboratory of the United States Geological Survey 1914–36. *U.S. Geol. Surv., Bull.*, **878**.
Wilkins, R. W. T., **Bird, J. R.** and **Ewald, A. H.**, 1981. Observations on deformation microstructures and fluid inclusions in proton-irradiated halite. *Neues Jahrb. Min., Abhandl.*, **141**, 240–57.
Williams-Jones, A. E. and **Samson, I. M.**, 1990. Theoretical estimation of halite solubility in the system $NaCl–CaCl_2–H_2O$: applications to fluid inclusions. *Can. Min.*, **28**, 299–304.
Wilson, A. A., 1993. The Mercia Mudstone Group (Trias) of the Cheshire Basin. *Proc. Yorks. Geol. Soc.*, **49**, 171–88.

Index

Mineral names in **bold** type are those described in detail; page numbers in **bold** type refer to the principal descriptions of the mineral. Entries other than mineral names are in *italic* type.

Alabaster, 40, 42, 56
Alforsite, 303
Alstonite, 264, 266, 268–270
Anglesite, 4, 5, 11–4, 30, 31, 34, 272, 275
Anglesobaryte, 7, 14
Anhydrite, 6–8, 14, 17, 20, 22, 30, 35, 36, 43, 45, 47–56, 60, 62, **74–94**, 123, 369, 371, 374, 377, 378
  Structure, 74
  Chemistry, 77
  Optics, etc., 80
  Paragenesis, 83
Ankerite, 6, 31, 98, 118, 122, 138, 140, 146, 172, 183, 189, 194, 195, 198, **219–231**, 259, 260, 274, 275
  Structure, 219
  Chemistry, 222
  Optics, etc., 227
  Paragenesis, 228
Antozonite, 357
Apatite, 128, 212, **297–334**, 337, 346, 364
  Structure, 297
  Chemistry, 302
  Optics, etc., 320
  Paragenesis, 322
Aragonite, 36, 54, 98, 105, 106, 108, 115, 118, 123–6, 201, 202, 209, 211, **235–52**, 253, 257–9, 263, 264–9, 272, 280, 283
  Structure, 235
  Chemistry, 238
  Optics, etc., 244
  Paragenesis, 247
Arrow-head twins, 59
Avogadrite, 5
Azurite, 98, 183, 260, **278–94**
  Structure, 288
  Chemistry, 291
  Optics, etc., 292
  Paragenesis, 293

Baryte, **3–29**, 30–7, 64, 67, 79, 81–3, 256, 260, 263, 266, 268, 269, 270, 322, 346, 364
  Structure, 3
  Chemistry, 5
  Optics, etc., 11
  Paragenesis, 14
Barytoanglesite, 6
Barytocalcite, 6, 23, 113, 114, 260, 264–6, 268–70
Barytocelestine, 6, 31, 32
Bassanite, 43, 47, 54, 55, 59, 66, 67, 84
Beef, 126
Benstonite, 264–6, 269, 270
Bischofite, 372
Blue John, 355, 357, 359, 362
Breunnerite, 137, 138, 141, 144
Burbankite, 260
Bütschliite, 282, 283, 285, 286

Calciorhodochrosite, 160
Calciostrontianite, 255, 260
Calcite, 6, 11, 15, 16, 23, 31, 33, 34, 36, 54, 57, 60, 79, 80, 83, 86, 97, **101–35**, 136, 137, 144, 146, 150, 152, 157–60, 163, 167, 171, 172, 178, 182, 183, 189–93, 197–202, 204–6, 208–13, 223, 228, 229, 235–49, 253, 259, 260, 265, 267, 269, 270, 277, 278, 280, 283, 284, 286, 293, 322, 325, 358, 362, 364
  Structure, 101
  Chemistry, 108
  Optics, etc., 117
  Paragenesis, 124
Calcite II, 105, 106, 121, 122
Calcite III, 105, 106, 121, 122
Calcite IV, 105, 106
Calcite V, 105, 106
Capillitite, 152
Carbonate-apatite, 297, 301, 302, 308, 310–3, 314, 317, 325
Carbonates–introduction, 95
Celestine, 4–7, 10–14, 19, 20, 23, **30–9**, 45, 54, 55, 81, 83, 253, 259, 260
  Structure, 30
  Chemistry, 31
  Optics, etc., 33
  Paragenesis, 35
Cerussite, 11, 98, 183, 239, 259, 260, **272–6**,
  Structure, 272
  Chemistry, 273
  Optics, etc., 273
  Paragenesis, 275
Chalybite, 163
Cheralite, 336, 337, 345, 346
Chlorapatite, 297–9, 300, 302, 303, 316, 320, 321, 323, 325

Collophane, 236
Cryolite, 362

Dahllite, 297, 302, **314**
Dead-burned gypsum, 48
Dead-burned magnesite, 143
Desert roses, 12, 56, 60
Dolomite, 11, 15, 80, 98, 106, 107, 110, 113, 118, 119, 122, 123, 128, 138, 140, 141, 144–6, 158, 172, **189–218**, 219–222, 225, 227–9, 247–9, 260, 265, 269, 280, 281
  Structure, 189
  Chemistry, 194
  Optics, etc., 205
  Paragenesis, 209
Dry bone, 178

Fairchildite, 282–6
Fascicular gypsum, 42
Fission tracks, in apatite, 318
Florencite, 319, 347
Fluocerite, 346
Fluorapatite, 297–334
Fluor-chlor-oxyapatite, 304–5
Fluorhydroxylapatite, 299
Fluorite, 6, 11, 15, 16, 19–21, 31, 128, 260, 269, 325, **355–68**
  Structure, 355
  Chemistry, 356
  Optics, etc., 358
  Paragenesis, 362
Francolite, 297, 302, 312–3, **314**, 315, 319, 326, 327

Gaspéite, 97, 139–41
Glauberite, 57, 83, 87
Glaucosphaerite, 291
Gregoryite, 283, 284, 287
Gypsum, 14, 17, 19, 20, 30, 34–7, **40–73**, 74, 123, 253, 369, 374, 377
  Structure, 41
  Chemistry, 45
  Optics, etc., 56
  Paragenesis, 61

Halite, 51, 54, 62, 63, 81, 84, 89, 362, **369–80**
  Structure, 369
  Chemistry, 370
  Optics, etc., 374
  Paragenesis, 376
Hemihydrate, 40, 43–5, 47–51, 54–6, 63, 76, 79, 80, 83
Hepatite, 11
Hokutolite, 5
Hoshiite, 140, 141
Huntite, 98, **277–81**
  Structure, 277
  Chemistry, 277
  Optics, etc., 278
  Paragenesis, 280
Huttonite, 336, 337, 345
Hydrohalite, 370, 372
Hydromagnesite, 60, **138**, 142, 209, 280
Hydroxylapatite, **297–334**
Hydrozincite, 183, 184

Iceland spar, 118, 119

Kerstenite, 5
Kieserite, 85
Knistersalz, 371
Kolwézite, 291
Kopi, 56
Kularite, 348
Kutnahorite, 98, 111, 152, 153, 159, 160, 195, 219, 221, 222, 224–8

Lansfordite, 138, 142

Magnesite, 97, 106, 107, 118, 120, 123, **136–49**, 154, 157, 158, 173, 178, 182, 183, 190–3, 197, 198, 200–2, 204, 208, 209, 211, 213, 228, 249, 277, 278, 280, 281
  Structure, 136
  Chemistry, 138
  Optics, etc., 143
  Paragenesis, 145
Malachite, 11, 54, 98, 183, 260, 275, **288–94**
  Structure, 288
  Chemistry, 291
  Optics, etc., 292
  Paragenesis, 293
Manganapatite, 321
Manganocalcite, 160
Mcquinessite, 291
Miltonite, 67
Minrecordite, 98, 180, 182, 198
Mirabilite, 54
Monazite, **335–51**
  Structure, 335
  Chemistry, 336
  Optics, etc., 345
  Paragenesis, 346
Monetite, 315
Monheimite, 179
Mossottite, 238
Mountain milk, 281

Nahcolite, 370
Nesquehonite, 138, 142
Norsethite, 204, 264, 265, 267, 269, 270
Nyerereite, **282–87**
  Structure, 282
  Chemistry, 283
  Optics, etc., 285
  Paragenesis, 286

Oligonspar, 166, 171
Otavite, 97

Paralstonite, 264, 266, 268, 270
Plaster of Paris, 48, 56
Polyhalite, 54, 87, 374
Ponite, 152
Protodolomite, 194

Radiofluorite, 359
Rhabdophane, 319, 347
Rhodochrosite, 11, 97, 111, 118, 120, 144, **150–62**, 178, 179, 182, 183, 197
Rock roses, 15
Rock salt (halite), 369, 371
Rosasite, 291

Saltspar, 378
Sand baryte, 15
Sand gypsum, 51
Satin-spar, 40, 42, 48, 65
Seed gypsum, 56
Selenite, 40, 48, 56, 60, 64
Shortite, 270, 283–5
Siderite, 11, 15, 80, 97, 112, 118, 137–40, 144, 150, 158, **163–77**, 178, 179, 182, 183, 195, 198, 200, 201, 211, 223, 228, 229, 260, 364
  Structure, 163
  Chemistry, 164
  Optics, etc., 171
  Paragenesis, 172
Smithsonite, 97, 117, 120, 144, 158, 164, 167, 172, **178–85**, 272, 275
  Structure, 178
  Chemistry, 179
  Optics, etc., 182
  Paragenesis, 183
Soluble anhydrite, 40, 44, 45, 47, 48, 50, 76, 79, 80
Strontianite, 33, 36, 98, 247, **253–62**, 263, 264, 266, 268, 272, 274, 275, 325
  Structure, 253
  Chemistry, 255
  Optics, etc., 258
  Paragenesis, 259
Strontiobaryte, 6, 8, 14
Swallow tail twins, 59
Svanbergite, 319
Sylvite, 370, 371, 373–6
Syngenite, 54
System:
  $BaCO_3$–$CaCO_3$, 113, 114, 255
  $BaCO_3$–$CaCO_3$–$SrCO_3$, 266
  $BaCO_3$–$PbCO_3$, 266
  $BaCO_3$–$SrCO_3$, 266
  $BaSO_4$–$CaSO_4$, 6
  $BaSO_4$–$SrSO_4$, 10, 32
  $CaCO_3$–$FeCO_3$, 112, 167
  $CaCO_3$–$FeCO_3$–$MgCO_3$, 197, 222, 224
  $CaCO_3$–K_2CO_3–Na_2CO_3, 128
  $CaCO_3$–$MgCO_3$, 110, 199, 201, 203
  $CaCO_3$–$MgCO_3$–$MnCO_3$, 195
  $CaCO_3$–$MnCO_3$, 112, 152, 158
  $CaCO_3$–$PbCO_3$, 241
  $CaCO_3$–$SrCO_3$, 113, 239, 255
  $CaMgSi_2O_6$–$Ca_5(PO_4)_3F$–$NaAlSiO_4$, 324
  CaO–CO_2, 128
  CaO–CO_2–H_2O, 115, 116
  CaO–MgO–SiO_2–CO_2–H_2O, 213
  Ca–O–H–CO_2, 291
  $CaSO_4$–H_2O, 48, 50, 52, 54, 55, 78
  $CaSO_4$–K_2SO_4–H_2O, 58
  $CaSO_4$–K_2SO_4–$MgSO_4$–H_2O, 58
  $CaSO_4$–$NaCl$–H_2O, 51, 52, 78
  $CaSO_4$–Na_2SO_4–H_2O, 58
  $FeCO_3$–$MgCO_3$, 138
  $FeCO_3$–$MnCO_3$, 154, 158, 167
  KCl–NaCl, 372
  $MgCO_3$–$MnCO_3$, 140, 152, 158
  MgO–CO_3–H_2O, 142
  MgO–SiO_2–H_2O–CO_2, 145
  $MnCO_3$–$ZnCO_3$, 158
  $NaCl$–H_2O, 373
  $PbCO_3$–$SrCO_3$, 256, 273

Tarnowitzite, 238, 239, 246

Vanadinite, 303
Vaterite, **107**, 108, 114, 115, 235, 243
Vibertite, 67

Wavellite, 319
Witherite 11, 23, 33, 79, 98, 256, 260, **263–71**, 272
  Structure, 263
  Chemistry, 265
  Optics, etc., 268
  Paragenesis, 269
Woodhouseite, 319

Xenotime, 346

Yttrofluorite, 346, 356, 359

Zemkorite, 282–5